essentials of

Sociology

essentials of
Sociology

4TH
EDITION

DAVID B.
Brinkerhoff

LYNN K.
White

SUZANNE T.
Ortega

UNIVERSITY OF NEBRASKA–LINCOLN

Wadsworth Publishing Company

I(T)P® An International Thomson Publishing Company

Belmont, CA • Albany, NY • Boston • Cincinnati • Johannesburg • London • Madrid
Melbourne • Mexico City • New York • Pacific Grove, CA • Scottsdale, AZ • Singapore • Tokyo • Toronto

Publisher: Eve Howard
Assistant Editor: Barbara Yien
Marketing Manager: Chaun Hightower
Editorial Assistant: Ari Levenfeld
Project Editor: Jerilyn Emori
Print Buyer: Karen Hunt
Permissions Editor: Robert Kauser
Production Coordinator: Electronic Publishing Services Inc., NYC
Interior and Cover Designer: Paul Uhl/Design Associates
Copy Editor: Salvatore Allocco, Electronic Publishing Services Inc., NYC
Illustrator: Electronic Publishing Services Inc., NYC
Cover Photo: Cotton R. Coulson, Courtesy National Geographic Society
Compositor: Electronic Publishing Services Inc., NYC
Printer: World Color Book Services/Taunton

Printed in the United States of America
1 2 3 4 5 6 7 8 9 10

For more information, contact Wadsworth Publishing Company, 10 Davis Drive, Belmont, CA 94002,
or electronically at http://www.wadsworth.com

International Thomson Publishing Europe
Berkshire House 168-173
High Holborn
London, WC1V 7AA, England

International Thomson Publishing Asia
60 Albert Street
#15–01 Albert Complex
Singapore 189969

Thomas Nelson Australia
102 Dodds Street
South Melbourne 3205
Victoria, Australia

International Thomson Publishing Japan
Hirakawacho Kyowa Building, 3F
2-2-1 Hirakawacho
Chiyoda-ku, Tokyo 102, Japan

Nelson Canada
1120 Birchmount Road
Scarborough, Ontario
Canada M1K 5G4

International Thomson Publishing
Southern Africa
Building 18, Constantia Park
240 Old Pretoria Road
Halfway House, 1685 South Africa

International Thomson Editores
Campos Eliseos 385, Piso 7
Col. Polanco
11560 México D.F. México

Library of Congress Cataloging-in-Publication Data

Brinkerhoff, David B.
    Essentials of sociology / David B. Brinkerhoff, Lynn K. White,
Suzanne T. Ortega. -- 4th ed.
      p.   cm.
    Includes bibliographical references (p.   ) and index.
    ISBN 0-534-55548-9 (pbk.)
    1. Sociology.   I. White, Lynn K.   II. Ortega, Suzanne Trager,
1951–   .   III. Title.
HM51.B8533   1999   98–15988
301--dc21

Brief Contents

Contents

CHAPTER 4   *The Individual in Society*   83

CHAPTER 5   # Deviance, Crime, and Social Control   112

CHAPTER 11   *Family and the Life Course*   **253**

CHAPTER 15

Social Movements, Technology, and Social Change   373

Preface

One of the most stimulating aspects of the discipline of sociology is that it deals with the major issues confronting our planet, our nation, and our lives. Perhaps more than any other discipline, sociology deals with the substance of ordinary life— getting a job, getting married (and staying that way), caring for children, and caring for aging parents. Sociologists also grapple with the major national and international problems of our times: homelessness, health care reform, environmental degradation, and poverty and dependence in the least-developed nations. An introductory textbook has both the challenge and the opportunity of giving students the sociological framework that will help them better understand how social structures shape the details of their own personal lives and the larger social world that surrounds them.

This fourth edition of *Essentials of Sociology*, has the same goal as the preceding editions: to provide a concise, balanced introduction to the field of sociology that is shorter and less expensive than a comprehensive text. Like the first three editions, the fourth edition provides a careful blend of theory and the latest research, combined with a series of examples, case studies, and applications that will help students develop the sociological imagination.

The fourth edition of *Essentials of Sociology* will be particularly helpful for instructors who teach short academic terms and have difficulty covering the material in a comprehensive text or for those who wish to supplement their text with readers, literature, or activities. In condensing the book, we have retained all the central concepts, theories, and research findings. We have also retained the examples that help students grasp the basics of the discipline and ensure that the book will be engaging and readable.

Changes in the Fourth Edition

This edition of *Essentials of Sociology* is a thoroughly revised version of the previous edition of *Essentials*. This volume has been thoroughly updated. We have carefully reviewed all the major journals and many of the specialty journals in order to provide students with the most current findings. Figures and graphs have been revised to incorporate the very latest data available.

Second, the book has been revised to take into account recent topical developments in the field. Over the last few years, the discipline of sociology has become more structural, more interested in all aspects of diversity—including issues of global interdependency—and more concerned with the social impact of technology. The fourth edition of *Essentials of Sociology* reflects these changes. To reflect the concern with social structure, a new chapter on social networks and groups has been added to the text. Coverage of diversity issues has been expanded by including two new chapters on gender and age inequalities. More cross-national examples have been incorporated throughout the text. Chapter 10, for instance, includes a case study of declining life expectancies in Eastern Europe and Russia. In Chapter 3, the Focus On box features an analysis of "the strength of weak ties" in China and Singapore, and the Focus On box in Chapter 13 addresses Sweden's democratic socialist system. Material on the new Human Development Index is included in Chapter 7. Finally, the social impact of technological change has been expanded throughout the text. In addition to its extended treatment in Chapter 15, technology receives attention in a new Focus on Technology feature incorporated in seven of the chapters of this fourth edition. For example, the Focus On box in Chapter 11 analyzes the relationship between technology, gender roles, and kinkeeping; Chapter 9 includes a Focus On box, highlighting the ways in which technology can be an equity issue; and, in Chapter 5, the impact of technology on law enforcement is discussed.

Plan of the Book

The fourth edition of *Essentials of Sociology* retains the pedagogical features that have made it successful. Each chapter contains at least one high-interest boxed insert. Clearly identified concepts, concept summaries, and chapter summaries continue to aid students in mastering the material. New Critical Thinking questions have been added to each chapter. Useful as the basis for group discussion or as individual writing assignments, these questions challenge students to apply sociological concepts and theory to problems relevant to their immediate lives.

Focus On

A boxed insert in each chapter introduces provocative and interesting issues. These Focus Ons fall into four different formats: Focus on American Diversity, Focus on a Global Perspective, Focus on Application, and Focus on Technology. To demonstrate to students the importance of understanding the increasing diversity of American society, the Focus on American Diversity boxes examine issues such as the measurement of IQ, gay and lesbian families, and gender differences in mathematics. The Focus on Global Perspective series introduces students to a comparative approach to social issues and to social science research and deals with topics such as women and development, international migration, and survey research in Nigeria. The Focus on Application boxes address important social issues and policy concerns such as the environmental movement and who benefits from current government subsidy and welfare programs. Finally, the Focus on Technology series links technological innovation to social change by analyzing the implications of technology for social relationships, social control, and social institutions.

Chapter Summaries

A short point-by-point summary lists the chief points made in each chapter. These summaries will aid the beginning student in studying the text and distinguishing the central from the supporting points.

Concept Learning Aids

Learning new concepts and new vocabulary is vital to developing a new perspective. In *Essentials of Sociology*, this learning is facilitated in four ways. When new concepts first appear in the text, they are printed in boldface type, and complete definitions are set out clearly in the margin. Whenever several related concepts are introduced (for example, pluralist, power elite, conflict/dialectic, and state autonomy models of American government), a concept summary is included as a text figure to summarize the definitions, give examples, and clarify differences. Third, a glossary appears at the end of the book for handy reference. Finally, Critical Thinking questions encourage students to make concepts and terms a working part of their vocabulary by using them to discuss a problem of personal or social relevance.

Internet Resources

An entirely new feature of this edition is the addition of suggested Internet sites at the end of each chapter. Some of the most relevant sites on the net for finding information related to issues discussed in *Essentials of Sociology* have been identified. For example, browsers are directed to sites that specialize in Marxism (Chapter 1), deafness (Chapter 2), international population data (Chapter 14), and gay and lesbian issues (Chapter 4).

InfoTrac College Edition Exercises

Another new feature to the fourth edition is a series of InfoTrac College Edition exercises included at the end of each chapter. InfoTrac College Edition is an on-line database of 600 periodicals, journals, and magazines. Readers are asked to use specific InfoTrac articles in formulating their responses to the Critical Thinking questions in the text.

Student Study Guide

The student *Study Guide* will be invaluable in helping students master the material. The study guide contains an outline of each chapter, practice questions in matching, completion, multiple-choice, and essay form.

Acknowledgments

In preparing this edition as in the earlier editions, we have accumulated many debts. We are especially grateful for the good-natured and generous advice of our colleagues at the University of Nebraska at Lincoln. Special thanks go to Robert Benford, Miguel Carranza, Jay Corzine, Julie Harms Cannon, Kurt Johnson, Jennifer Lehmann, Helen Moore, Wayne Osgood, Keith Parker, and Al Williams.

They were always willing to share their expert knowledge and to comment and advise on our own forays into their substantive areas. Thanks also to Tim Pippert, who created the InfoTrac College Edition exercises in this edition.

Special thanks go to the people at West and Wadsworth Publishing, including Clyde Perlee who first prompted us to become authors and Denise Simon, who was generous with encouragement and advice. We also wish to thank our current editor, Eve Howard. We benefitted greatly from her knowledge of what makes a college textbook usable. A special thanks also to her assistant, Barbara Yien. Our production editor, Jodi Isman, played a crucial role in turning our material into an attractive finished product. Her patience and her skill are deeply appreciated. Our copy editor, Salvatore Allocco, not only saved us from technical gaffes and inconsistencies, but performed the important service of checking the manuscript for problematic writing and examples. His queries improved our writing and our thinking. At all levels, the people at Wadsworth have been a pleasure to work with—ready to help us make our book the best possible, while leaving the substance and direction of the book in our hands.

We would like to express our gratitude to those people who reviewed the manuscript for us:

Jon Ianniti
State University of New York–Morrisville

James Lindberg
Montgomery College

Carol Jenkins
Glendale Community College

Deidre Tyler
Salt Lake Community College

John Leib
Georgia State University

Once again, we thank those people who reviewed the manuscript for previous editions of *Sociology,* and *Essentials of Sociology.* Their suggestions and comments made a substantial contribution to the project: Margaret Abraham, Hofstra University, New York; Paul J. Baker, Illinois State University; Robert Benford, University of Nebraska; John K. Cochran, Wichita State University, Kansas; Carolie Coffey, Cabrillo College, California; Paul Colomy, University of Akron, Ohio; Ed Crenshaw, University of Oklahoma; Raymonda P. Dennis, Delgado Community College, New Orleans; Lynda Dodgen, North Harris County College, Texas; David A. Edwards, San Antonio College, Texas; Laura Eells, Wichita State University, Kansas; William Egelman, Iona College, New York; Constance Elsberg, Northern Virginia Community College; Christopher Ezell, Vincennes University, Indiana; Joseph Faltmeier, South Dakota State University; Daniel E. Ferritor, University of Arkansas; Charles E. Garrison, East Carolina University, North Carolina; James R. George, Kutztown State College, Pennsylvania; Harold C. Guy, Prince George Community College, Maryland; Rose Hall, Diablo Valley College, California; Michael G. Horton, Pensacola Junior College, Florida; Cornelius G. Hughes, University of Southern Colorado; Jon Ianitti, SUNY Morrisville; William C. Jenné, Oregon State University; Dennis L. Kalob, Loyola University, New Orleans; Sidney J. Kaplan, University of Toledo, Ohio; Florence Karlstrom, Northern Arizona University; Diane Kayongo-Male, South Dakota State University; William Kelly, University of Texas; James A. Kithens, North Texas State University; Phillip R. Kunz, Brigham Young University, Utah; Billie J. Laney, Central Texas College; Charles Langford, Oregon State University; Mary N. Legg, Valencia Community College, Florida; Joseph J. Leon, California State Polytechnic University, Pomona; J. Robert Lilly, Northern Kentucky University; Jan Lin, University of Houston;

Richard L. Loper, Seminole Community College, Florida; Carol May, Illinois Central College; Rodney C. Metzger, Lane Community College, Oregon; Vera L. Milam, Northeastern Illinois University; James S. Munro, Macomb College, Michigan; Lynn D. Nelson, Virginia Commonwealth University; J. Christopher O'Brien, Northern Virginia Community College; Charles O'Connor, Bemidji State University, Minnesota; Jane Ollenberger, University of Minnesota–Duluth; Robert L. Petty, San Diego Mesa College, California; Ruth A. Pigott, Kearney State College, Nebraska; John W. Prehn, Gustavus Adolphus College, St. Peter, Minnesota; Adrian Rapp, North Harris County College, Texas; Mike Robinson, Elizabethtown Community College, Kentucky; Will Rushton, Del Mar College, Texas; Rita P. Sakitt, Suffolk Community College, New York; Martin Scheffer, Boise State University, Idaho; Richard Scott, University of Central Arkansas; Ida Harper Simpson, Duke University, North Carolina; James B. Skellenger, Kent State Unversity, Ohio; Ricky L. Slavings, Radford University, Virginia; John M. Smith, Jr., Augusta College, Georgia; James Steele, James Madison University, Virginia; Michael Stein, University of Missouri–St. Louis; Barbara Stenross, University of North Carolina; Jack Stirton, San Joaquin Delta College, California; Emil Vajda, Northern Michigan University; Steven L. Vassar, Mankato State University, Minnesota; Jane B. Wedemeyer, Sante Fe Community College, Florida; Thomas J. Yacovone, Los Angeles Valley College, California; David L. Zierath, University of Wisconsin.

The Study of Society

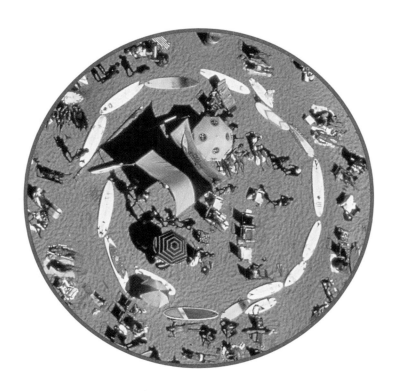

Outline

What Is Sociology?

Each of us starts the study of society with the study of individuals. We wonder why Theresa keeps getting involved with men who treat her badly, why Mike never learns to quit drinking before he gets sick, why our aunt puts up with our uncle, and why anybody likes the Spice Girls. We wonder why people we've known for years seem to change drastically when they get married or change jobs.

If Theresa were the only person who ever got into this predicament and if Mike were the only person who ever drank too much, then we might try to understand their behavior by peering into their personalities. We know, however, that there are thousands, maybe millions, of men and women who have disappointing romances and who drink too much. To understand Mike and Theresa, then, we must place them in a larger context and examine the forces that seem to compel so many people to behave in a similar way.

Sociologists tend to view these common human situations as if they were plays. They might, for example, title a common human drama *Boy Meets Girl.* Just as *Hamlet* has been performed all around the world for 400 years with different actors and different interpretations, *Boy Meets Girl* has also been performed countless times. Of course, the drama is acted out a little bit differently each time, depending on the scenery, the people in the lead roles, and the century—but the essentials are the same. Thus we can read nineteenth- or even sixteenth-century love stories and understand why those people did what they did. They were playing roles in a play that is still performed daily.

More formal definitions will be introduced later, but the metaphor of the theater can be used now to introduce two of the most basic concepts in sociology: role and social structure. By *role* we mean the expected performance of someone who occupies a specific position. Mothers have roles, teachers have roles, students have roles, and lovers have roles. Each position has an established script that suggests appropriate lines, gestures, and relationships with others. Discovering what each society offers as a stock set of roles is one of the major themes in sociology. Sociologists try to find the common roles that appear in society and to determine why some people play one role rather than another.

The other major sociological concept is *social structure,* which is concerned with the larger structure of the play in which the roles appear. What is the whole set of roles that appears in this play, and how are the roles interrelated? Thus the role of mother is understood in the context of the social structure we call the family. The role of student is understood in the context of the social structure we call education. Through these two major ideas, role and social structure, sociologists try to understand the human drama.

The Sociological Imagination

The **sociological imagination** is the ability to see the intimate realities of our own lives in the context of common social structures; it is the ability to see personal troubles as public issues.

The ability to see the intimate realities of our own lives in the context of common social structures has been called the **sociological imagination** (Mills 1959, 15). Sociologist C. Wright Mills suggests that the sociological imagination is developed when we can place such personal troubles as poverty, divorce, or loss of faith into a larger social context, when we can see them as common public issues. He suggests that many of the things we experience as individuals are really beyond our control. They have to do with society as a whole, its historical development, and the way it is organized. Mills gives us some examples of the differences between a personal trouble and a public issue:

> When, in a city of 100,000, only one man is unemployed, that is his personal trouble, and for its relief we properly look to the character of the man, his skills, and his immediate opportunities. But when in a nation of 50 million employees, 15 million men are unemployed, that is an issue, and we may not hope to find its solution within the range of opportunities open to any one individual. The very structure of opportunities has collapsed. Both the correct statement of the problem and the range of possible solutions require us to consider the economic and political institutions of the society, and not merely the personal situation and character of a scatter of individuals. . . .
>
> Consider marriage. Inside a marriage a man and a woman may experience personal troubles, but when the divorce rate during the first four years of marriage is 250 out of every 1000 attempts, this is an indication of a structural issue having to do with the institutions of marriage and the family and other institutions that bear on them. (Mills 1959, 9)

In everyday life, we do not define personal experiences in these terms. We frequently do not consider the impact of history and social structures on our own experiences. If a child becomes a drug addict, parents tend to blame themselves; if spouses divorce, their friends usually focus on their personality problems; if you flunk out of school, everyone will be likely to blame you personally. To develop the sociological imagination is to understand how outcomes such as these are, in part, a product of society and not fully within the control of the individual.

Some people flunk out of school, for example, not because they are stupid or lazy but because they are confused about just which play they are appearing in. The "this is the best time of your life" play calls for very different roles from the "education is the key to success" play. Other people may flunk out because they come from a social class that does not give them the financial or psychological support that they need. These students may be working 25 hours a week in addition to going to school; they may be going to school despite their family's indifference. In contrast, other students may find it difficult to fail: Their parents provide tuition, living expenses, a personal computer, a car, and moral support. As we will discuss in more detail in Chapter 12, parents' social class is one of the best predictors of who will fail and who will graduate. Success or failure is thus not entirely an individual matter; it is socially structured.

Sociological imagination, the ability to see our own lives and those of others as part of a larger social structure and a larger human drama, is central to sociology. Once we develop this imagination, we will be less likely to explain others' behavior through their personality and will increasingly look to the roles and social structures that determine behavior. We will also recognize that the solutions to many social problems lie not in changing individuals but in changing the social structures and roles that are available to them. Although poverty, divorce, illegitimacy, and racism are experienced as intensely personal hardships, they are unlikely to be reduced effectively through massive personal therapy. To solve these and many other social problems, we need to change social structures; we need to rewrite the play. Sociological imagination offers a new way to look at—and a new way to search for solutions to—the common troubles and dilemmas that face individuals.

These homeless people are obviously experiencing dire personal problems: no food, no home, no shelter, no money, no medical care. Unfortunately, there are somewhere between 300,000 and 3 million others who share the same circumstances—for roughly the same reasons. Although some homeless individuals may suffer from mental illness or substance abuse, the extent of homelessness largely reflects the lack of affordable housing and adequately paying jobs in the contemporary United States. Learning to see personal experiences and tragedies as part of larger patterns of social problems is a vital element of the sociological imagination.

THINKING

CRITICALLY

Which of your own personal troubles might reasonably be reframed as a public issue? Does such a reframing change the nature of the solutions you can see?

Sociology as a Social Science

Sociology is concerned with people and with the rules of behavior that structure the ways in which people interact. As one of the social sciences, sociology has much in common with political science, economics, psychology, and anthropology. All these fields share an interest in human social behavior and, to some extent, an interest in society. In addition, they all share an emphasis on the scientific method as the best approach to knowledge. This means that they rely on critical and systematic examination of the evidence before reaching any conclusions and their practitioners approach each research question from a position of moral neutrality—that is, they try to be objective observers. This scientific approach is what distinguishes the social sciences from journalism and other fields that comment on the human condition.

Sociology is a social science whose unique province is the systematic study of human social interaction. Its emphasis is on relationships and patterns of interaction—how these patterns develop, how they are maintained, and also how they change.

Sociology is the systematic study of human social interaction.

The Emergence of Sociology

Sociology emerged as a field of inquiry during the political, economic, and intellectual upheavals of the eighteenth and nineteenth centuries. Rationalism and science replaced tradition as methods of understanding the world, leading to changes in government, education, economic production, and even religion and family life. The clearest symbol of this turmoil is the French Revolution (1789), with its bloody uprising and rejection of the past.

Although less dramatic, the industrial revolution had an even greater impact. Within a few generations, traditional rural societies were replaced by industrialized urban societies. The rapidity and scope of the change resulted in substantial social disorganization. It was as if society had changed the play without bothering to tell the actors, who were still trying to read from old scripts. Although a few people prospered mightily, millions struggled desperately to make the adjustment from rural peasantry to urban working class.

The picture of urban life during these years—in London, Chicago, or Hamburg—was one of disorganization, poverty, and dynamic and exciting change. This turmoil and tragedy provided the inspiration for much of the intellectual effort of the nineteenth century: Charles Dickens's novels, Jane Addams's reform work, Karl Marx's revolutionary theory. It also inspired the scientific study of society. These were the years in which science was a new enterprise and nothing seemed too much to hope for. After electricity, the telegraph, and the X-ray, who was to say that science could not discover how to turn stones into gold or how to eliminate poverty or war? Many hoped that the tools of science could help in understanding and controlling a rapidly changing society.

The Founders: Comte, Martineau, Spencer, Marx, Durkheim, and Weber

Auguste Comte (1798–1857)

The first major figure to be concerned with the science of society was the French philosopher Auguste Comte. He coined the term *sociology* in 1839 and is generally considered the founder of this field.

Comte was among the first to suggest that the scientific method could be applied to social events (Konig 1968). The philosophy of positivism, which he developed, suggests

Auguste Comte, 1798–1857

that the social world can be studied with the same scientific accuracy and assurance as the natural world. Once the laws of social behavior were learned, he believed, scientists could accurately predict and control events. Although thoughtful people wonder whether we will ever be able to predict human behavior as accurately as we can predict the behavior of molecules, the scientific method remains central to sociology.

Another of Comte's lasting contributions was his recognition that an understanding of society requires a concern for both the sources of order and continuity and the sources of change. Comte called these divisions the theory of statics and the theory of dynamics. Although sociologists no longer use his terms, Comte's basic divisions of sociology continue under the labels of social structure (statics) and social process (dynamics).

Harriet Martineau (1802–1876)

Born into a wealthy English family, Harriet Martineau entered sociology at a time when few women were receiving any formal education. Initially, she helped to promote the development of the new discipline by translating Comte's work into English. A social activist and a sociologist in her own right, Martineau then used Comte's notions about the predictability and changeability of human society as a model for understanding and changing exploitative labor laws and the unfair treatment of women. Martineau traveled widely, writing about U.S. family customs, politics, religion, and race relations, while also serving as a forceful advocate for the abolition of slavery.

Harriet Martineau, 1802–1876

Herbert Spencer (1820–1903)

Another pioneer in sociology was the British philosopher-scientist Herbert Spencer, who advanced the thesis that evolution accounts for the development of social, as well as natural, life. Spencer viewed society as similar to a giant organism: Just as the heart and lungs work together to sustain the life of the organism, so the parts of society work together to maintain society. Spencer's analogy led him to some conclusions that seem foolish by modern standards, but they also led him to some basic principles that still guide the study of sociology.

One of the Spencer's guiding principles was that society must be understood as an adaptation to its environment. This principle of adaptation implies that to understand society, we must focus on processes of growth and change. It also implies that there is no "right" way for a society to be organized. Instead, societies will change as circumstances change.

Spencer's second major contribution was his concern with the scientific method. More than many scholars of his day, Spencer was aware of the importance of objectivity and moral neutrality in investigation. In essays on the bias of class, the bias of patriotism, and the bias of theology, he warned sociologists that they must suspend their own opinions and wishes when studying the facts of society (Turner & Beegley 1981).

Karl Marx (1818–1883)

A philosopher, economist, and social activist, Karl Marx was born in Germany to middle-class Jewish parents. Marx received his doctorate in philosophy at the age of 23, but because of his radical views he was unable to obtain a university appointment and spent most of his adult life in exile and poverty (Rubel 1968).

Marx was repulsed by the poverty and inequality that characterized the nineteenth century. Unlike other scholars of his day, he was unwilling to see poverty as either a natural or a God-given condition of the human species. Instead, he viewed

Herbert Spencer, 1820–1903

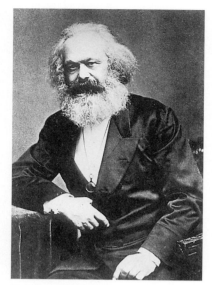

Karl Marx, 1818–1883

Economic determinism means that economic relationships provide the foundation on which all other social and political arrangements are built.

Dialectic philosophy views change as a product of contradictions and conflict between the parts of society.

Emile Durkheim, 1858–1917

poverty and inequality as human-made conditions fostered by private property and capitalism. As a result, he devoted his intellectual efforts to understanding—and eliminating—capitalism. Many of Marx's ideas are of more interest to political scientists and economists than to sociologists, but he left two enduring legacies to sociology: the theories of economic determinism and the dialectic.

ECONOMIC DETERMINISM. Marx began his analysis of society by assuming that the most basic task of any human society is to provide food and shelter to sustain itself. Marx argued that the ways in which society does this—its modes of production—provide the foundations on which all other social and political arrangements are built. Thus he believed that family, law, and religion all develop after and adapt to the economic structure; in short, they are determined by economic relationships. This idea is called **economic determinism.**

A good illustration of economic determinism is the influence of economic conditions on marriage choices. In traditional agricultural societies, young people often remain economically dependent upon their parents until well into adulthood because the only economic resource, land, is controlled by the older generation. In order to support themselves now and in the future, they must remain in their parents' good graces; this means they cannot marry without their parents' approval. In societies where young people can earn a living without their parents' help, however, they can marry when they please and whomever they please. Marx would argue that this shift in mate selection practices is the result of changing economic relationships. Because Marx saw all human relations as stemming ultimately from the economic systems, he suggested that the major goal of a social scientist is to understand economic relationships: Who owns what, and how does this pattern of ownership affect human relationships?

THE DIALECTIC. Marx's other major contribution was a theory of social change. Many nineteenth-century scholars applied Darwin's theories of biological evolution to society; they believed that social change was the result of a natural process of adaptation. Marx, however, argued that the basis of change was conflict, not adaptation. He argued that conflicts between opposing economic interests lead to change.

Marx's thinking on conflict was influenced by the German philosopher George Hegel, who suggested that for every idea (thesis), a counteridea (antithesis) develops to challenge it. As a result of conflict between the two ideas, a new idea (synthesis) is produced. This process of change is called the **dialectic** (see Figure 1.1).

Marx's contribution was to apply this model of ideological change to change in economic and material systems. Within capitalism, Marx suggested, the capitalist class was the thesis and the working class was the antithesis. He predicted that conflicts between them would lead to a new synthesis, a new economic system that would be socialism. Indeed, in his role as social activist, Marx hoped to encourage conflict and ignite the revolution that would bring about the desired change. The workers, he declared, "have nothing to lose but their chains" (Marx & Engels [1848] 1965).

Although few sociologists are revolutionaries, many accept Marx's ideas on the importance of economic relationships and economic conflicts. Much more controversial is Marx's argument that the social scientist should also be a social activist, a person who not only tries to understand social relationships but also tries to change them.

Emile Durkheim (1858–1917)

Like Marx, Durkheim was born into a middle-class family. While Marx was starving as an exile in England, however, Durkheim spent most of his career occupying a prestigious professorship at the Sorbonne. Far from rejecting society, Durkheim embraced it, and much of his outstanding scholarly energy was devoted to understanding the stability of society and the importance of social participation for individual happiness. Whereas the lasting legacy of Marx is a theory that looks for the conflict-laden and changing aspects of social practices, the lasting legacy of Durkheim is a theory that examines the positive contributions of social patterns. Together they allow us to see both order and change.

Durkheim's major works are still considered essential reading in sociology. These include his studies of suicide, education, divorce, crime, and social change. Two enduring contributions are his ideas about the relationship between individuals and society and the development of a method for social science.

One of Durkheim's major concerns was the balance between social regulation and personal freedom. He argued that community standards of morality, which he called the collective conscience, not only confine our behavior but also give us a sense of belonging and integration. For example, many people complain about having to dress up; they complain about having to shave their faces or their legs, having to wear a tie or pantyhose. "What's wrong with my jeans?" they want to know. At the same time, most of us feel a sense of satisfaction when we appear in public in our best clothes. We know that we will be considered attractive and successful. Although we may complain about having to meet what appear to be arbitrary standards, we often feel a sense of satisfaction in being able to meet those standards successfully. In Durkheim's words, "institutions may impose themselves upon us, but we cling to them; they compel us, and we love them" (Durkheim [1895] 1938, 3). This beneficial regulation, however, must not rob the individual of all freedom of choice.

In his classic study, *Suicide,* Durkheim identified two types of suicide that stem from an imbalance between social regulation and personal freedom. Fatalistic suicide occurs when society overregulates and allows too little freedom, when our behavior is so confined by social institutions that we cannot exercise our independence ([1897] 1951, 276). Durkheim gave as an example of fatalistic suicide in the very young husband who feels overburdened by the demands of work, household, and family. Anomic suicide, on the other hand, occurs when there is too much freedom and too little regulation, when society's influence does not check individual passions ([1897] 1951, 258). Durkheim said that this kind of suicide is most likely to occur in times of rapid social change. When established ways of doing things have lost their meaning, but no clear alternatives have developed, individuals feel lost. The high suicide rate of Native Americans (approximately twice that of Caucasian Americans) is generally attributed to the weakening of traditional social regulation.

Durkheim was among the first to stress the importance of using reliable statistics to examine theories of social life. Each of his works illustrates his ideal social scientist: an objective observer who only seeks the facts. As sociology became an established discipline, this ideal of objective observation replaced Marx's social activism as the standard model for social science.

Max Weber (1864–1920)

A German economist, historian, and philosopher, Max Weber (Veh-ber) provided the theoretical base for half a dozen areas of sociological inquiry. He wrote on religion, bureaucracy, method, and politics. In all these areas, his work is still valuable and insightful; it is covered in detail in later chapters. Three of Weber's more general contributions were an emphasis on the subjective meanings of social actions, on social as opposed to material causes, and on the need for objectivity in studying social issues.

Weber believed that knowing patterns of behavior was less important than knowing the meanings people attach to behavior. For example, Weber would argue that it is relatively meaningless to compile statistics such as one-half of all marriages contracted

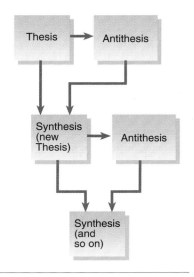

FIGURE 1.1
The Dialectic
The dialectic model of change suggests that change occurs through conflict and resolution rather than through evolution.

Max Weber, 1864–1920

today may end in divorce compared to only 10 percent in 1890 (Cherlin 1992). More critical, he would argue, is understanding how the meaning of divorce has changed in the past hundred years. Weber's emphasis on the subjective meanings of human actions has been the foundation of scholarly work on topics as varied as religion and immigration.

Weber trained as an economist, and much of his work concerned the interplay of things material and things social. He rejected Marx's idea that economic factors were the determinants of all other social relationships. In a classic study, *The Protestant Ethic and the Spirit of Capitalism,* Weber tried to show how social and religious values may be the foundation of economic systems. This argument is developed more fully in Chapter 12, but its major thesis is that the religious values of early Protestantism (self-discipline, thrift, and individualism) were the foundation for capitalism.

One of Weber's more influential ideas was his declaration that sociology must be **value-free.** Weber argued that sociology should be concerned with establishing what is and not what ought to be. Weber's dictum is at the heart of the standard scientific approach that is generally advocated by modern sociologists. Thus, although one may study poverty or racial inequality because of a sense of moral outrage, such feelings must be set aside to achieve an objective grasp of the facts. This position of neutrality is directly contradictory to the Marxian emphasis on social activism, and sociologists who adhere to Marxist principles generally reject the notion of value-free sociology. Most modern sociologists, however, try to be value-free in their scholarly work.

Value-free sociology concerns itself with establishing what is, not what ought to be.

Sociology in the United States

Sociology in the United States developed somewhat differently than it did in Europe. Although U.S. sociology has the same intellectual roots as European sociology, it has some distinctive characteristics. Three features that have characterized U.S. sociology from its beginning are a concern with social problems, a reforming rather than a radical approach to these problems, and an emphasis on the scientific method.

Baptism is a religious ritual common to most Christian faiths. We can study what baptism means in Christian theology, we can compile statistics on the percentage of the population that has been baptized, or we can follow Weber's emphasis on subjective meanings by asking what it means to the individuals who take part in it. The typical Presbyterian baptism in which an infant's head is sprinkled with a few drops of water during a formal service is quite different in symbolic meaning from this woman's baptism by immersion in the River Jordan.

One reason that U.S. sociology developed differently from European sociology is that our social problems differ. Slavery, the Civil War, and high immigration rates, for example, made racism and ethnic discrimination much more salient issues in the United States. One of the first sociologists to study these issues was W. E. B. DuBois. DuBois, who received his doctorate in 1895 from Harvard University, devoted his career to developing empirical data about African Americans and to using those data to combat racism. The work of Jane Addams, another early sociologist, founder of Hull House and recipient of the 1931 Nobel Peace Prize, nicely illustrates the reformist approach of much U.S. sociology. Addams and the other women who lived at Hull House used quantitative social science data to lobby successfully for legislation on safer working conditions, a better juvenile justice system, better public sanitation, and services for the poor (Deegan 1987).

As sociology became more established, it also became more conservative. In the years between the two World Wars, a new generation of sociology professors became convinced that social activism was incompatible with academic respectability. However, by the 1950s and into the 1960s, sociologists such as C. Wright Mills and Ralph Dahrendorf turned renewed attention to social problems and social conflict.

The first sociology course in the United States was taught at Yale University in 1876. By 1910, most colleges and universities in the United States offered sociology courses, although separate departments were slower to develop. Most of the courses were offered jointly with other departments, most often with economics but frequently with history, political science, philosophy, or general social science departments.

By 1960, almost all colleges and universities had departments of sociology, and by 1990, 120 of them offered doctoral programs. Higher-degree sociology programs are more popular in the United States than in any other country in the world. This is partly because sociology in the United States has always been oriented toward the practical as

W. E. B. DuBois (1868–1963)

Jane Addams (1860–1935)

well as the theoretical. The focus has consistently been on finding solutions to social issues and problems, with the result that U.S. sociologists not only teach sociology but also work in government and industry.

As recognition that the solution to problems such as AIDS, environmental degradation, poverty, and ethnic conflict requires international effort, U.S. sociologists are developing closer working relationships with their counterparts in agencies and universities throughout the world.

Current Perspectives in Sociology

As this brief review of the history of sociological thought has demonstrated, there are many ways of approaching the study of human social interaction. The ideas of Marx, Weber, Durkheim, and others have given rise to dozens of theories about human behavior. In this section, we bring together and summarize ideas that are the foundation of the three dominant theoretical perspectives in sociology today: structural functionalism, conflict theory, and symbolic interactionism.

Structural-Functional Theory

Structural-functional theory addresses the question of social organization and how it is maintained.

Structural-functional theory addresses the question of social organization and how it is maintained. This theoretical perspective has its roots in natural science and the analogy between society and an organism. In the analysis of a living organism, the scientist's task is to identify the various parts (structures) and to determine how they work (function). In the study of society, a sociologist with this perspective tries to identify the structures of society and how they function, hence the name *structural functionalism*.

The Assumptions Behind Structural Functionalism

In the sense that any study of society must begin with an identification of the parts of society and how they work, structural functionalism is basic to all perspectives. Scholars who use this perspective, however, are distinguished from other social analysts by their reliance on three major assumptions:

1. *Stability.* The chief evaluative criterion for any social pattern is whether it contributes to the maintenance of society.
2. *Harmony.* As the parts of an organism work together for the good of the whole, so the parts of society are also characterized by harmony.
3. *Evolution.* Change occurs through evolution—the adaptation of social structures to new needs and demands and the elimination of unnecessary or outmoded structures.

Because it emphasizes harmony and adaptation, structural functionalism is sometimes called consensus theory.

Structural-Functional Analysis

A structural-functional analysis asks two basic questions: What is the nature of this social structure (what patterns exist)? What are the consequences of this social structure (does it promote stability and harmony)? In this analysis, positive consequences are

MAP 1.1
Sociologists Around the World Who Are Members of the International Sociological Association, 1995

Sociology is a global discipline, although the vast majority of members in the International Sociological Association are citizens of wealthy, western nations. Many of these European and U.S. sociologists study developing regions of Asia, South America, or Africa. But indigenous peoples increasingly bring their own experiences to analyzing their societies, especially in issues concerning population and development.

SOURCE: Data supplied by the International Sociological Association.

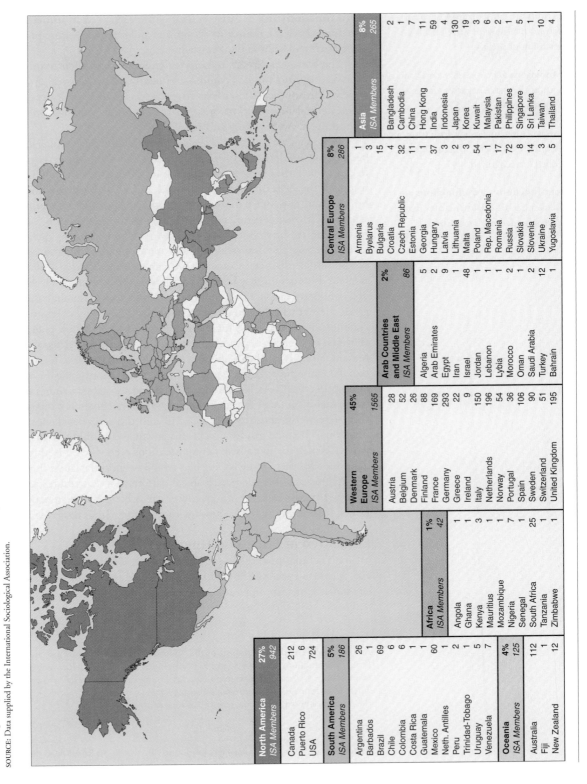

North America	27% 942
ISA Members	
Canada	212
Puerto Rico	6
USA	724

South America	5% 186
ISA Members	
Argentina	26
Barbados	1
Brazil	69
Chile	6
Colombia	6
Costa Rica	1
Guatemala	1
Mexico	60
Neth. Antilles	1
Peru	2
Trinidad-Tobago	1
Uruguay	5
Venezuela	7

Africa	1% 42
ISA Members	
Angola	1
Ghana	1
Kenya	3
Mauritius	1
Mozambique	1
Nigeria	7
Senegal	1
South Africa	25
Tanzania	1
Zimbabwe	1

Oceania	4% 125
ISA Members	
Australia	112
Fiji	1
New Zealand	12

Western Europe	45% 1565
ISA Members	
Austria	28
Belgium	52
Denmark	26
Finland	88
France	169
Germany	293
Greece	22
Ireland	9
Italy	150
Netherlands	196
Norway	54
Portugal	36
Spain	106
Sweden	90
Switzerland	51
United Kingdom	195

Arab Countries and Middle East	2% 86
ISA Members	
Algeria	5
Arab Emirates	2
Egypt	9
Iran	1
Israel	48
Jordan	1
Lebanon	1
Libya	1
Morocco	2
Oman	1
Saudi Arabia	2
Turkey	12
Bahrain	1

Central Europe	8% 286
ISA Members	
Armenia	1
Byelarus	3
Bulgaria	15
Croatia	4
Czech Republic	32
Estonia	11
Georgia	1
Hungary	37
Latvia	3
Lithuania	2
Malta	3
Poland	54
Rep. Macedonia	1
Romania	17
Russia	72
Slovakia	8
Slovenia	2
Ukraine	12
Yugoslavia	1

Asia	8% 265
ISA Members	
Bangladesh	2
Cambodia	1
China	7
Hong Kong	11
India	59
Indonesia	4
Japan	130
Korea	19
Kuwait	3
Malaysia	6
Pakistan	2
Philippines	1
Singapore	5
Sri Lanka	1
Taiwan	10
Thailand	4

Functions are consequences of social structures that have positive effects on the stability of society.

Dysfunctions are consequences of social structures that have negative effects on the stability of society.

Manifest functions or dysfunctions are consequences of social structures that are intended or recognized.

Latent functions or dysfunctions are consequences of social structures that are neither intended nor recognized.

called **functions** and negative consequences are called **dysfunctions.** A distinction is also drawn between **manifest** (recognized and intended) consequences and **latent** (unrecognized and unintended) consequences.

The basic strategy of looking for structures and their manifest and latent functions and dysfunctions is common to nearly all sociological analysis. Scholars from widely different theoretical perspectives use this framework for examining society. What sets structural-functional theorists apart from others who use this language are their assumptions about harmony and stability.

Many states are currently considering legislation that would allow women who have been victims of domestic violence to use the "battered women's syndrome" as a defense in cases where they subsequently assault or kill their abuser. Such laws would explicitly recognize the right of women who assault or kill an abusive partner to plead not guilty by reason of temporary insanity. What would be the consequences of this new social structure? Its manifest function (intended positive outcome) is, of course, to give legal recognition to the devastating long-term psychological consequences of domestic violence. The manifest dysfunction is that some offenders might use the battered women's syndrome defense as an excuse for a malicious, premeditated assault on a significant other. A latent dysfunction may be that women who are acquitted of legal charges on the basis of a temporary insanity plea could find it difficult to maintain custody of their children, given the stigma often attached to individuals with any diagnosis of mental disorder. Another latent outcome is more difficult to classify: The new policy may perpetuate the view that women are dependent on men. Is this persisting viewpoint a function or a dysfunction? This is a difficult question to answer from a neutral point of view, and it is here that the assumptions behind structural-functional theory guide the analysis. Following the assumption that the major criterion for judging a social structure is whether it contributes to the maintenance of society, structural-functional analysis has tended to call structures that preserve the status quo "functions" and those that challenge the status quo "dysfunctions." Because gender bias in this law may contribute to an established pattern of women remaining in abusive family situations even when it is physically or emotionally dangerous for them to do so, the bias would be judged a latent function (see Table 1.1).

As this example suggests, a social pattern that contributes to the maintenance of society may benefit some groups more than others. A pattern may be functional—that is, help maintain the status quo—without being either desirable or equitable.

Evaluation of Structural Functionalism

Structural-functional theory tends to produce a static and conservative analysis of social systems (Turner 1982). This tendency is not a requirement for functional analysis, but it is commonplace. For example, an enumeration of the ways in which the

TABLE 1.1

A Structural-Functional Analysis of the "Battered Women's Syndrome" as a Legal Defense
Structural-functional analysis examines the intended and unintended consequences of social structures. It also assesses whether the consequences are positive (functional) or negative (dysfunctional). There is no moral dimension to the assessment that an outcome is positive; it merely means that the outcome contributes to the stability of society.

	Manifest	Latent
Function	Gives legal recognition to the psychological consequences of domestic violence	Encourages the dependence of women on men
Dysfunction	May contribute to abuses of the criminal justice system	Makes it more difficult for victims of domestic violence to maintain custody

Structural-functional theory focuses on the benefits that social structures provide for individuals and society. It argues that the regularity and routine provided by such social structures as the family and government are as necessary as the regulations governing a school crossing. Theorists with this perspective often have a tendency to focus on the advantages rather than the disadvantages of particular social structures.

education system contributes to the maintenance of inequality (an argument outlined in Chapter 12) is not the same thing as saying that maintaining inequality is a good thing. The distinction is a fine one, however, and in general structural functionalism tends to be a more attractive perspective for those who want to preserve the status quo than for those who want to challenge it.

Conflict Theory

If structural-functional theory sees the world in terms of consensus and stability, then it can be said that conflict theory sees the world in terms of conflict and change. Conflict theorists contend that a full understanding of society requires a critical examination of the competition and conflict in society, especially of the processes by which some people are winners and others losers. As a result, **conflict theory** addresses the points of stress and conflict in society and the ways in which they contribute to social change.

Conflict theory addresses the points of stress and conflict in society and the ways in which they contribute to social change.

Assumptions Underlying Conflict Theory

Conflict theory is derived from Marx's ideas. The following are three primary assumptions of modern conflict theory:

1. *Competition.* Competition over scarce resources (money, leisure, sexual partners, and so on) is at the heart of all social relationships. Competition rather than consensus is characteristic of human relationships.
2. *Structural inequality.* Inequalities in power and reward are built into all social structures. Individuals and groups that benefit from any particular structure strive to see it maintained.
3. *Revolution.* Change occurs as a result of conflict between competing interests rather than through adaptation. It is often abrupt and revolutionary rather than evolutionary.

THINKING

CRITICALLY

Consider how a structural-functional analysis of gender roles might differ from a conflict analysis. Are men or women more likely to choose conflict theory?

Conflict Analysis

Like structural functionalists, conflict theorists are interested in social structures. The two questions they ask, however, are different. Conflict theorists ask: Who benefits from those social structures? How do those who benefit maintain their advantage?

A conflict analysis of modern education, for example, notes that the highest graduation rates, the best grades, and the highest monetary returns per year of education go to students from advantaged backgrounds. The answer to the question "Who benefits?" is that educational benefits go to the children of those who are already well off. Conflict theorists go on to ask how this situation developed and how it is maintained. Their answers (developed more extensively in Chapter 12) focus on questions such as how are educational resources (texts, teachers, school buildings) allocated by neighborhood and whether the curriculum is designed for one kind of child (white middle class) rather than other kinds. They also look for ways in which this system benefits the powerful—for example, by creating a class of nongraduates who can be hired cheaply.

Evaluation of Conflict Theory

Thirty years ago, sociology was dominated by structural-functional theory, but conflict theory has become increasingly popular. It allows us to ask many of the same questions as structural-functional theory (What is a certain social structure? What are its outcomes?), but it also encourages us to take a more critical look at outcomes; for example, this particular structure is functional for whom? Together the two perspectives provide a balanced view, allowing us to analyze the sources of both conflict and harmony, order and change.

Conflict theory tends to produce a critical picture of society, and the emphasis on social activism and social criticism that is at the heart of conflict theory tends to attract scholars who would like to change society. In general, conflict theorists place less emphasis than other sociologists on the importance of value-free sociology.

What did Jewish immigrants to New York City in the early twentieth century have in common with Latinas now living and working in Los Angeles? The answer is that most were/are poorly paid, with many finding work in garment industry sweatshops such as the one pictured here. Conflict theory takes a critical look at who benefits from the low wages of immigrants (and women) and at what social structures serve to maintain this centuries-old pattern.

Symbolic Interaction Theory

Both structural-functional and conflict theories focus on social structures and the relationships between them. What about the relationship between individuals and social structures? Sociologists who focus on the ways that individuals relate to and are affected by social structures generally use symbolic interaction theory. **Symbolic interaction theory** addresses the subjective meanings of human acts and the processes through which we come to develop and share these subjective meanings. The name of this theory comes from the fact that it studies the *symbolic* (or subjective) meaning of human *interaction*.

Symbolic interaction theory addresses the subjective meanings of human acts and the processes through which people come to develop and communicate shared meanings.

Assumptions Underlying Symbolic Interaction Theory

When symbolic interactionists study human behavior, they begin with three major premises (Blumer 1969):

1. *Symbolic meanings are important.* Any behavior, gesture, or word can have multiple interpretations (can symbolize many things). In order to understand human behavior, we must learn what it means to the participants.
2. *Meanings grow out of relationships.* When relationships change, so do meanings.
3. *Meanings are negotiated.* We do not accept others' meanings uncritically. Each of us plays an active role in negotiating the meaning that things will have for us.

THINKING

CRITICALLY

Can you think of situations where a change of friends, living arrangements, or jobs has caused you to have new interpretations of the events surrounding you?

Symbolic Interaction Analysis

These premises direct symbolic interactionists to the study of how individuals are shaped by relationships and social structures. For example, symbolic interactionists would be interested in how growing up in a large as opposed to a small family or in a working-class as opposed to an upper-class family affects individual attitudes and behaviors.

Symbolic interactionists are also interested in the active role of the individual in modifying and negotiating his or her way through these relationships. Why do two children raised in the same family turn out differently? The answer lies in part in the fact that each child experiences subtly different relationships and situations; the meanings that the youngest child derives from the family experience may be different from those the oldest child derives.

Most generally, symbolic interaction is concerned with how individuals are shaped by relationships. This question leads first to a concern with childhood and the initial steps we take to learn and interpret our social worlds. It is also concerned with later relationships with lovers and friends, employers, and teachers.

Evaluation of Symbolic Interaction Theory

The value of symbolic interaction is that it focuses attention on the personal relationships and encounters that are so important in our everyday lives. By showing how the relationships dictated by the larger social structure affect our subjective worlds, symbolic interactionists give us a more complete picture of these social structures.

Neither symbolic interactionism nor the conflict or structural-functional theories are complete in themselves. Symbolic interactionism focuses on individual relationships, and the other two theories focus largely on society. Together, however,

CONCEPT SUMMARY	*Major Theoretical Perspectives in Sociology*		
	STRUCTURAL FUNCTIONALISM	**CONFLICT THEORY**	**SYMBOLIC INTERACTIONISM**
Nature of society	Interrelated social structures that fit together to form an integrated whole	Competing interests, each seeking to secure its own ends	Interacting individuals and groups
Basis of interaction	Consensus and shared values	Constraint, power, and competition	Shared symbolic meanings
Major question	What are social structures? Do they contribute to social stability?	Who benefits? How are these benefits maintained?	How do social structures relate to individual subjective experiences?
Level of analysis	Social structure	Social structure	Interpersonal interaction

they provide a valuable set of tools for understanding the relationship between the individual and society.

Interchangeable Lenses

As this brief review of major theoretical perspectives illustrates, the field of sociology uses a variety of theoretical perspectives. These perspectives can be regarded as interchangeable lenses through which society may be viewed. Just as a telephoto lens is not always superior to a wide-angle lens, one sociological theory will not always be superior to another.

Occasionally, the same subject can be viewed through any of these perspectives. For example, one can examine prostitution through the theoretical lens of structural-functional, conflict, or symbolic interaction theory. Following are three snapshots of female prostitution using these perspectives.

The Functions of Prostitution

The functional analysis of female prostitution begins by examining its social structure. It identifies the recurrent patterns of relationships among pimps, prostitutes, and customers. Then it examines the consequences of this social structure. In 1961, Kingsley Davis listed the following outcomes of prostitution:

- It provides a sexual outlet for men who cannot compete on the marriage market—the physically or mentally handicapped or the very poor.
- It provides a sexual outlet for men who are away from home a lot, such as salesmen and sailors.
- It provides a sexual outlet for the kinky.

Provision of these services is the manifest or intended function of prostitution. Davis goes on to note that, by providing these services, prostitution has the latent function of protecting the institution of marriage from malcontents who, for one reason or another, do not receive adequate sexual service through marriage. Prostitution is the safety valve that makes it possible to restrict respectable sexual relationships (and hence child-bearing and childrearing) to marital relationships while still allowing for the variability of human sexual appetites.

Prostitution: Marketing a Scarce Resource

Conflict theorists analyze prostitution as part of the larger problem of unequal allocation of scarce resources. Women, they argue, have not had equal access to economic opportunity. In some societies, they are forbidden to own property; in others, they suffer substantial discrimination in opportunities to work and earn. Because of this inability to support themselves, women have had to rely on economic support from men. They get this support by exchanging the one scarce resource they have to offer: sexual availability. To a Marxist, it makes little difference whether a woman barters her sexual availability by the job (prostitution) or by contract (marriage); the underlying cause is the same.

Although most analyses of prostitution focus on adult women, the conflict perspective helps explain the growing problem of prostitution among runaway and homeless boys and girls. The young people have few realistic opportunities to support themselves by regular jobs: many are not old enough to work legally and, in any case, would be unable to support themselves adequately on the minimum wage. Their young bodies are their most marketable resource.

Prostitution: Learning the Trade

Symbolic interactionists who examine prostitution will take an entirely different perspective. They will want to know how prostitutes learn the trade and how they manage their self-concept so that they continue to think positively of themselves in spite of engaging in a socially disapproved profession. One such study was done by Barbara Heyl, who intensively interviewed a middle-aged woman who had spent her career first as a prostitute and then as a madam and trainer of prostitutes. Heyl found that much of the training in the prostitute's role consists of business training, not sex. They learn

Prostitution occurs in nearly every culture and every historical period. Consequently, sociologists have been interested in trying to explain why this pattern of social behavior is so persistent and what meaning it has to its participants. Conflict sociologists have focused on prostitution as an economic relationship created by differential access to legitimate rewards. Structural-functionalists have emphasized that prostitution is voluntary and benefits both prostitutes and society. Symbolic interactionists have analyzed the social setting where prostitution occurs, the gestures, vocabularies, and symbols used to transact business.

how to hustle—how to get the maximum amount of money for the minimum amount of work. In speaking of what her training produces, the madam says she is turning out professional hustlers, not whores. She is proud of her work. She says, "They find that I am teaching them how to make money, to dress tastefully, to conserve and be poised with men, to be knowledgeable about good hygiene, to have good working habits, such as punctuality, which will help them whether they stay in the rackets or not, and to have self-respect" (Heyl 1979, 105).

Summary

As these examples illustrate, many topics can be fruitfully studied with any of the three theoretical perspectives. Just as a photographer with only one lens can shoot almost any subject, the sociologist with only one perspective will not be unduly limited in what to examine. One will generally get better pictures, however, by selecting the theoretical perspective that is best suited to the particular subject. In general, structural functionalism and conflict theory are well suited to the study of social structures, or **macrosociology.** Symbolic interactionism is well suited to the study of the relationship between individual meanings and social structures, or **microsociology.**

> **Macrosociology** focuses on social structures and organizations and the relationships between them.

> **Microsociology** focuses on interactions among individuals.

The Science of Society

The things that sociologists study—for example, deviance, marital happiness, and poverty—have probably interested you for a long time. You may have developed your own opinions about why some people have good marriages and some have bad marriages or why some people break the law and others do not. Sociology is an academic discipline that uses the procedures of science to critically examine commonsense explanations of human social behavior. Science is not divorced from common sense but is an extension of it.

Defining Science

The ultimate aim of science is to better understand the world. Science directs us to find this understanding by observing and measuring what actually happens. This is not the only means of acquiring knowledge. Some people get their perceptions from the Bible or the Koran or the Book of Mormon. Others get their answers from their mothers or their husbands or their girlfriends. When you ask such people, "But how do you know that that is true?" their answer is simple. They say "My mother told me" or "I read it in *Reader's Digest.*"

> **Science** is a way of knowing based on empirical evidence.

    Science differs from these other ways of knowing in that it requires empirical evidence as a basis for knowledge; that is, it requires evidence that can be confirmed by the normal human senses. We must be able to see, hear, smell, or feel it. Before a social scientist would agree that she or he "knew" religious differences increased the likelihood of divorce, for example, she or he would want to see statistical evidence.

    Science has two major goals: accurate description and accurate explanation. In sociology, we are concerned with accurate descriptions of human interaction (how many people marry, how many people abuse their children, how many people flunk out of school). After we know the patterns, we hope to be able to explain them, to say *why* people marry, abuse their children, or flunk out of school.

The Research Process

At each stage of the scientific process, certain conventional procedures are used to ensure that a researcher's findings will be accepted as scientific knowledge. The procedures used in sociological research are covered in depth in classes on research methods, statistics, and theory construction. At this point, we merely want to introduce a few ideas that you must understand if you are to be an educated consumer of research results. We look at the general research process and review three concepts central to research: variables, operational definitions, and sampling.

Step One: Stating the Problem

The first step in the research process is the careful statement of the issue to be investigated. We may select a topic because of a personal experience or out of commonsense observation. For example, we may have observed that black Americans appear more likely to experience unemployment and poverty than white Americans. Alternatively, we might begin with a theory that predicts, for instance, that black Americans will have higher unemployment rates than white Americans because they have been discriminated against in schools and the workplace. In either case, we begin by reviewing the research of other scholars to help us specify exactly what it is that we want to know. If a good deal of research has already been conducted on the issue and good theoretical explanations have been advanced for some of the patterns, then a problem may be stated in the form of a **hypothesis**—a statement about empirical relationships that we expect to observe if our theory is correct. A hypothesis must be testable; that is, there must be some way in which data can help weed out a wrong conclusion and identify a correct one. For example, the hypothesis that mothers ought to stay home with their children cannot be tested; the hypothesis that children who spend their early years in day-care centers are emotionally less secure than those who stay at home with their mothers can be tested.

A **hypothesis** is a statement about relationships that we expect to find if our theory is correct.

Step Two: Gathering Data

To narrow the scope of a problem to manageable size, researchers focus on variables rather than people.

VARIABLES.   **Variables** are measured characteristics that vary from one individual or group to the next (Babbie 1995). If we wish to analyze differences in rates of black/white unemployment, we need information on two variables: race and unemployment. The individuals included in our study would be complex and interesting human beings, but for our purposes, we would be interested only in these two aspects of each person's life.

Variables are measured characteristics that vary from one individual or group to the next.

When we hypothesize a cause-and-effect relationship between two variables, the cause is called the **independent variable,** and the effect is called the **dependent variable.** In the example of the effects of day care on children's emotional security, attending day care is the independent variable, and emotional security is the dependent variable; that is, we hypothesize that security depends on whether or not a child attends a day-care program.

The **independent variable** is the variable that does the causing in cause-and-effect relationships.

The **dependent variable** is the effect in cause-and-effect relationships. It is dependent on the actions of the independent variable.

OPERATIONAL DEFINITIONS.   In order to describe a pattern or test a hypothesis, each variable must be precisely defined. Before we can describe racial differences in unemployment rates, for instance, we need to be able to decide whether an individual is unemployed. To assess the effects of day care on emotional security, we must be able

Operational definitions describe the exact procedures by which a variable is measured.

to sort children into at least two categories: those who attend day care and those who do not. The exact procedure by which a variable is measured is called an **operational definition.** Reaching general agreement about these definitions may pose a problem. For instance, people are typically considered to be unemployed if they are actively seeking work but cannot find it. This definition ignores all the people who became so discouraged in their search for work that they simply gave up. Obviously, including discouraged workers in our definition of the unemployed might lead to a different description of patterns of unemployment. In the day-care example, exposure to day care might be defined simply in terms of whether a child attends a day-care program, but it might also be defined in terms of the number of hours per week a child spends in a day-care facility or the number of years a child has been attending. Again, research results may vary depending on how the day-care experience is measured. Consumers of research should always check carefully to see what operational definitions are being used when they evaluate results.

SAMPLING.   It would be time-consuming, expensive, and probably nearly impossible to get information on race and employment status for all adults or on the emotional security and day-care experiences of all children. It is also unnecessary. The process of **sampling**—taking a systematic selection of representative cases from a larger population—allows us to get accurate empirical data at a fraction of the cost that examining all possible cases would involve.

A **sample** is a systematic selection of representative cases from the larger population.

Sampling involves two processes: (1) obtaining a list of the population you want to study and (2) selecting a representative subset or sample from the list. Selecting from the list is easy; choosing a relatively large number by a random procedure generally assures that the sample will be unbiased. The more difficult task is getting a list. A central principle of sampling is that a sample is only representative of the list from which it is drawn. If we draw a list of people from the telephone directory, then our sample can only be said to describe households listed in the directory; it will omit those with unlisted numbers, those with no telephones, and those who have moved since the directory was issued. The best surveys begin with a list of all the households or telephone numbers in a target region.

Step Three: Finding Patterns

The third step in the research process is to analyze the data, looking for patterns. If we study unemployment, for example, we will find that black Americans are more than twice as likely as white Americans to experience unemployment (U.S. Department of Labor 1997). This generalization notes a **correlation,** an empirical relationship between two variables—in this case, between race and employment.

Correlation occurs when there is an empirical relationship between two variables.

Step Four: Generating Theories

After a pattern is found, the next step in the research process is to explain it. As we will discuss in the section on survey research, finding a correlation between two variables does not necessarily mean that one variable causes the other. For example, even though there is a correlation between race and unemployment, not all black Americans are unemployed, and being black is not the only cause of unemployment. Nevertheless, if we have good empirical evidence that being black increases the probability of unemployment, the next task is to explain why that should be so. Explanations are usually embodied in a **theory,** an interrelated set of assumptions that explains observed patterns. Theory always goes beyond the facts at hand; it includes untested assumptions that explain the empirical evidence.

A **theory** is an interrelated set of assumptions that explains observed patterns.

In the unemployment example, one might theorize that the reason black Americans face more unemployment than whites is because many of today's black adults grew up in a time when the racial difference in educational opportunity was much greater than it is now. This simple explanation goes beyond the facts at hand to include some assumptions about how education is related to race and unemployment. Although theory rests on an empirical generalization, the theory itself is not empirical; it is . . . well, theoretical.

It should be noted that many different theories can be compatible with a given empirical generalization. We have proposed that education explains the correlation between race and unemployment. An alternative theory might argue that the correlation arises because of discrimination. Because there are often many plausible explanations for any correlation, theory development is not the end of the research process. We must go on to test the assumption of the theory by gathering new data.

The scientific process can be viewed as a wheel that continuously moves us from theory to data and back again (Figure 1.2). Two examples illustrate how theory leads to the need for new data and how data can lead to the development of new theory.

As we have noted, data show that unemployment rates are higher among black Americans than among white Americans. One theoretical explanation for this pattern links higher black unemployment to educational deficits. From this theory, we can deduce the hypothesis that blacks and whites of equal education will experience equal unemployment. To test this hypothesis, we need more data, this time about education and its relationship to race and unemployment.

A study by Melvin Thomas (1993) tests a closely related hypothesis. Thomas asked whether educational deficits explained why African Americans earned less income than whites. He found the hypothesis was not supported: Even if educational levels were equal, the odds were that whites would earn more than African Americans.

Jay Belsky (1990) has reviewed a decade of research on the effects of day care on children's socioemotional development. In reviewing the literature, Belsky notes that some studies show that children with day-care experience were more likely to be attached insecurely to their parents; other studies found that day care had no effects on security. The data show no clear pattern in part, Belsky hypothesizes, because researchers have failed to take into account other variables that may affect children's development—

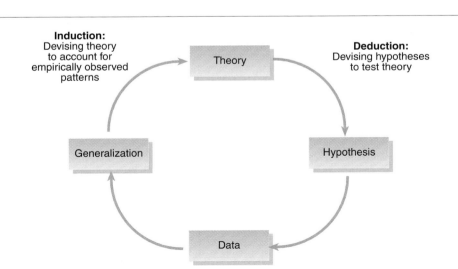

Induction:
Devising theory to account for empirically observed patterns

Deduction:
Devising hypotheses to test theory

Theory

Generalization

Hypothesis

Data

FIGURE 1.2
The Wheel of Science
The process of science can be viewed as a continuously turning wheel that moves us from data to theory and back again.
SOURCE: Adapted from Wallace, Walter. 1969. Sociological Theory. Chicago: Aldine.

Day care plays an increasingly important role in the lives of children, and researchers remain uncertain about its effects on relationships between parents and their offspring. One major hypothesis is that it is not whether or not children attend day care that matters for children's development but rather the quality of the day care provided.

Induction is the process of moving from data to theory by devising theories that account for empirically observed patterns.

Deduction is the process of moving from theory to data by testing hypotheses drawn from theory.

the quality of the day care, the quality of maternal employment, and whether children were in day care under one year of age.

Thomas's findings and Belsky's review can be the basis for revised theories. These new theories will again be subject to empirical testing, and the process will begin anew. In the language of science, the process of moving from data to theory is called **induction,** and the process of moving from theory to data is called **deduction.** These two processes and their interrelationships are also illustrated in Figure 1.2.

Three Strategies for Gathering Data

The theories and findings reported in this book represent a variety of research strategies. Three of the more general strategies are outlined here: experiments, survey research, and participant observation. In this section, we review each method and illustrate its advantages and disadvantages by showing how it would approach the test of a common hypothesis, namely, that alcohol use reduces grades in school.

The Experiment

The **experiment** is a method in which the researcher manipulates independent variables to test theories of cause and effect.

An **experimental group** is the group in an experiment that experiences the independent variable. Results for this group are compared with those for the control group.

A **control group** is the group in an experiment that does not receive the independent variable.

The **experiment** is a research method in which the researcher manipulates the independent variable to test theories of cause and effect. Sometimes, experiments are conducted in carefully controlled conditions in laboratories, but often they take place in normal classrooms and work environments. In the classic experiment, a group that experiences the independent variable, an **experimental group,** is compared with a **control group** that does not. If the groups are equal as far as everything else, a comparison between them will show whether experience with the independent variable is associated with unique change in the dependent variable.

An experiment designed to assess whether alcohol use affects grades, for example, would need to compare an experimental group that drank alcohol with a control group that did not. A hypothetical experiment might begin by observing student grades for several weeks until students' normal performance levels had been established. Then the class would be divided randomly into two groups. If the initial pool is large enough, we can assume that the two groups are probably equal on nearly everything. For example, we assume that both groups probably contain an equal mix of good and poor students, of lazy and ambitious students. The control group might be requested to abstain from alcohol use for five weeks, and the experimental group might be requested to drink three times a week during the same period. At the end of the five weeks, we would compare the grades of the two groups. Both groups might have experienced a drop in grades because of normal factors such as fatigue, burnout, and overwork. The existence of the control group, however, allows us to determine whether alcohol use causes a reduction in grades over and above that which normally occurs.

Experiments are excellent devices for testing hypotheses about cause and effect. They have three drawbacks, however. First, experiments are often unethical because they expose subjects to the possibility of harm. The study on alcohol use, for example, might damage student grades, introduce students to bad habits, or otherwise injure them. A more extreme example might be the question of whether people who were abused as children are more likely to abuse their own children. We could not set up an experiment in which one of two randomly assigned groups of children was beaten and the other not. Because of such ethical issues, many areas of sociological interest cannot be studied with the experimental method.

A second drawback to experiments is that subjects often behave differently when they are under scientific observation than they would in their normal environment. For example, although alcohol consumption might normally have the effect of lowering student grades, the participants in our study might find the research so interesting that their grades would actually improve. In this case, the subjects' knowledge that they are participating in an experiment affects their response to the independent variable. This response is called the **guinea-pig effect.** In sociology, it is often called the Hawthorne effect because it was first documented in a research project at the Hawthorne Electric plant.

The **guinea-pig effect** occurs when subjects' knowledge that they are participating in an experiment affects their response to the independent variable.

A final drawback to the experimental method is that laboratory experiments are often highly artificial. When researchers try to set up social situations in laboratories, they often must omit many of the factors that would influence the same behavior in a real-life situation. The result is often a very unnatural situation. Like the guinea-pig effect, this artificiality has the effect of reducing our confidence that the results that appear in the experiment can be generalized to the more complex conditions of the real world.

Because of these disadvantages and because of ethical limitations, relatively little sociological research uses the controlled experiment. The areas in which it has been most useful are the study of small-group interaction and the simulation of situations that seldom occur in real life.

The Survey

The most common research strategy in sociology is the survey. In **survey research,** the investigator asks a relatively large number of people the same set of standardized questions. These questions may be asked in a personal interview, over the telephone, or in a paper-and-pencil format. Because it asks the same questions of a large number of

Survey research is a method that involves asking a relatively large number of people the same set of standardized questions.

Incidence is the frequency with which an attitude or behavior occurs.

Trends are changes in a variable over time.

Differentials are differences in the incidence of a phenomenon across subcategories of the population.

A **cross-sectional design** uses a sample (or cross section) of the population at a single point in time.

The **panel design** follows a sample over a period of time.

people, it is an ideal method for furnishing evidence on incidence, trends, and differentials. Thus, survey data on alcohol use may allow us to say such things as the following: Eighty percent of the undergraduates of Midwestern State currently use alcohol (**incidence**); the proportion using alcohol has remained about the same over the last 10 years (**trend**); and the proportion using alcohol is higher for males than females (**differential**). Survey research is extremely versatile; it can be used to study attitudes, behavior, ideals, and values. If you can think of some way to ask a question about such matters, then you can study it with survey research.

Most surveys use what is called a **cross-sectional design;** they take a sample (or cross section) of the population at a single point in time and expect it to show some variability on the independent variable. Thus in our study of alcohol use we would take a sample of students, expecting to find that some of them drink and some do not. We could then compare these two groups to see which gets the best grades.

In 1997, we conducted such a survey of our undergraduate students. The results are displayed in Table 1.2. The table shows that students who drink alcohol report getting worse grades: Although 51 percent of the nondrinkers have GPAs over 3.5, only 28 percent of those who admit drinking do. At the other end of the grade distribution, drinkers were more likely than abstainers to report a grade point average of less than 2.5.

The difficulty with the cross-sectional design is that we cannot reach any firm conclusions about cause and effect. We cannot tell whether alcohol use causes bad grades or whether bad grades lead to alcohol use. A more striking problem is that we cannot be sure there is a cause-and-effect relationship at all. Because the drinkers and the abstainers were not randomly assigned to the two categories, the two groups differ on many other variables besides drinking. For example, the drinkers may have less conventional families, come from worse neighborhoods, or simply hate reading. It could be that one of these factors is causing the poor grades and that alcohol-related behavior is just coincidental.

A different strategy used in survey research is the **panel design,** which follows a sample over a period of time. During this period of time, some sample members will experience the independent variable, and we can observe how they differ from those who do not both before and after this experience. To use this design for examining the effect of alcohol use on grades would require surveying a group of young people at several points in time, say, from when they were 12 until they were 25. This design would not alter the fact that some people choose to drink and others do not, but it would let us look at the same people before and after their decision. It would allow us to see whether students' grades actually fell after they started to drink alcohol or whether they

TABLE 1.2
Cross-Tabulation of College Grades By Use of Alcohol
556 Undergraduate Students at a Midwestern State University, 1997.

College Grades	USE ALCOHOL	
	No	Yes
B+ or A	51%	28%
C+ or B	42	60
Below C+	7	12
Total	100%	100%
Number	114	452

were always poor students. The major disadvantage of a panel design is, of course, that it is expensive and time-consuming.

Another important drawback of survey research in general is that respondents may misrepresent the truth. Prejudiced people may tell you that they are unprejudiced, and only a small fraction of those who abuse their children are likely to admit it. This misrepresentation is known as **social-desirability bias**—the tendency for people to color the truth so that they appear to be nicer, richer, and generally more desirable than they really are. The consequences of this bias vary in seriousness depending on the research aim and topic. Obviously, it is a greater problem for such sensitive topics as drug use and prejudice.

Survey research is designed to obtain standard answers to standard questions. It is not the best strategy for studying deviant or undesirable behaviors or for getting at ideas and feelings that cannot easily be reduced to questionnaire form. An additional drawback of survey research is that it is designed to study individuals rather than contexts. Thus, it focuses on the individual alcohol user or abstainer rather than the setting and relationships in which drinking takes place. For these kinds of answers, we must turn to participant observation.

Participant Observation

Under the label **participant observation,** we classify a variety of research strategies—participating, interviewing, observing—that examine the context and meanings of human behavior. Instead of sending forth an army of interviewers, participant observers go out into the field themselves to see firsthand what is going on. These strategies are used more often by sociologists interested in symbolic interaction theory—that is, researchers who want to understand subjective meanings and personal relationships. The goals of this research method are to discover patterns of interaction and to find the meaning of the patterns for the individuals involved.

The three major tasks involved in participant observation are interviewing, participating, and observing. A researcher goes to the scene of the action, where she may interview people informally in the normal course of conversation, participate in whatever

Social-desirability bias is the tendency of people to color the truth so that they sound nicer, richer, and more desirable than they really are.

THINKING

CRITICALLY

Consider what study design you could *ethically* use to determine whether drinking alcohol, living in a sorority, or growing up with a single parent reduces academic performance.

Participant observation includes a variety of research strategies—participating, interviewing, observing—that examine the context and meanings of human behavior.

What is going on here? Survey research is not going to give you the answer. Not only is there no time to select a sample and draw up a questionnaire, the participants do not appear to be particularly cooperative respondents. When we want to study a social process or when we want to study deviant and uncooperative populations, participant observation is usually our best research strategy. In this case—a confrontation between white supremacists, the police, and anti-KKK protestors at the "Aryan Woodstock" festival—a researcher would need to be on the scene for as long as possible and then track down participants later to discuss their behavior and motivations.

they are doing, and observe the activities of other participants. Not every participant observation study includes all three dimensions equally. A participant observer studying alcohol use on campus, for example, would not need to get smashed every night. She would, however, probably do long, informal interviews with both users and nonusers, attend student parties and activities, and attempt to get a feel for how alcohol use fits in with certain student subcultures.

The data produced by participant observation are often based on small numbers of individuals who have not been selected according to random-sampling techniques. The data tend to be unsystematic and the samples not very representative; however, we do learn a great deal about the few individuals involved. This detail is often useful for generating ideas that can then be examined more systematically with other techniques. In this regard, participant observation may be viewed as a form of initial exploration of a research topic.

In some situations, however, participant observation is the only reasonable way to approach a subject. This is especially likely when we are examining undesirable behavior, real behavior rather than attitudes, or uncooperative populations. In the first instance, social-desirability bias makes it difficult to get good information about undesirable behavior. Thus, what we know about running a brothel (Heyl 1979) or the homeless (Snow & Anderson 1987) rests largely on the reports of participant observers. This style of research produces fewer distortions than would have occurred if a middle-class survey researcher dropped by to ask the participants about their activities.

In the second case, participant observation is well suited to studies of behavior—what people do rather than what they say they do. Behaviors are sometimes misrepresented in surveys simply because people are unaware of their actions or don't remember them very well. For example, individuals may believe they are not prejudiced; yet observational research will demonstrate that these same people systematically choose not to sit next to persons of another race on the bus or in public places. Sometimes, actions speak louder than words.

In the third case, we know that survey research works best with people who are predisposed to cooperate with authorities and who are relatively literate. Where either one or both of these conditions is not met, survey research may not be possible. For this reason there is little survey research on prison populations, juvenile gangs, preschoolers, or rioters. Participant observation is often the only means to gather data on these populations.

A major disadvantage of participant observation is that the observations and generalizations rely on the interpretation of one investigator. Because researchers are not robots, it seems likely that their findings reflect some of their own world view. This is a greater problem with participant observation than with survey or experimental work, but all social science suffers to some extent from this phenomenon. The answer to this dilemma is **replication,** redoing the same study with another investigator to see if the same results occur.

Replication is the repetition of empirical studies with another investigator or a different sample to see if the same results occur.

Alternative Strategies

The bulk of sociological research uses these three strategies. There are, however, a dozen or more other imaginative and useful ways of doing research, many of them involving the analysis of social artifacts rather than people. For example, a study of women's magazines of the nineteenth century illustrates changing attitudes towards spinsterhood during that period (Hickok 1981). A study of church paintings, epitaphs, and cemeteries over the centuries has shown how our ideas about death have changed (Aries

CONCEPT SUMMARY	*Research Methods*

CONTROLLED EXPERIMENTS

Procedure	Dividing subjects into two equivalent groups, applying the independent variable to one group only, and observing the differences between the two groups on the dependent variable
Advantages	Excellent for analysis of cause-and-effect relationships; can stimulate events and behaviors that do not occur outside the laboratory in any regular way
Disadvantages	Based on small, nonrepresentative samples examined under highly artificial circumstances; unclear that people would behave the same way outside the laboratory; unethical to experiment in many areas

SURVEY RESEARCH

Procedure	Asking the same set of standard questions of a relatively large, systematically selected sample
Advantages	Very versatile—can study anything that we can ask about; can be done with large, random samples so that results represent many people; good for incidence, trends, and differentials
Disadvantages	Shallow—does not get at depth and shades of meaning; affected by social-desirability bias; better for studying people than situations

PARTICIPANT OBSERVATION

Procedure	Observing people's behavior in its normal context; experiencing others' social settings as a participant; in-depth interviewing
Advantages	Seeing behavior in context; getting at meanings associated with behavior; seeing what people do rather than what they say they do
Disadvantages	Limited to small, nonrepresentative samples; dependent on interpretation of single investigator

1974). Studies of court records and government statistics have demonstrated incidence, trends, and differentials in many areas of sociological interest.

Sociologists: What Do They Do?

A concern with social problems has been a continuing focus of U.S. sociology. This is evident both in the kinds of courses that sociology departments offer (social problems, race and ethnic relations, crime and delinquency, for example) and in the kinds of research sociologists do.

The majority of U.S. sociologists are employed in colleges and universities, where they teach and do research. Much of this research is *basic sociology,* which has no immediate practical application and is motivated simply by a desire to describe or explain some aspect of human social behavior more fully. Even basic research, however, often has implications for social policy.

In addition to the pure research motivated by scholarly curiosity, an increasing proportion of sociologists are engaged in *applied sociology,* seeking to provide immediate practical answers to problems of government, industry, or individuals. The proportion of sociologists who are engaged in applied work has more than doubled in the last decade, from 9 percent in 1976 to more than 25 percent today. This increase is evident in government, business, and nonprofit organizations.

Working in Government

A long tradition of sociological work in government has to do with measuring and forecasting population trends. This work is vital for decisions about where to put roads and schools and when to stop building schools and start building nursing homes. In addition, sociologists have been employed to design and evaluate public policies in a wide variety of areas. In World War II, sociologists designed policies to increase the morale and fighting efficiency of the armed forces. During the so-called war on poverty in the 1960s, sociologists helped plan and evaluate programs to reduce the inheritance of poverty.

Sociologists work in nearly every branch of government. For example, sociologists are employed by the Centers for Disease Control (CDC), where they examine how social relationships are related to the transmission of AIDS, how intravenous drug users share needles, and how AIDS is transmitted along chains of sexual partners. While the physicians and biologists of CDC examine the medical aspects of AIDS, sociologists work at understanding the social aspects.

Working in Business

Sociologists are employed by General Motors and Pillsbury as well as by advertisers and management consultants. Part of their work concerns internal affairs (bureaucratic structures and labor relations), but much of it has to do with market research. Business and industry employ sociologists so that they can use their knowledge of society to predict which way consumer demand is likely to jump. For example, the greater incidence today of single-person households has important implications for life insurance companies, food packagers, and the construction industry. To stay profitable, companies need to be able to predict and plan for such trends. Sociologists are also extensively involved in the preparation of environmental impact statements, in which they try to assess the likely impact of, say, a slurry coal operation on the social and economic fabric of a proposed site.

Working in Nonprofit Organizations

Nonprofit organizations range from hospitals and clinics to social-activist organizations and private think tanks; sociologists are employed in all of them. Sociologists at Planned Parenthood, for example, are interested in determining the causes and consequences of teenage fertility, evaluating communication strategies that can be used to prevent teenage pregnancy, and devising effective strategies to pursue some of that organization's more controversial goals, such as the preservation of legal access to abortion on demand.

The training that sociologists receive has a strong research orientation and is very different from the therapy-oriented training received by social workers. Nevertheless, a thorough understanding of the ways that social structures impinge on individuals can be useful in helping individuals cope with personal troubles. Consequently, some sociologists also do marriage counseling, family counseling, and rehabilitation counseling.

a global perspective

SURVEY RESEARCH IN NIGERIA

"*H*ow do you feel about the current government?" "Do you think men and women ought to be treated equally?" "How many televisions do you own?" Without much effort, we can imagine places in the world where questions like these that are so commonplace to us would appear foolish and perhaps dangerous to ask.

In the United States and much of the developed world, we are accustomed to questions about the most intimate details of our lives. It is common for bureaucratic agencies to record data pertaining to our height, weight, IQ, taxes, fertility, and credit rating. Our acceptance of intrusive questioning is based on the assumption that this information is somehow necessary to good governance and the trust that our privacy will be protected. Because we are familiar with routine bureaucratic data gathering, it is relatively easy for survey researchers to enlist our cooperation. If called on the telephone, approximately 75 percent of persons in the United States will re-

spond to survey researchers' inquiries on topics ranging from politics to religion.

When survey research is conducted in Third World nations, many of these conditions do not hold. Often people there have good reason not to trust their governments. More generally, they simply are not accustomed to opening their private lives to the probing of bureaucratic agents.

One of the most common areas, where western survey research methods meet resistance from Third World peoples, is research on fertility and family planning. Agencies such as the United Nations and the U.S. Agency for International Development wish to know how many children women in Kenya, Nigeria, and other high-fertility nations *want* so that they can use the discrepancy between actual and desired child-bearing to develop contraceptive programs. Agnes Riedmann's (1993) analysis of a survey research project among the Yoruba of Nigeria highlights several cultural clashes that may impede such research efforts.

o   In many nonwestern cultures, fertility is "up to God." It would be unthinkable for individuals to put their own opinions forward. A question such as, "If you had

more money, would you rather have a new car or another child?" presumes that individuals have a choice about fertility and, moreover, that dollar values can be assigned to children.

o   Asking *women's* opinions is often considered indecent (unless one's husband is present) or at best a waste of time because women's opinions do not count.

o   There is profound distrust of strangers who ask personal questions. After submitting to an ill-understood interview on fertility, for example, one Yoruban respondent asked whether the police were now coming to take her away.

In many cases Yoruban respondents mocked, yelled at, and ran from interviewers. The persistent inquiry of strangers into their private lives seemed to some to be just one more instance of the crazy behavior of "oyinbos," or white people. Others speculate that it was a result of white people not having enough to do! If badgered into participation, Yorubans politely told interviewers what they thought they wanted to hear and sent them on their way.

Although mothers and mothering are universal, the meanings attached to children and to childbearing are not. Survey research on fertility and family planning may give seriously misleading results in nations such as Nigeria, where conception is viewed as being "up to God," not "up to the individual," and where answering questions about personal topics is considered indecent, if not downright dangerous.

Sociology in the Public Service

Although most sociologists are committed to a value-free approach to their work as scholars, many are equally committed to changing society for the better. They see sociology as a "calling—work that is inseparable from the rest of one's life and driven by a sense of moral responsibility for people's welfare (Yamane 1994). As a result, sociologists have served on a wide variety of public commissions and in public offices in order to effect social change. They work for change independently, too. For example, one sociologist, Claire Gilbert, publishes an environmental newsletter, *Blazing Tattles,* that reports adverse effects of pollution (Alesci 1994).

Perhaps the clearest example of sociology in the public service is the award of the 1982 Nobel Peace Prize to Swedish sociologist Alva Myrdal for her unflagging efforts to increase awareness of the dangers of nuclear armaments. Value-free scholarship does not mean value-free citizenship.

Summary

1. Sociology is the systematic study of social behavior. Sociologists use the concepts of role and social structure to analyze common human dramas. Learning to understand how individual behavior is affected by social structures is the process of developing the "sociological imagination."

2. The rapid social change that followed the industrial revolution was an important inspiration for the development of sociology. Problems caused by rapid social change stimulated the demand for accurate information about social processes. This social-problems orientation remains an important aspect of sociology.

3. Sociology has three major theoretical perspectives: structural-functional theory, conflict theory, and symbolic interaction theory. The three can be seen as alternative lenses through which to view society, with each having value as a tool for understanding how social structures shape human behavior.

4. Structural functionalism has its roots in evolutionary theory. It identifies social structures and analyzes their consequences for social harmony and the maintenance of society. Identification of manifest and latent functions and dysfunctions is part of its analytic framework.

5. Conflict theory developed from Karl Marx's ideas about the importance of conflict and competition in structuring human behavior. It analyzes social structures by asking who benefits and how are these benefits maintained. It assumes that competition is more important than consensus and that change occurs as a result of conflict and revolution rather than through evolution.

6. Symbolic interaction theory examines the subjective meanings of human interaction and the processes through which people come to develop and communicate shared symbolic meanings. Although structural functionalism and conflict theory study macrosociology, symbolic interactionism is a form of microsociology.

7. Sociology is a social science. This means it relies on critical and systematic examination of the evidence before reaching any conclusions and that it approaches each research question from a position of neutrality. This is called value-free sociology.

8. The four steps in the research process are stating the problem, gathering the data, finding patterns, and generating theory. These steps form a continuous loop, called the "wheel of science." The movement from data to theory is called induction, and the movement from theory to hypothesis to data is called deduction.

9. A design for gathering data depends on identifying the variables under study, agreeing on precise operational definitions of these variables, and obtaining a representative sample of cases in which to study relationships among the variables.

10. The experiment is a method designed to test cause-and-effect hypotheses. Although it is the best method for this purpose, it has three disadvantages: ethical problems, the guinea-pig effect, and highly artificial conditions. It is most often used for small-group research and for simulation of situations not often found in real life.

11. Survey research is a method that asks a large number of people a set of standard questions. It is good at describing incidence, trends, and differentials for random samples, but it is not as good at describing the contexts of human behavior or for establishing causal relationships.

12. Participant observation is a method in which a small number of individuals who are not randomly chosen are observed or interviewed in depth. The strength of this method is the detail about the contexts of human behavior and its subjective meanings; its weaknesses are poor samples and lack of verification by independent observers.

13. Most sociologists teach and do research in academic settings. A growing minority are employed in government and business, where they do applied research. Regardless of the setting, sociological theory and research have implications for social policy.

Sociology on the Net

As you begin your journey on the Internet, it is important that you are aware of the norms governing Internet behavior. The University of Michigan's Electronic Library has an excellent set of selections on the topic of Internet etiquette.

```
http://mel.lib.mi.us/internet/
INET-netiquette.html
```

Netiquette (Net One). Now open and read the section on the *Internet Revealed: Netiquette.* Browse through the selections and please note the 10 commandments for computer ethics. Please keep these in mind as you do the exercises in this text as well as all of your other computer work.

Now that you are familiar with some of the dos and don'ts of the Internet, it is time to explore some sociological sources that will give you some more insight into the life and times of Karl Marx.

```
http://csf.Colorado.EDU/psn/marx/index.html
```

Go to the section called *Bio Material* and open the first item entitled the *Marx/Engels Thumbnail Chronology.* Read the chronology and find out how Marx combined scholarship and activism in his own life. How wealthy was Marx and what kind of life did he lead? What were the circumstances of his death? You may wish to get a more personal feel for Marx and his family by scrolling down to the Marx and Engels Photo Gallery and browsing through the offerings.

FIND IT ON INFOTRAC COLLEGE EDITION
Chapter One includes a section titled "Interchangeable Lenses," in which prostitution is examined using the three major theoretical perspectives in sociology. After reading this section, use InfoTrac College Edition to look up the following article:

"The Division of Household Labor." Beth Anne Shelton and Daphnie John. *Annual Review of Sociology,* 1996, v22 p229.

(Hint: Enter the search term *household labor* using the Subject Guide.)

Using the material from the article, construct your own "interchangeable lenses" section in which you examine household labor from the **Marxist/socialist perspective** (a subcategory of conflict theory) and **exchange theory.** Note: Exchange theory is not spelled out in the article as directly as the Marxist/socialist perspective, but you will find ample discussion of it throughout the article.

Suggested Readings

Babbie, Earl. 1994. *The Sociological Spirit.* Belmont, Calif.: Wadsworth. From a dedicated sociologist who believes that the world and national problems that concern us most have their solutions in the realm addressed by sociology; a book of essays that not only introduces the sociological imagination but also motivates the reader to develop it.

Berger, Peter L. 1963. *Invitation to Sociology: A Humanistic Perspective.* Garden City, N.Y.: Doubleday Anchor. Thirty years old and a classic. A delightful, well-written introduction to what sociology is and how it differs from other social sciences. Blends a serious exploration of basic sociological understandings with scenes from everyday life—encounters that are easy to relate to and that make sociology both interesting and relevant.

Deegan, Mary Jo, ed. 1987. *Women in Sociology: A Bio-bibliographical Sourcebook.* New York: Greenwood Press. Not only a great sourcebook but also arresting reading. A volume that tells about often-forgotten women in sociology from the early nineteenth century until today.

Mills, C. Wright. (1959). *The Sociological Imagination.* New York: Oxford University Press. A penetrating account of how the study of sociology expands understanding of common experiences.

Neuman, W. Lawrence. (1994). *Social Research Methods: Qualitative and Quantitative Approaches.* Boston: Allyn and Bacon. A textbook for undergraduates that covers the major research techniques in sociology. Coverage is up-to-date, thorough, and readable.

Reinharz, Schulamit. (1992). *Feminist Methods in Social Research.* New York: Oxford University Press. A well-written review of the relationship between feminist research methods and conventional research design. Richly illustrated with examples from feminist scholarship.

CHAPTER 2

Culture and Society

Outline

Introduction to Culture

In Chapter 1 we said that sociology is concerned with analyzing the contexts of human behavior and how these contexts affect our behavior. Our neighborhood, our family, and our social class provide part of that context, but the broadest context of all is our culture. **Culture** is the total way of life shared by members of a society.

Culture resides essentially in nontangible forms such as language, values, and symbolic meanings, but it also includes technology and material objects. A common image is that culture is a "tool kit" that provides us with the equipment necessary to deal with the common problems of everyday life (Swidler 1986). Consider how culture provides patterned activities of eating and drinking. We share a common set of tools and technologies in the form of refrigerators, ovens, toasters, microwaves, and coffeepots. As the advertisers suggest, we share similar feelings of psychological release and satisfaction when, after a hard day of working or playing, we take a break with a cup of coffee or a cold beer. The beverages we choose and the meanings attached to them are part of our culture.

Culture can be roughly divided into two categories: material and nonmaterial. *Nonmaterial culture* consists of language, values, rules, knowledge, and meanings shared by the members of a society. *Material culture* includes the physical objects that a society produces—tools, streets, sculptures, and toys, to name but a few. These material objects depend on the nonmaterial culture for meaning. For example, Barbie dolls and figurines of fertility goddesses share many common physical features; their meaning depends on nonmaterial culture.

Culture is the total way of life shared by members of a society. It includes not only language, values, and symbolic meanings but also technology and material objects.

Theoretical Perspectives on Culture

Within sociology, there are two approaches to the study of culture (Wuthnow & Witten 1988). The first approach treats culture as the underlying basis of interaction. It accepts culture as a given and is more interested in how culture shapes us than in how culture itself is shaped. Scholars taking this approach have concentrated on illustrating how norms, values, and language guide our behavior. The second approach focuses on culture as a social product. It asks why particular aspects of culture develop. These scholars would be interested, for example, in why the content of commercial television is so different from the content of public television. How does the economic structure of television affect its products? How does the way arts are funded in the United States affect what art becomes accepted by the public? Generally, the first perspective on culture is characteristic of structural-functional theory, while conflict theorists are more interested in the determinants of culture. Because both the content of culture and the determinants of culture are of interest to us, we will use both perspectives.

THINKING CRITICALLY

What features of U.S. society might explain why children here are raised in small nuclear families rather than in more extended kin groups?

Bases of Human Behavior: Culture and Biology

Why do people behave as they do? What determines human behavior? To answer these questions, we must be able to explain both the varieties and the similarities in human behavior. Generally, we will argue that biological factors help explain what is common

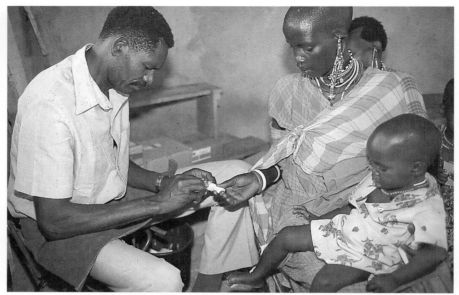

On the left, a traditional Brazilian medicine man uses herbs and religious ritual to cure ill-ness. In sharp contrast, on the right, the techniques of modern science are being used by a doctor in Kenya. Despite the obvious differences among cultures, a close look shows a great many underlying similarities. Although some rely on prayer and others on antibiotics, all cultures provide some routine set of mechanisms for dealing with illness and injury.

to humankind across societies, whereas culture explains why people and societies dif-fer from one another.

Cultural Perspective

Regardless of whether they are structural functionalists or conflict theorists, sociologists share some common orientations toward culture: Nearly all hold that culture is *prob-lem solving,* culture is *relative,* and culture is a *social product.*

Culture is Problem Solving

Regardless of whether people live in tropical forests or in the crowded cities of New York, London, or Tokyo, they confront some common problems. They all must eat, they all need shelter from the elements (and often from each other), and they all need to raise children to take their place and continue their way of life. Although these prob-lems are universal, the solutions are highly varied. For example, responsibility for child rearing may be assigned to the mother's brother, as is done in the Trobiand Islands; to the natural mother and father, as is done in the United States; or to communal nurs-eries, as is done in the Israeli kibbutz.

Whenever people face a recurrent problem, cultural patterns will have evolved to provide a ready-made answer. This does not mean that the answer provided is the best answer or the only answer or the fairest answer, but merely that culture provides a standard pattern for dealing with this common dilemma. One of the issues that di-vides conflict and functional theorists is how these answers develop. Functionalists argue that the solutions we use today have evolved over generations of trial and error and that they have survived because they work, because they help us meet basic needs. A conflict theorist would add that these solutions work better for some people than others. Conflict theorists argue that elites manipulate culture in order to rationalize and maintain solutions that work to their advantage. Scholars from both perspectives agree that culture provides ready-made answers for most of the recurrent situations that we face in daily life; they disagree in their answer to the question "Who benefits?" from a particular solution.

Culture is Relative

The solutions that each culture devises may be startlingly different. Among the Wodaabe of Niger, for example, mothers are not allowed to speak directly to their first- or second-born children and, except for nursing, are not even allowed to touch them. The babies' grandmothers and aunts, however, lavish affection and attention on them (Beckwith 1983). The effect of this pattern of child rearing is to emphasize loyalties and affections throughout the entire kin group rather than just to one's own children or spouse. This practice helps ensure that each new entrant will be loyal to the group as a whole. Is it a good or a bad practice? That is a question we can answer only by seeing how it fits in with the rest of the Wodaabe culture. Does it help the people meet recurrent problems and maintain a stable society? If so, then it works; it is functional. The idea that each cultural trait should be evaluated in the context of its own culture is called **cultural relativity.** A corollary of cultural relativity is that no practice is universally good or universally bad; goodness and badness are relative, not absolute.

> **Cultural relativity** requires that each cultural trait be evaluated in the context of its own culture.

This type of evaluation is sometimes a difficult intellectual feat. For example, no matter how objective we try to be, most of us believe that infanticide, human sacrifice, and cannibalism are absolutely and universally wrong. Such an attitude reflects **ethnocentrism**—the tendency to use the norms and values of our own culture as standards against which to judge the practices of others.

> **Ethnocentrism** is the tendency to view the norms and values of our own culture as standards against which to judge the practices of other cultures.

Ethnocentrism usually means that we see our way as the right way and everybody else's way as the wrong way. When American missionaries came to the South Sea Islands, they found that many things were done differently in Polynesian culture. The missionaries, however, were unable to view Polynesian folkways as simply different. If they were not like American practices, then they must be wrong and were probably wicked. As a result, the missionaries taught the islanders that the only acceptable way (the American way) to have sexual intercourse was in a face-to-face position with the man on top, the now-famous "missionary position." They taught the Polynesians that women must cover their breasts, that they should have clocks and come on time to appointments, and a variety of other Americanisms that the missionaries maintained to be morally right behavior.

Ethnocentrism is often a barrier to interaction of people from different cultures, leading to much confusion and misinterpretation. It is not, however, altogether bad. In the sense that it is pride in our own culture and confidence in our own way of life, ethnocentrism is essential for social integration. In other words, we learn to follow the ways of our culture because we believe that they are the right ways; if we did not share that belief, there would be little conformity in society. Ethnocentrism, then, is a natural and even desirable product of growing up in a culture. An undesirable consequence, however, is that we simultaneously discredit or diminish the value of other ways of thinking and feeling.

Culture is a Social Product

A final assumption sociologists make about culture is that culture is a social, not a biological, product. The immense cultural diversity that characterizes human societies is not the product of isolated gene pools, but of cultural evolution.

Some aspects of culture are produced deliberately. Shakespeare picked up paper and pen in order to write *Hamlet;* some advertising executive worked to invent the Energizer bunny. Governments, bankers, and homeowners deliberately commission the designing of homes, offices, and public buildings, and people buy publishing empires so that they can spread their own version of the truth. Other aspects of culture—such as our culture's ideas about right and wrong, its dress patterns, and its language—develop gradually out of social interaction. But all of these aspects of culture are human products; none of them is instinctive. People *learn* culture and, as they use it, they modify it and change it.

Culture depends on a unique human attribute: the capacity for language. Only after language is invented can bits of practical knowledge (such as "fire is good" and "don't use electricity in the bathtub") or ideas (such as "God exists") be transmitted from one generation to the next. Inventions, discoveries, and forms of social organization are socially bestowed and intentionally passed on so that each new generation potentially elaborates on and modifies the accumulated knowledge of the previous generations. In short, culture is cumulative only because of language.

Because of language, human beings are not limited to the slow process of genetic evolution in adapting to their circumstances. Cultural evolution is a uniquely human way for a species to adapt to its environment. Whereas biological evolution may require literally hundreds of generations to adapt the organism fully to new circumstances, cultural evolution allows the changes to be made within a short period of time. In this sense, cultural evolution is an extension of biological evolution, one that speeds up the processes of change and adjustment to new circumstances in the environment (van den Berghe 1978).

Biological Perspective

As the continued popularity of *National Geographic* attests, the wide diversity of human cultures is a continuous source of fascination. Costumes, eating habits, and living arrangements vary dramatically. It is tempting to focus on the exotic variety of human behavior and to conclude that there are no limits to what humankind can devise. A closer look, however, suggests that there are some basic similarities in cultures around the world—the universal existence of the family, religion, aggression, and warfare. When we focus on these universals, cultural explanations are likely to be supplemented with biological explanations.

Sociobiology is the study of the biological basis of all forms of human behavior.

Within the past two decades sociology has witnessed renewed interest in the role of evolution and biology in human behavior. A relatively new field, **sociobiology,** is the study of the biological basis of all forms of human behavior (E. O. Wilson 1978). Sociobiology makes the assumption that humans and all other lifeforms developed through evolution and natural selection. According to this perspective, change in a species occurs primarily through one mechanism: Some individuals are more successful at reproduction than others. As the offspring of successful reproducers increase in number relative to those of the less successful reproducers, the species comes to be characterized by the traits that mark successful reproducers.

Who are the successful reproducers? They are those who have more children and raise more of them until they are old enough to reproduce themselves (Daly & Wilson 1983). Among the characteristics of human society that are thought to be related to these reproductive strategies are altruism and male/female differences in mating behavior, parenting, and aggression. For example, sociobiologists suggest that parents who are willing to make sacrifices for their children, occasionally even giving their life for them, are more successful reproducers; by ensuring their children's survival, they increase the likelihood that their own genes will contribute to succeeding generations. Thus, sociobiologists argue that we have evolved biological predispositions toward altruism (an unselfish concern for others), *but only insofar as our own kin are concerned.*

Sociobiology provides an interesting theory about how the human species has evolved over tens of thousands of years. Most of the scholars who study the effect of

Although mothers and mothering are universal, the meanings attached to children and to childbearing are not. Survey research on fertility and family planning may give seriously misleading results in nations such as Nigeria, where conception is viewed as being "up to God," not "up to the individual," and where answering questions about personal topics is considered indecent, if not downright dangerous.

biology on human behavior, however, are concerned with more contemporary questions such as "How do hormones, genes, and chromosomes affect human behavior?" Joint work by biologists and social scientists is helping us to understand how biological and social factors work together to determine human behavior.

This approach is nicely illustrated in a recent set of studies by Udry, who asks how the biological changes that accompany puberty interact with social structures to determine adolescent sexuality. Using blood and urine analyses to determine hormone levels, Udry finds that boys and girls with higher levels of testosterone report higher levels of sexual behavior and sexual interest. He also finds that strict family supervision, especially having a father in the home, can override the effect of hormonal change among girls (Udry 1988). This suggests that social *and* biological factors play a role in adolescent sexuality.

We are at the same time animals and social products. It is a false dichotomy to ask whether culture or biology determines behavior; instead, our behavior represents an intersection between the two. One leading sociologist has argued, for example, that "ignorance of biological processes may doom efforts" to create equality between women and men or to decrease the burden that child care poses for women (Rossi 1984, 11). Only by recognizing and taking into account the joint effects of culture and biology can we have the complete picture of the determinants of human behavior.

The Carriers of Culture

In the following sections, we review three vital aspects of nonmaterial culture—language, values, and norms—and show how they shape both societies and individuals.

Language

The essence of culture is the sharing of meanings among members of a society. The chief mechanism for this sharing is a common language. Language is the ability to communicate in symbols—orally or in writing.

THE FAR SIDE   By GARY LARSON

"And now, Randy, by use of song, the male sparrow will stake out his territory . . . an instinct common in the lower animals."

Reprinted by permission of Universal Press Syndicate.

What does *communicate with symbols* mean? It means, for example, that when you see the combinations of circles and lines that appear on your textbook page as the word *orally,* you are able to understand that it means "speaking aloud." On a different level, it means that the noise we use to symbolize "dog" brings to your mind a four-legged domestic canine. Almost all communication is done through symbols. Even the meanings of physical gestures such as touching or pointing are learned as part of culture.

Scholars of sociolinguistics (the relationship of language to society) point out that language has three distinct relationships to culture: Language embodies culture, language is a framework for culture, and language is a symbol of culture (Fishman 1985a, 1985b).

Language as an Embodiment of Culture

Language is the carrier of culture; it embodies the values and meanings of a society as well as its rituals, ceremonies, stories, and prayers. Until you share the language of a culture, you cannot participate in it (Fishman 1985b).

A corollary is that loss of language may mean loss of a culture. Currently, many Native American languages have fewer than 40 speakers, most of whom are more than 50 years old. When these people die, they will take their language with them, and important aspects of those Native American cultures will be lost. This vital link between language and culture is why many Jewish parents in the United States send their children to Hebrew school on Saturdays. This is why U.S. law requires that people must be able to speak English before they can be naturalized as U.S. citizens. To participate fully in Jewish culture requires some knowledge of Hebrew; to participate in U.S. culture requires some knowledge of English.

Language as Framework

The **linguistic relativity hypothesis** argues that the grammar, structure, and categories embodied in each language affect how its speakers see reality.

Language gives us capabilities, but it also shapes and confines us. The **linguistic relativity hypothesis** associated with Whorf (1956) argues that the grammar, structure, and categories embodied in each language affect how its speakers see reality and that, therefore, reality is different for speakers of, say, English and Lakota (Sioux). The argument is that our thinking and perception are in some ways fashioned by our linguistic capacities (Fishman 1985b).

We can see the shaping qualities of language in the development of written language. Literacy allows us to communicate with those whom we cannot see; it takes communication out of the face-to-face context and makes it impersonal (Cicourel 1985). This means that literate individuals have the capacity to experience the world beyond themselves in a way not open to the nonliterate. Written language thus expands and changes the world we experience. Writing also encourages a different mode of thinking, an analytic approach in which the written word can be set down, modified, and studied in a way that oral communication cannot. The bureaucratic form of organization and the development of science would both be impossible without written language.

Because language shapes how we perceive reality, one way to change perceptions is to change the words we use. For example, in the last few years, a growing number of black leaders have begun to use "African American" instead of "black." This shift in language usage symbolizes a distinction based on cultural heritage (African) rather than on race (color). It moves us away from thinking about physical differences to thinking about cultural differences. Obviously, this shift has important implications for understanding the causes of current racial inequalities and for framing policy responses. Because most of the research reported in this text is based on respondents identifying themselves by race (black or white), however, we generally use the racial labels in reporting research results.

Language as a Symbol

A common language is often the most obvious outward sign that people share a common culture. This is true of national cultures such as French and Italian and subcultures such as youth. A distinctive language symbolizes a group's separation from others while it simultaneously symbolizes unity within the group of speakers (Cobarrubias 1983). For this reason, groups seeking to mobilize their members often insist on their own distinct language.

Values

After language, the most central and distinguishing aspect of culture is **values,** shared ideas about desirable goals. Values are typically couched in terms of whether a thing is good or bad, desirable or undesirable (Williams 1970, 27). For example, many people in the United States believe that a happy marriage is desirable. In this case and many others, values may be very general. They do not, for example, specify what a happy marriage consists of.

Values are shared ideas about desirable goals.

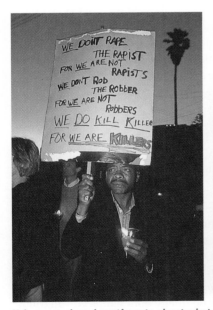

Values are shared sentiments about what is good and just; norms are the specific guidelines a culture provides for reaching values. As recent elections and public opinion polls show, most people in the United States value security and a crime-free society. These two men are probably no exception. At one level, their disagreement is probably less about the value they place on personal safety than about the rules of conduct that should be used to ensure it. At another level, these men may disagree about many other important values—the meaning of justice or the fundamental nature of human beings, for instance. Value conflicts cannot be resolved by scientific evidence. When sociologists study values, they do not address issues of right or wrong, but instead consider questions such as "What types of people are most likely to hold a particular value?" or "What factors cause people to support one set of norms for achieving a value versus another?"

This Tennessee barbershop provides explicit written instruction about appropriate language. For the most part, norms that govern daily life are not as clear as this. Nevertheless, most of us pick up a pretty good understanding of these and other norms by observing how others behave and watching their reactions to our behavior. Negative sanctions such as stares of surprise of consternation, grimaces of distaste, and shrugs of disdain will tell us we have crossed a sensitive line and violated a norm.

Norms are shared rules of conduct that specify how people ought to think and act.

Folkways are norms that are the customary, normal, habitual ways a group does things.

Mores are norms associated with fairly strong ideas of right or wrong; they carry a moral connotation.

Some cultures value tenderness, others toughness; some value cooperation, others competition. Nevertheless, because all human populations face common dilemmas, certain values tend to be universal. For example, nearly every culture values stability and security, a strong family, and good health. There are, however, dramatic differences in the guidelines that cultures offer for pursuing these goals. In societies like ours, an individual may try to ensure security by putting money in the bank or investing in an education. In many traditional societies, security is maximized by having a large number of relatives. In societies such as that of the Kwakiutl of the Pacific Northwest, security is achieved, not by saving your wealth, but by giving it away. The reasoning is that all of the people who accept your goods are now under an obligation to you. If you should ever need help, you would feel free to call on them and they would feel obliged to help. Thus, although many cultures place a value on establishing security against uncertainty and old age, the specific guidelines for reaching this goal vary. The guidelines are called norms.

Norms

Shared rules of conduct are **norms.** They specify what people *ought* to or *ought not* to do. The list of things we ought to do sometimes seems endless. We begin the day with an "I'm awfully tired, but I ought to get up," and many of us end the day with "This is an awfully good show, but I ought to go to bed." In between, we ought to brush our teeth, eat our vegetables, work hard, love our neighbors, and on and on. The list is so extensive that we may occasionally feel that we have too many obligations and too few choices. Of course, some pursuits are optional and allow us to make choices, but the whole idea of culture is that it provides a blueprint for living, a pattern to follow.

Norms vary enormously in their importance both to individuals and to society. Some, such as fashions, are powerful while they last but are not central to society's values. Others, such as those supporting monogamy and democracy, are central to our culture. Generally, we distinguish between two kinds of norms: folkways and mores.

Folkways

The word **folkways** describes norms that are simply the customary, normal, habitual ways a group does things. Folkways is a broad concept that covers relatively permanent traditions (such as fireworks on the Fourth of July) as well as short-lived fads and fashions (such as wearing baseball caps backwards or sneakers with "pumps").

A key feature of all folkways is that no strong feeling of right or wrong is attached to them. They are simply the way people usually do things. For example, if you choose to have hamburgers for breakfast and oatmeal for dinner, you will be violating U.S. folkways. If you sleep on the floor or dye your hair purple, you will be deviating from the usual pattern. If you violate folkways, you may be regarded as eccentric, weird, or crazy, but you will not be regarded as immoral or criminal.

Mores

Some norms are associated with strong feelings of right and wrong. These norms are called **mores** (more-ays). Whereas eating oatmeal for dinner may only cause you to be considered crazy (or lazy), other things that you can do will really offend your neighbors.

CONCEPT SUMMARY | *Values, Norms, and Laws*

CONCEPT	DEFINITION	EXAMPLE FROM MARRIAGE	RELATIONSHIP TO VALUES
Values	Shared ideas about desirable goals	It is desirable that marriage include physical love between wife and husband.	
Norms	Shared rules of conduct	Have sexual intercourse regularly with each other, but not with anyone else.	Generally accepted means to achieve value
Folkways	Norms that are customary or usual	Share a bedroom and a bed; kids sleep in a different room.	Optional but usual means to achieve value
Mores	Norms with strong feelings of right and wrong	Thou shalt not commit adultery.	Morally required means to achieve value
Laws	Formal standards of conduct, enforced by public agencies	Illegal for husband to rape wife; sexual relations must be voluntary.	Legally required means; may or may not be supported by norms

---

If you eat your dog or spend your last dollar on liquor when your child needs shoes, you will be violating mores. At this point, your friends and neighbors may decide that they have to do something about you. They may turn you in to the police or to a child protection association; they may cut off all interaction with you or even chase you out of the neighborhood. Not all violations of mores result in legal punishment, but all result in such informal reprisals as ostracism, shunning, or reprimand. These punishments, formal and informal, reduce the likelihood that people will violate mores.

Laws

Rules that are enforced and sanctioned by the authority of government are **laws.** Very often the important mores of society become laws and are enforced by agencies of the government. If the laws cease to be supported by norms and values, they are either stricken from the record or become dead-letter laws, no longer considered important enough to enforce. Not all laws, of course, are supported by public sentiment; in fact, many have come into existence as the result of lobbying by powerful interest groups. Laws regulating marijuana use in the United States, for example, owe their origins to lobbying by the liquor industry. Similarly, laws requiring the wearing of seat belts are not a response to social norms. In these cases, laws are trying to create norms rather than respond to them.

Social Control

From our earliest childhood, we are taught to observe norms, first within our families and later within peer groups, at school, and in the larger society. After a period of time, following the norms becomes so habitual that we can hardly imagine living any other way—they are so much a part of our lives that we may not even be aware of them as constraints. We do not think, "I ought to brush my teeth or else my friends and family will shun me;" instead we think, "It would be disgusting not to brush my teeth, and I'll hate myself if I don't brush them." For thousands of generations, no human

THINKING CRITICALLY

Can you think of an example from U.S. culture where values, norms, and laws are not consistent with each other? What consequences, if any, are these inconsistencies likely to have?

Laws are rules that are enforced and sanctioned by the authority of government. They may or may not be norms.

considered it disgusting to go around with unbrushed teeth. For most people in the United States, however, brushing their teeth is so much a part of their feeling about themselves, about who they are and the kind of person they are, that they would disgust themselves by not observing the norm.

Through indoctrination, learning, and experience, many of society's norms come to seem so natural that we cannot imagine acting differently. No society relies completely on this voluntary compliance, however, and all encourage conformity by the use of **sanctions**—rewards for conformity and punishments for nonconformity. Some sanctions are formal, in the sense that the legal codes identify specific penalties, fines, and punishments that are to be meted out to individuals found guilty of violating formal laws. Formal sanctions are also built into most large organizations to control absenteeism and productivity. Some of the most effective sanctions, however, are informal. Such positive sanctions as affection, approval, and inclusion encourage normative behavior, whereas such negative sanctions as a cold shoulder, disapproval, and exclusion discourage norm violations.

Despite these sanctions, norms are not always a good guide to what people actually do, and it is important to distinguish between normative behavior (what we should do) and actual behavior. For example, our own society has powerful mores supporting marital fidelity. Yet research has shown that nearly half of all married men and women in our society have committed adultery (Thompson 1983). In this instance, culture expresses expectations that differ significantly from actual behavior. This does not mean the norm is unimportant. Even norms that a large minority, even a majority, fail to live up to are still important guides to behavior. The discrepancy between actual behavior and normative behavior—termed *deviance*—is a major area of sociological research and inquiry (see Chapter 5).

Subcultures and Countercultures

Sharing a culture does not mean there is complete homogeneity. When segments of society face substantially different social environments, subcultures grow up to help them adapt to these unique problems. **Subcultures** share in the overall culture of society, but also maintain a distinctive set of values, norms, lifestyles, and even language. In the United States, for example, some of the most prominent subcultures are based on region. Empirical studies show that people who live in the South are more likely to attend church frequently, have strong religious affiliations, and hold traditional attitudes (Hurlbert 1989). Compared to people who live in other regions of the United States, southerners are also more likely to own guns and approve of using violence to settle interpersonal quarrels (Ellison 1991). In addition, they have a distinctive speech pattern that serves as a symbol of their group membership. So even though Southerners share much of the larger U.S. culture, they also have their own unique cultural legacy.

Countercultures are groups that have values, interests, beliefs, and lifestyles that conflict with those of the larger culture. This theme of conflict is clear among one recent U.S. countercultural group—punkers. Some punkers are just part-timers who dye their hair purple and listen to death rock but nevertheless manage to go to school or hold a job. True punkers, however, angrily reject all contact with straight society. They refuse to work or take organized charity; they live angry and sometimes hungry lives on the streets. Asked about the kind of music he liked, one punker said, "I like hard core a lot. I don't like idiotic stuff though like Venom and stuff. I like lyrics that actually say something and music that makes you want to beat somebody up" (Baron 1989, 30).

Sanctions are rewards for conformity and punishments for nonconformity.

Subcultures are groups that share in the overall culture of society but also maintain a distinctive set of values, norms, lifestyles, and even language.

Countercultures are groups having values, interests, beliefs, and lifestyles that are opposed to those of the larger culture.

An Example: Deafness as Culture

Most of us view deafness as undesirable, even catastrophic (Dolnick 1993). At best, we see being deaf as a medical condition to be remedied. However, some spokespersons in the deaf community maintain that deafness is not a disability but a culture (Dolnick 1993). According to them, the essence of deafness is not the inability to hear but a culture based on their shared language, American Sign Language (ASL). ASL is not just a way to "speak" English with one's hands but is a language of its own, complete with its own rules of grammar, puns, and poetry. Furthermore, it is a language that is learned and shared. Where hearing babies begin to jabber nonsense syllables, for example, deaf babies of parents who sign begin to "babble" nonsense signs with their fingers (Dolnick 1993). This shared language encourages, in turn, shared values and a positive group identity. Studies show, for instance, that many deaf people would not choose to join the "hearing" culture even if they could. Asked whether she would prefer to hear, Roslyn Rosen, president of the National Association of the Deaf, answered that she doesn't want to be "fixed." Further, some deafness-as-culture activists see the use of an inner-ear operation that can reverse some children's deafness as a form of "child abuse" or "genocide." Partly as a result of their common language and cultural identity, deaf people tend to marry each other—another step toward solidifying community and culture. Interestingly, disputes have recently erupted among deaf persons about who is to be considered more authentically deaf. To be the deaf child of deaf parents has higher status within deaf culture because this is as deaf as one can be.

Thinking of deafness as culture illustrates many of the points made earlier. For instance, culture is problem solving, and deaf culture embodies a way to solve the human problem of communication. Using American Sign Language shapes deaf people's experience, reminding them of their common values, norms, and cultural identity.

Society

Culture is a way of life. In some places, it cuts across national boundaries and takes in people who live in two, three, or four nations. In other places, two distinct cultures

How do you view deafness?

A **society** is the population that shares the same territory and is bound together by economic and political ties.

(English and French in Canada, for example) may coexist within a single national boundary. For this reason, we distinguish between cultures and societies. A **society** is the population that shares the same territory and is bound together by economic and political ties. Often the members share a common culture but not always.

Cultural Variation and Change

Culture provides solutions to common and not-so-common problems. The solutions devised are immensely variable. Among the reasons for this variability are environment, isolation, technology, and dominant cultural themes.

Environment

Why are the French different from Australian aborigines, the Finns different from the Navajo? One obvious reason is the very different environmental conditions to which they must adapt. These different environmental conditions determine which kinds of economies can flourish and, to a significant extent, the degree of scarcity of abundance.

Isolation

When a culture is cut off from interaction with other cultures, it is likely to develop unique norms and values. Where conditions of isolation preclude contact with others, a culture continues on its own course, unaltered and uncontaminated by others. Since the middle of the eighteenth century, however, almost all cultures in the world, near and far, have been influenced by and have in turn influenced Western European culture.

Technology

The tools that a culture has available will affect its norms and values, its economic and social relationships. The effect of the automobile on U.S. dating and courtship customs is an obvious example.

Dominant Cultural Themes

Cultures generally contain dominant themes that give a distinct character and direction to the culture; they also create, in part, a closed system. New ideas, values, and inventions are usually accepted only when they fit into the existing culture or represent changes that can be absorbed without too greatly distorting existing patterns. The Native American hunter, for example, was pleased to adopt the rifle as an aid to the established cultural theme of hunting. Western types of housing and legal customs regarding land ownership, however, were rejected because they were alien to a nomadic and communal way of life.

An Example: American Culture

U.S. culture is a unique blend of complex elements. It is a product of the United States' environment, its immigrants, its technology, and its place in history. This culture closely resembles the cultures of two close cousins, Australia and Canada, with which it shares vital characteristics. All three are new countries settled by diverse groups of immigrants, yet dominated by English culture; all three had uncrowded spaces and a sense of frontier; and all three now offer high levels of industrialization and wealth. Yet U.S. culture is distinguishable from the cultures of these first cousins as well as from those of more

At least since the middle of the eighteenth century, industrialization and colonialism have extended Western culture to previously isolated societies. The diffusion of modern technology has been particularly rapid in cases where the new tools enhance a society's ability to meet basic human needs at the same time that they are consistent with existing cultural patterns. Leaders, regardless of time, place, or the cultural bases of their authority, share a common need to communicate effectively with large numbers of followers.

distant cultural relatives in Europe. Although U.S. culture has changed in recent years, the changes are generally consistent with earlier cultural themes.

In 1970, one analyst of U.S. mores concluded that three values were central to understanding U.S. culture (Williams 1970):

- *Work.* Work is a good thing in itself, over and above the need to earn a living.
- *Achievement.* To strive to better yourself and get ahead is a good thing; being content with your present lot is considered rather lazy.
- *Morality.* People in the United States wish to be regarded as just, egalitarian, and generous. Behaving in accord with these principles is considered more desirable than acting in your own interest.

To a significant extent, these values continue to guide people of the United States. However, there are signs that the relative importance of these older values has declined. Results of a 1996 survey are presented in Figure 2.1. These data show that people of the United States consistently rank their family as the thing for which they are most thankful, followed closely by good health. Job and career is a distant third. These survey results and other indicators suggest that U.S. values are undergoing the following changes: a growing emphasis on personal well-being, a reduced emphasis on work, and a greater emphasis on consumerism. However, none of these values is completely new to U.S. culture. Daniel Boone and countless frontier explorers, for example, sought personal fulfillment as they abandoned their families to see what was on the other side of the mountain, and people of the United States have long been distinguished by their wealth of material goods.

U.S. norms have also changed. Premarital sex, childless marriages, working mothers, and interracial marriages are increasingly being accepted by the U.S. public. In

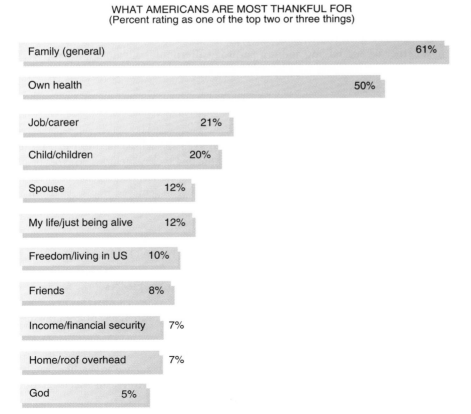

WHAT AMERICANS ARE MOST THANKFUL FOR
(Percent rating as one of the top two or three things)

Family (general)	61%
Own health	50%
Job/career	21%
Child/children	20%
Spouse	12%
My life/just being alive	12%
Freedom/living in US	10%
Friends	8%
Income/financial security	7%
Home/roof overhead	7%
God	5%

FIGURE 2.1
American Values in 1996
SOURCE: The Gallup Poll, November 1996.

general, U.S. norms now allow a wider variation in the means people use to achieve the values of self-fulfillment or a happy family life, but these changes do not represent a massive departure from earlier values. Across the United States, there remains a broad consensus that achievement, belief in God, and strong family ties are good things.

Social Structure and Institutions

Earlier we likened culture to a tool kit that would help us adapt to social and physical dilemmas. This tool kit is not merely an unorganized grab bag of norms and values. Instead it is organized by social structures and institutions. In the following sections, we examine these structures and how they enable us to make effective use of the tools that culture provides.

Social Structures

Many of our daily encounters occur in patterns. Every day we interact with the same people (our family or co-workers) or with the same kinds of people (sales clerks). These patterned relationships, these common dramas of everyday life, are called social structures. Each of these dramas has a set of actors (mother/child or buyer/seller) and a set of norms that defines appropriate behavior for each actor.

A **social structure** is a recurrent pattern of relationships.

Formally, a **social structure** is a recurrent pattern of relationships. Social structures can be found at all levels in society. Baseball games, friendship networks, families, and large corporations all fall into patterns that are repeated day after day. Some of these patterns are reinforced by formal rules or laws, but many more are maintained by force of custom.

The patterns in our lives are both constraining and enabling (Giddens 1984). If you would like to be free to work to your own schedule, you will find the nine-to-five, Monday-to-Friday work pattern a constraint. On the other hand, preset patterns provide convenient and comfortable ways of handling many aspects of life. They help us to get through crosstown traffic, find dating partners and spouses, and raise our children.

Whether we are talking about a Saturday afternoon ballgame, families, or the workplace, social structures can be analyzed in terms of three concepts: status, role, and institution.

Status

A **status** is a specialized position within a group.

The basic building block of society is the **status**—a specialized position in a group. Sociologists who want to study the status structure of a society include two types of statuses: achieved and ascribed. An **achieved status** is a position that a person can obtain in a lifetime. Being a father or president of the PTA is an achieved status. An **ascribed status** is a position fixed by birth or inheritance; it is unalterable in a person's lifetime. Sex and race are examples of ascribed statuses.

An **achieved status** is optional, one that a person can obtain in a lifetime.

An **ascribed status** is fixed by birth and inheritance and its unalterable in a person's lifetime.

Sociologists who analyze the status structure of a society are typically concerned with four related issues (Blau 1987; Blau & Schwartz 1984): (1) identifying the number and types of statuses that are available in a society; (2) assessing the distribution of people among these statuses; (3) determining how the consequences—the rewards, resources, and opportunities—are different for people who occupy one status rather than another; and (4) ascertaining what combinations of statuses are likely or even possible.

A Case Study: The Social Structure of Race

To illustrate how our lives are structured by status membership, we apply this approach to one particular ascribed status and ask how being African American affects relationships and experiences in the United States.

To begin: How many racial statuses are there in the United States? The 1990 census asked us to *identify* ourselves as belonging to one of five racial categories: white, black, Native American, Asian, or other. Although the labels may change, the same question appears on many of the other forms you will fill out as well as on almost every social survey. The apparent consensus on which racial statuses are significant and the nearly universal concern about them should alert us to the possible consequences that racial status has in our daily lives.

However, it is not just the number of statuses that has consequences. The numerical *distribution* of the population among racial statuses also encourages or discourages certain patterns of behavior. In 1990, for example, the census identified 2 million black residents in New York City, but only 12 black residents of Worland, Wyoming. In New York City, there are two whites for every black; in Worland, the ratio is 442 to 1. This means that one-half of all whites in New York could marry or be best friends with a black person; in Worland, it is impossible for more than a few whites to be this closely linked to a black resident.

Of course, numbers alone do not tell the whole story. By nearly every measure that one might choose, there is substantial inequality in the rewards, resources, and opportunities available to black and white people in the United States. Black unemployment is twice that of whites; the infant mortality rate is twice as high; the likelihood of being murdered is six times higher. Obviously, racial status has enormous *consequences* on the structures of daily experiences.

Although inequality persists, racial status does not correspond as directly with occupational and educational statuses as it once did, and different *combinations* of statuses are possible. Forty years ago, being black meant having much less education and very different kinds of jobs than whites. Today, knowing a person's ascribed status (race) is not such an accurate guide to his or her achieved statuses (education or occupation). Nevertheless, 30 percent of all nurse's aides and orderlies but only 5 percent of all physicians in the United States are black (U.S. Bureau of the Census 1996). The processes through which this continued overlap of racial, political, and economic statuses is maintained are discussed further in Chapter 8.

Roles

The status structure of a society provides the broad outlines for interaction. These broad outlines are filled in by **roles,** sets of norms that specify the rights and obligations of each status. To use a theatrical metaphor, the status structure is equivalent to the cast of characters, while roles are equivalent to the scripts that define how the characters ought to act, feel, and relate to one another. This language of the theater helps to make a vital point about the relationship between status and role: People occupy statuses, but they play roles. This distinction is helpful when we analyze how structures work in practice—and why they sometimes don't work. A man may occupy the status of father, but he may play the role associated with it very poorly.

A **role** is a set of norms specifying the rights and obligations associated with a status.

Structures in Action

Social structure is a concept designed to account for the patterns that appear in human behavior. Taken to extremes, it suggests a machinelike conformity in which we play the parts we are assigned. In reality, social structures are much more dynamic. Within the general pattern, there are many variations. Some of these variations are socially structured; others are due to individual actors negotiating their own roles. In order to fulfill role obligations, a person must have adequate resources to do so; a college student cannot be successful if he or she cannot afford to buy the books and has no time to study. People must also be motivated to meet their obligations by a system of sanctions; when there is little reward for good role performance, people become lax about meeting their obligations. Competing role demands can create additional difficulties. Finally, the norms of any social structure are never complete. For example, although the norms associated with the family provide the broad outlines of acceptable behavior, they leave many details unresolved. Being honest and standing by your family are both norms, but what do you do when telling the truth gets your brother in trouble? What do you do when something comes up that isn't in the script? Say, your mother wants to get a divorce and move in with you. Although social structure gives us a great deal of guidance about appropriate behavior, there are many occasions in everyday life when we must improvise and negotiate. This means that there will be variability in the way people play their parts.

Institutions

Social structures vary in scope and importance. Some, such as those that pattern the Friday night poker game, have limited application. The players could change to Saturday night or up the ante, and it would not have a major effect on the lives of anyone other than members of the group. If a major corporation changed seniority or family leave policies, it would have somewhat broader consequences, affecting not only employees of that firm but setting a precedent for the way other complex organizations might come to do business. Changes in other structures, however, have the power to shape the basic fabric of our lives. We call these structures social institutions.

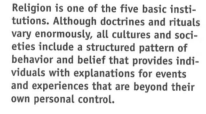

Institutions are enduring social structures that meet basic human needs.

An **institution** is an enduring and complex social structure that meets basic human needs. Its primary features are that it endures for generations; includes a complex set of values, norms, statuses, and roles; and addresses a basic human need. Embedded in the statuses and roles of the family institution, for example, are enduring patterns for dating and courtship, child rearing, and care of the elderly. Because the institution of family is composed of many separate family groups, however, the exact rules and behaviors surrounding dating or elder care will vary.

Nevertheless, institutions provide routine patterns for dealing with predictable problems of social life. Because these problems tend to be similar across societies, we find that every society tends to have the same types of institutions.

Religion is one of the five basic institutions. Although doctrines and rituals vary enormously, all cultures and societies include a structured pattern of behavior and belief that provides individuals with explanations for events and experiences that are beyond their own personal control.

Basic Institution

There are five basic social institutions:

- The family, to care for dependents and rear children.
- The economy, to produce and distribute goods.
- Government, to provide community coordination and defense.
- Education, to train new generations.
- Religion, to supply answers about the unknown or unknowable.

These institutions are basic in the sense that every society provides *some* set of enduring social arrangements designed to meet these important social needs. These arrangements may vary from one society to the next, sometimes dramatically. Government institutions may be monarchies or democracies or dictatorships. But a stable social structure that has the responsibility for meeting these needs is common to all societies.

In simple societies, all of these important social needs—political, economic, education, and religious—are met through one major social institution, the family or kinship group (Adams 1971). Social relationships based on kinship obligations serve as a basis for organizing production, reproduction, education, and defense.

As societies grow larger and more complex, the kinship structure is less able to furnish solutions to all the recurrent problems. As a result, some activities are gradually transferred to more specialized social structures outside the family. The economy, education, religion, and government become fully developed institutionalized structures that exist separately from the family. (The institutions of the contemporary United States are the subjects of Chapters 11 to 13).

As the social and physical environment of a society changes and the technology for dealing with that environment expands, the problems that individuals have to face change. Thus, institutional structures are not static; new structures emerge to cope with new problems.

Among the more recent social structures to be institutionalized in Western society are medicine, science, sport, the military, law, and the criminal justice system. Each of these areas can be viewed as an enduring social structure, complete with interrelated statuses and a unique set of norms and values.

Institutional Interdependence

Each institution of society can be analyzed as an independent social structure, but none really stands alone. Instead, institutions are interdependent; each affects the others and is affected by them.

In a stable society, the norms and values embodied in the roles of one institution will usually be compatible with those in other institutions. For example, a society that stresses male dominance and rule by seniority in the family will also stress the same norms in its religious, economic, and political systems. In this case, interdependence reinforces norms and values and adds to social stability.

Sometimes, however, interdependence is an important mechanism for social change. Because each institution affects and is affected by the others, a change in one tends to lead to change in the others. Changes in the economy lead to changes in the family; changes in religion lead to changes in government. For example, when years of school becomes more important than hereditary position in determining occupation, hereditary position will also be endangered in government, the family, and religion.

Institutions: Agents of Stability or Oppression?

Sociologists use two major theoretical frameworks to approach the study of social structures: structural functionalism and conflict theory. The first focuses on the part that institutions play in creating social and personal stability; the second focuses on the role of institutions in legitimizing inequality (Eisenstadt 1985).

A STRUCTURAL-FUNCTIONAL VIEW OF INSTITUTIONS.   Institutions provide ready-made patterns to meet most recurrent social problems. These ready-made patterns regulate human behavior and are the basis for social order. Because we share the same patterns, social life tends to be stable and predictable.

Structural-functional theorists point out that stability and predictability are important both for societies and individuals. The classic expression of this point of view is from Durkheim, who argued that social institutions create a "liberating dependence." By furnishing patterned solutions to our most pressing everyday problems, they free us for more creative efforts. They save us from having to reinvent the social equivalent of the wheel each generation and thus they facilitate our daily lives. Moreover, because these patterns have been sanctified by tradition, we tend to experience them as morally right. As a result, we find satisfaction and security in social institutions.

A CONFLICT VIEW OF INSTITUTIONS.   By the very fact that they regulate human behavior and direct choices, institutions also constrain behavior. By producing predictability, they reduce innovation; by providing security, they reduce freedom. Some regard this as a necessary evil, the price we pay for stability.

This benign view of constraint is challenged by conflict theorists. They accept that institutions meet basic human needs, but they wonder, Why *this* social pattern rather than another? Why *this* family system instead of another? To explain why one social arrangement is chosen over another, conflict theorists ask, "Who benefits?"

From the viewpoint of conflict theory, institutions represent a camouflaged form of inequality—one that is supported by norms and tradition. Because institutions have existed for a long time, we tend to think of our familial, religious, and political systems as not merely one way of fulfilling a particular need but as the only acceptable way. Just as a tenth-century Mayan may have thought "Of course, virgins should be sacrificed if the crops are bad," so we tend to think "Of course, women earn less than men." In both cases, the cloak of tradition obscures our vision of inequality or oppression, making inequality seem normal and even desirable. As a result, institutions stifle social change and help maintain inequality.

Summary

Institutions create order and stability; in doing so, they suppress change. In creating order, they preserve the status quo. In regulating, they constrain. In this sense, both conflict and structural-functional theories are right; they simply place a different value judgment on stability and order. The two theoretical perspectives prompt us to ask somewhat different questions about social structures. Structural functionalism prompts us to ask how an institution contributes to order, to stability, and to meeting the needs of society and the individual. Conflict theory prompts us to ask which groups are benefiting the most from the system and how they are seeking to maintain their advantage. Both are worthwhile questions, and both will be addressed when we look at social structures ranging from deviance to the economy (in Chapters 5–13).

Culture and Social Structure

A major theoretical debate in sociology has revolved around the question of which comes first, culture or social structure? As noted in Chapter 1, Karl Marx believed that the structure of economic institutions determines every other facet of social life, including culture. In contrast, Max Weber believed that values were important causes of social structure. Most sociologists agree with both points of view. On Marx's side, by regulating access to social rewards, social structures clearly reinforce certain values and norms and discourage and inhibit others. However, social institutions exist only to the extent that they are expressed in the roles that individuals play. Thus, as Weber would have it, changes in cultural values and norms eventually lead to changes in individual behavior and therefore to changes in the statuses and roles that make up the larger social structure.

A Case Study: Values, Institutions, and the Environment

Each day, a minimum of 140 plant and animal species are condemned to extinction. Each year, about 17 million hectares (more than 65,000 square miles) of forest vanish, the equivalent of an area about half the size of Finland. The ozone layer over the heavily populated latitudes of the Northern Hemisphere is thinning twice as fast as scientists thought just a few years ago (Postel 1992), and this loss is believed to explain the recent dramatic increase in the incidence of skin cancer. On Indian reservations and in many minority and poor communities, exposure to toxic waste has led to a surge in the number of children born with major birth defects. The question is, of course, how did things get this bad and what can we do about it?

Writing in 1967, history professor Lynn White, Jr. argued that the roots of the modern ecological crisis lie in Western traditions of technology and science. White believed that a technology capable of efficiently exploiting natural resources (and other human beings, for that matter) began its unparalleled growth in medieval Europe at least partly because of the Judeo-Christian philosophy that man is the master of nature rather than simply a part of it. White argues that the religious values of Hindus, Native Americans, and traditional African cultures are quite different on this key matter, and that, at least partly for this reason, their social structures have been less ecologically destructive. The traditional Hindu and African belief has been that human beings are connected to all elements of nature by insoluble spiritual and psychological bonds ([King] Moshoeshoe 1993; Singh 1993); Native American philosophy maintains that no action should be taken without assessing its consequences for the next seven gen-

There is no question that human societies affect and are affected by the natural environment, and there is no question that the effects of human social activity on the local environment are often devastating. The only real question is whether attitudes or social structures are more to blame. Is the diffusion of Western culture or the diffusion of Western economic and technological structures the primary cause of destruction such as this shown in Indonesia? The answer to this question is important because it provides a starting point for the even more important discussion of how we can save the planet. Should we start by trying to change people's beliefs, lifestyles, and consumption patterns, or should we begin by trying to change social structures—transportation systems or recycling efforts, for instance?

erations. These beliefs call into question the wisdom of the wasteful lifestyles that have accompanied technological and institutional developments in the West.

In Western culture, belief in individualism, materialism, "progress and growth," and the idea that government should have a very limited role in controlling business complements a religious ideology that tends to emphasize humankind's dominion over all other creatures. As Weber noted, these beliefs are important cultural correlates of capitalist economic structures; as ecofeminists point out, they also tend to be associated with male-dominated family structures. Even though scholars debate which came first—technology and social institutions or values—and theologians question just how conducive Western religious values have been toward the destruction of the environment, research clearly shows that commitment to these dominant political and economic values is associated with lower levels of environmental concern among United States citizens (Dunlap & Van Liere 1984).

Despite their different histories, cultural traditions, and social structures, many people living in both the industrialized West and the developing nations of Africa and South America are now starting to show a high level of concern for environmental deterioration and widespread support for environmental protection. People in poorer countries recognize that overpopulation creates ecological problems; residents of wealthier nations realize that technology and wasteful lifestyles have devastating consequences for the environment. Furthermore, results from an international survey indicate that a very high percentage of respondents throughout the world also say they are willing to pay higher prices for goods and services if doing so will protect the environment (Dunlap, Gallup, Jr., & Gallup 1993).

If cultural values toward the environment have changed, why hasn't there been a corresponding change in social structures and social institutions? Part of the answer is that current, environmentally harmful ways of doing business are profitable. Another part of the answer is that the attitudes of private citizens have changed faster than business and governmental leaders realize. However, a final part of the answer is that even when individual values change, it is hard to change behavior if supporting institutional structures are not in place. It is difficult, for instance, to conserve energy and reduce pollution if there is no system of mass transportation in place that would allow people to get to work or school on time without relying on a private automobile. No matter how committed one is to the principle of recycling, it is difficult to do so if there are no groups or organizations that routinely pick up or process recyclables. Change in cultural values or beliefs does lead to institutional change, but the process is slow, often depending on changes in one set of statuses and roles, one cluster of groups and group memberships, at a time.

THINKING

CRITICALLY

You are responsible for reducing air pollution in Los Angeles by at least 50% during the next five years. What would you do and where would you start? With culture or with social structure?

Types of Societies

Institutions give a society a distinctive character. In some societies, the church has been dominant; in others, it has been the family or the economy. Whatever the circumstance, recognizing the institutional framework of a society is critical to an understanding of how it works.

The history of human societies is the story of ever-growing institutional complexity. In simple societies, we often find only one major social institution—the family or kin group. Many modern societies, however, have as many as a dozen institutions.

What causes this expansion of institutions? The triggering event appears to be economic. When changes in technology, physical environment, or social arrangements increase the level of economic surplus, the possibilities of institutional expansion arise (Lenski 1966). In this section we sketch a broad outline of the institutional evolution that accompanied three revolutions in production.

Hunting, Fishing, and Gathering Societies

The chief characteristic of hunting, fishing, and gathering societies is that they have subsistence economies. This means that in a good year they produce barely enough to get by; that is, they produce no surplus. In some years, of course, game and fruit are plentiful, but in many years scarcity is a constant companion.

The basic units of social organization are the household and the local clan, both of which are based primarily on family bonds and kinship ties. Most of the activities of hunting and gathering are organized around these units. A band rarely exceeds 50 people in size and tends to be nomadic or seminomadic. Because of their frequent wanderings, members of these societies accumulate few personal possessions.

The division of labor is simple, based on age and sex. The common pattern is for men to participate in hunting and deep-sea fishing and for women to participate in gathering, shore fishing, and preserving. Aside from inequalities of status by age and sex, few structured inequalities exist in subsistence economies. Members possess little wealth; they have few, if any, hereditary privileges; and the societies are almost always too small to develop class distinctions. In fact, a major characteristic of subsistence societies is that individuals are homogeneous, or alike. Apart from differences occasioned by age and sex, members generally have the same everyday experiences.

Horticultural Societies

The first major breakthrough from subsistence economy to economic surplus was the development of agriculture. When people began to plant and cultivate crops, rather than harvesting whatever nature provided, stable horticultural societies developed. The technology was often primitive—a digging stick, occasionally a rudimentary hoe—but it produced a surplus.

The regular production of more than the bare necessities revolutionized society. It meant that some people could take time off from basic production and turn to other pursuits: art, religion, writing, and frequently warfare. Of course, the people who participate in these alternate activities are not picked at random; instead, a class hierarchy develops between the peasants, who must devote themselves full-time to food production, and those who live off their surplus.

Because of relative abundance and a settled way of life, horticultural societies tend to develop complex and stable institutions outside the family. Some economic activity is carried on outside the family, a religious structure with full-time priests may develop, and a stable system of government—complete with bureaucrats, tax collectors, and a hereditary ruler—often develops. Such societies are sometimes very large. The Inca empire, for example, included an estimated population of more than four million.

Agricultural Societies

Approximately 5,000 to 6,000 years ago, a second agricultural revolution occurred, and the efficiency of food production was doubled and redoubled through better technology. The advances included the harnessing of animals, the development, in time, of metal tools, the use of the wheel, and improved knowledge of irrigation and fertilization. These changes dramatically altered social institutions.

The major advances in technology meant that even more people could be freed from direct production. The people not tied directly to the land congregated in large urban centers and developed a complex division of labor. Technology, trade, reading and writing, science, and art grew rapidly as larger and larger numbers of people were able to devote full time to these pursuits. Along with greater specialization and occupational diversity came greater inequality. In the place of the rather simple class structure of horticultural societies, a complex class system developed, with merchants, soldiers, scholars, officials, and kings—and, of course, the poor peasants on whose labor they all ultimately depended. This last group still contained the vast bulk of the society, probably at least 90 percent of the population (Sjoberg 1960).

CONCEPT SUMMARY	*Types of Societies*

HUNTING, FISHING, AND GATHERING

Technology:	Very simple—arrows, fire, baskets
Economy:	Bare subsistence, no surplus
Settlements:	None or very small (bands of less than 50 people)
Social organization:	All resting within family
Examples:	Plains Indians, Eskimos

HORTICULTURAL SOCIETIES

Technology:	Digging sticks, occasionally blade tools
Economy:	Simple crop cultivation, some surplus and exchange
Settlements:	Semipermanent—some cities; occasionally kingdoms
Social organization:	Military, government, religion becoming distinct institutions
Examples:	Mayans, Incas, Egyptians under the pharaohs

AGRICULTURAL SOCIETIES

Technology:	Irrigation, fertilization, metallurgy, animal power used to increase agricultural productivity
Economy:	Largely agricultural, but much surplus; increased market exchange and substantial trade
Settlements:	Permanent—urbanization becoming important, empires covering continents
Social organization:	Educational, military, religious, and political institutions are well developed
Examples:	Roman empire, feudal Europe, Chinese empire

INDUSTRIAL SOCIETIES

Technology:	New energy sources (coal, gas, electricity) leading to mechanization of production
Economy:	Industrial—few engaged in agriculture or direct production; much surplus; fully developed market economy
Settlements:	Permanent—urban living predominating, nation states
Social organization:	Complex set of interdependent institutions
Examples:	Contemporary United States, Europe, Japan

One of the common uses to which societies put their new leisure and other new technology was warfare. With the domestication of the horse (cavalry) and the invention of the wheel (chariot warfare), military technology became more advanced and efficient. Military might was used as a means to gain even greater surplus through conquering other peoples. The Romans were so successful at this that they managed to turn the peoples of the entire Mediterranean basin into a peasant class that supported a ruling elite in Italy.

Industrial Societies

The third major revolution in production was the advent of industrialization 200 years ago in Western Europe. The substitution of mechanical, electrical, and fossil-fuel energy for human and animal labor caused an explosive growth in productivity, not only of goods but also of knowledge and technology. In the space of a few decades, agricultural societies were transformed. The enormous increases in energy, technology, and knowledge freed the bulk of the work force from agricultural production, and increasingly also from industrial production. The overall effect on society has been to transform its political, social, and economic character. Old institutions such as education have expanded dramatically, and new institutions such as science, medicine, and law have emerged.

When Institutions Die: The Tragedy of the Ojibwa

The story of modernization is the story of institutional change, of changes in the ways our production, reproduction, education, and social control are socially structured. Sometimes these institutional changes take place in harmony so that institutions continue to support one another and to provide stable patterns that meet ongoing human needs. On other occasions, however, old institutions are destroyed before new ones can evolve. When this happens, societies and the individuals within them are traumatized; societies and people fall apart.

In 1985, Anastasia Shkilnyk chronicled just such a human tragedy when she described the plight of the Ojibwa Indians of western Ontario in the book *A Poison Stronger Than Love*. Although the details are specific to the Ojibwa, her story is helpful in understanding what happened to Native Americans in general and other traditional societies when rapid social change altered social institutions.

In 1976, Anastasia Shkilnyk was sent by the Canadian Department of Indian Affairs to Grassy Narrows, an Ojibwa community of 520 people, to advise the band on how to alleviate economic disruption caused by mercury poisoning in nearby lakes and rivers. At the request of the chief and council, she stayed for two and a half years to assist the band in developing socioeconomic projects and in preparing the band's case for compensation for the damages they suffered from mercury pollution and misguided government policy.

A Broken Society

Grassy Narrows in the 1970s was a community destroyed. Drunken six-year-olds roamed winter streets when the temperatures were 40 below. The death rate for both children and adults was very high compared to the rest of Canada. Nearly three-quarters of all deaths were linked directly to alcohol and drug abuse. A quote from Shkilnyk's journal evokes the tragedy of life in Grassy Narrows.

"Time Zones" by Frank Bigbear, Jr.

Friday. My neighbor comes over to tell me that last night, just before midnight, she found four-year-old Dolores wandering alone around the reserve, about two miles from her home. She called the police and they went to the house to investigate. They found Dolores's three-year-old sister, Diane, huddled in a corner crying. The house was empty, bare of food, and all the windows were broken. The police discovered that the parents had gone to Kenora the day before and were drinking in town. Both of them were sober when they deserted their children.

It's going to be a bad weekend. The police also picked up an eighteen-month-old baby abandoned in an empty house. No one seemed to know how long it had been there. . . . The milk in the house had turned sour. The baby was severely dehydrated and lying in its own vomit and accumulated excrement. Next door, the police found two people lying unconscious on the floor.

The adults in the community were not alcoholics of the kind that had physiological reactions if they did not drink regularly. Rather, they were binge drinkers. When wages were paid or the welfare checks came, many drank until they were unconscious and the money was gone. Shkilnyk estimates that on one occasion when $20,000 in wages and $5,000 in social assistance was paid out on one day, within one week $14,000 had been spent on alcohol. The children felt as much despair as their parents and sought similar forms of escape. Often the children waited until their parents had drunk themselves unconscious and then drank the liquor that was left. If they could not get liquor, they sniffed glue or gasoline.

Yet 20 years before, the Ojibwa had been a thriving people. How was a society so thoroughly destroyed?

Ojibwa Society Before 1963

The Ojibwa have been in contact with whites for two centuries. In 1873, they signed the treaty that defines their relationship with the government of Canada. In the treaty, the government agreed to set aside reserves for the Indians, to give the Indians the right to pursue their traditional occupations of hunting, fishing, and trapping and to provide schools.

Generally, this arrangement does not seem to have been disruptive to the Ojibwa way of life. Their reserve was in an area they had traditionally viewed as their own on the banks of the English River in northern Ontario. Despite the development of logging and mining in the areas around them, they had very little contact with the white community except for an annual ceremonial visit on Treaty Day.

The Ojibwa were a hunting and gathering people; the family was the primary social institution. A family group consisted of a man and his wife, their sons and their wives and children, or of several brothers and their wives and children. The houses or tents of this family group would all be clustered together, perhaps as far as a half mile from the next family group.

Economic activities were all carried out by family groups. These activities varied with the season. In the late summer and fall, there was blueberry picking and harvesting wild rice. In the winter, there was hunting and trapping. In all of these endeavors, the entire family participated, everybody packing up and going to where the work was. The men would trap and hunt, the women would skin and prepare the meat, the old people would come along to take care of the children and teach them. They used their reserve only as a summer encampment. From late summer until late spring, the family was on the move.

Besides being the chief economic and educational unit, the family was also the major agent of social control. Family elders enforced the rules and punished those who violated them. In addition, most religious ceremonials were performed by family elders. Although a loose band of families formed the Ojibwa society, each family group was largely self-sufficient, interacting with other family groups only to exchange marriage partners and for other ceremonial activities.

The earliest influence of white culture did not disrupt this way of life particularly. The major change was the development of boarding schools, which removed many Indian children from their homes for the winter months. When they returned, however, they would be accepted back into the group and educated into Indian ways by their grandparents. The boarding schools took the children away, but did not disrupt the major social institutions of the society they left behind.

The Change

In 1963, the government decided that the Ojibwa should be brought into modern society and given the benefits thereof: modern plumbing, better health care, roads, and the like. To this end, they moved the entire Ojibwa community from the old reserve to a government-built new community about 4 miles from their traditional encampment. The new community had houses, roads, schools, and easy access to "civilization." The differences between the new and the old were sufficient to destroy the fragile interdependence of Ojibwa institutions.

An Application

ALCOHOLISM AND NATIVE AMERICANS

focus on

U nfortunately, the Ojibwa are not an exceptional case. Their tragedy has been played out in tribe after tribe, band after band, all over North America. Compared to the national average, Native Americans are (Podolsky 1986/87)

○ 3.8 times more likely to die from alcoholism
○ 4.5 times more likely to have cirrhosis of the liver
○ 5.5 times more likely to die in alcohol-related motor vehicle accidents

To paraphrase C. Wright Mills (Chapter 1), when one or two individuals abuse alcohol, this is an individual problem, and for its relief we rightfully look to clinicians and counselors. When large segments of a population have alcohol problems, this is a public issue and must be addressed at the level of social structure.

The Problem
There is enormous variety among Native American groups on this as well as other indicators. Some Native

American groups are almost entirely alcohol-free; in others, alcoholism touches nearly every family. Nevertheless, the overall level of alcohol abuse is very high (McConnell 1973). As a result, 95 percent of Native Americans report that alcohol abuse is a major problem of their people (Shafer 1989).

High levels of alcohol use are a health problem, an economic problem, and a social problem. Among the related issues are child and spouse abuse, unemployment, teenage pregnancy, birth defects, nonmarital births, and divorce. How can these interrelated problems be addressed?

Two Contrasting Approaches: Clinical Versus Structural
The clinical approach to treating alcoholism is based on traditional medical and psychological models: Take one patient at a time and treat his or her symptoms. Many clinical treatment programs use Alcoholics Anonymous as a framework for treatment: Individuals are encouraged to admit that they are alcoholics and to seek the support and help of the AA group and of God in conquering alcoholism. Even when faced with society-

wide alcoholism such as exists in some Native American communities, the clinical model still focuses on such solutions as more counselors and better diagnostic care (Weibel-Orlando 1986/87).

A sociological approach to understanding alcoholism on such a scale begins instead with the premise that any widely recurring pattern must be socially structured. According to this perspective, the appropriate strategy for reducing alcoholism among Native Americans is to ask what social structures encourage alcohol use. Conversely, why don't social structures reward sobriety?

The answer depends on one's theoretical framework. Structural functionalists focus on the destruction of Native American culture and the absence of harmony between their institutions and those of white society. Conflict theorists see the current situation as the result of a violent conflict over scarce resources, a conflict in which victorious Europeans systematically stripped Native Americans of their means of economic production and hence destroyed their society (Fisher 1987).

First, all the houses were close together in neat rows, assigned randomly without regard for family group. As a result, the kin group ceased to exist as a physical unit. Second, the replacement of boarding schools with a community school meant that a parent (the mother) had to stay home with the children instead of going out on the trap line. As a result, the adult woman overnight became a consumer rather than a producer, shattering her traditional relationship with her husband and community. As a consequence of the women's and children's immobility, men had to go out alone on the trap line. Because they were by themselves rather than with their family, the trapping trips were reduced from six to eight weeks to a few days, and trapping ceased to be a way of life for the whole family. The productivity of the Ojibwa reached bottom with the government order in May 1970 to halt all fishing because of severe mercury contamination of the water. Then the economic contributions of men as well as women were sharply curtailed; the people became heavily dependent on the government rather than on themselves or on each other.

What happened was the total destruction of old patterns of doing things—that is, of social institutions. The relationships between husbands and wives were no longer clear. What were their rights and obligations to each other now that their

Regardless of theoretical position, however, the obvious fact is that Native Americans are severely economically disadvantaged. Unemployment is often a way of life, and only 25 percent of the adults on some reservations are employed. Lack of employment is a critical factor in alcoholism in all populations. Having a steady, rewarding job is an incentive to stay sober; it also reduces the time available for drinking, which is essentially a leisure-time activity. Even clinically oriented scholars conclude that "occupational stability is a major factor in maintaining sobriety" (Ronan and Reichman 1986). From this perspective, the solution to high levels of Native American alcoholism must include changing economic institutions to provide full employment, as well as hiring more AA group leaders.

Summary

In many ways, fighting alcoholism is like fighting smallpox. We cannot eradicate the disease by treating people after they have it; we have to prevent its occurence. When alcoholism is epidemic in a community, it requires community-wide efforts at prevention. Statuses, roles, and institutions must be reformed so that people have a reason to stay sober.

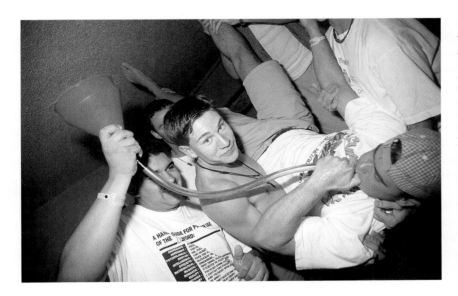

Native Americans are hardly the only subculture in American society to have problems with alcohol abuse. Another is young men. Heavy drinking and drinking games are especially characteristic of the high school and college years, and the expectation of drinking is built into many collegiate activities, such as parties and spring break. Although raising the drinking age has made much of this illegal, it is still a recurrent pattern supported by subcultural norms and values.

joint economic productivity was at an end? What were their rights and obligations to their children when no one cared about tomorrow?

The Future

In 1985 the Ojibwa finally reached an out-of-court settlement with the federal and provincial governments and the mercury-polluting paper mill. The $8.7 millon settlement was in compensation for damages to their way of life arising from government policies and mercury pollution. The band is using some of this money to develop local industries that will provide an ongoing basis for a productive and thriving society. Today, Ojibwa society has begun the process of healing and recovery.

Summary

Institutions offer stable patterns of responding to stable problems. Some of these stable problems are straightforward and obvious, such as finding enough to eat. Other problems are more subtle but just as important: having something to do each day that is meaningful and having bonds of obligation and exchange with others in the community.

In the case of the Ojibwa Indians and other indigenous peoples of North America, welfare and the supermarket can take care of the first problem, but they cannot replace the second. Stable social structures that define our roles relative to others in our environment, that assure us of the continuity of the past and the future, are an essential aspect of human society. It is often true, as conflict theorists stress, that any given institutional arrangement benefits one group over another. It is also true, as the structural functionalists stress, that some institutional arrangement is better than none.

Summary

1. Culture is a design for living that provides ready-made solutions to the basic problems of a society. It can be conceived of as a tool kit of material and nonmaterial components that help people adapt to their circumstances.

2. Language, or symbolic communication, is a central component of culture. Language embodies culture, serves as a framework for perceiving the world, and symbolizes common bonds.

3. Values spell out the goals that a culture finds worth pursuing, and norms specify the appropriate means to reach them.

4. The cultures of large and complex societies are not homogeneous. Subcultures and countercultures with distinct lifestyles and folkways develop to meet unique regional, class, and ethnic needs.

5. People of the United States say that they value the family, good health, a good self-image, and a sense of accomplishment. Over the last generation, their values have changed with more emphasis being placed on self-fulfillment and consumerism and less emphasis on work.

6. Culture is learned, taught, normative, problem solving, and relative.

7. The most important factors accounting for cultural variation are the physical and natural environment, isolation from other cultures, level of technological development, and cultural themes.

8. The analysis of social structure—recurrent patterns of relationships—revolves around three concepts: status, role, and institution. Statuses are specialized positions within a group and may be of two types: achieved or ascribed. Roles define how status occupants ought to act and feel.

9. Because societies share common human needs, they also share common institutions. The common institutions are family, economy, government, education, and religion. Each society has some enduring social structure to perform these functions for the group.

10. Institutions are interdependent; none stands alone, and a change in one results in changes in others. Institutions regulate behavior and maintain the stability of social life across generations. Conflict theorists point out that these patterns often benefit one group more than others.

11. An important determinant of institutional development is the ability of a society to produce an economic surplus. Each major improvement in production has led to an expansion in social institutions.

Sociology on the Net

In your text, the authors discuss "deaf culture." They state that through interaction, deaf persons influence each other—and the hearing as well—about their values, attitudes, and norms. One place where the deaf culture is available to all is on the Internet.

```
http://dww.deafworldweb.org/
```

Once you reached the home page, click on the search button. Click in the top box and when the cursor appears type *deaf and culture* in the space provided. Activate the search button at the bottom of the screen. Browse through the selections. Note the selection on guest comments and how the Internet has helped transmit the deaf culture to both the deaf and hearing alike. Browse a bit more and note the DWW news as well as the deaf culture of the bi-week.

If culture is problem solving, how does this site help the deaf solve their problems? As you read about cochlear implants and other deaf issues, how might the deaf culture promote ethnocentrism? How might being a member of

the hearing promote ethnocentrism about these same issues and topics?

FIND IT ON INFOTRAC COLLEGE EDITION

Your textbook has given you a "standard" definition of *culture,* based on the functionalist perspective. In contemporary
conversation, we use the term *culture* in many different ways.

Using InfoTrac College Edition, look up the following article:

"The Culture of 'Culture.'" Christopher Clausen,
The Wilson Quarterly, Autumn 1994, v18 n4 p132.

(Hint: Enter the search term *Christopher Clausen,*
using the Subject Guide.)

How does Clausen's analysis compare with the standard
definition?

Suggested Readings

Bell, Robert W., and Bell, Nancy (eds.). 1989. *Sociobiology
and the Social Sciences.* Lubbock, Tex.: Texas Tech Press. A
collection of essays applying the sociobiological perspective
to topics such as family violence, sexual attraction, and single parenting.

Bellah, Robert N., and Associates. 1985. *Habits of the Heart:
Individualism and Commitment in American Life.*
Berkeley: University of California Press. A critical look at
contemporary U.S. values. The authors argue that an emphasis on individual fulfillment at the expense of commitment is damaging to society and to personal
integration.

Kephart, William M., and Zellner, William W. *Extraordinary
Groups: The Sociology of Unconventional Life-Styles* (5th ed.).
New York: St. Martin's Press. A fascinating tour of some of
the most interesting subcultures and countercultures in the
United States—both past and present: the Amish, gypsies,
Father Divines, and Jehovah's Witnesses. Painless sociol-
ogy—applies basic concepts and theory to truly extraordi-
nary groups.

Lenski, Gerhard. 1966. *Power and Privilege: A Theory of Social
Stratification.* New York: McGraw-Hill. A major work distinguishing the fundamental characteristics of different
types of societies, particularly in terms of socially structured
inequality.

Shkilnyk, Anastasia. 1985. *A Poison Stronger Than Love: The
Destruction of an Ojibwa Community.* New Haven: Yale. An
ethnographic community study that focuses on the social
structures of a Native American community in Canada. A
powerful illustration of the extent to which individual well-
being depends on stable institutions.

Smith, Robert J., and Wiswell, Ella Lury. 1982. *The Women of
Suye Mura.* Chicago: University of Chicago Press. A companion volume to John Embree's now standard *Suye Mura:
A Japanese Village.* Smith and Wiswell's book can stand
alone as a unique account of the lives of rural Japanese
women in the mid-1930s. Also makes an interesting contrast to the more standard focus on social organization and
men's lives found in Embree's work.

Trask, Haunani-Kay. 1993. *From a Native Daughter:
Colonialism and Sovereignty in Hawaii.* Monroe, Maine:
Common Courage Press. A fiery yet factually supported
description of U.S. involvement in changing all the institutions of Native Hawaiians.

CHAPTER 3

Groups, Networks, and Organizations

Sociology is the study of relationships. We are concerned with how relationships are formed and the consequences these relationships have for the individual and the community. In this chapter we review the basic types of human relationships, from small and intimate groups to large and formal organizations, and discuss some of the consequences of these relationships.

Group Processes

Some relationships are characterized by harmony and stability; others are made stressful by conflict and competition. We use the term **social processes** to describe the types of interaction that go on in relationships. This section looks closely at four social processes that regularly occur in human relationships: exchange, cooperation, competition, and conflict.

Social processes are the forms of interaction through which people relate to one another; they are the dynamic aspects of society.

Exchange

Exchange is voluntary interaction in which the parties trade tangible or intangible benefits with the expectation that all parties will benefit. A wide variety of social relationships include elements of exchange. In friendships and marriages, exchanges usually include such intangibles as companionship, moral support, and a willingness to listen to the other's problems. In business or politics, an exchange may be more direct; politicians, for example, openly acknowledge exchanging votes on legislative bills—I'll vote for yours if you'll vote for mine.

Exchange is voluntary interaction from which all parties expect some reward.

Exchange relationships are based on the expectation that people will return favors and strive to maintain a balance of obligation in social relationships. This expectation is called the **norm of reciprocity** (Gouldner, 1960). If you help your sister-in-law move, then she is obligated to you. Somehow she must pay you back. If she fails to do so, the social relationship is likely to end, probably with bad feelings. A corollary of the norm of reciprocity is that you avoid accepting favors from people with whom you do not wish to enter into a relationship. For example, if someone you do not know very well volunteers to type your term paper, you will probably be suspicious. Your first thought is likely to be, "What is this guy trying to prove? What does he want from me?" If you do not want to owe this person a favor, you will say that you prefer to type your own paper. Nonsociologists might sum up the norm of reciprocity by concluding that there's no free lunch.

The **norm of reciprocity** is the expectation that people will return favors and strive to maintain a balance of obligation in social relationships.

Exchange is one of the most basic processes of social interaction. Almost all voluntary relationships are entered into as situations of exchange. In traditional U.S. families, these exchanges were clearly spelled out. He supported the family, which obligated her to keep house and look after the children; or, conversely, she bore the children and kept house, which obligated him to support her.

An exchange relationship persists only if each party to the interaction is getting something out of it. This does not mean that the rewards must be equal; in fact, rewards are frequently very unequal. You have probably seen play groups, for example, where one child is treated badly by the other children and is permitted to play with them only if he agrees to give them his lunch or allows them to use his bicycle. If this boy has no one else to play with, however, he may find this relationship more rewarding than the alternative of playing alone. The continuation of very unequal exchange relationships usually rests on a lack of desirable alternatives (Emerson 1962).

Cooperation

Cooperation is interaction that occurs when people work together to achieve shared goals.

Cooperation occurs when people work together to achieve shared goals. Exchange is a trade: I give you something and you give me something else in return. Cooperation is teamwork. It is characteristic of relationships where people work together to achieve goals that they cannot achieve alone. Consider, for example, a four-way stop. Although it may entail some waiting in line, in the long run we will all get through more safely and more quickly if we cooperate and take turns. Most continuing relationships have some element of cooperation. Spouses cooperate in raising their children; children cooperate in tricking their substitute teachers.

Competition

It is not always possible for people to reach their goals by exchange or cooperation. If your goal and my goal are mutually exclusive (for example, I want to sleep and you want to play your stereo), we cannot both achieve our goals. Similarly, in situations of scarcity, there may not be enough of a desired good to go around. In these situations, social processes are likely to take the form of either competition or conflict.

Competition is a struggle over scarce resources that is regulated by shared rules.

A struggle over scarce resources that is regulated by shared rules is **competition** (Friedsam 1965). The rules usually specify the conditions under which winning will be considered fair and losing will be considered tolerable. When the norms are violated, competition may erupt into conflict.

Competition is a common form of interaction in U.S. society. Jobs, grades, athletic honors, sexual attention, marriage partners, and parental affection are only a few of the scarce resources for which individuals or groups compete. In fact, it is difficult to identify many social situations that do not entail competition. One positive consequence of competition is that it stimulates achievement and heightens people's aspirations. It also, however, often results in personal stress, reduced cooperation, and social inequalities (elaborated on in Chapter 6–10).

Because competition often results in change, groups that seek to maximize stability often devise elaborate rules to avoid the appearance of competition. Competition is particularly problematic in such informal groups as friendships and marriages. Friends who want to stay friends will not compete for valued objects; they might compete over bowling scores, but they won't compete for the same promotion. Similarly, couples who value their marriage will not compete for their children's affection or loyalty. To do so would be to destroy the marriage.

Conflict

Conflict is a struggle over scarce resources that is not regulated by shared rules; it may include attempts to destroy or neutralize one's rivals.

When struggle over scarce resources is not regulated by shared rules, **conflict** occurs (Coser 1956, 8). Because no tactics are forbidden and anything goes, conflict may include attempts to neutralize, injure, or destroy one's rivals. Conflict creates divisiveness rather than solidarity.

When conflict takes place with outsiders, however, it may enhance the solidarity of the group. Whether we're talking about warring superpowers or warring street gangs, the us-against-them feeling that emerges from conflict with outsiders causes group members to put aside their jealousies and differences to work together. Groups from nations to schools have found that starting conflicts with outsiders is a useful device for restraining the more quarrelsome members of their own group. Critics of U.S. foreign policy point out that many of our foreign adventures (invading Grenada, bombing Libya) seem to be timed to take people's minds off divisive domestic issues such as the Iran-Contra scandals.

Summary

Exchange, cooperation, competition, and even conflict are important aspects of our relationships with others. Few of our relationships involve just one type of group process. Even friendships usually involve some competition as well as cooperation and exchange; similarly, relationships among competitors often involve cooperation.

We interact with people in a variety of relationships. Some of these relationships are temporary and others permanent; some are formal and others informal. In the rest of this chapter, we discuss the wide variety of relationships we have under three general categories: groups, social networks, and organizations.

Groups

What is a group? A **group** is two or more people who interact on the basis of a shared social structure and recognize mutual dependency. Groups may be large or small, formal or informal; they range from a pair of lovers to IBM. In all of them, members share a social structure specifying statuses, roles, and norms, and they share a feeling of mutual dependency.

The distinctive characteristics of a group stand out when we compare the group to two collections of people that do not have these characteristics. An **aggregate** is people who are temporarily clustered together in the same location (for example, all the people on a city bus, those attending a movie, or shoppers in a mall). Although these people may share some norms (such as walking on the right when passing others), they are not mutually dependent. In fact, most of their shared norms have to do with procedures to maintain their independence despite their close physical proximity. The other nongroup is a **category**—a collection of people who share a common characteristic. Dorm residents, Greeks, bald-headed men, and Samoans are categories of people. Most of the people who share category membership will never meet, much less interact.

The distinguishing characteristics of groups hint at the rewards of group life. Groups are the people we take into account and the people who take us into account. They are the people with whom we share norms and values. Thus, groups are a major source of solidarity and cohesion, reinforcing and strengthening our integration into society. The benefits of group life range all the way from sharing basic survival and problem-solving techniques to satisfying personal and emotional needs.

How Groups Affect Individuals

When a man opens a door for a woman, do you see traditional courtesy or intolerable condescension? When you listen to Snoop Doggy Dog or Bad Religion, do you hear good music or mindless noise? Like taste in music, many of the things we deal with and believe in are not true or correct in any absolute sense; they are simply what our groups have agreed to accept as right.

The tremendous impact of group definitions on our own attitudes and perceptions has been cleverly documented in a classic experiment by Asch (1955). In this experiment, the group consists of nine college students, all apparently unknown to each other. The experimenter explains that the task at hand is an experiment in visual judgment. The subjects are shown two cards similar to those in Figure 3.1 and are asked to

Many of the good things in life are in short supply: They are scarce resources. Some of the most serious struggles take place over intangible rewards such as respect, prestige, and honor. When the struggle is regulated by norms that specify the rules of fair play, as in a soccer game, we call it competition. When anything goes, we call it conflict.

A **group** is two or more people who interact on the basis of shared social structure and recognize mutual dependency.

An **aggregate** is people who are temporarily clustered together in the same location.

A **category** is a collection of people who share a common characteristic.

FIGURE 3.1
The Cards Used in Asch's Experiment
In Asch's experiment, subjects were instructed to select the line on Card B that was equal in length to the line on Card A. The results showed that many people will give an obviously wrong answer in order to conform to the group.
SOURCE: From "Opinions and Social Pressure," by Solomon E. Asch. Copyright © November, 1955 by Scientific American, Inc. All rights reserved.

Card A

Card B

judge which line on Card B is most similar to the line on Card A. This is not a difficult task; unless you have a bad squint or have forgotten your glasses, you can tell that Line 2 most closely matches the line on the first card.

The experimental part of this research consists of changing the conditions of group consensus under which the subjects make their judgments. Each group must make 15 decisions and, in the first few trials, all of the students agree. In subsequent trials, however, the first eight students all give an obviously wrong answer. They are not subjects at all but paid stooges of the experimenter. The real test comes in seeing what the last student—the real subject of the experiment—will do. Will he go along with everybody else, or will he publicly set himself apart? Photographs of the experiment show that the real subjects wrinkled their brows, squirmed in their seats, and gaped at their neighbors; in 37 percent of the trials the naïve subject publicly agreed with the wrong answer, and 75 percent gave the wrong answer on at least one trial.

In the case of this experiment, it is clear what the right answer should be. Many of the students who agreed with the wrong answer probably were not persuaded by group opinion that their own judgment was wrong, but they decided not to make waves. When the object being judged is less objective, however—for example, whether Janet Jackson is better than Hootie and the Blowfish—the group is likely to influence not only public responses but also private views. Whether we go along because we are really convinced or because we are avoiding the hassles of being different, we all have a strong tendency to conform to the norms and expectations of our groups. Thus, our group memberships are vital in determining our behavior, perceptions, and values.

Interaction in Small Groups

We spend much of our lives in groups. We have work groups, family groups, and peer groups. In class we have discussion groups, and everywhere we have committees. This section reviews some of the more important factors that affect the kind of interaction we experience in small groups.

In this sequence of pictures a subject (in the middle) shows the strain that comes from disagreeing with the judgments of other members of the group. Some subjects in the Asch experiment disagreed on all 12 trials of the experiment. However, 75 percent of the experimental subjects agreed with the majority on at least one trial. Subjects who initially yield to the majority find it increasingly difficult to make independent judgments as the experiment progresses.

Size

The smallest possible group is two people. As the group grows to three, four, and more, its characteristics change. With each increase in size, each member has fewer opportunities to share opinions and contribute to decision making or problem solving. In many instances, the larger group will be better equipped for solving problems and finding answers, but this practical utility may be gained at the expense of individual satisfaction. Although there will be more ideas to consider, the likelihood that our own ideas will be influential diminishes. As the group gets larger, interaction becomes more impersonal, more structured, and less personally satisfying.

Proximity

Dozens of laboratory studies demonstrate that interaction is more likely to occur among group members who are physically close to one another. This effect is not limited to the laboratory.

In a classic demonstration of the role of proximity in group formation and interaction, Festinger and his associates (1950) studied a married-student housing project. All of the residents had been strangers to one another before being arbitrarily assigned to the housing unit. The researchers wanted to know what factors influenced friendship choices within the project. The answer: physical proximity. Festinger found that people were twice as likely to choose their next-door neighbors as friends as they were to choose people who lived only two units away. In general, the greater the physical distance, the less likely friendships were to be formed. An interesting exception to this generalization is that people who lived next to the garbage cans were disproportionately likely to be chosen as friends. Why? Because many of their neighbors passed by their units daily and therefore had many chances to interact and form friendships.

Communication Patterns

Interaction of group members can be either facilitated or retarded by patterns of communication. Figure 3.2 shows some common communication patterns for five-person groups. The communication structure allowing the greatest equality of participation is the *all-channel network*. In this pattern, each person can interact with every other person with approximately the same ease. Each participant has equal access to the others and an equal ability to become the focus of attention.

The other two common communication patterns allow for less interaction. In the *circle pattern*, people can speak only to their neighbors on either side. This pattern reduces interaction, but it does not give one person more power than others. In the *wheel pattern*, on the other hand, not only is interaction reduced but a single, pivotal individual gains greater power in the group. The wheel pattern is characteristic of the traditional classroom. Students do not interact with one another; instead they interact directly only with the teacher, thereby giving that person the power to direct the flow of interaction.

Communication structures are often created, either accidentally or purposefully, by the physical distribution of group members. The seating of committee members at a round table tends to facilitate either an all-channel or a circle pattern, depending on the size of the table. A rectangular table, however, gives people at the ends and in the middle of the long sides an advantage. They find it easier to attract attention and are apt to be more active in interactions and more influential in group discussions. Consider the way communication is structured in the classes and groups you participate in. How do seating structures encourage or discourage communication?

THINKING CRITICALLY

As people around the world have been brought together by computer networks, something we might call "electronic proximity" has developed. Do you think electronic proximity will have the same effects on friendship as physical proximity does?

FIGURE 3.2
Patterns of Communication
Patterns of communication can affect individual participation and influence. In each figure the circles represent individuals and the lines are flows of communication. The all-channel network provides the greatest opportunity for participation and is more often found in groups where status differences are not present or are minimal. The wheel, by contrast, is associated with important status differences within the group.

All-channel

Circle

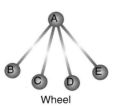
Wheel

Cohesion

Cohesion refers to the degree of attraction members feel to the group.

One of the important dimensions along which groups vary is their degree of **cohesion** or solidarity. A cohesive group is characterized by strong feelings of attachment and dependency. Because its members feel that their happiness or welfare depends on the group, the group may make extensive claims on the individual members (Hechter 1987). Cohesive adolescent friendship groups, for example, can enforce dress codes and standards of conduct on their members.

Marriages, churches, and friendship groups differ in their cohesiveness. What makes one marriage or church more cohesive than another? Among the factors that contribute to cohesion are small size, similarity, frequent interaction, long duration, and a clear distinction between insiders and outsiders (Homans 1950; Hechter 1987). Although all marriages in our society have the same size (two members), a marriage where the partners are more similar, spend more time together, and so on will generally be more cohesive than one where the partners are dissimilar and see each other for only a short time each day.

Social Control

Small groups rarely have access to legal or formal sanctions; yet they exercise profound control over individuals. The basis of this control is fear—fear of not being accepted by the group (Douglas 1983). The major weapons that groups use to punish nonconformity are ridicule and contempt, but their ultimate sanction is exclusion from the group. From "you can't sit at our lunch table anymore" to "you're fired," exclusion is one of the most powerful threats we can make against others. This form of social control is most effective in cohesive groups, but Asch's experiment shows that fear of rejection can induce conformity to group norms even in artificial lab settings.

Decision Making

One of the primary research interests in the sociology of small groups is how group characteristics (size, cohesion, and so on) affect group decision making. This research has focused on a wide variety of actual groups: flight crews, submarine crews, and juries, to name a few (Davis & Stasson 1988).

Generally, groups strive to reach consensus; they would like all their decisions to be agreeable to every member. As the size of the group grows, consensus requires lengthy and time-consuming interaction so that everybody's objections can be clearly understood and incorporated. Thus, as groups grow in size, they often adopt the more expedient policy of majority rule. This policy results in quicker decisions but often at the expense of individual satisfaction. It therefore reduces the cohesiveness of the group.

Choice Shifts

One of the most consistent findings of research is that it is seldom necessary to resort to majority rule in small groups. Both in the laboratory and in the real world, there is a strong tendency for opinions to converge. One of the classic experiments on convergence was done more than 50 years ago by Sherif (1936). In this experiment, strangers were put into a totally dark room. A dot of light was flashed onto the wall, and each participant was asked to estimate how far the light moved during the experimental period. After the first session, the participants recorded their own answers and then shared them with the other participants. There was quite a bit of variation in the estimates. Then they did the experiment again. This time there was less difference. After four trials, all participants agreed on an estimate that was close to the average of the initial estimates. (The dot of light was, in fact, stationary.)

The convergence effect has been demonstrated in dozens of studies since. Convergence, however, is not always to a middle position. Sometimes, the group reaches consensus on an extreme position. This is called the *risky shift* when the group converges on an adventurous option and the *tame shift* when the choice is extremely conservative. Sometimes these choice shifts depend on persuasive arguments put forward by one or more members, but often they result from general norms in the group that favor conservatism over risk (Davis & Stasson 1988). For example, one might expect the PTA steering committee to choose the safest opinion while members of a terrorist group would choose the riskiest option.

A special case of choice shift is *groupthink*. Groupthink occurs when pressures to agree are so strong that they stifle critical thinking. In such situations, people do not change; they merely hide their real opinions in order to be supportive. Irving Janis (1982) has documented the role of groupthink in a variety of twentieth-century political decisions. For example, in 1962 President Kennedy and his advisers rashly decided to invade Cuba. This so-called Bay of Pigs invasion ended in a disastrous rout for U.S. troops; it was poorly planned and probably foolish in any case. Afterward, nearly every member of the advisory group admitted that he had thought it was a dumb idea but had hesitated to say so (Schlesinger 1965). As this example illustrates, groupthink often results in bad decisions. Research shows that better decisions usually result when a persistent minority forces the majority to consider the minority's objections (Nemeth 1985).

Summary

Whether the small group arises spontaneously among neighbors or school-children, or whether it is a committee appointed to solve a community problem, the operation of the group depends on the quality of interaction among the members. Research suggests that interaction will be facilitated by small size, open communication networks, similarity, and physical proximity. When these circumstances align to produce high levels of interaction, individual satisfaction, group cohesion, and social control all tend to increase.

Types of Groups

Some groups have a more important impact than others on our lives. All of you, for example, probably belong to a family group as well as to the student body of your college or university. Except for an occasional student activist, membership in the family is far more important than membership in the student body and will have a more lasting effect. Sociologists call small, intimate, and lasting groups *primary groups;* they call large, impersonal groups *secondary groups.*

Primary Groups

Primary groups are characterized by intimate, face-to-face interaction (Cooley [1909] 1967). These groups represent our most complete experiences in group life. The closest approximation to an ideal primary group is probably the family, followed by adolescent peer groups and adult friendships. The relationships formed in these groups are relatively permanent and constitute a basic source of identity and attachment.

One of the most powerful mechanisms of social control is the threat of exclusion from valued groups. None of us likes to be rejected, and most of us will go to considerable lengths to avoid the threat of exclusion. This means that we conform: we dress as other group members do, think as they think, and do as they do. This desire to please others in our intimate social circles is the most powerful pressure for conformity. Compared to this, formal sanctions such as fines and jail sentences are relatively ineffective.

Primary groups are groups characterized by intimate, face-to-face interaction.

The ideal primary group tends to have the following characteristics: (1) personal and intimate relationships, (2) face-to-face communication, (3) permanence, (4) a strong sense of loyalty or we-feeling, (5) small size, (6) informality, and (7) traditional or nonrational decision making (Rogers 1960). In addition to the family, friendship networks, co-workers, and gangs may be primary groups. Groups such as these are major sources of companionship, intimacy, and belongingness, conditions that strengthen our sense of social integration into society.

Secondary Groups

Secondary groups are groups that are formal, large, and impersonal.

By contrast, **secondary groups** are formal, large, and impersonal. Whereas the major purpose of many primary groups is simply to provide companionship, secondary groups usually form to accomplish some specific task. The perfect secondary group is entirely rational and contractual in nature; the participants interact solely to accomplish some purpose (earn credit hours, buy a pair of shoes, get a paycheck). Their interest in each other does not extend past this contract. If you have ever been in a lecture class of 300 students, you have firsthand experience of a classic secondary group. The interaction is temporary, anonymous, and formal. Rewards are based on universal criteria, not on such particularistic grounds as your effort or need. The Concept Summary shows the important differences between primary and secondary groups.

Comparing Primary and Secondary Groups

Primary and secondary groups serve very different functions for individuals and societies. From the individual's point of view, the major purpose of primary groups is **expressive** activity, giving individuals social integration and emotional support. Your family, for example, usually provides an informal support group that is bound to help you, come rain or shine. You should be able to call on your family and friends to bring you some soup when you are down with the flu, to pick you up in the dead of night when your car breaks down, and to listen to your troubles when you are blue.

Expressive describes activities or roles that provide integration and emotional support.

Because we need primary groups so much, they have tremendous power to bring us into line. From the society's point of view, this is the major function of

Many of the groups that we participate in combine characteristics of primary and secondary groups. The elementary school and its classrooms are secondary groups: they are rationally designed and formally organized to meet specific instrumental goals. On the other hand, they also have some of the characteristics of primary groups, including the development of personal relationships, many of which will last for 6, 12, or even 40 years.

CONCEPT SUMMARY	*Differences Between Primary and Secondary Groups*	
	PRIMARY GROUPS	**SECONDARY GROUPS**
Size	Small	Large
Relationships	Personal, intimate	Impersonal, aloof
Communication	Face to face	Indirect—memos, telephone, etc.
Duration	Permanent	Temporary
Cohesion	Strong sense of loyalty, we-feeling	Weak, based on self-interest
Decisions	Based on tradition and personal feelings	Based on rationality and rules
Social structure	Informal	Formal—titles, officers, charters, regular meeting times, etc.

primary groups: They are the major agents of social control. The reason most of us don't shoplift is because we would be mortified if our parents, friends, or co-workers found out. The reason most soldiers go into combat is because their buddies are going. We tend to dress, act, vote, and believe in ways that will keep the support of our primary groups. In short, we conform. The law would be relatively helpless at keeping all the millions of us in line if we weren't already restrained by the desire to stay in the good graces of our primary groups. One corollary of this, however, which Chapter 5 addresses, is that if our primary groups accept shoplifting or street fighting as suitable behaviors, then our primary-group associations may lead us into deviance rather than conformity.

The major functions of secondary groups are **instrumental** activities, the accomplishment of specific tasks. If you want to build an airplane, raise money for a community project, or teach introductory sociology to 2,000 students a year, then secondary groups are your best bet. They are responsible for building our houses, growing and shipping our vegetables, educating our children, and curing our ills. In short, we could not do without them.

Instrumental describes activities or roles that are task oriented.

The Shift to Secondary Groups

In preindustrial society, there were few secondary groups. Vegetables and houses were produced by families, not by Del Monte or Georgia Pacific. Parents taught their own children, and neighbors nursed one another's ills. Under these conditions, primary groups served both expressive and instrumental functions. As society has become more industrialized, more and more of our instrumental needs are the obligation of some secondary group rather than of a primary group.

In addition to losing their instrumental functions to secondary groups, primary groups have suffered other threats in industrialized societies. In the United States, for example, approximately 21 percent of all households move each year (nearly 30 percent in the western states). This fact alone means that our ties to friends, neighborhoods, and co-workers are seldom really permanent. People change jobs, spouses, and neighborhoods. One consequence of this breakdown of traditional primary groups is that many people rely on secondary contacts even for expressive needs; they may hire a counselor rather than call their neighbor, for example.

THINKING CRITICALLY

Can you think of a situation in your life in which the primary source of social control was a secondary rather than a primary group?

Many scholars have suggested that these inroads on the primary group represent a weakening of social control; that is, the weaker ties to neighbors and kin means that people feel less pressure to conform. They don't have to worry about what the neighbors will say because they haven't met them; they don't have to worry about what mother will say because she lives 2,000 miles away and what she doesn't know won't hurt her.

There is apparently some truth in this suggestion, and it may be one of the reasons that small towns with stable populations are more conventional and have lower crime rates than do big cities with more fluid populations—an issue addressed more fully in Chapter 14.

Social Networks

A **social network** is an individual's total set of relationships.

Each of us has memberships in a variety of primary and secondary groups. Through these group ties we develop a **social network.** This social network is the total set of relationships we have. It includes our family, our insurance agent, our neighbors, our classmates and co-workers, and the people who belong to our clubs. Through our social network, we are linked to hundreds of people in our communities and perhaps across the country.

Strong and Weak Ties

Strong ties are relationships characterized by intimacy, emotional intensity, and sharing.

Weak ties are relationships with friends, acquaintances, and kin that are characterized by low intensity and intimacy.

Although our insurance agent and our mother are both part of our social network, there is a qualitative difference between them. We can divide our social networks into two general categories of intimacy: strong ties and weak ties. **Strong ties** are relationships characterized by intimacy, emotional intensity, and sharing. We have strong ties with the people we would confide in, for whom we would make sacrifices, and whom we expect to make sacrifices for us. **Weak ties** are relationships that are characterized by low intensity and intimacy (Granovetter 1973). Co-workers, neighbors, fellow club members, cousins, and in-laws generally fall in this category.

Your social network does not include everybody that you have ever interacted with. Many interactions, such as those with some classmates and neighbors, are so superficial that they cannot truly be said to be part of a relationship at all. Unless contacts develop into personal relationships that extend beyond the simple exchange of services or a passing nod, they would not be included in your social network.

Research suggests that social networks are vital for integration into society, encouraging conformity, and building a firm sense of self-identity. Because of their importance for the individual and society, documenting the trends in social networks has been a major focus of sociological study.

The Relationship Between Ties and Groups

The distinction between strong and weak ties obviously parallels the distinction between primary and secondary groups. The difference between these two sets of concepts is that strong and weak apply to one-to-one relationships while primary and secondary apply to the group as a whole. We can have both strong and weak ties within the primary as well as the secondary group.

For example, the family is obviously a primary group; it is relatively permanent, with strong feelings of loyalty and attachment. We are not equally intimate with every family member, however. We may be very close to our mother but estranged from our

A critical part of our social network is our strong ties—the handful of people to whom we feel intense loyalty and intimacy. For many people, family is an important source of strong ties. Although we may not be close to everyone in our family, there are usually a few family members to whom we feel very close. Women are somewhat more apt than men to choose their strong ties from among their family. Many, like these two sisters, will find that these ties form a lifelong bond that will provide a sense of continuity over the entire life course.

brother. Similarly, although the school as a whole is classified as a secondary group, we may have developed an intimate relationship, a strong tie, with one of our schoolmates. Strong and weak are terms used to describe the relationship between two individuals; primary and secondary are characteristics of the group as a whole.

Strong Ties

Although we have a large research literature on strong ties, there is little consensus on how to define them operationally. As a result, different studies yield different pictures of the extent of strong ties. For example, one recent study asked a national sample to name those individuals they could "discuss important matters with." Using this definition, the average individual had only three strong ties (Marsden 1987). Another major study used a definition that included every adult you lived with, engaged in social activities with, or would borrow from or confide in. This study found that the average person had 15–19 ties (Fischer 1982).

Despite these differences, all studies agree on the factors that affect the number and composition of strong ties. The most important of these factors is education. People with more education have more strong ties, have a greater diversity of strong ties, and are less reliant on kin. The number of ties also varies by residence and age. Urban residents have more strong ties than rural residents, perhaps in part because they have a greater variety from which to choose. Older respondents consistently report the fewest strong ties. Gender does not appear to make much difference in the number of strong ties that people have, but it does affect the source: Women's ties are more likely than men's to be drawn from the kin group (Marsden 1987; Moore 1990).

Voluntary Associations

In addition to relationships formed through family and work, many of us voluntarily choose to join other groups and associations. We may join the PTA, a bowling team, the Elks, or the Sierra Club. These groups, called **voluntary associations,** are nonprofit organizations designed to allow individuals an opportunity to pursue their shared interests collectively. They vary considerably in size and formality. Some—for example,

Voluntary associations are nonprofit organizations designed to allow individuals an opportunity to pursue their shared interests collectively.

a global perspective

FINDING JOBS, CHANGING JOBS IN SINGAPORE AND CHINA

*H*ow will you go about looking for a job after you graduate? After you have read over all of the classified ads and gone to the Career Planning and Placement Office, what will you do next? For most of us, the answer seems obvious: ask friends and family if they are aware of any challenging, well-paying, secure job opportunities for persons with just your credentials. Research conducted in the United States suggests that this just might be the wrong strategy. Mark Granovetter (1974; 1994) and others have shown that while individuals are likely to use their personal network of **strong** ties to search for work, they are more likely to actually find work through their **weak** ties. How does this work? This is how Granovetter explains "the strength of weak ties." The people to whom we are most strongly tied are a lot like us; our friends and family tend to be interested in the same things we are, share similar educational and economic

backgrounds, and know the same people we do. Consequently, they are unlikely to be able to give us much information about job possibilities that we didn't already know or couldn't find out. In contrast, our weak ties—people with whom we interact less frequently, whom we know less well—are more likely to have information about jobs that weren't listed in the newspaper and that our friends didn't know about. Furthermore, because our weak ties are most likely to bridge us to people with higher status than our own, weak ties give us access to information about more prestigious jobs and to people who may, in fact, have some influence in hiring decisions for them (Lin 1990). But are weak ties equally important in finding work in non-market economies or in cultures where individual social networks are much more directly linked to larger social structures?

In the late 1980s and early 1990s, Yanjie Bian and Soon Ang (1997) went to Tianjin, China, and Singapore to study this question. Tianjin is the third largest city in China (see Map 3.1) and in the late 1980s had still not begun the transition to a more Westernized market economy. Jobs in Tianjin and elsewhere in

China were assigned by the government, and although new workers could express their job preferences, these preferences did not necessarily affect their final assignments. In Singapore the situation was somewhat different. A city-state with 78 percent of its population of Chinese origin, labor shortages in Singapore meant that workers had a very broad range of jobs from which to choose. Because advertising for new employees was costly, Singaporean employers had a strong stake in trying to identify potential employees who would be loyal to the company and stay with them for a long time.

In both Tianjin and Singapore, individuals rely much more directly and explicitly on their social networks for favors of all kinds than is often true in the United States. The guiding principle for individual social networks is *guanxi* (pronounced gwan-she). *Guanxi* literally means "relationship" or "relation" and refers to a set of interpersonal connections that are based on the **reciprocal** exchange of favors and on mutual trust. The reciprocal obligations built into *guanxi* relationships lead to emotional attachments, of course, but in addition there is a strong moral dimension; failure

the Elks and the PTA—are very large and have national headquarters, elected officers, formal titles, charters, membership dues, regular meeting times, and national conventions. Others—for example, bowling teams and quilting groups—are small, informal groups that draw their membership from a local community or neighborhood.

Voluntary associations are an important mechanism for enlarging our social networks. Most of the relationships we form in voluntary associations will be weak ties, but voluntary associations can also be the means of introducing us to people who will become close friends and intimates.

Voluntary associations perform an important function for individuals. Studies document that people who participate in them generally report greater satisfaction and personal happiness, longer life, greater self-esteem, more political effectiveness, and a greater sense of community (Hanks 1981; Knoke 1981; Litwak 1961; Moen, Dempster-McLain, & Williams 1989; Pollock 1982). The correlation between high participation and greater satisfaction does not necessarily mean that joining a voluntary association is the road to happiness. At least part of the relationship is undoubtedly due to the fact that it is precisely those happy persons who feel politically effective to reciprocate causes people to lose face—*mianzi* (Bian 1997; Bian and Ang 1997). By this definition, *guanxi* relations are strong ties.

So what did Bian and Ang find? In contrast to the United States, finding work and changing work in Singapore and China depends more on strong ties than on weak ones. In China, the trust embedded in *guanxi* allows people to "bend government rules" in helping job seekers find more desirable employment; in Singapore, employers have more faith in the future loyalty of potential employees who are personally recommended to them through *guanxi* networks. Nevertheless, in both China and Singapore, the best strategy of all appears to be one that involves both weak ties and *guanxi*. Getting help in the job search process depends first and foremost on ties of trust and reciprocity. The highest level job placements occur, however, when a job seeker's strong ties "intercede" with someone of higher status; this typically occurs when the person of higher status is strongly tied to the "intermediary" but only weakly tied to the job seeker, him- or herself. What is true in China and Singapore may also be true in the United States. Looking for work is not nearly as hard when friends have well-placed friends.

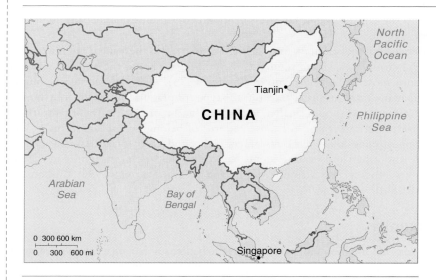

MAP 3.1
Tianjin, China.

and attached to their communities who seek out voluntary associations. It also appears to be true, however, that greater participation can be an avenue for achievement and lead to feelings of integration and satisfaction.

The Mediation Hypothesis

An important characteristic of voluntary associations is that they combine some of the features of primary and secondary groups—for example, the companionship of a small group and the rational efficiency of a secondary group. Some scholars have therefore suggested that voluntary associations mediate (provide a bridge) between primary and secondary groups (Pollock 1982). They allow us to pursue instrumental goals without completely sacrificing the satisfactions that come from participation in a primary group. Through participation in voluntary associations, we meet some of our needs for intimacy and association while we achieve greater control over our immediate environment. Take, for example, the sportsman who wishes to protect both wildlife habitat and the right to have guns. This individual can write letters to his member of Congress, but he will believe, rightly, that as an individual he is unlikely to have much clout. If this same individual joins with others in, say, the National Wildlife Federation or the National Rifle Association, he will have the enjoyment of associating with other like-minded individuals and the satisfaction of knowing that a paid lobbyist is representing his opinions in Washington. In this way, voluntary associations provide a bridge between the individual and large secondary associations.

Correlates of Membership Participation

Most people in the United States belong to at least one voluntary association, and approximately one-fourth participate in three or more. Among those who report membership, a large proportion are passive participants—they belong in name only. They buy a membership in the PTA when pressured to do so, but they don't go to meetings. Similarly, anyone who subscribes to *Audubon* magazine is automatically enrolled in the local Audubon Club, but few subscribers become active members. Because so many of our memberships are superficial, they are also temporary. Most people in the United States, however, maintain continuous membership in at least one association.

Membership in voluntary associations exhibits much the same pattern we noted earlier for strong ties. Urban residents, the middle-aged, and the well educated are the groups most frequently involved (Tomeh 1973). Although women belong to about as many associations as men do, the types of organizations they join are different; women tend to join smaller, less formal, and more expressive organizations. The most important reason for these patterns of belonging appears to be the different opportunities that men and women, higher and lower social classes, the young and old, and urban and rural residents have to participate meaningfully in the full range of voluntary associations (McPherson & Smith-Lovin 1986; Williams & Ortega 1986).

Community

In everyday life, we hear a lot of talk about the benefits of having a "sense of community." In mourning the contemporary loss of community, commentators seem to regard community as a good thing, but they usually aren't very specific about what community is. More importantly, they aren't very specific about what community was.

According to sociologists, a strong community is characterized by dense, cross-cutting social networks (Wellman & Berkowitz 1988). A community is strongest when all members are linked to one another through complex overlapping ties. Many scholars believe that communities of the past were characterized by just such networks. However, one recent study challenges this assumption. Using network data collected in 1939 from one neighborhood in Bloomington, Indiana, Campbell concludes that ". . . this neighborhood, and perhaps others like it, was not significantly more sociable than contemporary settings" (1990, 151).

Furthermore, studies show that network ties need not be strong to have important consequences for individuals and the community. For example, research shows that neighbors help each other in many ways, even though they share few confidences (Wellman & Wortley 1990). Studies also show that as the proportion of people who simply know one another increases, deviance and fear of crime are reduced, better control is exercised over children, and the weak and handicapped are more likely to be cared for (Freudenberg 1986). Finally, a growing body of literature shows that voluntary associations promote community cohesion and solidarity by creating a network of weak ties among members (McPherson 1983; Sampson 1988).

Strong communities are built by dense networks of weak ties. When everybody knows everybody else or at least knows them through a cousin or neighbor, then community norms are reinforced and the community is more likely to be able to work together. This is most easily accomplished in small, stable communities where the residents share much in common.

Summary

Even though the structure of our social networks may have changed over time, research shows that people remain importantly connected to neighbors, friends, and community; studies also demonstrate that the size and diversity of our social networks have a variety of significant consequences. For example, research indicates that many people first hear about jobs and career opportunities through weak ties (Granovetter 1974; Lin, Walter, & Vaughn 1981; McPherson & Smith-Lovin 1982). In this and other instances, the more people you know, the better off you are.

THINKING

CRITICALLY

Suppose you were trying to get help for a family member's substance-abuse problem. Who would you turn to, your strong or your weak ties?

Complex Organizations

Few people in our society escape involvement in large-scale organizations. Unless we are willing to retreat from society altogether, a major part of our lives is organization bound. Even in birth and death, large, complex organizations (such as hospitals and vital statistics bureaus) make demands on us. Throughout the in-between years, we are constantly adjusting to organizational demands (Kanter 1981).

Sociologists use the term **complex organizations** to refer to large, formal organizations with complex status networks (Brinkerhoff & Kunz 1972). Examples include universities, governments, business corporations, and voluntary associations such as the Masons and the Catholic church.

These complex organizations make a major contribution to the overall quality of life within society. Because of their size and complexity, however, they don't supply the cohesion and personal satisfaction that smaller groups do. They may make their members feel as if they are simply cogs in the machine rather than important people in their own right. This is nowhere more true than in a bureaucracy.

Complex organizations are large, formal organizations with complex status networks.

Bureaucracy

Bureaucracy is a special type of complex organization characterized by explicit rules and a hierarchical authority structure, all designed to maximize efficiency.

Bureaucracy is a special type of complex organization characterized by explicit rules and a hierarchical authority structure, all designed to maximize efficiency. In popular usage, bureaucracy often has a negative connotation: red tape, silly rules, and unyielding rigidity. In social science, however, it is simply an organization in which the roles of each actor have been carefully planned to maximize efficiency.

The Classic View

Most large, complex organizations are also bureaucratic: IBM, General Motors, U.S. Steel, the Catholic church, colleges, and hospitals. The major characteristics of bureaucracies were outlined 80 years ago by Max Weber (1910/1970a):

1. *Division of labor and specialization.* Bureaucratic organizations employ specialists in each position and make them responsible for specific duties. Job titles and job descriptions specify who is to do what and who is responsible for each activity.
2. *Hierarchy of authority.* Positions are arranged in a hierarchy so that each position is under the control and supervision of a higher position. Frequently referred to as chains of command, these lines of authority and responsibility are easily drawn on an organization chart, often in the shape of a pyramid.
3. *Rules and regulations.* All activities and operations of a bureaucracy are governed by abstract rules or procedures. These rules are designed to cover almost every possible situation that might arise: hiring, firing, and the everyday operations of the office. The object is to standardize all activities.
4. *Impersonal relationships.* Interactions in a bureaucracy are supposed to be guided by the rules rather than by personal feelings. Consistent application of impersonal rules is intended to eliminate favoritism and particularism.
5. *Careers, tenure, and technical qualifications.* Candidates for bureaucratic positions are almost always selected on the basis of technical qualifications such as high scores on civil service examinations, education, or experience. Once selected for a position, persons advance in the hierarchy by means of achievement and seniority.
6. *Efficiency.* Bureaucratic organizations coordinate the activities of a large number of people in the pursuit of organizational goals. All have been designed to maximize efficiency. From the practice of hiring on the basis of credentials rather than personal contacts to the rigid specification of duties and authority, the whole system is constructed to keep individuality, whim, and particularism out of the operation of the organization.

Organizational Culture

Weber's classic theory of bureaucracy almost demands robots rather than individuals. A list of rules that covered every possible situation would be unwieldy and impossibly long. Not surprisingly, therefore, we find that few organizations try to be totally bureaucratic. Instead, organizations strive to create an atmosphere of goodwill and common purpose among their members so that they all will apply their ingenuity and best efforts to meeting organizational goals (DiTomaso 1987). This goodwill is as essential to efficiency as are the rules. In most organizations, in fact, working exactly according to the rules is considered a form of sabotage. For example, unions of public employees (such as the police) that cannot legally strike engage in "working to the rule" as a form of protest: They follow every little nit-picking rule and fill out every form carefully. The result is usually a sharp slowdown in work and general chaos.

Sociologists use the term *organizational culture* to refer to the pattern of norms and values that affects how business is actually carried out in an organization. The key to a successful organizational culture is cohesion, and most organizations strive to build cohesion among their members. They do this by encouraging interaction among employees (providing lunchrooms, sponsoring after-hours sports leagues, having company picnics and newsletters, and developing unifying symbols, such as mascots, company colors, or uniforms). This is most clearly apparent when you think about the large bureaucratic organization represented by a university, but it is also characteristic of transnational corporations.

In many organizations, the formal rules have very little to do with the day-to-day activity of the members. In situations as varied as classrooms and shop floors, people evolve their own way of doing business and may have little use for the formal rules (Ouichi & Wilkins 1985); in fact, they may not even know what the rules are. What determines whether an organization works by the rules or not?

A major factor affecting the degree of bureaucratization in an organization is the degree of uncertainty in the organization's activities. When activities tend to be routine and predictable, the organization is likely to emphasize rules, central planning, and hierarchical chains of command. When activities change rapidly in unpredictable ways, there is more emphasis on flexibility and informal decision making (Simpson 1985). This explains why, for example, classrooms tend to be less bureaucratic and ball-bearings factories more bureaucratic.

Criticisms of Bureaucracies

Bureaucracy is the standard organizational form in the modern world. Organizations from churches to governments are run along bureaucratic lines. Despite the widespread adoption of this organizational form, it has several major drawbacks. Three of the most widely acknowledged are as follows:

1. *Ritualism.* Rigid adherence to rules may mean that a rule is followed regardless of whether it helps accomplish the purpose for which it was designed. The rule becomes an end in itself rather than a means to an end. For example, individuals may interpret a rule stating that the workday ends at 5 P.M. to mean that they cannot work later even if they want to. Overemphasis on rules can stifle initiative and prevent the development of more efficient procedures (Blau & Meyer 1971).
2. *Alienation.* The stress on rules, hierarchies, and impersonal relationships can sharply reduce the cohesion of the organization. This has several drawbacks: It reduces social control, it increases turnover, and it reduces member satisfaction and commitment. All of these factors may interfere with the organization's ability to reach its goals.
3. *Structured inequality.* Critics charge that the modern bureaucracy with its multiple layers of authority is a profoundly antidemocratic organization. In fact, the whole purpose of the bureaucratic form is to concentrate power in one or two decision makers whose decisions are then passed down as orders to subordinates below. Some observers believe that the amassing of concentrated power in the name of efficiency and rationalism is incompatible with a democratic society (Perrow 1986).

Many bureaucratic organizations use the company picnic as a strategy to build a positive organizational culture. They hope that the personal ties built during such informal occasions will motivate employees to go beyond the formal rules and to give their best efforts to promoting the company's and their employer's interests. In this picture, Steve Jobs (formerly of Apple Computer and now head of NeXT Computers) appears at a NeXT Company picnic barefooted. This just-one-of-the-guys informality and family atmosphere may mask the fact that such picnics are good for business.

THINKING

CRITICALLY

From your experience, what are some functions of bureaucracy? What are some dysfunctions? Would you classify the dysfunctions as manifest or latent? Why?

The Japanese Model of Organization

Within the last two decades, Japan has come to dominate dozens of areas of international trade, from automobiles to video recorders. The astounding success of Japanese manufacturers has led to intense scrutiny of Japanese management practices in order to discover their secret (Dore 1973; Marsh & Mannari 1976; Masatsugu 1982).

Because Japanese firms, like their U.S. counterparts, are large and complex, they are run as bureaucracies. They have hierarchies of authority, lots of rules, hiring on the basis of technical qualifications, and many of the other features of classic bureaucratic theory. In addition, however, they have some distinct characteristics that may represent the complex organization of the future. Among these are several practices that move the Japanese bureaucracy in the direction of the collectivist organization. A recent review by Lincoln and McBride (1987) suggests the following unique features of Japanese organizations:

1. *Permanent employment.* Japanese organizations have a strong norm of permanent employment. On leaving school, an employee enters a firm and remains there until retirement. The company makes a lifetime commitment to the employee, and the employee also makes a lifetime commitment to the firm.
2. *Internal labor markets.* Higher-level positions are filled from below from within the same firm. Thus, people can make a lifetime commitment to the same firm and still have the opportunity for promotion and upward mobility. This system differs from the common U.S. practice of pursuing careers by changing employers. Because of lifetime commitment and internal labor markets, seniority in the firm and age are highly correlated—and generally highly rewarded.
3. *Relative absence of white-collar/blue-collar distinction.* In Japan, all workers are expected to be committed to the firm; thus all workers work on salary rather than for an hourly wage. There are no separate cafeterias and parking lots for the supervisors and for the line workers. This also means that the internal labor market crosses the white-collar/blue-collar line: Although one might start as a blue-collar worker, seniority is likely to move one into management ranks.
4. *Participatory decision making.* Unlike the classic bureaucratic model where decisions flow in only one direction (down), a more circular pattern of decision making is the norm in Japanese organizations. Suggestions may originate among workers' circles and then rise to the top (called bottom-up management), or they may originate at the top but be implemented only after there is substantial consensus among the workers. This participatory decision-making process is yet another mechanism that reduces the distinction between management and line workers and increases the cohesion and commitment of the employees.

THINKING

CRITICALLY

Why do you suppose union leaders and managers each often oppose participatory decision making? What other obstacles do you think there will be to U.S. firms implementing Japanese management techniques?

When U.S. scholars first started to look at Japanese organizations, they often attributed the unique features of these organizations to Japan's feudal heritage. Increasingly, however, scholars view Japan's organizational style as the wave of the future rather than a holdover from the past. Unlike the classic bureaucratic model, which pretty much ignores the human component of the workplace, the Japanese organizational form recognizes that complex organizations are *human groups* rather than machines. Thus, it takes into consideration such issues as cohesion and group dynamics. The realization that even corporations such as IBM are groups and are affected by group processes is having a substantial effect on the evolution of bureaucracy in the United States and around the world.

Small-group responsibility at all levels is a key part of Japanese-style organizations. This is very different from a classic bureaucracy, where the people at the bottom are expected to follow the rules established by their superiors rather than think for themselves. The bottom-up style of management characteristic of Japanese organizations gives workers a greater investment in their product, and may be a key to higher-quality products.

Summary

1. Relationships are characterized by four basic social processes: exchange, cooperation, competition, and conflict.

2. Groups are distinguished from aggregates and categories in that members take one another into account and their interaction is shaped by shared expectations.

3. Group interaction is affected by group size and the proximity and communication patterns of group members. The amount of interaction in turn affects group cohesion, the amount of social control the group can exercise over members, and the quality of group decisions.

4. A fundamental distinction between groups is the extent to which they are primary or secondary. Primary groups are essential to individual satisfaction and integration; they are also the primary agents of social control in society. Secondary groups are generally task-oriented and perform instrumental functions for societies and individuals.

5. Each person has a social network that consists of both strong and weak ties. The number of these ties is generally greater for individuals who are urban, middle-aged, and highly educated.

6. In the United States weak ties are important in finding new jobs; in China and Singapore, strong ties appear to provide better help to job seekers.

7. Voluntary associations may mediate between the primary and secondary group, providing a bridge that links the individual to larger groups. Voluntary associations combine some of the expressive functions of primary groups with the instrumental functions of secondary groups.

8. A bureaucracy is a rationally designed organization whose goal is to maximize efficiency. The chief characteristics of a bureaucracy are division of labor and specialization, a hierarchy of authority, a system of rules and regulations, impersonality in social relations, and emphasis on careers, tenure, and technical qualifications.

9. Although most contemporary organizations are built on a bureaucratic model, many are far less rational and impersonal than the classic model suggests. All effective bureaucracies rely on an organizational culture to inspire employees to give their best efforts and to help meet organizational goals.

10. Japanese organizations fit somewhere between the extreme rationalism of the classic bureaucracy and the people-centered collectivist organizations. Japanese organizations try to maximize employees' commitment to the firm through four mechanisms: lifetime employment, internal labor markets, little distinction between blue- and white-collar workers, and participatory decision making.

Sociology on the Net

A community is strongest when everybody knows everybody else and when all members are linked to one another through complex and overlapping ties. In the past we would think of neighborhoods where the residents knew each other and talked on porches and over back fences. Perhaps the community of the future will be a human community linked by computers in an "electronic village." Several U.S. communities are electronically networked. One of the most advanced is the Blacksburg Electronic Village in Blacksburg, Virginia.

        http://www.bev.net

Browse through this electronic village by starting with the sections for newcomers. The Welcome Page and Introduction will get you off to a good start. After you are familiar with this site, visit the section called *Starting a Village*. How did this community become an electronic village? Are all segments of the community joined with links? Who is in and who is left out? Could a system like this work in your home town?

FIND IT ON INFOTRAC COLLEGE EDITION

In the text's discussion of social networks, the importance of strong ties is discussed. One way to develop strong ties and a sense of community with your neighbors is through the redevelopment of neighborhoods. Using InfoTrac College Edition, look up the following article:

> "Building a Better World, One Neighborhood at a Time." Russell Mawby. *Alternatives Journal,* April–May 1996, v22 n2 p4. (Hint: Enter the search term *Russell Mawby,* using the Subject Guide.)

Do you see cohousing as an effective way to encourage a strong sense of community among neighbors? What problems might arise from such living arrangements? Would you consider living in the type of neighborhood Mawby is describing?

Suggested Readings

Douglas, Tom. 1983. *Groups: Understanding People Gathered Together.* London: Tavistock. An engaging book that focuses on the similarity of group processes in teams, families, and other small groups.

Fischer, Claude S. 1982. *To Dwell among Friends: Personal Networks in Town and City.* Chicago: University of Chicago Press. A report of Fischer's research in northern California, including an excellent overview of sociological concerns about social networks.

Fischer, Claude S. 1992. *America Calling: A Social History of the Telephone to 1940.* Berkeley: University of California Press. Fischer's research on social networking has motivated him to investigate the social history of the telephone in the United States, a cultural artifact that facilitates networking in modern society.

Hall, Richard H. (ed.). 1995. *Complex Organizations.* Brookfield, Ver.: Dartmouth University Press. A collection of theoretical and research-based essays on various aspects of complex organizations in industrialized societies today.

Marsden, Peter V. 1992. "Social Network Theory." Pp. 1887–1894 in Edgar F. Borgatta and Marie L. Borgatta (eds.) *Encyclopedia of Sociology,* vol. 4. New York: Macmillan.

Nestmann, Frank, and Hurrelmann, Klaus, eds. 1994. *Social Networks and Social Support in Childhood and Adolescence.* New York: Walter de Gruyter. A research-based evaluation of the way that networks and social support can be used to positively influence children and teenagers.

CHAPTER 4

The Individual in Society

Outline

The previous chapters of this book have focused on macrosociology—analysis of cultures, institutions, social structures, groups, and organizations. This focus on structures should not obscure the fact that the heart of sociology is a concern with *people.* Sociology is interesting and useful to the extent that it helps us explain why people do what they do. It should let us see ourselves, our family, and our acquaintances in a new light.

In this chapter we deal directly with individuals. We begin with an overview of the ways in which individuals are molded by social structures and conclude by looking at how individuals respond to and experience social structure.

The Self and Self-Concept

From the small lump of flesh that is the newborn infant develops a complex and fascinating human being, a human being who is simultaneously much like every other human being and at the same time exactly like no other.

> The **self** is a complex whole that includes unique attributes and normative responses. In sociology, these two parts are called the I and the me.

Each individual **self** may be thought of as a combination of unique attributes and normative responses. Within sociology these two parts of the self are called the I and the me (Mead 1934).

> The **I** is the spontaneous, creative part of the self.

The **I** is the spontaneous, creative part of the self; the **me** is the self as social object, the part of the self that responds to others' expectations. In English grammar, *me* is used when we speak of ourselves as the object of others' actions (She sent me to the office); *I* is used when we speak of ourselves as the actor (I acted up). Sociological use follows this convention.

> The **me** represents the self as social object.

As this description of the self implies, the two parts may pull us in different directions. For example, many people face a daily conflict between their I and their me when the alarm clock goes off in the morning—the I wants to roll over and go back to sleep, but the me knows it is supposed to get up and go to class. Some of these conflicts are resolved in favor of the me and some in favor of the I. Daily behavior, however, is viewed as the result of an ongoing internal dialogue between the I and the me.

> The **self-concept** is the self we are aware of. It is our thoughts about our personality and social roles.

The self is enormously complex, and we are often not fully aware of our own motives, capabilities, and characteristics. The self that we are aware of is our **self-concept.** It consists of our thoughts about our personality and social roles. For example, a young man's self-concept might include such qualities as young, male, Methodist, good athlete, student, shy, awkward with girls, responsible, American. His self-concept includes all the images he has of himself in the dozens of different settings in which he interacts.

Learning To Be Human

What is human nature? Are we born with a tendency to be cooperative and sharing or with a tendency to be selfish and aggressive? The question of the basic nature of humankind has been a staple of philosophical debate for thousands of years. It continues to be a topic of debate because it is so difficult (some would say impossible) to separate the part of human behavior that arises from our genetic heritage from the part that is developed after birth. The one thing we are sure of is that nature is never enough.

The Necessity of Nurture

Each of us begins life with a set of human potentials: the potential to talk, to walk, to love, and to learn. By themselves, however, these natural capacities are not enough to enable us to join the human family. Without nurture—without love and attention and hugging—the human infant is unlikely to survive, much less prosper. The effects of neglect are sometimes fatal and, depending on severity and length, almost always result in retarded intellectual and social development.

How can we determine the importance of nurture? There are a few case studies of tragically neglected children, but luckily the instances are rare. Some of the first clinical evidence on the effects of limited social interaction on human development was provided by René Spitz's (1945) study of an orphanage where each nurse was in charge of a dozen or more infants. Although the children's physical needs were met, the nurses had little time to give individual attention to each child.

Children who spend the first years of their lives in this type of institutional environment are devastated by the experience. Because of limited personal attention, such children withdraw from the social world; they seldom cry and are indifferent to everything around them. The absence of handling, touching, and movement is the major cause of this retarded development. In time, the children become increasingly retarded intellectually and more susceptible to disease and death. Of the 88 children Spitz studied, 23 died before reaching the age of two and a half. Even if they live, Spitz found, socially deprived children are likely to become socially crippled adults.

Deprivation can also occur in homes where parents fail to provide adequate social and emotional stimulation. Children who have their physical needs met but are otherwise ignored by their parents have been found to exhibit many of the same symptoms as institutionalized infants. These and other studies of the effects of isolation and deprivation on children suggest that children need intensive interaction with others to survive and develop normally. Much of the evidence for this conclusion, however, is derived from atypical situations in which unfortunate children have been subjected to extreme and unusual circumstances. To test the limits of these findings and to examine the reversibility of deprivation effects, researchers have turned to experiments with monkeys.

Monkeying with Isolation and Deprivation

For more than 20 years, researchers have been experimenting with mother deprivation and isolation of infant monkeys. In a classic series of experiments, Harry Harlow and his associates raised infant monkeys in total isolation. The infants lived in individual cages with a mechanical mother figure that provided milk. Although the infant monkeys' nutritional needs were met, their social needs were not. As a result, both their physical and social growth suffered. They exhibited such bizarre behavior as biting themselves and hiding in corners. As adults, these monkeys refused to mate; if artificially impregnated, the females would not nurse or care for their babies (Harlow & Harlow 1966). These experiments provide dramatic evidence about the importance of being with others; even apparently innate behaviors such as sexuality and maternal behavior must be developed through interaction. On the bright side, the monkey experiments affirm that some of the ill effects of isolation and deprivation are reversible. The

focus on

American Diversity

GENDER DIFFERENCES IN MATHEMATICS

Fierce debates about the relative importance of nature versus nurture no longer preoccupy most social scientists. In one area of research, however, the nature/nurture controversy remains heated—the apparent male advantage in mathematics. Studies show that boys outperform girls on standardized math tests such as the SAT, that by high school boys substantially outnumber girls in advanced mathematics courses, and that in adulthood women remain substantially underrepresented in such occupations as engineering that depend heavily on a foundation in mathematics. For some scholars, the pattern seems so consistent that they have begun to wonder if gender differences in mathematical skills aren't biologically based, after all.

Neuroscientists are beginning to unravel the relationship between exposure to male or female hormones during pregnancy and the characteristic differences in the structure of adult men and women's brains. Fetal exposure to testosterone, for instance, is associated with right-brain dominance. This association may help to explain not only why more men are left-handed than women but why they tend to have better visual-spatial skills. Conversely, the fact that language centers are typically located in the left hemisphere of the brain may account for women's tendency towards better verbal skills.

From findings such as these, researchers reason that gender differences in mathematical ability are at least partially a result of biology. Statistically speaking, however, the differences are small. Because the differences within each sex are so much larger than the differences between them, critics of the biological perspective argue that hormones can explain only a very small part of overall variation in mathematical aptitude. This leaves a great deal of room for the influence of social factors. Evidence for this point of view comes from two lines of research.

First, sociologists Richard Felson and Lisa Trudeau (1991) have shown in a study of children in Grades 5 through 12 that, contrary to results based only on the SAT, girls actually outperform boys on most tests of mathematical performance, especially those that are designed to correspond to the curriculum. Although boys are more likely to *choose* advanced mathematics courses in high school, by many other criteria girls are ahead. Girls' grades in mathematics are higher than boys' and, for required mathematics courses, they are more likely than boys to be placed in the advanced tracks. Based on these results, Felson and Trudeau argue that males do not have a general advantage in mathematical skills, but instead a very specific one—probably the type of spatial ability usually measured by mental rotation tests.

Second, studies clearly show that the male advantage in mathematical performance has been declining dur-

---

young monkeys experienced almost total recovery when placed in a supportive social environment (Suomi et al. 1972).

Summary

Although it is dangerous to generalize from monkeys to humans, the evidence from the monkey experiments supports the observations about human infants: Physical and social development depend on interaction with others. Even being a monkey does not come naturally. Walking, talking, loving, and laughing all depend on sustained and intimate interaction with others. Clearly, our identities, even our lives, are socially bestowed and socially sustained (Berger 1963, 98).

Socialization

Socialization is the process of learning the roles, statuses, and values necessary for participation in social institutions.

The process of learning the roles, statuses, and values necessary for participation in social institutions is called **socialization.** Socialization is a lifelong process. It begins with learning the norms and roles of our family and our subculture and making these part

ing the last 20 years (Hyde et al. 1990). One possible explanation for this pattern is that boys and girls are now being socialized more similarly, thereby reducing the traditional male advantage in math. Biopsychologist Janice Juraska has demonstrated that female rats have fewer nerve connections than males into the hippocampus, a brain region associated with spatial relations and memory. However, when the cages of female rats were "enriched" with stimulating toys, the females developed more neural connections. As Juraska says, "Hormones do affect things—it's crazy to deny that, but there is no telling which way sex differences might go if we completely changed the environment" (as quoted in Gorman 1992).

On the average, males outperform females on mathematical tasks that require the kind of spatial skill measured by mental rotation tests. Although this difference may have a genetic basis, many scholars believe that the difference stems primarily from socialization. Boys play the kind of games and are given the kinds of toys, for instance, that are most likely to develop visual-spatial abilities (Entwisle, Alexander, & Olson 1994). Consequently, the newest methods of teaching mathematics focus on providing comparable "hands-on" experiences in the classroom. Changes in teaching and parenting are helping to reduce the gender gap in mathematical performance.

of our self-concept. As we grow older and join new groups and assume new roles, we learn new norms and redefine our self-concept.

Types of Socialization

Sociologists generally distinguish three types of socialization: primary socialization, anticipatory socialization, and resocialization.

Primary Socialization

Early childhood socialization is called **primary socialization.** It is primary in two senses. It occurs first, and it is most critical for later development. During this period, children develop personality and self-concept, acquire motor abilities, reasoning, and language skills, and are exposed to a social world consisting of roles, values, and norms.

During the period of primary socialization, children are expected to learn and embrace the norms and values of society. They learn that conforming to rules is an important key to gaining acceptance and love, first from their family and then from a larger network. Because young children are so dependent on the love and acceptance of their family, they are under especially strong pressure to conform to their family's expectations. This is a

Primary socialization is personality development and role learning that occurs during early childhood.

critical step in their becoming conforming members of society. If this learning does not take place in childhood, then conformity is exceptionally difficult to develop in later life.

Anticipatory Socialization

Anticipatory socialization is role learning that prepares us for roles we are likely to assume in the future.

As we progress from infancy to old age, we must continually shed old roles and adopt new ones. Many of these role changes are relatively easy because of **anticipatory socialization**—role learning that prepares us for roles we are likely to assume in the future. Because of this socialization, most of us are more-or-less prepared for the responsibilities we will face as spouses, parents, and workers. Goals have been established, skills acquired, and attitudes developed that prepare us to accept and even embrace adult roles.

With disconcerting frequency, however, individuals find that their anticipatory socialization is incomplete and idealized. Thus, people who marry at 18 to escape the hassles of living with their parents may not be fully aware of the different kinds of hassles that come with maintaining a home, a spouse, and a child. The statement of a 33-year-old automobile painter, the father of three, married 13 years, illustrates the point:

> I had to work from the time I was thirteen and turn over most of my pay to
> my mother to help pay the bills. By the time I was nineteen, I had been
> working for all those years and I didn't have anything—*not a thing*. I used
> to think a lot about how when I got married, I would finally get to keep my
> money for myself. I guess that sounds a little crazy when I think about it
> now because I have to support the wife and kids. I don't know *what* I was
> thinking about, but I never thought about that then. (Rubin 1976, 56–57)

As this quotation suggests, it is seldom possible to prepare fully for a new role. No matter how many books we read or how many other people we observe, we still find that a new job, a new spouse, or a new child requires some on-the-job training.

Resocialization

Resocialization occurs when we abandon our self-concept and way of life for a radically different one.

The most extreme example of role change comes about when we abandon our self-concept and way of life for a radically different one. This is called **resocialization.** Changing the social behavior, values, and self-concept acquired over a lifetime of experience is difficult, and few people undertake the change voluntarily.

A drastic example of resocialization occurs when people become permanently disabled. Those who become paralyzed experience intense resocialization to adjust to their handicap. All of a sudden, their social roles and capacities are changed. Their old self-concept no longer covers the situation. They may have lost bladder and bowel control, be severely limited in their ability to get around, or be incapable of full sexual functioning. If they are single, they must face the fact that they are unlikely ever to marry and become parents; if they are older, they have to reevaluate their adequacy as spouses or parents. These changes require a radical redefinition of self. If self-esteem is to remain high, priorities will have to be rearranged and new, less active roles given prominence.

Resocialization may also be deliberately imposed by society. When an individual's behavior leads to social problems—as is the case with habitual criminals, problem alcoholics, and the mentally disturbed—society may decree that the individual must abandon the old identity and accept a more conventional one.

TOTAL INSTITUTIONS.   Generally speaking, a radical change in self-concept requires a radical change in environment. Drug counseling one night a week is not likely to drastically alter the self-concept of a teenager who spends the rest of the week among kids who are constantly wasted. Thus, the first step in the resocialization process is to isolate the individual from the past environment.

This is most efficiently done in **total institutions**—facilities in which all aspects of life are strictly controlled for the purpose of radical resocialization (Goffman 1961a). Monasteries, prisons, boot camps, and mental hospitals are good examples. Within them, past statuses are wiped away. Social roles and relationships that formed the basis of the previous self-concept are systematically eliminated. New statuses are symbolized by regulation clothing, rigidly scheduled activity, and new relationships. Inmates are encouraged to engage in self-analysis and self-criticism, a process intended to reveal the inferiority of past perspectives, attachments, and statuses.

Total institutions are facilities in which all aspects of life are strictly controlled for the purpose of radical resocialization.

A CASE STUDY: PRISON BOOT CAMPS AND RESOCIALIZATION.

Polls repeatedly show that most people and politicians in the United States want to get tough with criminals. For some, getting tough on crime means meting out harsher punishments—locking up more people for longer periods of time. However, at least one recent study (Newcomb 1994) shows that almost three times as many people believe that prevention and rehabilitation are more important in controlling crime than longer periods of incarceration. Advocates have proposed as much as $1 billion in new financing for prison boot camps or "shock incarceration programs." Perhaps their popularity arises out of the fact that among all of the proposed crime control measures, they alone simultaneously satisfy the public's urge to punish criminals with its desire to reform them.

Like all resocialization programs, prison boot camps begin with the premise that inmates must radically alter their lifestyles, values, and beliefs. To become productive members of society, inmates must come to reject their old identity and often their old relationships. How are such goals accomplished with groups of young lawbreakers, who though usually convicted of a nonviolent first felony offense, often have a long history of prior delinquent acts?

First, new prisoners are segregated or isolated from competing social environments. Shaved heads and being called names rather than by name remind prisoners that they are leaving one self-concept behind and taking on another. Second, all aspects of daily life are strictly controlled; activities are closely regulated and interaction with other inmates is strictly controlled. Days typically begin with an hour of strenuous calisthenics in the early morning dark and proceed through long hours of military drilling, hard physical labor, drug counseling, and study. Meals last for eight minutes and absolutely no talking is allowed. Prisoners never walk; they run. They are taught to stop and stand at attention before approaching a prison official and then never to look him or her in the eye (*Corrections Today* 1990).

To many, boot camps sound like a great idea. They promise to teach young nonviolent offenders self-discipline, respect for the law, and the value of work by placing them in a rigorous military-style setting while simultaneously satisfying the public's apparent need "for seeing some civility pounded into thugs who terrorize their neighborhoods" (Katel, Liu, & Cohn 1994).

There is some evidence that shock incarceration strategies work. Offenders leave prison boot camps with stronger positive attitudes about their future than they had when they entered; in contrast, prison inmates' attitudes become even more negative during their time in prison (McKenzie, Shaw, & Gowdy 1993). The long-term success of these programs is, however, far more questionable. Camp graduates are just as likely as prison parolees to end up back in trouble with the law. In a study of inmates released from Louisiana prison camps, 37 percent were arrested at least once during their first year of freedom, compared with only 25.7 percent of parolees (Katel, Liu, & Cohn 1994).

Uniforms, shaved heads, harsh discipline, and rigorous physical demands are all designed to encourage young prison boot-camp inmates to cast off their old deviant identity and adopt a new conformist self-concept. Whether it takes place in the context of a religious cult or in the quasi-military setting of this prison, research shows that resocialization often fails; it is difficult for anyone to sustain a change in identity when it conflicts with the expectations of old friends and perhaps even some family members.

Given the disparity between the norms espoused in the prison camps and in the peer groups to which most inmates return, it is little wonder that the rate of recidivism is so high. Proponents of shock incarceration and critics alike agree that without educational programs, drug and vocational counseling, and a well-designed program of postincarceration supervision, the changes in self-concept and values that prison boot camps can generate will not be sufficient to enable offenders to overcome the difficulties they face successfully when they return to their home environment. (MacKensie, Brame, McDowall, & Souryal 1995).

Agents of Socialization

Socialization is a continual process of learning. Each time we encounter new experiences, we are challenged to make new interpretations of who we are and where we fit into society. This challenge is most evident when we make major role transitions—when we leave home for the first time, join the military, change careers, or get divorced, for example. Each of these shifts requires us to expand our skills, adjust our attitudes, and accommodate ourselves to new social roles. Child psychologists have noted that these periods of transition tend to herald both intellectual and moral growth in the youngster. They constitute a crisis that challenges our old assumptions about ourselves and prompts a fundamental reappraisal of who we are.

Learning takes place in many contexts. We learn at home, in school and church, on the job, from our friends, and from television. These agents of socialization have a profound effect on the development of personality, self-concept, and the social roles we assume. Each of these agents of socialization is discussed more fully in later chapters. They are introduced here to illustrate the importance of social structures for learning.

Family

The most important agent of socialization is the family. As the tragic cases of child neglect and the monkey experiments so clearly demonstrate, the initial warmth and nurturance we receive at home are essential to normal cognitive, emotional, and physical development. In addition, our parents are our first teachers. From them we learn to tie our shoes and hold a pencil, and from them we also learn the goals and aspirations that will stay with us for the rest of our lives.

The activities required to meet the physical needs of a newborn provide the initial basis for social interaction. Feeding and diaper changing give opportunities for cuddling, smiling, and talking. These nurturant activities are all vital to the infant's social and physical development; without them, the child's social, emotional, and physical growth will be stunted (Gardner 1972; Lynch 1979; Provence & Lipton 1962; Spitz 1945).

In addition to these basic developmental tasks, the child has a staggering amount of learning to do before becoming a full member of society. Much of this early learning occurs in the family as a result of daily interactions: The child learns to talk and communicate, to play house, and to get along with others. As the child becomes older, teaching is more direct, and parents attempt to produce conformity and obedience, impart basic skills, and prepare the child for events outside the family.

One reason the family is the most important agent of socialization is that the self-concept formed during childhood has lasting consequences. In later stages of development, we pursue experiences and activities that integrate and build on the foundations established in the primary years. Although the personality and self-concept are not rigidly fixed in childhood, we are strongly conditioned by childhood experiences (Mortimer & Simmons 1978).

The family is also an important agent of socialization in that the parents' religion, social class, and ethnicity influence the child's social roles and self-concept. They influence the expectations that others have for the child, and they determine the groups with which the child will interact outside the family. Thus, the family's race, class, and religion shape the child's initial experiences in the neighborhood, at school, and at work.

Schools

In Western societies, schooling has become institutionalized as the natural habitat for children. The central function of schools in industrialized societies is to impart specific skills and abilities necessary for functioning in a highly technological society.

Schools do much more than teach basic skills and technical knowledge, however; they also transmit society's central cultural values and ideologies. Unlike the family, in which children are treated as special persons with unique needs and problems, schools expose children to situations in which the same rules, regulations, and authority patterns apply to everyone. In schools, children first learn that levels of achievement affect status in groups (Parsons 1964, 133). In this sense, schools are training

THINKING CRITICALLY
List some specific ways you think that a family's social class might influence the way a child is socialized. Can you think of any ways that living in the city versus living in the country might matter?

grounds for roles in the workplace, the military, and other bureaucracies in which relationships are based on uniform criteria.

Peers

Compulsory education and the late age at which most youths become full-time workers have contributed to the emergence of a youth subculture in modern societies. In recent years, this development has been accelerated by the tendency for both parents to work outside the home, creating a vacuum that may be filled by peer interaction.

What are the consequences of peer interaction for socialization and the development of the self-concept? Because kids who hang out together tend to dress and act a lot alike, many observers have concluded that peer pressure creates conformity; they have also assumed that conformity to peer values and lifestyles is a frequent source of family conflict, as parents and friends compete for influence. A fairly recent study suggests, however, that the effects of peer pressure have been substantially overestimated. First, it appears that peer similarity precedes group membership; in other words, kids hang around together because they share attitudes (Dornbusch 1989). Second, adolescents are generally more concerned about their parents' opinions than their friends'. Even if they engage in behavior with their friends that their parents would disapprove of, they usually do so only if they think their parents won't find out (Dornbusch 1989).

There are, however, areas of a child's development where peer group socialization has an important influence (Gecas 1981). Because the judgments of one's peers are unclouded by love or duty, they are particularly important in helping us get an accurate picture of how we appear to others. In addition, the peer group is often a mechanism for learning social roles and values that adults don't want to teach. For example, much sexual knowledge and social deviance is learned in the peer group.

Mass Media

Throughout our lives we are bombarded with impersonal messages from radios, magazines, films, and billboards. The most important mass medium for socialization, however, is undoubtedly television. Nearly every home has one, and the average person in the United States spends many hours a week watching it.

The transformation of physical appearance is a powerful symbolic way of leaving one's self-concept behind and taking on another. In adolescence, the peer group becomes an increasingly important arena for trying on various identities and making decisions about what to wear, whom to date, and what clubs to join. Although teenagers may disdain parental advice on matters of aesthetics—music or hair—research shows that parents remain a major influence in terms of decisions about core values, finances, education, or a career.

The effect of television viewing on learning is vigorously debated, and the evidence is somewhat contradictory. The most universally accepted conclusion is that the mass media can be an important means of supporting and validating what we already know. Through a process of selective perception, we tend to give special notice to material that supports our beliefs and self-concept and to ignore material that challenges us.

Television, however, may play a more active part than this. Studies suggest that characters seen regularly on television can become role models, whose imagined opinions become important as we develop our own roles. For example, a working mother might ask, now how would Murphy Brown handle this? Material on television may supplement the knowledge our own experience gives us about U.S. roles and norms. These findings imply that the content of television can have an important influence.

Religion

In every society, religion is an important source of individual direction. The values and moral principles in religious doctrine give guidance about appropriate roles and behaviors. Often the values we learn through religion are compatible with the ideals we learn through other agents of socialization. For example, the golden rule ("Do unto others as you would have them do unto you") that U.S. children learn in Sunday school is likely to fit in easily with similar messages they have heard at home and at school.

The role of religion, however, cannot be reduced to a mere reinforcer of society's norms and values. As we point out in Chapter 12, religious ideals have the power to change societies and the individuals in them. Moreover, even within modern U.S. society, there are important differences in the messages delivered by, say, Mormon, Jewish, and Baptist religions. These differences account for some significant variability in socialization experiences.

Workplace

Almost all of us will spend a significant portion of our adult lives working outside the home for wages or salaries. The environments in which we work, however, are very different. Some of us will work with machines, others with ideas; some will work with people, others on people. Work is found in cities, factories, offices, and fields. Much of it is impersonal, monotonous, and regulated by time clocks; but some is highly personal, challenging, and flexible.

Long-term research by Kohn and his associates indicates that the nature of our work affects our self-concept and behavior. The amount of autonomy, the degree of supervision and routinization, and the amount of cognitive complexity demanded by the job have important consequences. If your work demands flexibility and self-discipline, you will probably come to value these traits elsewhere—at home, in government, and in religion. If your work instead requires subordination, discipline, and routine, you will come to find these traits natural and desirable (Kohn & Schooler 1983).

Because of the salience of our work-role identity, studies demonstrate that losing one's job can be a major blow to the self-concept. Although many people who lose their jobs can protect their self-esteem by blaming the economy or the government, studies also show that unemployment increases depression and physical illness (Hamilton, Broman, Hoffman, & Renner 1990) and reduces the sense of personal competence (Gecas 1989).

Symbolic Interaction Theory

Symbolic interaction theory addresses the subjective meanings of human acts and the processes through which people come to develop and communicate shared meanings.

The **interaction school of symbolic interaction** focuses on the active role of the individual in creating the self and self-concept.

The **structural school of symbolic interaction** focuses on the self as a product of social roles.

Socialization provides a broad overview of the processes through which we come to share roughly the same norms and values as other members of our culture. But how does this learning take place? What are the processes through which, as children and adults, we translate our social experiences into part of our self and self-concept? This question is addressed by the **symbolic interaction theory.** As noted in Chapter 1, this theory addresses the subjective meanings of human acts and the processes through which we come to develop and share these subjective meanings.

Over the years, two distinct schools have developed within this perspective: the interaction school and the structural school (Biddle 1986; Turner 1985). The **interaction school of symbolic interaction** focuses on the active role of the individual in creating the self and self-concept. The **structural school of symbolic interaction** focuses on the self as a product of social roles. We review each separately and then discuss their similarities and differences.

The Interaction School

The major premise of the interaction school is that people are actively involved in creating and negotiating their own roles and self-concept. Although each of us is born into an established social structure with established expectations for our behavior, nevertheless we have opportunities to create our own self-concept. The concepts of *looking-glass self* and *role taking* illustrate how this process works.

Looking-Glass Self

Charles Horton Cooley (1902) provided a classic description of how we develop our self-concept. He proposed that we learn to view ourselves as we think others view us. He called this the **looking-glass self.** According to Cooley, there are three steps in the formation of the looking-glass self:

The **looking-glass self** is the process of learning to view ourselves as we think others view us.

1. We imagine how we appear to others.
2. We imagine how others judge our appearance.
3. We develop feelings about and responses to these judgments.

For example, an instructor whose students openly talk to one another or doze during class and who frequently finds himself talking to a half-empty room is likely to gather that his students think he is a bad teacher. He need not, however, accept this view of himself. The third stage in the formation of the looking-glass self suggests that the instructor may either accept the students' judgment and conclude that he is a bad teacher or reject their judgment and conclude that the students are simply not smart enough to appreciate his profound remarks. Our self-concept is not merely a mechanical reflection of those around us; rather it rests on our interpretations of and reactions to those judgments. We are actively engaged in defining our self-concept, using past experiences as one aid in interpreting others' responses. A person who considers herself witty will assume that others are laughing with her, not at her; someone used to making clumsy errors, however, will form the opposite interpretation from the laughter.

We also actively define our self-concept by choosing among potential looking glasses; that is, we try to choose roles and associates supportive of our self-concept (Gecas & Schwalbe 1983). The looking glass is thus a way of both forming and maintaining self-concept.

As Cooley's theory indicates, symbolic interaction considers subjective interpretations to be extremely important determinants of the self-concept. It is not only others' judgments of us that matter; our subjective interpretation of those judgments is equally important. This premise of symbolic interactionism is apparent in W.I. Thomas's classic statement: If people "define situations as real, they are real in their consequences" (Thomas & Thomas 1928, 572). People interact through the medium of symbols (words, gestures) that must be subjectively interpreted. The interpretations have real consequences—even if they are *mis*interpretations.

Role Taking

The most influential contributor to symbolic interaction theory during this century is George Herbert Mead. Mead argued that we learn social norms through the process of **role taking.** This means imagining ourselves in the role of the other in order to determine the criteria others will use to judge our behavior. This information is used as a guide for our own behavior.

> **Role taking** involves imagining ourselves in the role of the other in order to determine the criteria others will use to judge our behavior.

According to Mead, role taking begins in childhood, when we learn the rights and obligations associated with being a child in our particular family. To understand what is expected of us as children, we must also learn our mother's and father's roles. We must learn to see ourselves from our parents' perspective and to evaluate our behavior from their point of view. Only when we have learned their roles as well as our own will we really understand what our own obligations are.

Mead maintained that children develop their role knowledge by playing games. When children play house, they develop their ideas of how husbands, wives, and children relate to one another. When the little boy comes in saying "I've had a hard day; I hope it's not my turn to cook dinner," or when the little girl warns her dolls not to play in the street and to wash their hands before eating, they are testing their knowledge of family role expectations.

In the very early years, role playing and role taking are responsive to the expectations of **significant others**—role players with whom we have close personal relationships. Parents, siblings, and teachers, for example, are decisive in forming a child's self-concept. As children mature and participate beyond this close and familiar network, the process of role taking is expanded to a larger network that helps them understand

> **Significant others** are the role players with whom we have close personal relationships.

Anticipatory socialization prepares us for the roles we will take in the future. Children everywhere, whether they dress up in their parents' old clothes or enact their fantasies through dolls or trucks, play out their visions of how mommies and daddies ought to behave. As Mead suggests, this role taking is an essential way that children learn and practice acceptable adult behavior. No matter how much they play or how well they know the expectations of their significant others, however, people are never fully prepared for their new roles. A new spouse or a new child will almost certainly require some on-the-job training.

what society in general expects of them. They learn what the bus driver, their neighbors, and their employers expect. Eventually, they come to be able to judge their behavior not only from the perspective of significant others but also from what Mead calls the **generalized other**—the composite expectations of all the other role players with whom they interact. Being aware of the expectations of the generalized other is equivalent to having learned the norms and values of the culture. One has learned how to act like an American or a Pole or a Nigerian.

The **generalized other** is the composite expectations of all the other role players with whom we interact; it is Mead's term for our awareness of social norms.

Saying that everyone learns the norms and values of the culture does not mean that everyone will behave alike or that everyone will follow the same rules. My significant others are not the same as yours. Not only will our family experiences differ depending on the culture and subculture in which we are reared, but as we get older, we have some freedom in choosing those whose expectations will guide our behavior. Although we may know perfectly well what society in general expects of us, we may choose to march to the beat of a different drummer.

The Negotiated Self

Role taking and the looking-glass self are ways in which the individual can become an active agent in the construction of his or her own self-concept. The self that emerges is a negotiated self, a self that we have fashioned by selectively choosing looking glasses and significant others.

The idea of negotiation suggests that we have an end in view. What is that end? An important one is to protect and enhance our self-esteem. **Self-esteem** is the evaluative part of the self-concept; it is our judgment of our worth compared with others. Because we would all like to think well of ourselves, we strive to negotiate a self-concept that reinforces that image.

Self-esteem is the evaluative component of the self-concept; it is our judgment about our worth compared with others'.

Summary

According to the interaction school of symbolic interaction, the individual takes an active role in negotiating the self and self-concept. These are not imposed by others or by the social structure; rather they are negotiated by the individual during the

Perhaps no idea is so widespread in U.S. society as the idea that participating in sports is a good way of socializing young people to important American values. Through sports, children may learn the value of teamwork and the importance of giving their best and trying to be winners. Critics suggest, however, that sport also promotes sexism by emphasizing traditional gender roles, particularly seen in the aggression often associated with male-dominated sports. Although many parents undoubtedly still feel that sports are more important for their sons than their daughters, female sports participation has increased substantially over the last 20 years, and many activities are now coed. As a result, sports may not reinforce gender role differences as much as they once did.

process of interacting with others. An important goal in this process is the enhancement of self-esteem.

The Structural School

The structural school of symbolic interaction differs from the interaction school by stressing the importance of institutionalized social structures. Unlike the interaction school, which gives the individual a great deal of freedom in negotiating a self-concept, the structural school assumes that individuals are constrained and shaped in important ways by society.

Scholars from the structural school focus on institutionalized social roles such as working, parenting, and going to school, and stress the profound ability of these roles to shape both our behavior and our personality. The key concept in structural theory is **role identity,** the image we have of ourself in a specific social role (Burke 1980, 18). For example, a woman who is a professor, a mother, and an aerobics student will have a different role identity in each setting. According to structural theorists, her self-concept will be a composite of these multiple identities (Stryker 1981).

The idea of role identities draws heavily on the analogy of life as a stage. As we move from scene to scene, we change costumes, get a new script, and come out as a different character. A young man may play the role of dutiful son at home, party animal at the dorm, and serious scholar in the classroom. None of these images is necessarily false; but because the roles are difficult to carry on simultaneously, the young man does not try to do so. Most of us engage in this practice without even thinking about it. We adjust our vocabularies, topics of conversation, and apparent values and concerns automatically as we talk to elderly relatives, our friends, and our co-workers. In this sense, we are like the elephant described by the six blind men; someone trying to find the Real Us might have a hard time reconciling the different views that we present.

Identity Salience Hierarchies

The concept of role identity implies that we play one role at a time and ignore the obligations associated with other roles. In practice, however, competing demands often force us to choose between different selves. If your boss tells you a racist joke, do you laugh courteously or do you tell him you are offended? Which self will take priority?

One mechanism for making such choices is the establishment of an **identity salience hierarchy,** a ranking of your various role identities in order of their importance to you (Callero 1985). Whenever two roles come into conflict, you simply follow the role that ranks highest on your list. Research shows that in ranking role identities, we give preference to roles that provide us with the most self-esteem. This reflects two things: the social status associated with a role and our skill in performing it.

Most women, for example, who are both professors and aerobics students will rank being a professor higher in their identity salience hierarchy. A woman who was a very bad professor and a very good aerobics student, however, might reverse the order. The relative ranking of mother and professor might be more problematic. Undoubtedly, some women would rank the professor role higher than the mother role; such women would follow the norms of their job in cases of role conflicts. Differences in identity salience hierarchies help explain why two people with the same sets of roles will behave differently. Some professor/mothers will go to the office on Saturday and some will stay at home; these choices reflect differences in their identity salience hierarchies (Serpe 1987).

Role identity is the image we have of ourself in a specific social role.

An **identity salience hierarchy** is a ranking of an individual's various role identities in order of their importance to him or her.

THINKING CRITICALLY
What role identity is most important to you now? Do you think that might change over the course of your life? Why or why not?

CONCEPT SUMMARY	*The Two Schools of Symbolic Interaction*	
	INTERACTION SCHOOL	**STRUCTURAL SCHOOL**
The self-concept is . . .	Negotiated	Determined by roles
The individual is . . .	Active in creating self-concept; has more freedom to choose self	Less active in creating self-concept; has less freedom to choose self
The self-concept is developed through . . .	Role taking (taking the role of others)	Performing institutionalized roles
Roles are . . .	Negotiated	Allocated
Major concepts	Looking-glass self, role taking, and self-esteem	Role identity and identity salience hierarchy

Summary

The structural school of symbolic interaction suggests that the self and self-concept depend on the roles we play and, consequently, on the social structures we are a part of. This helps to explain why two friends who go in very different directions after high school, say, one into the army and the other into a rock band, develop into such very different people despite their initial similarities. We grow into our roles, and the parts we play eventually become a part of our self-concept.

Conclusion

Despite their differences (displayed in the Concept Summary), the structural and interaction schools have much in common. Scholars from both schools accept the general premises of symbolic interaction theory. Reworded from Chapter 1, these premises are that:

1. In order to understand human behavior, we must learn what it means to the individual actors.
2. Meanings depend on relationships.
3. Actors negotiate their self-concept (at the least, by choosing the salience hierarchy of their role identities).

Both those theorists who stress the importance of social structure and those who stress the importance of individual choice agree that interaction with others is central to understanding individual behavior.

The Sociology of Everyday Life

Why do people do what they do? The answer depends in part on which social roles they are playing, but it also depends on the situation and on the individual role player. Two people playing the role of physician will do it differently; the same individual will play the role differently depending on the patient's attitude and the circumstances. Social structure explains the broad outlines of why we do what we do, but it doesn't deal with specific concrete situations. This is where the sociology of everyday life comes in. The **sociology of everyday life** focuses on the social processes that structure our experience in ordinary face-to-face situations.

The **sociology of everyday life** focuses on the social processes that structure our experience in ordinary face-to-face situations.

Assumptions About Everyday Life

The sociology of everyday life is closely identified with the interaction school of symbolic interaction theory. Like that school, it stresses the importance of subjective meanings assigned to symbolic communications and the active part of the individual in negotiating roles and identities. Scholars who use the everyday life (EDL) perspective, however, emphasize four additional assumptions: the problematic nature of culture, the dialectic, biography, and thick description.

The Problematic Nature of Culture

In Chapter 2, we likened culture to a design for living and to a tool kit. If you consider these two images, you will see that they offer subtly different meanings of culture. The "design for living" image suggests a set of blueprints that need only to be correctly followed; the "tool kit" image, on the other hand, suggests a more dynamic approach. You do not follow a tool kit, you *use* it.

The sociology of everyday life is based on the tool kit image. It assumes that day-to-day behavior is not a matter of following clear cultural scripts but of improvising, negotiating, and adjusting to the general outline. From this point of view, culture is problematic.

Culture does furnish a great many rules and rituals, but it is not always clear which rules apply when. Being honest and standing by your friends are both norms; what do we do when telling the truth gets a friend in trouble? This predicament emphasizes that conformity to cultural norms is problematic: It requires a continual stream of choices (Oberschall & Leifer 1986).

The Dialectic

The rules that govern our behavior may contradict each other; they may also contradict our own wishes.

Scholars from the EDL perspective take seriously the I/me split of the self proposed by Mead. They recognize that, on the sidelines of every social encounter, the I stands ready to assert itself—to barge in and do something impulsive and perhaps selfish. Any concrete situation can be envisioned as a negotiation between these two parts of ourself.

This negotiation can be viewed as a dialectic, a process of conflict between individual freedom and social constraint (Bensman & Lilienfeld 1979). Here is a commonplace example: You are hurrying down the sidewalk, late for class, and you pass an acquaintance with a quick "Hi there, Lori, how are you?" Instead of replying as expected, Lori stops and proceeds to tell you that she is really depressed because she has just heard that her mother has cancer. So now what do you do? One aspect of your self (the me) tells you that you must pause and show interest and concern. Your I, the spontaneous, impulsive part of your self, wants to keep hurrying to class. After all, you hardly know Lori, much less her mother. In this case and in many others, your behavior represents a dialectic, a conflict between social convention and individual impulses.

A dialectic is not a simple contradiction but a process in which opposing forces engage, meet, and produce change. Viewing interaction episodes as a dialectic leads to the proposition that social interactions are never completely programmed by social structures. The outcome of a conflict is never fully predictable.

Biography

Each of us possesses a personal history, a biography, that makes us unique. Although we share many things, none of our experiences exactly duplicates the experiences of another. Although we may all face similar social constraints as in the awkward episode of Lori's mother, we will each bring a different self to the dialectic. Because of this, no two encounters are identical.

The uniqueness of each encounter does not mean that there is no patterned regularity in everyday life. Almost all of our daily encounters follow recognizable routines that allow us to interact without awkwardness, even with strangers. These routines explain the similarities in our encounters; biography explains why each one is just a little different from the rest.

Thick Description

The sociology of everyday life uses a methodological technique called thick description. Unlike thin description (who did what when), *thick description* tells us why the actors did what they did and what it meant to them. Thick description requires that we get into the actors' conceptual world, that we understand what is going on inside their heads (Geertz 1973).

The switch from thin to thick description brings about an important change in the stance the observers take toward reality. Scholars such as Durkheim and Comte assumed that an objective reality exists and that the goal of science is to put aside all personal, or subjective, ideas so that this objective reality can be studied. Sociologists of everyday life, on the contrary, assume that there are multiple subjective realities and that the world as I see it is just as real as the world as you see it. Thick description is the attempt to understand the subjective social worlds of individual actors in specific situations (Lincoln & Guba 1985).

Summary

The sociology of everyday life is in many ways the application of the interaction school to the mundane encounters of daily life. It stands apart from other branches of sociology, however, because of its emphasis on the problematic nature of culture, the dialectic, biography, and the reality of subjective worlds.

Managing Everyday Life

Much of our daily life is covered by routines. The most important routines we use for interaction with others are carried out through talk. We all learn dozens of these verbal routines and can usually pull out an appropriate one to suit each occasion. Small rituals such as

> "Hello. How are you?"
>
> "Fine. How are you?"
>
> "Fine."

What's going on here? In order to plan an effective course of action, one of the first issues to resolve in any social setting is defining just what kind of a situation it is. Next we have to figure out who the relevant players are. Our strategy for responding to the disaster would certainly be different if we believed that this Los Angeles scene had resulted from a tornado rather than from the riots.

will carry us through dozens of encounters every day. If we supplement this ritual with a half a dozen others, such as "thanks/you're welcome" and "excuse me/okay," we will be equipped to meet most of the repetitive situations of everyday life. Nevertheless, each encounter is potentially problematic, and successful interaction requires selecting the appropriate ritual plus the skill and motivation to carry it out.

At the beginning of any encounter, individuals must resolve two issues: (1) What is going on here—what is the nature of the action? (2) What identities will be granted—who are the actors? All action depends on our answers to these questions. Even the decision to ignore a stranger in the hallway presupposes that we have asked and answered these questions to our satisfaction. How do we do this?

Frames

The first step in any encounter is to develop an answer to the question, what is going on here? The answer forms a frame, or framework, for the encounter. A **frame** is roughly identical to a definition of the situation, a set of expectations about the nature of the interaction episode that is taking place.

A **frame** is an answer to the question, what is going on here? It is roughly identical to a definition of the situation.

All face-to-face encounters are preceded by a framework of expectations—how people will act, what they will mean by their actions, and so on. Even the most simple encounter, say, approaching a salesclerk to buy a package of gum, is covered by dozens of expectations: We expect the salesclerk will speak English, will wait on the person who got to the counter first, will charge exactly the sum on the package and will not try to barter with us, will not comment on the fact that we are overweight or need a haircut. These expectations—the frame—give us guidance on how we should act and allow us to evaluate the encounter as normal or as deviant.

Our frames will be shared with other actors in most of our routine encounters, but this is not always the case. We may simply be wrong in our assessment of what is going on, or other actors in the encounter may have an entirely different frame. The final frame that we use to define the situation will be the result of a negotiation between the actors.

Identity Negotiation

After we have put a frame on an encounter, we negotiate an answer to the second question: What identities will be granted? This question is far more complex than attaching names to the actors. Because each of us has a repertoire of roles and identities from which to choose, we are frequently uncertain about which identity an actor is presenting *in this specific situation*.

To some extent, identities will be determined by the frame being used. If a student visit to a professor's office is framed as an academic tutorial, then the professor's academic identity is the relevant one. If the visit is framed as a social visit, then other aspects of the professor's identity (hobbies, family life, and so on) become relevant.

In many routine encounters, identities are not problematic. Although confusion about identities is a frequent device in comedy (for example, many of Shakespeare's comedies), in real life, a few verbal exchanges are usually sufficient to resolve confusion about the actors' identities. In some cases, however, identity definitions are a matter of serious conflict. For example, I want to be considered a status equal, but you wish to treat me as an inferior.

Resolving the identity issue involves negotiations about both your own and the other's identity. How do we negotiate another's identity? We do so by trying to manipulate others into playing the roles we have assigned them. Mostly we handle this through talk. For example, "Let me introduce Mary, the computer whiz," sets up a different encounter than "Let me introduce Mary, the best party giver I know." Of course, others may reject your casting decisions. Mary may prefer to present a different identity than you have suggested. In that case, she will begin to try to renegotiate her identity.

Even those who have very high self-esteem seek out ways to enhance and maintain their credit with others. Unfortunately, few of us receive as many ribbons as this mouse breeder, and even he probably finds few opportunities to show off all of his awards. Most of us, then, have to seek somewhat more subtle ways to demonstrate what wonderful people we are.

THINKING

CRITICALLY

Describe a time when you disagreed with someone about his or her identity. What kind of situation was it and why was identity problematic? In the end, whose definition of identity was accepted? Why do you think that was?

Identity issues can become a major hidden agenda in interactions. Imagine an incompetent man talking to a competent woman. If the man finds this situation uncomfortable, he may try to define it as a man/woman encounter rather than a competence encounter. He may start with techniques such as "How do you, as a woman, feel about this?" To reinforce this simple device, he will probably follow up with remarks such as "You're so small, it makes me feel like a giant." He may interrupt her by remarking on her perfume. He may also use a variety of nonverbal strategies such as stretching his arm across the back of her chair to assert dominance. Through such strategies, actors try to negotiate both their own and the other's identity.

Dramaturgy

Dramaturgy is a version of symbolic interaction that views social situations as scenes manipulated by the actors to convey the desired impression to the audience.

The management of everyday life is the focus of a sociological perspective called dramaturgy. **Dramaturgy** is a version of symbolic interaction that views social situations as scenes manipulated by the actors to convey the desired impression to the audience.

The chief architect of the dramaturgical perspective is Erving Goffman (1959, 1963a). To Goffman, all the world is a theater. The theater has both a front region, the stage, where the performance is given, and a back region, where rehearsals take place and behavior inappropriate for the stage may appear. For example, waiters at expensive restaurants are acutely aware of being on stage and act in a dignified and formal manner. Once in the kitchen, however, they may be transformed back into rowdy college kids.

The ultimate back region for most of us, the place where we can be our real selves, is at home. Even here, however, front-region behavior is called for when company comes. ("Oh yes, we always keep our house this clean.") On such occasions, a married couple functions as a team in a performance designed to manage their guests' impressions. People who were screaming at each other before the doorbell rang suddenly start called each other "dear" and "honey." The guests are the audience, and they too play a role. By seeming to believe the team's act (Goffman calls this "giving deference"), they contribute to a successful visit.

Nonverbal Cues

A successful act requires careful stage management. Not only must the script be right, but costumes and props and body gestures must support our act.

DRESS.   All of us who have stood in front of our closet wondering what to wear have faced the dilemma of how to present ourself. Although most of us have many outfits that meet the necessary criteria of covering our bodies decently and dealing with the climate, we choose among potential outfits based on the impression we would like to create. Concepts such as dressing for success are examples of this. From punk haircuts to cowboy boots, the way we dress furnishes important auxiliary information to our audience; it can either support our act or discredit it.

BODY LANGUAGE.   The cues we give through facial expressions, eye contact, and posture can enhance, reinforce, or even contradict the meaning of a verbal communication. For example, when words of affection are accompanied by hugs or caresses, the verbal statements are enhanced and reinforced. When words of encouragement and interest are delivered without looking up from the newspaper, however, the verbal statements are contradicted by actions. Whether intended or unintended, the impression given belies and discredits the verbal communication.

Although President Bill Clinton undoubtedly spends more time in a suit than in his running clothes, this attire symbolized one of Clinton's salient role identities: "amateur athlete." All of us use clothing as a prop to support the roles we play. When we change from high heels to high-tops, we signal a change in identity. All of us try to affect others' opinions of us by the way we dress. Would world leaders listen as carefully to President Clinton if he showed up for the summit in his jogging shorts instead of a sports coat? Would you?

In other cases, gestures and posture may substitute entirely for verbal communication. Anybody who has watched a football game will be familiar with the small drama where a receiver who drops an important pass falls to the ground and drums the turf with his hands. This communicates that he took his failure seriously and he is angry with himself. In effect, it says, "You don't need to bawl me out, Coach, I know I goofed and I feel really bad about it and I'll try my best not to do it again."

CUING STATUS. Nonverbal cues are used extensively to send out messages that would be socially awkward if expressed verbally. Sexual interest is one of these messages. Another is a message about social status. It is difficult to say outright, "You know that I'm more important than you, don't you?" Thus, such messages are sent through nonverbal cues.

Studies of status encounters in laboratories show that the following nonverbal strategies are used to signal dominance: looking directly into the other's eyes while listening and speaking, taking up a lot of physical space (sprawling in one's chair, putting a foot on another chair, leaning an arm across the back of another chair), choosing a seat at the end of the table, interrupting others, and being the first person to speak (Ridgeway, Berger, & Smith 1985). In these last two instances, the timing of the speech acts has an effect independent of their symbolic content. The effectiveness of these strategies is confirmed if we consider the inferences we would make about the status of a person who did just the opposite: who wouldn't look at us, scrunched up in a corner of the chair, chose an unassuming position at the table, and was hesitant to speak.

Although some nonverbal communication appears to have universal meaning, most nonverbal communication must be framed before it can be accurately understood. A woman who takes a long glance up and down a man's body shows that she is interested in his body. Only knowledge of the context will tell you what that means. Is she a doctor looking at him with clinical interest? A tailor? A cannibal? A woman looking for a sexual encounter? An employer sizing up a laborer? As this suggests, a variety of meanings can be associated with the same behavior. The correct interpretation depends on the social context, interactions, and biographies of the members.

Whose Definition: Ex-Wife at the Funeral

How we act in any encounter depends in large part on the frame we have developed and the identities we and the other participants have worked out. One of the many processes of negotiation that goes on in social encounters is determining whose definition of the situation will be accepted. The person who wins this negotiation gains a powerful advantage. A paper by Riedmann (1987) on the role of the ex-wife at the funeral is an insightful illustration of this process.

In this case, a 44-year-old man dies suddenly and accidentally at the home he shares with his new wife of six months. Immediately after the accident, his 20-year-old daughter, who is staying with them, calls her mother to break the news. The ex-wife is terribly upset: She went steady with Bill from age 16 to 22 and was married to him for 20 years; even after the divorce and his remarriage, they remained friends and had lunch frequently. Although they could no longer live together, they shared a concern for their two children and half a lifetime of memories. His parents were almost as close to her as her own; she had spent many summer afternoons with his brothers and their children.

When informed of his death, therefore, she felt bereaved. Her first response was "What? Our dad is dead?" Her definition of the situation was that she had had a death in her family. Acting on this definition, she went over to the home of her parents-in-law. Here she came up against a different definition. From the in-laws' position, it was a death in their family but not in her family. Although they did not say so, they wanted her to leave quickly before the current wife came over.

Over the next few days, the ex-wife again and again ran into the dilemmas posed by contrasting definitions of the situation. The obituary did not mention her among the relatives left behind. When she went to view the body at the mortuary, she was denied admittance on the grounds that only members of the immediate family were eligible. When the minister asked everybody but the immediate family to leave the graveyard, she was supposed to leave. Although the ex-wife defined Bill's death as a death in her family, her definition was not commonly shared.

The ex-wife and the new wife formed teams, each trying to pursue its own definition of the situation. The "old family team" (made up of the ex-wife, one of her children, her relatives, and one brother-in-law) and the "new family team" competed to be recognized as the "real family team." At the funeral service itself, each team held down one corner of the lobby. When family and friends entered the lobby, they were faced with the predicament of which team to console first. Because many of Bill's old friends knew the ex-wife well and the new wife not at all, the "old family team" may be considered to have won this round.

Because of legal relationships, in the end the ex-wife was relatively powerless to impose her definition of the situation. The mortician and the priest were paid agents of the "new family team." The ex-wife was ultimately forced to reframe the situation and to recognize her powerlessness to impose her definition of the situation on others.

Identity Work

If we view day-to-day encounters as a series of negotiations, we may be satisfied with simple descriptions of what people usually do in such encounters. Most social scientists, however, wish to go beyond this and ask why people do what they do. The answer most often supplied by scholars studying everyday behavior is that people are trying to enhance their self-esteem. One commentator noted wryly that this research assumes an "approval-starved person hot on the trail of a compliment" (Schneider 1981). More generally, we assume that social approval is one of the most important rewards that human interaction has to offer and we try to manage our identities so that this approval is maximized.

Managing identities to support and sustain our self-esteem is called identity work (Snow & Anderson 1987). It consists of two general strategies: avoiding blame and gaining credit (Tedeschi & Riess 1981).

Avoiding Blame

There are many potential sources of damage to our self-esteem. We may have lost our job or flunked a class; on a more mundane level, we may have said the most embarrassing and stupidest thing imaginable, been unintentionally rude to an older relative, or otherwise made a fool of ourselves. When we behave in ways that make us look bad, or when we fear we are on the verge of saying or doing something embarrassing, then it is important to try to protect our identities.

Most of this work is done through talk. C. Wright Mills (1940, 909) noted that we learn the vocabulary for making excuses at the same time we learn the norms themselves. We learn what the rules are, and we simultaneously learn what kinds of accounts will justify our violations of those rules. If we can successfully explain away our rule breaking, we can present ourselves as a person who normally obeys norms and who deserves to be thought well of by ourselves and others. The two basic strategies we use to avoid blame are accounts and disclaimers.

Accounts

Many of the rule breakings that go on in everyday life are of a minor sort that can be explained away. We do this by giving **accounts,** explanations of unexpected or untoward behavior. Accounts are of two sorts—excuses and justifications (Scott & Lyman 1968). **Excuses** are accounts in which an individual admits that the act in question is bad, wrong, or inappropriate but claims he or she couldn't help it. **Justifications** are accounts that explain the good reasons the violator had for choosing to break the rule; often these are appeals to some alternate rule (1968, 47).

One of the most fertile fields for both excuses and justifications occurs in the student role. Students are expected to study and turn papers in on time. Many times, however, they don't. Kathleen Kalab, a sociology professor at Western Kentucky University, asked her students to explain in writing why they missed class. The answers she received show how students try to explain away their norm violations so that they can preserve a positive identity. For example, one student offered the following *excuse:*

> I am so sorry I missed your class, among others, Tuesday the [date]. The reason I missed is a simple yet probably unacceptable reason, I slept until 1:45 P.M. Not only did I sleep through Sociology, I also slept through these things: Geography, trash pick-up at Poland Hall, maintenance workers sawing down a tree outside my window, my alarm clock (I believe), and many other loud and interesting things. Please see your way clear of forgiving me for this horrible and strange event. (Kalab 1987, 79)

Other students *justified* sacrificing student performance in order to be good sons or employees. For example,

> Sorry I wasn't here but Friday my parents called and our entire herd of cattle broke out and were all over the county and then I got to castrate two 500-lb. bull calves. So needless to say I was extremely busy. (Kalab 1987, 75)

Or

> I am really sorry for missing your class Friday. The reason being I was part of my Reserve unit's advance party to Fort Knox. I didn't know this until late Thursday night. So please excuse me for not being there. (Kalab 1987, 76)

Accounts such as these are verbal efforts to resolve the discrepancy between what has happened and what can legitimately be expected. If they are accepted, self-identity is preserved and interaction can proceed normally.

Disclaimers

A person who recognizes that he or she is likely to violate expectations may preface that action with a **disclaimer,** a verbal device used, in advance, to defeat any doubts and negative reaction that might result from conduct (Hewitt & Stokes 1975, 3). Students commonly begin a query with "I know this is a stupid question, but. . . ." The disclaimer lets the hearer know that the speaker knows the rules, even though he or she doesn't know the answer.

Accounts are explanations of unexpected or untoward behavior. They are of two sorts: excuses and justifications.

Excuses are accounts in which one admits that the act in question is wrong or inappropriate but claims one couldn't help it.

Justifications are accounts which explain the good reasons the violator had for choosing to break the rule; often they are appeals to some alternate rule.

Disclaimers are verbal devices employed in advance to ward off doubts and negative reactions that might result from one's conduct.

Disclaimers occur before the act; accounts occur after the act. Nevertheless, both are verbal devices we use to try to maintain a good image of ourselves, both in our own eyes and in the eyes of others. They help us to avoid self-blame for deviant behavior, and they try to reduce the blame that others might attribute to us. If we are successful in this identity work, we can retain fairly good reputations despite a few failures in meeting our social responsibilities.

Gaining Credit

Our compliment-starved actor will not be content just to be thought a good sort of person; she wants all the credit she can get. This means that she will employ a variety of verbal devices to associate herself with positive outcomes. Just as there are a variety of ways to avoid blame, there are many ways to claim credit. One way is to link yourself spatially or verbally to situations or individuals that have high status. This ranges from making a $1,000 donation to a political fund-raiser so that you can have a photograph of yourself with the president to aligning yourself verbally with "our" team (when it is winning).

Claiming credit is a strategy that requires considerable tact. Bragging is generally considered inappropriate, and if you pat yourself too hard on the back, you are likely to find that others will refuse to do so. The trick is to find the delicate balance

Because the classroom is an explicitly competitive setting, it is an arena that demands a lot of identity work. Every day there are papers, grades, tests, and performances that may reflect either poorly or positively on our self-esteem and our social roles. Whether we squeal with delight when informed of a good grade or just tap our papers meaningfully on our desks, most of us try to get as much credit as we can for our good work.

where others are subtly reminded of your admirable qualities without your actually having to ask for or demand praise.

Again the classroom and the negotiation of student identity provide good examples of how we go about doing positive identity work. Daniel and Cheryl Albas (1988) studied the ways their students at the University of Manitoba managed identity when their papers were returned. "Aces" are students who have done very well on examinations, and their task is to claim all the credit they can without bragging or being condescending to classmates who have done poorly, the "bombers." The Albases note that aces differ in sophistication; all leave their examinations and grades prominently displayed on their desks, and some simply announce their score to others. In the most successful strategy, however, the ace begins by complaining about how hard the test was and then asks how other students did. Only when asked does the ace give his or her own score. Another strategy that enhances the ace's credit while protecting the bomber's identity is to blame both success and failure on luck. When talking to a bomber, an ace may say that "it was only luck that I studied the stuff that was on the exam." This strategy gives the ace double credit: for being a good student and for being a nice person.

A Case Study of Identity Work: The Homeless

A particularly useful illustration of identity work can be seen in individuals who have what Goffman (1961b) calls *spoiled identities*—identities that are not merely low in status but are actively rejected by society. Examples include the severely handicapped, traitors, and, in some communities, people with AIDS. How do people with spoiled identities sustain their self-esteem?

In 1961, Goffman argued that two general strategies for protecting one's self-esteem in the face of a spoiled identity are (1) physically withdrawing from interaction with higher-status others and (2) trying to pass as a member of some higher-status group. People with AIDS might use these strategies successfully; the homeless cannot. They cannot pass because their clothing and circumstances firmly declare their current status. Physical withdrawal is also difficult: By definition, the homeless have nowhere to go. Lacking these two options, what strategies do these women and men use to negotiate a positive identity?

A study among the homeless in Austin, Texas, investigated such identity mechanisms. David Snow and Leon Anderson (1987) ate at the Salvation Army and hung out under bridges and at the plasma center. They listened to homeless people talk about themselves, trying to discover the processes that these people used to negotiate their identities and protect their self-esteem.

Procedures and Context

Snow and Anderson used the strategies of participant observation described in Chapter 1 to study homeless people. The two sociologists dressed in their oldest clothes and hung around just to see what was happening. Although their study deals only with the homeless in one city, they argue that their results are probably applicable to the homeless everywhere. Not only are the homeless a mobile population that drifts from city to city and state to state, but "aside from variations in climate and the availability of free shelter and food, most aspects of life on the street are quite similar from one city to another. We think it is therefore reasonable to expect considerable similarity in basic patterns and process of identity construction and avowal among the homeless" (1987, 1341).

Snow and Anderson did not take notes or use tape recorders while on the street, but they took copious notes after each shift of observation. Although they announced their true identities to some of the street people they talked with, for the most part they passed as fellow homeless. They asked few direct questions, but used an interviewing style that they call "interviewing by comment." For example, instead of asking what a person is doing on the streets, they would say something like "I haven't seen you around." This indirect method of interviewing allowed them to see what the street person would volunteer about himself or herself.

Results

As they analyzed the kinds of stories they heard from the street people in Austin, Snow and Anderson found that they fell into three general categories: role distancing, role embracement, and storytelling. All are verbal strategies that help the homeless maintain their self-esteem and develop a positive identity.

Role Distancing

Whether we're homeless or not, we occasionally find ourselves playing a role that does us no credit. Students, for example, often have low-status jobs such as working at fast-food outlets or even sorting laundry in their dormitories. When we have roles that do us no credit, we engage in a process of *role distancing,* explaining to anyone who will listen that this is just temporary and not a reflection of who we really are. We reject this role identity.

Snow and Anderson found that role distancing was one of the most common forms of identity work among the homeless. For example, a 24-year-old man who has been on the street for only a few weeks says, "I'm not like the other guys who hang out down at the 'Sally' [Salvation Army]. If you want to know about street people, I can tell you about them; but you can't really learn about street people from studying me, because I'm different" (1987, 1349).

Role Embracement

Some of the street people that Snow and Anderson studied had been on the streets for years and recognized that this might be a permanent way of life. They coped with this by developing romantic notions of "brethren of the road" and cultivating nicknames such as Boxcar Billy and Panama Red. These people would introduce themselves as bums and tramps; it was one of their salient role identities.

Although it might not be obvious to outsiders, there are degrees of success in performing the role of street person. Those who had embraced the role took pride in their skill at such activities as panhandling and "dumpster diving" (going through trash cans for food, clothing, recyclables, and so on). They also took pride in their independence from conventional constraints and their ability to get food and shelter without toeing the line. These people looked down their nose at those who depended on the Salvation Army to take care of them; a real street person could take care of himself or herself.

Storytelling

A final form of identity work that the homeless engage in is storytelling. This is most common among those who have been on the street less than four years

This photograph was taken outside the Salvation Army in Austin, Texas, where David Snow and Leon Anderson did their participant observation study of the homeless. Their research shows that many of the people who hang out at the "Sally" manage to negotiate a positive identity despite being homeless. Getting by without a job, car, money, or house requires a lot of ingenuity, and some long-term homeless take pride in their ability to fend for themselves.

and who have not yet embraced a street identity. These stories have the purpose of establishing a positive, nonstreet identity by telling others how important one once was or how rich one will be or could easily be. One homeless man who was hanging around a transient bar bumming cigarettes and beers announced often that he had been offered a job as a mechanic with Harley-Davidson at $18.50 per hour. Others told stories about the future—how they were going to set up their own business, inherit a lot of money, or buy extravagant presents for their families.

Some of these stories were within the realm of reason, but most were not. Either the story itself was extraordinarily improbable (one man claimed to have been a guard on the nonexistent Alaska-Siberia border), or the stories changed so much from day to day that they were no longer believable. Although these unbelievable stories were generally saved for strangers or chance acquaintances, Snow and Anderson found that tactful audiences could be relied upon to give deference to the act of storytelling even if they knew perfectly well that it was all a story. On the street, as well as in more conventional settings, an unspoken norm is that one doesn't ask too many questions or challenge a person's stories about himself or herself. In this way, the street people work together to support one another's identity claims.

Conclusion

The homeless people that Snow and Anderson studied had all the ingredients for the formation of a negative identity: They were hungry, poor, ragged, homeless, frequently drunk and sick, and unemployed. It is not too surprising that some of them had negative self-images. More surprisingly, many managed to feel good about themselves. Somehow they had found it possible to construct a positive identity. Their success in doing so reaffirms the assumption made by the interaction school: Even in the face of a spoiled identity, we can negotiate a positive self-concept.

Where This Leaves Us

In the 1950s, structural-functional theory dominated sociology, and a great deal of emphasis was placed on the power of institutionalized norms to *determine* behavior. Durkheim, with his views on positivism and constraint, was a favorite classic theorist. Similarly, during this period, the structural school of symbolic interaction theory clearly dominated the interaction school, and theorists stressed the power of institutionalized roles to *determine* behavior and personality.

Beginning in the 1960s, sociologists grew increasingly concerned that this view of human behavior reflected an "oversocialized view of man" (Wrong 1961). In 1967, Garfinkel signaled rebellion against this perspective when he argued that the deterministic model presented people as "judgmental dopes" who couldn't do their own thinking.

In the last decade, sociological thinkers have increasingly tended to view social behavior as more negotiable and less rule bound (Perrow 1986). This change is most obvious in the sociology of everyday life, but it is also evident in most other areas of sociology. Studies of mental hospitals, businesses, and complex organizations now suggest that the behavior of actors may be best understood as a game in which each player chooses a strategy to maximize her or his self-interest (Crozier & Friedberg 1980). Even bureaucracies are not seen as wholly deterministic. Employees stress some rules, ignore others, and reinterpret the rest (Fine 1984).

This does not mean that the rules don't make a difference. Indeed, they make a great deal of difference, and there are obvious limits to the extent to which we can negotiate given situations. As W. I. Thomas noted in 1923, "The child is always born into a group of people among whom all the general types of situations which may arise have already been defined and corresponding rules of conduct developed, and where he has not the slightest chance of making his definitions and following his wishes without interference" (cited in Shalin 1986).

The perspective of life as problematic and negotiable is a useful balance to the role of social structure in determining behavior. Our behavior is neither entirely negotiable nor entirely determined.

Summary

1. The self is a combination of unique qualities and shared norms. The two key components of the self are the *I* and the *me*.

2. Although biological capacities enter into human development, our identities are socially bestowed and socially sustained. Without human relationships, even our natural capacities would not develop.

3. Socialization, the process of learning the norms and values necessary for participation in social institutions, occurs at three basic levels: primary socialization, anticipatory socialization, and resocialization.

4. The family is the major agent of socialization. It is responsible for the early nurturance that helps the infant develop into a human being, and it is central in laying the foundation of the self-concept. In addition, the family provides a background (social class, religion, and so on) that determines much of the child's other interactions. Other important agents of socialization include peers, schools, mass media, the workplace, and religion.

5. Symbolic interaction is the dominant theoretical framework in sociological studies of human development. It emphasizes that learning takes place through subjectively interpreted interaction, and that the development of the self and self-concept depend on the quality of our relationships with others.

6. The interaction school gives the individual a very active role in constructing the self. The idea of the negotiated self suggests that we use selective interpretations to construct and maintain our self-concept with a special eye to enhancing our self-esteem.

7. The structural school stresses the profound impact of conventional roles and statuses on the development of the self-concept. The concept of role identity suggests that our self and self-concept depend in important ways on the situations or roles in which we find ourselves.

8. The sociology of everyday life is a perspective that analyzes the patterns of human social behavior in concrete encounters in daily life. Four assumptions guide theory and research in this approach to studying human interaction: culture is problematic; individuals experience a dialectic between freedom and social constraint; each individual is unique, possessing a biography unlike any other individual; and understanding everyday life can be accomplished best through thick description.

9. Deciding how to act in a given encounter requires answering two questions: What is going on here? What identities will be granted? These issues of framing and identity resolution may involve competition and negotiation between actors or teams of actors.

10. Dramaturgy is a perspective pioneered by Erving Goffman. It views the self as a strategist who is choosing roles and setting scenes to maximize self-interest.

11. The desire for approval is an important factor guiding human behavior. To maximize this approval, people engage in active identity work to sustain and support their self-esteem. This work takes two forms: avoiding blame and gaining credit.

Sociology on the Net

In managing everyday life, we constantly engage in framing our encounters and identity negotiation. The frame defines the situation and in identity negotiation, we negotiate our own identity as well as the identity of others. These activities are carried out by individuals and groups alike. The case of homosexuality is an excellent example. America is a homophobic nation with a strong hatred and dislike of homosexuals. This is especially the case in the religious institution.

```
http://www.whidbey.net/~dcloud/fbns/
articlesfbns.htm
```

You have reached the Fundamentalist Baptist News Service. Scroll down through the sections on recent and past articles and select out several articles on homosexuality. What kind of a frame is being used here? How is this organization trying to negotiate the identities of homosexuals? What identity do they negotiate for themselves? You might also look for articles dealing with contemporary music.

Certainly, most gay and lesbians do not want to accept the identities negotiated for them by most religious organizations. In response to the identities being thrust upon them, a number of gays and lesbians have formed their own church.

        http://www.ufmcc.com

This web site is maintained by the Universal Fellowship of Metropolitan Community Churches. Click on the *Enter Our World* highlight. A good place to begin your browsing is in the section entitled *Who the UFMCC Is . . . What Do We Do?* Now go to the section on *Homosexuality and the Bible* and continue with the discussion of *Homosexuality: Not a Sin . . . Not a Sickness.* You may wish to follow this discussion of the Bible by following the links at the end of the discussion. How do these discussions constitute identity negotiation? What

kind of an identity is being negotiated? How does your faith look at this issue?

FIND IT ON INFOTRAC COLLEGE EDITION

Your text includes a brief discussion of the importance of family in a child's socialization. An article by Susan D. Witt, in *Adolescence,* gives more attention to the issue of gender-role socialization. Using InfoTrac College Edition, look up the following article:

> "Parental Influence on Children's Socialization to Gender Roles." Susan D. Witt. *Adolescence,* Summer 1997, v32 n126 p253.
>
> (Hint: Enter the search term *Susan D. Witt,* using the Subject Guide.)

Discuss how the majority of children in the United States are socialized in relation to gender roles. What are the benefits to raising children with a more androgynous gender-role orientation? Comment on the challenges of raising a child with an androgynous gender-role orientation.

Suggested Readings

Barnes, J. A. 1994. *A Pack of Lies: Toward a Sociology of Lying.* New York: Cambridge University Press. An interesting book on lying in everyday life. Barnes would argue that almost all of us lie once in a while, especially in socially "ambiguous domains." Barnes also looks at lying among politicians, bureaucrats, and medical practitioners.

Benedict, Ruth. 1961. *Patterns of Culture.* Boston, Mass.: Houghton Mifflin. Originally published in 1934. A classic that draws on several different cultures to illustrate how behavior and personality are consistent with the culture in which a person is reared. The emphasis is on the continuity of socialization.

Goffman, Erving. 1959. *The Presentation of Self in Everyday Life.* New York: Doubleday. The book in which Goffman lays out the basic ideas behind dramaturgy. Each of Goffman's books is enjoyable reading and easily accessible to the average undergraduate.

Goffman, Erving. 1961. *Asylums.* Garden City, N.Y.: Anchor/Doubleday. A penetrating account of total institutions and the significance of social structure in producing conforming behavior. Primarily an analysis of mental hospitals and mental patients, although the analysis is applicable to other total institutions.

Heusmann, L. Rowell, and Malamuth, Neil (eds.). 1986. "Media Violence and Antisocial Behavior," *Journal of Social Issues,* special issue, vol. 42. A collection of essays that spans the range from pornography to children's cartoons. Included are several articles discussing intervening strategies and regulatory issues.

Kohn, Melvin, L., and Schooler, Carmi, and Associates. 1983. *Work and Personality: An Inquiry into the Impact of Social Stratification.* Norwood, N.J.: Ablex. A summary of 20 years of research on the impact of work roles on personality and the self-concept. It provides strong evidence in favor of the structural school.

Snow, David A., and Anderson, Leon. 1993. *Down on Their Luck: A Case Study of Homeless Street People.* Berkeley: University of California Press. A comprehensive look at the structural causes and the personal costs of homelessness. Excellent example of a study that uses both qualitative and quantitative methods.

Deviance, Crime, and Social Control

Outline

Conformity, Nonconformity, and Deviance

In providing a blueprint for living, our culture supplies sets of norms and values that structure our behavior. They tell us what we ought to believe in and what we ought to do. Because we are brought up to accept them, for the most part we do what we ought to do and think as we ought to think. Only "for the most part," however, because none of us follows all the rules all the time.

Previous chapters concentrated on how norms and values structure our lives and how we learn them through socialization. This chapter considers some of the ways individuals break out of these patterns—from relatively unimportant eccentric behaviors to serious violations of others' rights.

Social Control

An understanding of deviance and nonconformity requires that we first consider what brings about conformity. The forces and processes that encourage conformity are known as **social control.** Social control takes place at three levels:

1. Through self-control, we police ourselves.
2. Through informal controls, our friends and intimates reward us for conformity and punish us for nonconformity.
3. Through formal controls, the state or other authorities discourage nonconformity.

Self-control occurs because individuals **internalize** the norms and values of their group. They make conformity to these norms part of their self-concept. Thus, most of us do not murder, rape, or rob, not simply because we are afraid the police would catch us but because it never occurs to us to do these things; they would violate our sense of self-identity. A powerful support of self-control is **informal social control,** self-restraint exercised because of fear of what others will think. Thus, even if your own values did not prevent you from cheating on a test, you might be deterred by the thought of how embarrassing it would be to be caught. Your friends might sneer at you or drop you altogether; your family would be disappointed in you; your professor might publicly embarrass you by denouncing you to the class. If none of these considerations is a deterrent, you might be scared into conformity by the thought of **formal social controls,** administrative sanctions such as fines, expulsion, or imprisonment. Cheaters, for example, face formal sanctions such as automatic failing grades and dismissal from school.

Whether we are talking about cheating on examinations or murder, social control rests largely on self-control and informal social controls. Few formal agencies have the ability to force compliance to rules that are not supported by individual or group values. Sex is a good example. In many states, sex between unmarried persons is illegal, and you can be fined or imprisoned for it. Even if the police devoted a substantial part of their energies to stamping out illegal sex, however, they would probably not succeed. In contemporary United States, a substantial proportion of unmarried people are not embarrassed about having sexual relations; they do not care if their friends know

Social control consists of the forces and processes that encourage conformity, including self-control, informal control, and formal control.

Internalization occurs when individuals accept the norms and values of their group and make conformity to these norms part of their self-concept.

Informal social control is self-restraint exercised because of fear of what others will think.

Formal social controls are administrative sanctions such as fines, expulsion, or imprisonment.

about it. In such conditions, formal sanctions cannot enforce conformity. Prostitution, marijuana use, seat-belt laws—all are examples of situations where laws unsupported by public consensus have not produced conformity.

Deviance Versus Nonconformity

Deviance refers to norm violations that exceed the tolerance level of the community and result in negative sanctions.

People may break out of cultural patterns for a variety of reasons and in a variety of ways. Whether your nonconformity is regarded as deviant or merely eccentric depends on the seriousness of the rule you violate. If you wear bib overalls to church or carry a potted palm with you everywhere, you will be challenging the rules of conventional behavior. Probably nobody will care too much, however; these are minor kinds of nonconformity. We speak of **deviance** when norm violations exceed the tolerance level of the community and result in negative sanctions. Deviance is behavior of which others disapprove to the extent that they believe something ought to be done about it (Archer 1985).

Deviance as Relative

Defining deviance as behavior of which others disapprove has an interesting implication: It is not the act that is important but the audience. The same act may be deviant in front of one audience but not another, deviant in one place but not another.

Few acts are intrinsically deviant. Even taking another's life may be acceptable in war, police work, or self-defense. Whether an act is regarded as deviant often depends on the time, the place, the individual, and the audience. For this reason, sociologists stress that *deviance is relative.* Some examples: Alcohol use is deviant for adolescents but not for adults; having two wives is deviant for the United States but not for Nigeria; carrying a gun to town is deviant in the late twentieth century but was not in the nineteenth century; wearing a skirt is deviant for a U.S. male but not for a U.S. female.

The sociology of deviance has two concerns: why people break the rules of their time and place and the processes through which the rules become established. In the following sections, we review several major theories of deviance before looking at crime rates in the United States.

Theories About Deviance

There are a dozen or more theories about deviant behavior. For argument's sake, we present them in three groups according to our familiar theoretical framework: structural-functional theories, symbolic interaction theories, and conflict theories.

Structural-Functional Theories

In Chapter 1, we said that the basic premise of structural-functional theory is that the parts of society work together like the parts of an organism. From this point of view, deviance is alien to society, an indication that the parts are not working right.

This perspective was first applied to the explanation of deviance by Durkheim in his classic study of suicide (1897/1951). Durkheim was trying to explain why people in industrialized societies are more likely to commit suicide than are people in other societies. He suggested that in traditional societies the rules tend to be well known and widely supported. As a society grows larger, becomes more heterogeneous, and experiences rapid social change, the norms of society may be unclear or no longer applicable to current conditions. Durkheim called this situation **anomie;** he believed that it was a major cause of suicide in industrializing nations.

Anomie is a situation where the norms of society are unclear or no longer applicable to current conditions.

The anomie idea was broadened to apply to all sorts of deviant behavior in Robert Merton's (1957) **strain theory.** Strain theory suggests that deviance results when culturally approved goals cannot be reached by culturally approved means. This is most likely in the case of our strong cultural emphasis on economic success and achievement. The goals of educational and economic achievement are widely shared. The means to live up to these goals, however, are not. In particular, Merton argued, people from the lower social classes have less opportunity to become successful. They find that the norms about achievement are not applicable to their situation.

Of course, not all people who find society's norms inapplicable to their situation will turn to a life of crime. Merton identifies four ways in which people adapt to situations of anomie (see the Concept Summary): innovation, ritualism, retreatism, and rebellion. The mode of adaptation depends on whether an individual accepts or rejects society's cultural goals and accepts or rejects appropriate ways of achieving them.

People who accept both society's goals and its norms about how to reach them are conformists. Most of us conform most of the time. When people cannot successfully reach society's goals using society's rules, however, deviance is a likely result. One form deviance may take is innovation; people accept society's goals but develop alternative means of reaching them. Innovators, for example, may pursue academic achievement through cheating, athletic achievement through steroids, or economic success by becoming gangsters. In these instances, deviance rests on using illegitimate means to accomplish socially desirable goals.

Other people who are blocked from achieving socially desired goals respond by rejecting the goals themselves. Ritualists slavishly go through the motions prescribed by society, but their goal is security, not success. Their major hope is that they will not be noticed. Thus, they do their work carefully, even compulsively. Although ritualists may appear to be overconformers, Merton says they are deviant because they have rejected our society's values on achievement and upward mobility. They have turned their back on normative goals but are clinging desperately to procedure. Retreatists, by contrast, adapt by rejecting both procedures and goals. They are society's dropouts: the vagabonds, drifters, and street people. The final mode of adaptation—rebellion—involves the rejection of society's goals and means and the adoption of alternatives that challenge society's usual patterns. Rebels are the people who start communes or revolutions to create an alternative society. Unlike retreatists, they are committed to working toward a different society.

The basic idea of Merton's theory is that, in complex and rapidly changing societies, there are dislocations between ends and means that encourage individuals to commit acts that are defined as deviant (Douglas & Waksler 1982). This theory explicitly defines deviance as a social problem rather than a personal trouble; it is a property of the social structure, not of the individual. As a consequence, the solution to deviance lies not in reforming the individual deviant but in reducing the mismatch between structural goals and structured means.

Two basic criticisms of strain theory have emerged. First, conflict theorists object to its structural-functional roots. Strain theory suggests that deviance results from a lack of integration among the parts of a social structure (norms, goals, and resources); it is viewed as an abnormal state produced by extraordinary circumstances. Conflict theorists, however, see deviance as a natural and inevitable product of competition in a society in which groups have different access to scarce resources. They suggest that the ongoing processes of competition should be the real focus of deviance studies (Lemert 1981).

Strain theory suggests that deviance occurs when culturally approved goals cannot be reached by culturally approved means.

Getting your nose pierced isn't against the law and your neighbors probably won't reject you. On the other hand, your mother and father are not likely to be pleased about it. Body piercing is an example of behavior that steps out of conventional rules without crossing over into deviance. It is an example of nonconformity, but it does not violate any major norms or arouse too much public disapproval.

CONCEPT SUMMARY	*Types of Strain Deviance*

Merton's strain theory of deviance suggests that deviance results whenever there is a disparity between goals and the institutionalized means available to reach them. Individuals caught in this dilemma may reject the goals or the means or both. In doing so, they become deviants.

MODES OF ADAPTATION	CULTURAL GOALS	INSTITUTIONAL MEANS
CONFORMITY	Accepted	Accepted
DEVIANCE		
Innovation	Accepted	Rejected
Ritualism	Rejected	Accepted
Retreatism	Rejected	Rejected
Rebellion	Rejected/replaced	Rejected/replaced

Second, critics question Merton's assertion that deviance is more characteristic of lower-class people. There is evidence that most lower-class people are able to adjust their goals downward sufficiently so that they can be reached by respectable means (Simons & Gray 1989). In addition, there is overwhelming evidence that many highly successful individuals adjust their goals so far upward that they cannot reach them by legitimate means. The savings and loan scandals of the late 1980s and early 1990s, which revealed that men who earned millions a year were cheating to earn still more, are clear evidence that the means-versus-goals discrepancy is not limited to the lower class.

In spite of these criticisms, sociologists continue to find strain theory both interesting and useful as an explanation of deviance. It underscores the sociological view that society, not the individual, is an important cause of deviant behavior.

Symbolic Interaction Theories

Symbolic interaction theories of deviance suggest that deviance is learned through interaction with others and involves the development of a deviant self-concept. Deviance is not believed to be a direct product of the social structure but of specific face-to-face interactions. This argument takes three forms: differential association theory, deterrence theories, and labeling theory.

Differential Association Theory

In the late 1940s, Edwin Sutherland developed a theory to explain the common observation that kids who grow up in neighborhoods where there are many delinquents are more likely to be delinquent themselves. **Differential association theory** argues that people learn to be deviant when more of their associates favor deviance than favor conformity.

How does differential association encourage deviance? There are two primary mechanisms. First, if our interactions are mostly with deviants, we may develop a biased image of the generalized other. We may learn that, of course, everybody steals or, of course, the ability to beat other people up is the most important criterion for judging a person. The norms that we internalize may be very different from those of conventional society. The second mechanism has to do with reinforcements. Even if we learned conventional norms, a deviant subculture will not reward us for following them.

THINKING

CRITICALLY

Can differential association theory explain why some boys/girls who grow up in bad neighborhoods do not become delinquents? Can you?

Differential association theory argues that people learn to be deviant when more of their associates favor deviance than favor conformity.

In fact, a deviant subculture may reward us for violating the norms. Through these mechanisms, we can learn that deviance is acceptable and rewarded.

Differential association theory stems largely from the structural school of symbolic interaction. People develop a deviant identity because they are thrust into a deviant subculture. The situation determines the identity.

Deterrence Theories

Many contemporary scholars use some form of deterrence theory to explain deviance. **Deterrence theories** suggest that deviance results when social sanctions, formal and informal, provide insufficient rewards for conformity. Deterrence theories combine elements of structural-functional and symbolic interaction theories. Although they place the primary blame for deviance on an inadequate (dysfunctional) sanctioning system, they also assign the individual an active role in choosing whether to deviate or conform. This theory assumes that the actor assesses the relative balance of positive and negative sanctions and makes a cost/benefit decision about whether to conform or be deviant (Piliavin et al. 1986; Paternoster 1989). When social structures do not provide adequate rewards for conformity, a larger portion of the population will choose deviance.

Empirical studies show that three kinds of rewards are especially important in deterring deviance: instrumental rewards, family ties, and self-esteem.

INSTRUMENTAL REWARDS. Unemployment and low wages are among the very best predictors of crime rates at any age (Devine, Shaley, & Smith 1988; Crutchfield 1989). People with no jobs or with dead-end jobs have little to lose and perhaps much to gain from deviance. People who have or can look forward to good jobs, on the other hand, are likely to conclude that they have too much to lose by being deviant.

FAMILY TIES. Consistent evidence shows that young people with strong bonds to their parents are more likely to conform (Hirschi 1969; Messner & Krohn 1990). Parents are in a very strong position to exert informal sanctions that encourage conformity, and young people who are close to their parents are vulnerable to these informal sanctions. There is an important corollary to this rule; deviant parents often produce delinquent children, in large part because they use harsh and inconsistent punishment and form only weak family attachments. Interestingly, family ties have even stronger effects on delinquency in China than in the West probably because in China the family remains the most important institution of social control (Zhang & Messner 1995).

SELF-ESTEEM. On a more symbolic level, deterrence theory suggests that people choose deviance or conformity depending on which will do the most to enhance their self-esteem (Kaplan, Martin, & Johnson 1986). For most of us, self-esteem is enhanced by conformity; we are rewarded for following the rules. People whose efforts are not rewarded, however, many find deviance an attractive alternative in their search for positive feedback. Especially among lower-class boys, delinquency has been found to be a means of improving self-esteem (Rosenberg, Schooler, & Schoenbach 1989).

According to deterrence theorists, positive sanctions give individuals a "stake in conformity"—something to lose, whether it's a job, parental approval, or self-esteem. When social structures fail to reward conformity, individuals have less to lose by choosing deviance.

Deterrence theories suggest that deviance results when social sanctions, formal and informal, provide insufficient rewards for conformity.

Differential association theory points out that people who grow up in crime-ridden neighborhoods are more likely to grow up to be criminals themselves. This is true both because there are peers who can teach a person how to commit crime and because he or she receives more rewards for violating conventional norms than for following them. It is easy to see how differential association theory might apply to gang members such as these, but can you think of ways it might also help to explain white-collar crime such as the widespread practice by physicians and hospitals of overbilling insurance companies and Medicare?

Labeling Theory

A third theory of deviance that combines symbolic interaction and conflict theories is labeling theory. **Labeling theory** is concerned with the processes by which the label *deviant* comes to be attached to specific people and specific behaviors. This theory takes to heart the maxim that deviance is relative. As the chief proponent of labeling theory puts it, "Deviant behavior is behavior that people so label" (Becker 1963, 90).

The process through which a person becomes labeled as deviant depends on the reactions of others toward nonconforming behavior. The first time a child acts up in class, it may be owing to high spirits or a bad mood. This impulsive act is *primary deviance*. What happens in the future depends on how others interpret the act. If teachers, counselors, and other children label the child a troublemaker *and* if she accepts this definition as part of her self-concept, then she may take on the role of a troublemaker. Continued rule violation because of a deviant self-concept is called *secondary deviance*.

This explanation of deviance fits in neatly with the structural school of symbolic interactionism. Deviance becomes yet another role identity that is integrated into the self-concept.

POWER AND LABELING.   A crucial question for labeling theorists is the process by which an individual comes to be labeled *deviant*. Many labeling theorists take a conflict perspective when answering this question. They assume that one of the strategies groups use in competing with one another is to get the other groups' behavior labeled as deviant while protecting its own behavior. Naturally, the more power a group has, the more likely it is to be able to brand its competitors deviant. This, labeling theorists allege, explains why lower-class deviance is more likely to be subject to criminal sanctions than is upper-class deviance.

In a classic study, Becker (1963) describes how this competition between interest groups caused marijuana users in the United States to be labeled *deviant*. Before 1937, marijuana use was not illegal in the United States. In 1937, however, a powerful vested-interest group, the Federal Bureau of Narcotics, campaigned to have it declared illegal. (Since Prohibition had ended, the bureau either had to find a new enemy or go out of business.) In conjunction with another vested-interest group, the Consolidated Brewers,

Labeling theory is concerned with the processes by which labels such as *deviant* come to be attached to specific people and specific behaviors.

Although anything is possible, one would be pretty safe in predicting that these young women will manage to stay out of trouble. Why? Because they receive significant rewards for conventional behavior. They have medals to display and friends and family who admire them. People who receive many rewards for following the rules are much less tempted by opportunities for deviance.

the FBN launched a major media campaign to stigmatize marijuana use by associating marijuana with violence and other criminal behaviors. As a result of its successful campaign, it created a new group of deviants. Becker refers to those who are in a position to create and enforce new definitions of morality as **moral entrepreneurs.**

Moral entrepreneurs are people who are in a position to create and enforce new definitions of morality.

FROM SIN TO SICKNESS.   Labeling theory's emphasis on subjective meanings gives us a framework for understanding the changing definitions of deviance. In recent years, there has been an increasing tendency for behaviors that used to be labeled *deviant* to be labeled *illnesses* instead. For example, many now consider alcoholism to be a disease. When a form of deviance comes to be viewed as illness, social reaction changes. It is no longer appropriate to put people in jail for being public drunks; instead they are put in hospitals. Physicians and counselors, rather than judges and sheriffs, treat them. Other forms of deviance, such as child abuse, gambling, murder, and rape, also may be regarded as forms of mental illness that are better treated by physicians than sheriffs (Link 1987; Rosecrance 1985). At present we have reached few firm decisions on these issues. The public seems to believe that although some murderers, rapists, and so on are mentally ill and should be treated by physicians, others are just bad and should be put in jail.

Individuals who acquire *sick* rather than *bad* labels are entitled to treatment rather than punishment and are allowed to absolve themselves from blame for their behavior (Conrad & Schneider 1980). As you might expect, people in positions of power are more apt to be successful in claiming the sick label. For example, the upper-class woman who shoplifts is likely to be labeled *neurotic,* whereas the lower-class woman who steals the same items is likely to be labeled *shoplifter.* The middle-class boy who acts up in school may be defined as hyperactive, the lower-class boy as a troublemaker.

EVALUATION.   Labeling theory combines elements of symbolic interaction and conflict theory. The deviant label becomes part of the self-concept, affecting further interaction. But this deviant label is imposed by powerful others rather than being self-selected.

The theory has become extremely popular in the last 20 years and has been applied in diverse situations (Chapter 12 discusses labeling in the schools, for example). It does, however, have some important limitations. One critic sums these up by saying that labeling theory gives the impression that this innocent guy was just walking along when, wham, society stuck the label *deviant* on him, and after that he had no choice but to cause trouble (Akers 1968).

More formally, labeling theory is criticized because (1) it doesn't explain primary deviance; (2) its emphasis on the relativity of deviance suggests that the only thing wrong with murder or assault is that someone arbitrarily called it deviant; and (3) it cannot explain repeated deviance by those who haven't been caught, that is, labeled.

Conflict Theory

Conflict theory proposes that competition and class conflict within society create deviance. Class conflict affects deviance in two ways (Archer 1985): (1) Class interests determine which acts are criminalized and how heavily they are punished. (2) Economic pressures lead to offenses, particularly property offenses, among the poor.

Defining Crime

The conflict perspective on defining crime has already been described in the section on labeling theory. Marxists argue that the law is a weapon used by the ruling class to maintain the status quo (Liska, Chamlin, & Reed 1985). This interpretation fits in

Posters such as this one were widely distributed in the 1930s in the successful attempt to criminalize marijuana use. Who sponsored this campaign? The answer is on the bottom line of the poster: the Consolidated Brewers Association of America. It doesn't take a financial genius to figure out why the breweries opposed marijuana use. This is one of the more obvious examples of the extent to which what is legal and what is illegal depends more on politics and economics than on unambiguous moral codes.

CONCEPT SUMMARY | *Theories of Deviance*

	MAJOR QUESTION	MAJOR ASSUMPTION	CAUSE OF DEVIANCE	MOST USEFUL FOR EXPLAINING DEVIANCE OF
Structural-Functional Theory				
Strain theory	Why do people break rules?	Deviance is an abnormal characteristic of the social structure	A dislocation between the goals of society and the means to achieve them	The working and lower classes who cannot achieve desired goals by prescribed means
Symbolic Interaction Theories				
Differential association theory	Why is deviance more characteristic of some groups than others?	Deviance is learned like other social behavior	Subcultural values differ in complex societies; some subcultures hold values that favor deviance. These are learned through socialization.	Delinquent gangs and those integrated into deviant subcultures and neighborhoods
Deterrence theories	When is conformity not the best choice?	Deviance is a choice based on cost/benefit assessments	Failure of sanctioning system (benefits of deviance exceed the costs)	All groups, but especially those lacking a "stake in conformity"
Labeling theory	How do acts and people become labeled *deviant?*	Deviance is relative and depends on how others label acts and actors	People whose acts are labeled *deviant* and who accept that label become career deviants	The powerless who are labeled *deviant* by more powerful individuals
Conflict theory	How does unequal access to scarce resources lead to deviance?	Deviance is a normal response to competition and conflict over scarce resources	Inequality and competition	All classes: Lower class is driven to deviance to meet basic needs and to act out frustration; upper class uses deviant means to maintain their privilege

with the general Marxist notion that all social institutions, including law, have been created to rationalize and support the current distribution of economic resources.

Supporters of this position note that we spend more money deterring muggers than embezzlers. We impose much more severe sentences for street crimes than corporate crimes. We are more likely to arrest those who assault members of the ruling class (well-off whites) than we are to arrest those who assault the powerless (nonwhites and the poor) (Smith 1987). Finally, even when people from the upper and lower classes commit similar crimes, those from the lower class are more likely to be arrested, prosecuted, and sentenced (Williams & Drake 1980). The system clearly seems to benefit the upper classes.

Many Marxists deny that crime is more prevalent among the poor. They argue that in fact the well-off are the least conforming group in the population; it is the rich rather than the working class who flout convention (Sorokin & Lundin 1959). Because of their control of the labeling apparatus—the state, the schools, the courts—the upper class has been able to avoid deviant labels.

Lower-Class Crime

Although the preceding view of the way crime is defined would be accepted by all Marxists, some believe that the lower class really is more likely to commit criminal acts.

One Marxist criminologist has declared that crime is a rational response for the lower class (Quinney 1980). These criminologists generally seem to agree with Merton's strain theory that a means/ends discrepancy is particularly acute among the poor and that it may lead to crime (Greenberg 1985). They believe, however, that this is a natural condition of an unequal society rather than an unnatural condition.

Summary

There are many theories of deviance in the field of sociology. These reflect differences in basic theoretical assumptions as well as differences in the kinds of deviance they try to explain (see the Concept Summary). All, however, are sociological, not psychological or biological, theories: They place the reasons for deviance within the social structure rather than within the individual. In the following sections, we apply these theories as we review major differentials in U.S. crime rates.

Crime

Most of the behavior that is regarded as deviant or nonconforming is subject only to informal social controls. **Crimes** are acts that are subject to legal penalties. Most, though not all, crimes violate social norms and are subject to informal as well as legal sanctions. In this section, we briefly define the different types of crimes, look at crime rates in the United States, and examine the findings about who is most likely to commit these crimes.

Crimes are acts that are subject to legal or civil penalties.

Index Crimes: Major Crimes Involving Violence or Property

Each year the federal government publishes the *Uniform Crime Report* (UCR), which summarizes crimes known to the police for eight major index crimes (U.S. Department of Justice 1996a):

- *Murder and nonnegligent manslaughter.* Overall, murder is a rare crime; yet some segments of society are touched by it much more than others; more than 48 percent of all murder victims in 1995 were African American and 77 percent were male.
- *Rape.* Rape accounts for about 5 percent of all violent crimes, and reported rapes have doubled in the last two decades. Nearly 97,500 women were raped in 1995.
- *Robbery.* Robbery is defined as taking or attempting to take anything of value from another person by force, threat of force, violence, or by putting the victim in fear. Unlike simple theft or larceny, robbery involves a personal confrontation between the victim and the robber and is thus a crime of violence.
- *Assault.* Aggravated assault is an unlawful attack for the purpose of inflicting severe bodily injury. Kicking and hitting are included in assault, but in 74 percent of the cases assault involves a weapon. In 1995, however, the number of assaults involving firearms decreased by 9 percent.
- *Burglary, larceny-theft, motor-vehicle theft,* and *arson* are the four property crimes included in the UCR. (Arson has only been added recently and is not covered in the trend data in Figure 5.1.) Property crimes are much more common than crimes of violence and account for 87 percent of the crimes covered in the UCR.

It is a common public perception that crime rates are much higher than they used to be. The accuracy of this perception depends on the time frame one uses and the specific

FIGURE 5.1
Changes in Violent Crime Rates, 1973–1994

Violent crime rose during the 1970s. After a dramatic slump in the early 1980s, violent crime rates rose through the early 1990s. Assault and rape rates are near record levels.

SOURCE: U.S. Department of Justice 1995a.

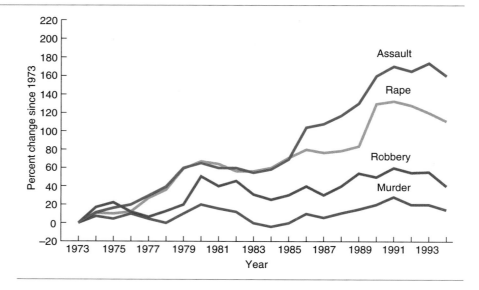

crime. Figures 5.1 and 5.2 depict trends in seven index crimes over the last 20 years. They show that crime rates are indeed higher than they were in 1973—in the case of rape and assault, twice as high. After a general slump in crime rates during the first years of the 1980s, all the UCR index crimes except burglary began to rise. In 1992, crime began to decrease and this trend has continued through to present.

Victimless Crimes

Victimless crimes such as drug use, prostitution, gambling, and pornography are voluntary exchanges between persons who desire goods or services from each other.

The so-called **victimless crimes**—such as drug use, prostitution, gambling, and pornography—are voluntary exchanges between persons who desire goods or services from one another (Schur 1979). They are called victimless crimes because participants in the exchange typically do not see themselves as being victimized or as suffering from the transaction: There are no complaining victims.

FIGURE 5.2
Changes in Property Crimes, 1973–1994

Property crime increased throughout the 1970s. Although all property crime rates dropped in the early 1980s, larceny-theft and motor-vehicle theft rates have returned to or exceeded their previous high levels. Burglary rates remain substantially lower than they were 10 years ago.

SOURCE: U.S. Department of Justice 1995a.

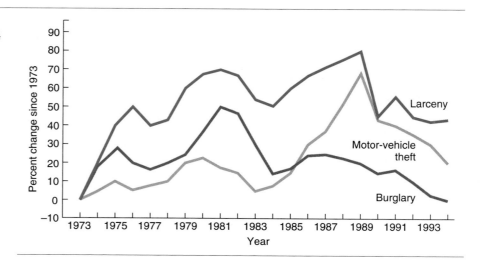

There is substantial debate about whether these crimes are truly victimless. Many argue that prostitutes, drug abusers, and pornography models *are* victims (Chapman & Gates 1978; Dworkin 1981): Even though there is an element of choice in the decision to engage in these behaviors, individuals are usually forced or manipulated into it by their disadvantaged class position. Others believe that such activities are legitimate areas of free enterprise and free choice and that the only reason these acts are considered illegal is because of some self-righteous busy-bodies (Jenness 1990).

Because there are no complaining victims, these crimes are difficult to control. The drug user is generally not going to complain about the drug pusher, and the illegal gambler is unlikely to bring charges against a bookie. In the absence of a complaining victim, the police must find not only the criminal but also the crime. Efforts to do so are costly and divert attention from other criminal acts. As a result, victimless crimes are irregularly and inconsistently enforced, most often in the form of periodic crackdowns and routine harassment.

White-Collar Crimes

Crime committed by respectable people of high social status in the course of their occupation is called **white-collar crime** (Sutherland 1961).

White-collar crime occurs at several levels. It is committed, for example, by employees against companies, by companies against employees, by companies against customers, and by companies against the public (for example, by dumping toxic wastes into the air, land, or water). When companies are the perpetrators white-collar crime is often referred to as *corporate crime*. Sometimes, corporate crime more closely parallels racketeering and organized crime than it does anything else. In the wake of a bailout that will cost taxpayers billions of dollars, investigators now believe that many failed savings-and-loan institutions were created for the sole purpose of generating corporate and personal profit from illegal activity. By definition, the only thing distinguishing this type of corporate crime from organized crime is that the thrift embezzlers wore Brooks Brothers suits and expensive white shirts while "racketeers," at least in popular mythology, often wear black shirts and white ties (Calavita & Pontell 1993).

Because most white-collar crime goes unreported, its total economic cost is difficult to assess. However, most scholars and law enforcement officials believe that the dollar loss due to corporate crime dwarfs that of street crime. In addition to the economic cost, there are social costs as well. Exposure to repeated tales of corruption tends to breed distrust and cynicism and, ultimately, to undermine the integrity of social institutions. If you think that all members of Congress are crooks, then you quit voting. If you think that every police officer can be bought, then you cease to respect the law. Thus, the costs of such crime go beyond the actual dollars involved in the crime itself.

The reasons for white-collar crime tend to be about the same as for street crimes: People want more than they can legitimately get and think the benefits of a crime outrun its potential costs (Coleman 1988). Differential association also plays a role. In some corporations, organizational culture winks at or actively encourages illegal behavior. Speaking of the insider trading scandals that have rocked Wall Street recently, one observer notes:

> You gotta do it. . . . Everybody else is. [It] is part of the business. . . . You work at a deli, you take home pastrami every night for free. It's the same thing as information on Wall Street. . . . I know you want to help your mother and provide for your family. This is the way to do it. Don't be a schmuck. Nobody gets hurt (cited in Reichman 1989).

Jim McDougal was convicted of fraud in the Arkansas Whitewater scandal. Sentenced to 2 years in prison, 1 year of house arrest, and several million dollars in restitution, this sounds like a stiff penalty for a crime few people are sure they understand. However, McDougal and his associates probably made a substantial profit from their illegal activities, and his prison sentence does not compensate taxpayers for the government dollars that will ultimately be spent either on prosecuting the crime or on bailing this and other failed savings and loans out. Although it may, in fact, be more costly than street crime, white-collar crime rarely provokes the same gut reaction as robbery or burglary, and there is less public demand to get tough on white-collar criminals.

White-collar crime is crime committed by respectable people of high status in the course of their occupation.

THINKING

CRITICALLY

Why do you think most Americans view street crime as more serious than corporate crime? What would a conflict theorist say? A structural functionalist?

The magnitude of white-collar crime in our society makes a mockery of the idea that crime is predominantly a lower-class phenomenon. Instead, it appears that people of different statuses simply have different opportunities to commit crime. Those in lower statuses are hardly in the position to engage in price fixing, stock manipulation, or tax evasion. They are in a position, however, to engage in high-risk, low-yield crimes such as robbery and larceny. In contrast, higher status individuals are in the position to engage in low-risk, high-yield crimes (Schur 1979). Because of the complexity of the transactions involved, white-collar crime is difficult to detect. Even if detected, white-collar offenders usually receive more lenient treatment than street criminals. For example, a recent study found that health-care professionals guilty of Medicaid fraud were much less likely to be incarcerated than persons charged with grand theft, despite the fact that the dollar losses from the Medicaid crimes, on the average, were much greater (Tillman & Pontell 1992).

Marxist critics argue that the absence of white-collar crime statistics from the UCR and the relative absence of white-collar criminals from our prisons reflect the fundamental class bias in our criminal justice system (Braithwaite 1985).

Correlates of Crime: Age, Sex, Class, and Race

Only 21 percent of the crimes reported in the UCR are cleared by an arrest. This means that the people arrested for criminal acts represent only a sample of those who commit reported crimes; they are undoubtedly not a random sample. The low level of arrests coupled with the low levels of crime reporting warn us to be cautious in applying generalizations about arrestees to the larger population of criminals. With this caution in mind, we note that the persons arrested for criminal acts are disproportionately male, young, and from minority groups. Figure 5.3 shows the pattern of arrest rates in 1995 by sex and age. As you can see, crime rates, especially for men, peak sharply during ages 15–24; during these peak crime years, young men are about four times more likely to be arrested than women of the same age. Minority data are not available by age and sex, but the overall rates show that African Americans and Hispanics are over three times more likely than whites to be arrested.

What accounts for these differentials? Can the theories reviewed earlier help explain these patterns?

FIGURE 5.3
Arrest Rates by Age and Sex, 1994
Arrest rates in the United States and most other nations show strong and consistent age and sex patterns. Arrest rates peak sharply for young people ages 15–24; at all ages, men are about four times more likely than women to be arrested.

SOURCE: Arrest rate data are from Uniform Crime Reports 1995, Ts 40–41; Population base data come from the U.S. Statistical Abstract 1995, T. 21.

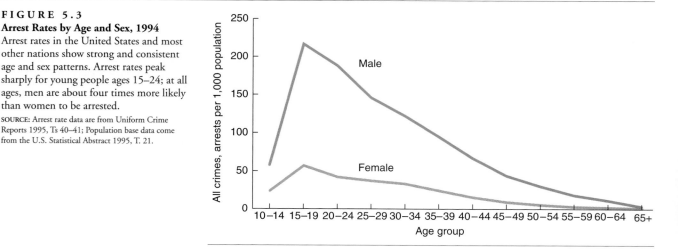

Age Differences

The age differences in arrest rates noted in Figure 5.3 are characteristic of nearly every nation in the world that gathers crime statistics (Hirschi & Gottfredson 1983). A great deal of controversy exists over the reasons for the very high arrest rates of young adults, but deterrence theories have the most promise for explaining this age pattern.

In many ways, adolescents and young adults have less to lose than other people. They don't have a "stake in conformity"—a career, a mortgage, a credit rating (Steffensmeier et al. 1989). When young people do have jobs, and especially when they have good jobs, their chances of getting into trouble are much smaller (Allan & Steffensmeier 1989).

Delinquency is basically a leisure-time activity. It is strongly associated with spending large blocks of unsupervised time with peers (Agnew & Petersen 1989; Osgood & Wilson 1989). When there is "nothing better to do," a substantial portion of young people will get in trouble. On the other hand, deviance is deterred by spending a lot of time with one's parents or with conforming peers (Gardner & Shoemaker 1989).

Sex Differences

The sex differential in arrest rates has both social and biological roots. Women's smaller size and lesser strength make them less likely to use violence or personal confrontation; they have learned that, for them, this is an ineffective strategy. Evidence linking male hormones to aggressiveness indicates that biology also may be a factor in women's lower inclination to engage in violent behavior.

Among social theories of deviance, deterrence theory seems to be the most effective in explaining these differences. Generally, girls are supervised more closely than boys, and they are subject to more social control; this is especially true in the lower class (Hagan, Gillis, & Simpson 1985; Heidensohn 1985; Thompson 1989). Whereas parents may let their boys wander about at night unsupervised, they are much more likely to insist on knowing where their daughters will be and with whom they will be associating. The greater supervision that girls receive increases their bonds to parents and other conventional institutions; it also reduces their opportunity to join gangs or other deviant groups.

These explanations raise questions about whether changing roles for women will affect women's participation in crime. Will increased equality in education, labor-force participation, smoking, drinking, also show up in greater equality of criminal behavior? The evidence shows only a modest tendency for this to happen. It is true that the crime rate for women has increased faster than the crime rate for men in a few areas (vagrancy, aggravated assault, gambling, and some property crimes). Female drug-related arrest rates have also increased faster than men's, but their arrests are primarily for possession and not for drug dealing. Indeed, research shows that women may be even less involved in the lucrative crack-selling marketplace of the late 1980s and may be even more vulnerable than they were in the heroin markets of even a few years earlier (Maher & Daly 1996). Furthermore, female arrest rates overall may actually still be lower now than they were in the mid-1800s (Boritch & Hagan 1990).

Differences by Social Class

The effect of social class on crime rates is complex. Although sociologists have historically held that social class is an important correlate of criminality (Braithwaite 1981; Elliott & Ageton 1980; Thornberry & Farnworth 1982), some studies have found that the relationship is not very strong and in some cases is nonexistent (Hirschi 1969; Johnson 1980; Krohn et al. 1980; Tittle & Meier 1990). Much of the inconsistency appears to center on difficulties in measuring both social class and crime.

Braithwaite's (1981) review of more than 100 studies leads to the conclusion that lower-class people commit more of the direct interpersonal types of crimes normally handled by the police than do people from the middle class. These are the types of crimes reported in the UCR. Middle-class people commit more of the crimes that involve the use of power particularly in the context of their occupational roles: fraud, embezzlement, price fixing, and other forms of white-collar crime. There is also evidence that the social-class differential may be greater for adult crime than for juvenile delinquency (Thornberry & Farnworth 1982).

Nearly all the deviance theories we have examined offer some explanation of the social-class differential. Strain theorists and some Marxists suggest that the lower class is more likely to engage in crime because of blocked avenues to achievement. Deterrence theorists attribute greater crime among the lower class to the fact that these people may be receiving fewer rewards from conventional institutions such as school and the labor market. All of these theories accept and seek to explain the social-class pattern found in the UCR, where, indeed, the lower class is overrepresented. Labeling and Marxist theories, on the other hand, argue that this overrepresentation is not a reflection of underlying social-class patterns of deviance but of bias in the law and within social control agencies (Williams & Drake 1980). Evidence suggests, for instance, that the disproportionately high lower-class homicide rates found in most modern societies result from governmental failure to provide the least privileged with the same legal means of conflict resolution as is typically provided to the social elite (Cooney 1997). However, overrepresentation of the lower class also reflects the particular mix of crimes included in the UCR; if embezzlement, price fixing, and stock manipulations were included in the UCR, we might see a very different social-class distribution of criminals.

Differences by Race

Although African Americans compose only 12 percent of the population, they make up 37 percent of those arrested for murder, 41 percent of those arrested for rape, and 38 percent of those arrested for assault. Hispanics, who compose about 8 percent of the total population, represent about 28 percent of those imprisoned for violent crimes. These strong differences in arrest and imprisonment rates are explained in part by social-class differences between minority and Anglo populations. Even after this effect is taken into account, however, African and Hispanic Americans are still much more likely to be arrested for committing crimes.

The explanation for this is complex. As we will document in Chapter 8, race continues to represent a fundamental cleavage in U.S. society. The continued and even growing correlation of race with unemployment, inner-city residence, and female-headed households reinforces the barriers between African Americans and whites in U.S. society. An international study confirms that the larger number of overlapping dimensions of inequality, the higher the "pent-up aggression which manifests itself in diffuse hostility and violence" (Messner 1989). The root cause of higher minority crime rates, from this perspective, is the low quality of minority employment—a factor that leads directly to unstable families and neighborhoods (Sampson 1987; Sampson & Groves 1989).

Poverty and segregation combine to put African American children in the worst neighborhoods in the country, neighborhoods where getting into trouble is a way of life and where conventional achievement is remote (Matsueda & Heimer 1987). Differential association theory thus explains a great deal of the racial difference in arrest rates. Deterrence theory is also important. African American children are much more likely to live in a fatherless home and thus lack an important social bond that might deter deviant behavior. In addition to these factors that may increase the propensity to deviance among minorities, there is also evidence that whether we are talking about troublemaking in school, stealing cars, or petty theft, minority-group members are more likely than Anglos to be labeled *deviant* and, if apprehended by the police, more apt to be cited, prosecuted and convicted (Unnever, Frazier, & Henretta 1980; Peterson 1988).

The Sociology of Law

We have reviewed theories about deviance and examined current findings about crime and criminals. In the last sections of this chapter, we will look at the formal mechanisms of social control. Among the questions of interest are: Why punish? What is a just punishment? How does the criminal justice system work? Can we reduce crime? We begin by taking a broad theoretical overview of the sociology of law.

Theories of Law

The cornerstone of the formal control system is the law. Generally, law is seen as serving three major functions: It provides formal sanctions to encourage conformity and discourage deviance, it helps settle disputes, and it may be an instrument for social change (Vago 1989). Beyond this simple summary, there is substantial discussion among scholars about how the law operates.

Most citizens, and probably even most sociologists, take a general structural-functional approach to law (Rich 1977). By clearly spelling out expected behaviors and punishing violators, the law helps maintain society. Although some may benefit more than others from particular laws and laws may be unequally enforced, law itself is a good thing, a benefit to society.

Conflict theorists, of course, take a somewhat different position: They suggest that the legal apparatus was designed to maintain and reproduce the system of inequality. Law, in this view, is a tool used by elites to dominate and control the lower orders (Chambliss 1978).

Both perspectives have obvious merit. Although law does serve the general interest by maintaining order, it is not surprising to find that it serves some interests better than others. The relationship between law and inequality is a central concern of sociologists of law.

What Is Justice?

Justice is an enormously difficult concept to define. Some argue that justice is served when everyone is treated equally, for example, when everyone who commits first-degree murder gets 30 years, no exceptions and no parole. Others believe that justice should be more flexible. They believe that circumstances should make a difference.

In this as in many other fields, Max Weber's contributions are insightful. Weber ([1914] 1954) distinguished between two types of legal procedures: rational and substantive. Rational law is based on strict application of the rules, regardless of fairness in specific cases. Substantive law, on the other hand, takes into account the unique circumstances of the individual case. For example, although the penalty for motor-vehicle homicide might be three years, substantive law might levy a lower penalty on a grief-stricken father who has killed his child than on someone who has killed a stranger.

Studies of actual sentencing outcomes suggest that law tends to be much more substantive than rational. If law were rational, one would expect that sentences would be highly correlated with the nature of the crime; they are not. If law were a tool of the elite, one would expect sentences to be affected by the race and class of the offender and the victim. There is evidence that death penalty decisions *are* affected by

Police work involves a great deal of discretion. In most situations, the police officer is out on her or his own, away from supervision and direction, and must make snap decisions about whether to pursue violations or disregard them. Because we recognize that there are not enough police officers to pursue every violation, we hope officers will use good judgment about what kinds of violations and violators are worth pursuing, and will not let prejudice or bias affect their decisions.

Street-level justice consists of the decisions the police make in the initial stages of an investigation.

the race of the victim; convicted murderers, for example, are less likely to get the death penalty if the victim was black (Radelet 1989). The most general conclusion, however, is that sentencing has only a rough association with the crime or the characteristics of the victim or offender. Decisions seem to depend more on the individual judge than on any characteristic of a particular case.

The Criminal Justice System

The formal mechanisms of social control mentioned at the beginning of the chapter are administered through the criminal justice system. In the United States, this system consists of a vast network of agencies set up to deal with persons who deviate from the law: police departments, probation and parole agencies, rehabilitation agencies, criminal courts, jails, and prisons.

The Police

Police officers occupy a unique and powerful position in the criminal justice system because they are empowered to make arrests in a context of low visibility: Often there are no witnesses to police encounters with suspected offenders. Although they are supposed to enforce the law fully and uniformly, everyone realizes that this is neither practical nor possible. In 1995 there were 2.3 full-time law enforcement officers for every 1,000 persons in the nation (U.S. Department of Justice 1996a). This means that the police ordinarily must give greater attention to more serious crimes. Minor offenses and ambiguous situations are likely to be ignored.

Police officers have a considerable amount of discretionary power in determining the extent to which the policy of full enforcement is carried out. Should a drunk and disorderly person be charged or sent home? Should a juvenile offender be charged or reported to parents? Should a strong odor of marijuana in an otherwise orderly group be overlooked or investigated? The decisions the police make in the initial stages of an investigation are called **street-level justice.** Unlike the justice meted out in courts, street-level justice is relatively invisible and thus hard to evaluate (Smith & Visher 1981).

The Courts

Once arrested, an individual starts a complex journey through the criminal justice system. This trip can best be thought of as a series of decision stages. There is considerable attrition as defendants pass from arrest to prosecution to sentencing and punishment. Even in felony cases, as many as 40 to 50 percent of those arrested will not be prosecuted because of problems with evidence or witnesses (Brossi 1979). At the same time, approximately 90 percent of all convictions are the result of pretrial negotiations (Figueira-McDonough 1985). This means that only about 10 percent of criminal convictions are processed through public trials. Thus, the pretrial phases of prosecution are often more crucial to arriving at judicial decisions of guilt or innocence than are court trials themselves. Like the police, prosecutors have considerable discretion in deciding whom to prosecute and on what charges.

Throughout the entire process, the prosecution, the defense, and the judges participate in negotiated plea bargaining. The accused is encouraged to plead guilty in the

technology

DOES IT TAKE A ROCKET SCIENTIST TO DO POLICE WORK?

When a young Manhattan Beach police officer was killed during what seemed to be a routine traffic stop, videotapes from the surveillance cameras at an ATM near the crime scene could provide investigators with blurry clues only. At a nearby aerospace lab, some of America's brightest scientists and engineers were hard at work developing systems that could produce clear images of objects hundreds and thousands of miles away. When one of the scientists heard about the case, he offered his company's image enhancement services to police investigators. Ultimately, aerospace engineers, using technology developed for the space program, helped clarify the images taken at the ATM and, in turn, were able to identify the car used by the killer, giving police a real break in the case (Preimsberger 1996).

While such specific partnerships between cops and rocket scientists remain relatively rare, technology is increasingly changing the way that law enforcement work is done. In 1964, only one city, St. Louis, had a police computer system; by 1968, ten states had statewide criminal justice information systems. Today, police departments in nearly every city with a population of over 50,000 people have computer support services (Siegel 1995), and many are acquiring even more sophisticated systems. The Dick Tracy or Dragnet image of the detective—notepad and pencil in hand, carefully sketching where the body lay and where the bullets entered the wall—is increasingly being replaced by one of a police technician, using a palmtop computer and a computer-aided drafting system to create a three-dimensional reconstruction of the crime scene (Breuninger 1995). ID imaging now permits many police officers to obtain photo identification and fingerprint images without the need for transporting a suspect to the station. Prints and mugshots can be immediately cross-checked over wide-ranging networks to find out whether the suspect has a previous record or an outstanding warrant in almost any jurisdiction. With integrated ballistics identification systems, police officers can determine whether a spent bullet or cartridge casing originated from a weapon that has previously been used in the commission of a crime; even when no one is hurt, information about the cas-

ings and bullets can be maintained in the system, allowing detectives to track an individual's activity over time. This aspect of computerized ballistics systems is particularly useful to police in their efforts to monitor gang activity (Boyle & O'Connor 1996). The list of future technological applications in law enforcement is long: Currently being considered are helmet-mounted facial identification systems, miniaturized chemical detection sensors, suspect disabling systems, and motion detectors that could be used to alert police and the community to unusual or illegal activities in after-hours parks or abandoned buildings.

It is hard to imagine that in the 1920s this was the type of police work August Vollmer had in mind as he issued the first calls for the modernization and professionalization of police work. In fact, some critics claim that the emphasis on professionalization started by Vollmer and his disciples has led modern law enforcement to focus too much on hardware and not enough on police–community relations (Siegel 1995). Such critics might find it ironic, indeed, that the proliferation of one type of technology—videocams—led to the apprehension of a rather unexpected group of offenders: the police officers involved in the Rodney King beating.

interest of getting a lighter sentence, a reduced charge, or, in the case of multiple offenses, the dropping of some charges. In return, the prosecution is saved the trouble of assembling evidence sufficient for a jury trial.

Punishment Rationales

Any assessment of prisons and punishment must come to grips with the issue: Why are we doing this? Before we can assess the adequacy of punishment, we need to be clear about its purpose. Traditionally there have been five major rationalizations for punishment (Conrad 1983):

1. *Retribution.* Society punishes offenders to revenge the victim and society as a whole; this is a form of revenge and retaliation.
2. *Reformation.* Offenders are not punished but rather are corrected and reformed so that they will become conforming members of the community.

3. *Specific deterrence.* Punishment is intended to scare offenders so that they will think twice about violating the law again.
4. *General deterrence.* By making an example of offenders, society scares the rest of us into following the rules.
5. *Prevention.* By incapacitating offenders, society keeps them from committing further crimes.

THINKING

CRITICALLY

Devise a strategy for deterring white-collar or corporate crime.

Today, social control agencies in the United States represent a mixture of these different philosophies and practices. However, as a result of crowded court dockets and prisons, in the 1980s and 1990s some scholars have argued that the goal of prisons and community corrections has shifted from the punishment or rehabilitation of individual criminals to the identification and management of unruly "high risk" groups (Feeley & Simon 1992).

Prisons

For most people, getting tough on crime means locking criminals up and throwing away the key. Presidential politics, the increasing strength of the Republican Party, the rise of conservative religious denominations, and overall public opinion have contributed to rapid expansion in the number of law enforcement officers, paramilitary police units, and incarcerations (Curry 1996; Jacobs & Helms 1996; 1997; Kraska and Kappeler 1997). More than 80 percent of the U.S. public, for instance, wants to make it much harder for people convicted of drug dealing to get parole (U.S. Department of Justice 1996b). In response to political pressures and public demand, the rate of imprisonment has risen sharply in the last 10 years (see Figure 5.4).

As a result, prison populations are soaring. In 1994, there were 1,016,760 people in state and federal prisons—more than three times the number in 1980. Disproportionately, these prisoners are young men who are uneducated, unskilled, poor, and black. The latest figures show that 22 percent are under age 25, 47 percent are black, and only slightly more than half have graduated from high school. Unemployment rates prior to arrest are nearly four to five times the national average (U.S. Department of Justice 1996b; U.S. Census Bureau 1996a).

The sharp increase in the use of prison to control crime has resulted in a crisis in prison conditions. Many facilities are housing twice as many inmates as they were

FIGURE 5.4

Rates of Imprisonment per 1,000 Index Crimes and per 100,000 Population

Since 1970, there has been a sharp upturn in our use of prison sentences to control crime. Both the number of prisoners per crime and the number of prisoners per 100,000 population have doubled since 1970. The consequence is a "crisis in penalty." The economic and social costs of imprisonment may be so high that we must consider alternative strategies.

SOURCE: U.S. Department of Justice 1989b, 6: U.S. Bureau of the Census 1989a, 183, U.S. Department of Justice 1993b, 608; U.S. Bureau of the Census, 1996 Statistical Abstract, no. 350.

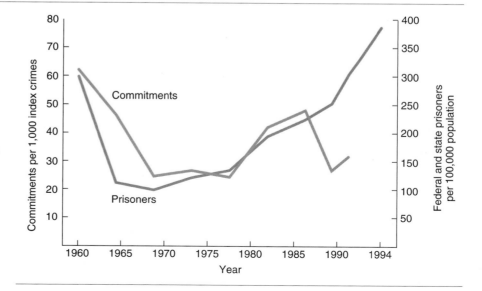

An Application

focus on

IS CAPITAL PUNISHMENT RACIST?

In 1972 in the case of *Furman v. Georgia,* three black defendants appealed their death sentences to the U.S. Supreme Court on the grounds that capital punishment, at least in cases of rape and murder, constituted cruel and unusual punishment. Their argument was that other defendants, many of whom were white, committed equally or more serious crimes and were not sentenced to death. There was, in fact, good statistical support for their claim that capital punishment is racist. Between 1930 and 1967, 54 percent of all civil executions involved nonwhite offenders. Eighty-nine percent of those executed for rape during this time period were black (Radelet 1981). These figures are clearly much higher than the percentage of nonwhites in the general population and also much higher than the percentage of nonwhites among convicted rapists. Between 1945 and 1965, 13 percent of the blacks convicted of rape were sentenced to death, whereas only 2 percent of the convicted whites received a similar sentence (Wolfgang & Reidel 1973). In a 5–4 decision, the Supreme Court agreed with the defendants, holding that the uncontrolled discretion of judges and juries in capital cases denied defendants constitutionally guaranteed rights to due process (Bell 1992).

The *Furman* decision put a temporary stop to capital punishment, but states attempted to solve the problem of disparity in sentencing by passing capital punishment statutes that gave judges and juries less discretion. Have the new, less discretionary laws eliminated earlier racial discrepancies in the administration of the death penalty?

> **Studies continue to show that, even among cases where legal factors are similar, race is a strong determinant of who is sentenced to death.**

Studies conducted in the post-*Furman* era continue to show that, even among cases where legal factors are similar, race is a strong determinant of who is sentenced to death. Fifty-six percent of those executed between 1977 and 1990 were white, 39 percent were black, and 5 percent were Hispanic (Culver 1992). New research shows, however, that the race of the victim is at least as important as that of the defendant. In a rigorous study of sentencing patterns in Georgia, Baldus and his associates found that after statistically controlling for 39 nonracial variables, defendants charged with killing white victims were 4.3 times more likely to

receive a death sentence than those charged with killing a black (Baldus, Pulaski, & Woodworth 1986; Radelet 1989).

On the basis of evidence such as this, Warren McCleskey, a black man convicted and sentenced to death for the 1978 murder of a white police officer appealed to the U.S. Supreme Court, arguing that his sentence was a product of racial discrimination. Speaking for the majority, Supreme Court Justice Powell found that statistical evidence of racial disparity in sentencing was insufficient to "demonstrate a constitutionally significant risk of racial bias." While the Court admitted "some risk of racial prejudice influencing a jury's decision in a criminal case," the level of risk documented by social science research was not "unacceptable." The Court reasoned that the criminal justice system would be immobilized by the "impossible" task of making itself even-handed, if the only test of constitutionality was statistical evidence of arbitrary sentencing. Significantly, the court never disputed the results of the Baldus study and never denied the influence of racial prejudice on capital sentencing in Georgia. Instead, they seemed to accept discrimination as an unfortunate cost of doing law enforcement's business (Bell 1992, 332–337).

---

designed to hold. These overcrowded conditions have been shown to be the chief determinant of violence in prisons (Gaes & McGuire 1985). As a result, prisons in more than 30 states are under court order to reduce crowding and improve conditions. This is an enormously expensive undertaking. The costs are so huge that one observer has called this a "crisis of penalty" (Young, cited in Currie 1989). Most observers agree that "a total commitment to the incarceration of all adult felons . . . cannot be sustained in practice" (Mushane et al. 1989, 137). Do we really want to spend billions and billions of dollars to build more prisons to warehouse a growing proportion of those convicted of crime? Do we need to?

A growing number of empirical studies demonstrate that the certainty of getting caught has more deterrent effect on crime than the length of the sentence (Klepper & Nagin 1989). These findings suggest that we are pursuing the wrong strategy. Rather than building more prisons to warehouse criminals for longer periods of time, we need

Prisons are total institutions where inmates are assigned numbers, wear identical uniforms, live in identical cells, and follow the same routines. They are also environments full of anger, hatred, violence, boredom, and insecurity. In this totally negative environment, prisons become warehouses for the deviant and the violent. They are unlikely environments for rehabilitation.

to put more money into law enforcement. Today, most experts agree that increasing the certainty that criminals will be caught will reduce crime more than will clobbering the few that we do catch.

Community-Based Corrections

Another approach to solving the prison crisis is to change the way we punish convicted criminals. Only one quarter of the convicted offenders under the jurisdiction of social control agencies are actually in jail or prison. The other three quarters are on probation or parole. The public has been generally negative about probation and parole, believing—often rightly—that probation has meant giving criminals a "slap on the wrist" and parole has meant letting criminals out without effective supervision.

As the cost of imprisoning larger numbers of people balloons to crisis proportions, there has been increased interest in effective community-based corrections. New intensive supervision probation (ISP) programs are being used across the country. They include curfews, mandatory drug testing, supervised halfway houses, mandatory community service, frequent reporting and unannounced home visits, restitution, electronic surveillance, and split sentences (incarceration followed by supervised probation) (Lurigio 1990). These programs are alleged to cost only half as much as imprisonment and to be more likely to result in rehabilitation (Mushane et al. 1989). Such claims are still being evaluated.

Social Change

The conservative approach to confronting crime has generally been to increase penalties for convicted criminals. This approach dominated the 1970s and 1980s, which is why prison populations have soared. An alternative approach, which may be gaining renewed momentum, is to address the social problems that give rise to crime. A leading criminologist advocates four major strategies for reducing crime (Currie 1989):

1. Reduce inequality and social impoverishment.
2. Replace unstable, low-wage, dead-end jobs with decent jobs.
3. Enact a supportive national family policy.
4. Increase the economic and social stability of communities.

These strategies would require a massive commitment of energy and money. They are not only expensive but also politically risky. While law-and-order advocates want to get tough on crime by sending more criminals to jail, a policy incorporating Currie's four strategies would channel dollars and beneficial programs into high-crime neighborhoods. Such a policy calls for teachers, not police officers, and good jobs rather than more prisons.

Observers from all sociological perspectives and all political parties recognize that social control is necessary. They recognize that soaring rape and assault rates and the explosion of drug-related crimes are serious problems that must be addressed. The issue is how. The sociological perspective suggests that crime can be addressed most effectively by examining social institutions rather than individual criminals.

Summary

1. Most of us conform most of the time. We are constrained to conform through three types of social control: (a) self-restraint through the internalization of norms and values, (b) informal social controls, and (c) formal social controls.
2. Nonconformity occurs when people violate expected norms of behavior. Acts that go beyond eccentricity and challenge important norms are called deviance. Crimes are a specific kind of deviance for which there are formal sanctions.
3. Deviance is relative. It depends on society's definitions, the circumstances surrounding an act, and the particular groups or subcultures one belongs to.
4. All three major theoretical perspectives in sociology have implications for the explanation of deviance. Structural functionalists use strain theory to blame deviance on social disorganization; symbolic interactionists propose differential association, deterrence, and labeling theories, which lay the blame on interaction patterns that encourage a deviant self-concept; Marxists and other conflict theorists find the cause of deviance in inequality and class conflict.
5. Most crimes are property crimes rather than crimes of violence. Although crime rates dropped in the early 1980s,

index crimes generally have been rising in the last few years. In 1992 all crimes except rape and assault decreased for the first time in almost a decade.
6. Many arrests are for victimless crimes—acts for which there is no complainant. Such crimes are the most difficult and costly to enforce.
7. The high incidence of white-collar crimes, those committed in the course of one's occupation, indicates that crime is not merely a lower-class behavior.
8. Males, minority-group members, lower-class people, and young people are disproportionately likely to be arrested for crimes. Some of this differential is due to their greater likelihood of committing a crime, but it is also explained partly by their differential treatment within the criminal justice system.
9. The sociology of law is concerned with how law is established and how it works in practice. Law reflects economic and political institutions and operates differently depending on social class. In practice, legal decisions are highly variable rather than determined by formal rules.

10. The criminal justice system includes the police, the courts, and the correctional system. Considerable discretion in the execution of justice is available to authorities at each of these levels.

11. The United States faces a "crisis of penalty," as our "get-tough" approach to crime is populating prisons far beyond capacity. Evidence suggests that longer sentences may not be necessary. Alternatives to imprisonment include community-based corrections and social change to reduce the causes of crime.

Sociology on the Net

We are bombarded every day with crime statistics and news reports of criminal behavior. It seems that we are surrounded by crime and it seems to be getting worse. For a real look at what is happening let's abandon the tabloid atmosphere of the mass media for the United States Justice Department for a clearer look at the statistics.

http://www.usdoj.gov

Scroll down the page and open the *Topical Index* section. Browse through the different topics being sure to read the *Press Release, Statistics and the Violence Against Women Home Page.* What kinds of crime does the government seek to eliminate? What crimes are the most common and which are relatively uncommon? What is the government doing to combat violence against women?

Now let's take a good look at the bad guys and gals by returning to the *Topical Index* and opening the section on *Fugitives—Wanted.* Who are the bad guys? Why aren't there more women on these lists?

There are many ways for society to label deviance and crime. One way is to label criminal behavior as illness. The hospital and the treatment center replace the jail and treatment replaces punishment. Drunks used to be tossed into jail. Now they are taken to a detoxification center. This change is referred to as the medicalization of deviance. Today, the treatment of chemical dependency is big business. One of the older and more respected treatment facilities is Hazelden.

http://www.hazelden.com

Begin your tour of this web site by clicking on the *Visit Us* section. Next, return to the home page and open the section entitled *Resource Center* and browse through the topics. You might even try your luck at the Quiz. What is the Hazelden mission? How does Hazelden refer to the people who come for treatment and successfully complete the program? What kinds of deviance are treated at Hazelden? What is the difference between this approach and the one taken at a prison?

FIND IT ON INFOTRAC COLLEGE EDITION

Capital punishment (the death penalty) is a hotly debated issue in the United States today. Using InfoTrac College Edition, look up the following two articles:

"The Case Against the Death Penalty." Eric M. Freedman. *USA Today (Magazine),* March 1997, v125 n2622 p48.

"Death Penalty is a Deterrent." George E. Pataki. *USA Today (Magazine),* March 1997 v125, n2622 p52.

(Hint: Both articles can be found among the first 40 hits after doing a Key Word search using the term *capital punishment.*)

Which argument do you believe to be more accurate? Why? Can you suggest any other strategies designed to deter capital crimes that could be used in conjunction with, or in replacement of, capital punishment?

Suggested Readings

Ben-Yehuda, Nachman. 1985. *Deviance and Moral Boundaries.* Chicago: University of Chicago Press. A wide-ranging coverage of deviance from witchcraft and the occult to cheating in science. A nice balance to the usual emphasis on criminal deviance.

Chambliss, William J., and Seidman, Robert B. 1982. *Law, Order, and Power.* Reading, Mass.: Addison-Wesley. A major text from the conflict perspective.

Cohen, Stanley. 1985. *Visions of Social Control: Crime, Punishment, and Classification.* Cambridge, England: Polity Press. A thoughtful book that considers the rationales that we use to justify punishment and the evidence supporting them. Cohen draws a provocative distinction between doing justice and doing good.

Elliott, Delbert, Huizinga, David, and Ageton, Suzanne. 1985. *Explaining Delinquency and Drug Use.* Beverly Hills: Sage. A research report on this most familiar form of deviance. It includes a good introductory discussion of symbolic interactionist theories of deviance.

Vago, Steven. 1989. *Law and Society.* (2d ed.) Englewood Cliffs, N.J.: Prentice-Hall. Covers the sociology of law, including a historical treatment of theories of the law as well as contemporary issues.

Stratification

Structures of Inequality

Inequality exists all around us. Maybe your mother loves your sister more than you or your brother received a larger allowance than you did. This kind of inequality is personal. Sociologists study a particular kind of inequality called stratification. **Stratification** is an institutionalized pattern of inequality in which social statuses are ranked on the basis of their access to scarce resources.

If your parents gave your brother more money because they decided he was nicer than you, this inequality was not stratification. Inequality becomes stratification when two conditions exist:

1. The inequality is *institutionalized,* backed up by long-standing social norms about what ought to be done.
2. The inequality is based on membership in a status (such as oldest son or blue-collar worker) rather than on personal attributes.

The scarce resources that we focus on are generally of three types: material wealth, prestige, and power. When inequality in one of these dimensions is supported by widely accepted and long-standing social norms and when it is based on status membership, then we speak of stratification.

Types of Stratification Structures

Stratification is present in every society that we know. All societies have norms specifying that some categories of people ought to receive more wealth, power, or prestige than others. There is, however, wide variety in the ways in which inequality is structured.

A key difference among structures of inequality is whether the categories used to distribute unequal rewards are based on ascribed or achieved statuses. As noted in Chapter 2, *ascribed statuses* are those that are fixed by birth and inheritance and are unalterable during a person's lifetime. *Achieved statuses* are optional ones that a person can obtain in a lifetime. Being African American or female, for example, is an ascribed status; being an ex-convict or a physician is an achieved status.

Every society uses some ascribed and some achieved statuses in distributing scarce resources, but the balance between them varies greatly. Stratification structures that rely largely on ascribed statuses as the basis for distributing scarce resources are called **caste systems;** structures that rely largely on achieved statuses are called **class systems.**

Caste Systems

In a caste system, whether you are rich or poor, powerful or powerless, depends entirely on who your parents are. Whether you are lazy and stupid or hardworking and clever makes little difference. Your parents' position determines your own.

This system of structured inequality reached its extreme form in nineteenth-century India. The level of inequality in India was not very different from that in many European nations at the time, but the system for distributing rewards was markedly different. The Indian population was divided into castes, roughly comparable to occupation groups, that differed substantially in the amount of prestige, power, and wealth

Stratification is an institutionalized pattern of inequality in which social statuses are ranked on the basis of their access to scarce resources.

Caste systems rely largely on ascribed statuses as the basis for distributing scarce resources.

Class systems rely largely on achieved statuses as the basis for distributing scarce resources.

THINKING

CRITICALLY

Can you think of any ways in which the U.S. system of stratification resembles a caste system?

How much should a person be paid for removing garbage? Removing garbage is both unpleasant and vital; in large metropolitan areas, garbage strikes have brought cities such as New York to a virtual halt. Nevertheless, because almost anyone can haul garbage, structural-functional theory suggests that wages will still be quite low. Conflict theorists, on the other hand, argue that the low wages of garbage collectors have more to do with their lack of power than with their lack of skill.

they received. The distinctive feature of the caste system is that caste membership is unalterable; it marks one's children and one's children's children. The inheritance of position was ensured by rules specifying that all persons should (1) follow the same occupation as their parents, (2) marry within their own caste, and (3) have no social relationships with members of other castes (Weber [1910] 1970b).

Class Systems

In a class system, achieved statuses are the major basis of unequal resource distribution. Occupation remains the major determinant of rewards, but it is not fixed at birth. Instead, you can achieve an occupation far better or far worse than that of your parents. The amount of rewards you receive is influenced by your own talent and ambition or their lack.

The primary difference between caste and class systems is not the level of inequality but the opportunity for achievement. The distinctive characteristic of a class system is that it permits **social mobility**—a change in social class. Technically, mobility that occurs from one generation to the next is **intergenerational mobility.** Change in occupation and social class during an individual's own career is **intragenerational mobility.** Both kinds of mobility may be downward as well as upward.

Social mobility is the process of changing one's social class.

Intergenerational mobility is the change in social class from one generation to the next.

Intragenerational mobility is the change in social class within an individual's own career.

Even in a class system, ascribed characteristics have an influence. Whether you are male or female, Hispanic or non-Hispanic, Jewish or Protestant is likely to influence which doors are thrown open and which barriers have to be surmounted. Nevertheless, these factors are much less important in a class than in a caste society. Because class systems predominate in the modern world, the rest of this chapter is devoted to them.

Classes—How Many?

A class system is an ordered set of statuses. Which statuses are included, and how are they divided? Two theoretical answers to these questions are presented here.

Marx

The **bourgeoisie** is the class that owns the tools and materials for their work—the means of production.

Karl Marx (1818–1883) believed that there were only two classes. We could call them the haves and the have-nots; Marx called them the bourgeoisie (boor-zhwah-zee) and the proletariat. The **bourgeoisie** are those who own the tools and materials necessary for their work—the means of production; the **proletariat** do not. They must therefore support themselves by selling their labor to those who own the means of production. In Marx's view, **class** is determined entirely by one's relationship to the means of production.

The **proletariat** is the class that does not own the means of production. They must support themselves by selling their labor to those who own the means of production.

Class refers to a person's relationship to the means of production.

Relationship to the means of production obviously has something to do with occupation, but it is not the same thing. According to Marx, your college instructor, the manager of the Sears store, and the janitor are all proletarians because they work for someone else. Your garbage collector is probably also a proletarian who sells his labor to an employer; if he owns his own truck, however, your garbage collector is a member of the bourgeoisie. The key factor is not income or occupation but whether individuals control their own tools and their own work.

False consciousness is a lack of awareness of one's real position in the class structure.

Class consciousness occurs when people are aware of their relationship to the means of production and recognize their true class identity.

Marx, of course, was not blind to the fact that in the eyes of the world managers of Sears stores are regarded as more successful than truck-owning garbage collectors. Probably managers think of themselves as being superior to garbage collectors. In Marx's eyes, this is **false consciousness**—a lack of awareness of one's real position in the class structure. Marx, a social activist as well as a social theorist, hoped that managers and janitors could learn to see themselves as part of the same oppressed class. If they developed **class consciousness**—an awareness of their true class identity—he believed a revolutionary movement to eliminate class differences would be likely to occur.

Weber: Class, Status, and Power

Status is social honor, expressed in lifestyle.

Several decades after Marx wrote, Max Weber developed a more complex system for analyzing classes. Instead of Marx's one-dimensional ranking system, which provided only two classes, Weber proposed three independent dimensions on which people are ranked in a stratification system (see Figure 6.1). One of them, as Marx suggested, is class. The second is **status,** or social honor, expressed in lifestyle. Unlike people united by a common class, people united by a common lifestyle form a community. They invite one another to dinner, marry one another, engage in the same kinds of recreation, and generally do the same things in the same places. The third dimension is **power,** the ability to direct others to act against their wishes, the ability to overcome resistance.

Power is the ability to direct others' behavior even against their wishes.

Weber argued that although status and power often follow economic position, they may also stand on their own and have an independent effect on social inequality. In particular, Weber noted that status often stands in opposition to economic power, depressing the pretensions of those who "just" have money. Thus, for example, a member of the Mafia may have a lot of money, may in fact own the means of production (a brothel, a heroin manufacturing plant, a casino)—but he will not have social honor.

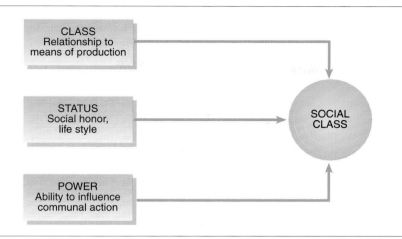

FIGURE 6.1
Weber's Model of Social Class
Weber identifies three independent dimensions of stratification. This multidimensional concept is sometimes called social class.

Most sociologists use some version of Weber's framework to guide their examination of stratification systems. Rather than speaking of class (the Marxian dichotomy), we speak of *social* class. **Social class** is a category of people who share roughly the same class, status, and power and who have a sense of identification with one another. When we speak of the upper class or of the working class, we are speaking of a social class in this sense.

Social class differs from *class* in two ways. First, it recognizes the importance of status and power as well as class. Second, it includes the element of self-awareness. Although people may be ignorant of their class situation, they are usually well aware of their social-class position, often using it as an important means to map the social world and their own place in it. People recognize that they are similar to others of their own social class but different in important ways from those in other social classes. The manager of the Sears store and the garbage collector, for example, are likely to be aware that they are members of the middle and working class, respectively.

Social class is a category of people who share roughly the same class, status, and power and who have a sense of identification with each other.

Inequality in the United States

Stratification exists in all societies. In Britain, India, and China alike, social structures ensure that some social classes routinely receive more rewards than do others. This section considers how the system works in the United States.

Measuring Social Class

If you had to rank all the people in your classroom by social class, how would you do it? There are many different strategies you could use: their incomes, their parents' incomes, the size of their savings accounts, or the way they dress and the cars they drive. Some students would score highly no matter how you ranked them, but others' scores might be very sensitive to your measurement procedure. The same thing is true when we try to rank people in the United States; the picture of inequality we get depends on our measurement procedure.

Self-Identification

A direct way of measuring social class is simply to ask people what social class they belong to. Given our definition of *social class* as a self-aware group, people should be able

THINKING

CRITICALLY

To what social class do you belong?
How do you know?

to tell you which social class they are in. Sure enough, when we ask people "Which of the following social classes would you say you belong to?" fewer than 1 percent say they don't know. Hardly anybody says, "What do you mean, 'social class'?" and even fewer tell you that we don't have social classes in the United States. The concept of social class is meaningful to most people in the United States, and they have an opinion about where they fit in the hierarchy.

The results of a 1986 survey are presented in Figure 6.2. As you can see, only tiny minorities see themselves as belonging to the upper and lower classes, and the bulk of the population is split nearly evenly between working- and middle-class identification. Studies show that the difference between working- and middle-class identification has important consequences, affecting what church you go to, how you vote, and how you raise your children. Some of these differences are discussed in later parts of this book.

Socioeconomic Status (SES)

Socioeconomic status (SES) is a measure of social class that ranks individuals on income, education, occupation, or some combination of these.

An alternative way to measure social class is by **socioeconomic status (SES),** which ranks individuals on income, education, occupation, or some combination of these. SES measures do not produce self-aware social-class groupings, but result in a ranking of the population from high to low on criteria such as years of school completed, family income, or occupation.

Many scholars use occupation alone as their indicator of social-class position. The device most often used to rank occupations is the Occupational Prestige Scale. The scale is based on survey research in which large random samples are given lists of occupations and asked "to pick out the statement that best gives your own personal opinion of the general standing that such a job has: excellent, good, average, somewhat below average, or poor." The prestige of an occupation rests on the overall evaluation that sample respondents give to each occupation. Repeated tests have shown that this procedure yields consistent results; the same ordering of occupations has been demonstrated on U.S. samples since 1927 as well as in other Westernized societies, from urban Nigeria to Great Britain (Hodge, Siegel, & Rossi 1964; Hodge, Treiman, & Rossi 1966). In spite of the fact that the question is specifically about *men* who hold these occupations, occupations are ranked the same way for women, too (Bose & Rossi 1983). Thus, we can be confident that the scale produces a reliable ordering of occupations (see Table 6.1 for a partial list of ranked occupations).

Economic Equality

FIGURE 6.2
Social-Class Identification in the United States
Social class is a very real concept to most Americans. They are aware of their own social-class membership: They feel that, in a variety of important respects, they are similar to others in their own social class and different from those in other social classes. The great majority of Americans place themselves in either the working or the middle class.
SOURCE: James Allen Davis and Tom N. Smith, 1986.

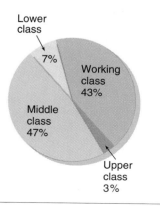

Lower class

7%

Working class 43%

Middle class 47%

Upper class 3%

All contemporary class systems have very high levels of inequality. In the United States, income inequality has been substantial since the beginning of the republic. Despite wars on poverty, large-scale increases in educational attainment, and a fourfold increase in the number of two-earner households, inequality in the distribution of household income has changed little. The poorest 20 percent continue to receive only 4 percent of all personal income, whereas the richest 20 percent receive almost 49 percent—or 11 times more (see Figure 6.3).

The inequality documented by income distribution is actually an underestimate of inequality. If we measure material inequality by the distribution of *wealth*—all that the person and the person's family have accumulated over the years (savings, investments, homes, land, cars, and other possessions)—we find that the richest 20 percent of households hold between 75 and 80 percent of all wealth. Historical research suggests that this unequal distribution of wealth is a longstanding pattern in the United States, dating back at least to 1810.

TABLE 6.1
Occupational Prestige Ratings

Occupation	Score	Occupation	Score	Occupation	Score
Physician	82	Social worker	52	Barber	38
College Professor	78	Funeral director	52	Jeweler	37
Judge	76	Computer specialist	51	Watchmaker	37
Lawyer	76	Stockbroker	51	Bricklayer	36
Physicist	74	Reporter	51	Airline steward	36
Dentist	74	Office manager	50	Meter reader	36
Banker	72	Bank teller	50	Mechanic	35
Aeronautical engineer	71	Electrician	49	Baker	34
Architect	71	Machinist	48	Shoe repairer	33
Psychologist	71	Police officer	48	Bulldozer operator	33
Airline pilot	70	Insurance agent	47	Bus driver	32
Chemist	69	Musician	46	Truck driver	32
Minister	69	Secretary	46	Cashier	31
Civil engineer	68	Foreman	45	Salesclerk	29
Biologist	68	Real estate agent	44	Meat cutter	28
Geologist	67	Fire fighter	44	Housekeeper	25
Sociologist	66	Postal clerk	43	Dock worker	24
Political scientist	66	Advertising agent	42	Gas station attendant	22
Mathematician	65	Mail carrier	42	Cab driver	22
Secondary school teacher	63	Railroad conductor	41	Elevator operator	21
Registered nurse	62	Typist	41	Bartender	20
Pharmacist	61	Plumber	41	Waiter	20
Veterinarian	60	Farmer	41	Farm laborer	18
Elementary school teacher	60	Telephone operator	40	Domestic servant	18
Accountant	57	Carpenter	40	Garbage collector	17
Librarian	55	Welder	40	Janitor	17
Statistician	55	Dancer	38	Shoe shiner	9

SOURCE: James A. Davis and Tom W. Smith, National Data Program for the Social Sciences: General Social Survey, Cumulative File, 1972–1982 (Ann Arbor, Mich.: Inter-University Consortium for Political and Social Research, 1983), Appendix F. Reprinted with the permission of the Inter-University Consortium for Political and Social Research.

The Consequences of Social Class

The following chapters point out the influence of social class in a number of areas—among them religious affiliation and participation, divorce, prejudice and discrimination, and work satisfaction. Here it suffices to say that almost every behavior and attitude we have is related to our social class. Do you prefer bowling to tennis? What kind of movies do you like (they call them films in the upper class)? Would you rather drink beer or sherry? These choices and nearly all the others you make are influenced by your social class. Knowledge of a person's social class will often tell us more about an individual than any other single piece of information. This is why "What do you do for a living?" almost always follows "Glad to meet you."

Can Money Buy Happiness?

Some social-class differences are merely subcultural differences in tastes and lifestyles. If you prefer football or even all-star wrestling to the symphony, there is no objective

FIGURE 6.3
Income Inequality in the United States, 1995
Distributions of income in the United States show little change in income inequality since World War II. Over 48 percent of the total income in the United States goes to the richest 20 percent of the population, whereas the poorest 20 percent of the population consistently receives about 4 percent.

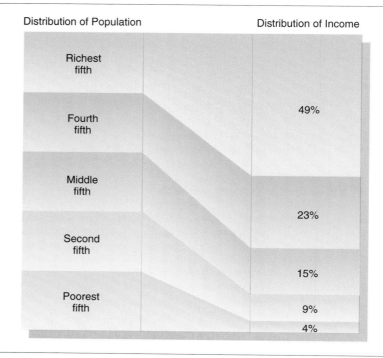

Distribution of Population | Distribution of Income

Richest fifth

Fourth fifth — 49%

Middle fifth — 23%

Second fifth — 15%

Poorest fifth — 9%

4%

way to say that your tastes are worse than those who subscribe to elite culture; they are just different. On many dimensions, however, social-class differences are more meaningful. Consider the following:

- People with incomes below $10,000 a year have nearly three times as many disability days as those with incomes of $35,000 or more (U.S. Bureau of the Census 1996a).
- People with incomes of less than $7,500 a year are more than two times as likely to have been the victim of a violent crime as those with incomes over $75,000 (U.S. Bureau of the Census 1996a).
- Infants whose mothers fail to graduate from high school are twice as likely to die before their first birthday as infants born to mothers with college degrees (Bertoli et al. 1984).
- People who fail to graduate from high school are twice as likely to be divorced within the first five years of marriage as those who complete at least one year of college (Martin & Bumpass 1989).

On these and many other indicators, the better off are not just different, they are also healthier and happier. The differences between social classes go far beyond tastes in recreation to fundamental differences in the quality of life.

Change and Continuity in the Consequences of Social Class

Levels of income inequality have changed little since World War II, and differences in wealth have been remarkably constant over 180 years. Nevertheless, there have been some striking changes in the consequences of social class. We can see the clearest evidence in two studies of Middletown (Muncie, Indiana), the first done in 1924–25. After the original classic investigation, the researchers concluded that social class

> is the most significant single cultural factor tending to influence what one does all day long throughout one's life: whom one marries; when one gets

THINKING

CRITICALLY

Can you think of any aspect of individual behavior or of social life that is likely to be unaffected by social class?

up in the morning; whether one belongs to the Holy Roller or Presbyterian church; or drives a Ford or a Buick; whether or not one's daughter makes the desirable high school Violet Club; or one's wife meets with the Sew We Do Club or with the Art Students' League; whether one belongs to the Odd Fellows or to the Masonic Shrine; whether one sits about evenings with one's necktie off; and so on indefinitely throughout the daily comings and goings of a Middletown man, woman, or child (cited in Caplow & Chadwick 1979).

In 1972, another team of investigators went back to see how Middletown had changed in the ensuing 50 years. The second study found a dramatic decline in social-class differences (see Table 6.2). The working class still got up a little earlier and still placed less stress on independence, but there was a marked convergence on all measures except the directly economic one—percentage unemployed.

As the differentials reported on disabilities, infant mortality, victimization, and divorce demonstrate, social class still does make a big difference. Nevertheless, the difference is less than it was 65 years ago. To some extent, this is a result of the major increases in real income that have been experienced in this country since 1924. The increases have been particularly important for those who were barely keeping their heads above water. Although the cars, televisions, and homes of the working class are not of the same quality as those of the middle class, the working class does have them. An additional factor in reducing some of the major differences in life chances is the extension of public services. Public schools, the GI Bill, and veterans' benefits have helped reduce some of the more severe consequences of lower social class. As we will document later in this chapter, however, some pockets of truly disadvantaged people still exist.

TABLE 6.2
Changes in Lifestyles by Social Class in Middletown Between 1924 and 1972
Differences in social class declined sharply in Middletown between 1924 and 1972. With the exception of unemployment, the lifestyles of working-class families and business-class families were much more similar in 1972 than they were in 1924.

	1924	1972
Percentage of families rising before 6 A.M. on workdays		
Business class	15%	31%
Working class	93	38
Percentage of families where husband unemployed in last year		
Business class	1	4
Working class	28	25
Percentage of families with a working wife		
Business class	3	42
Working class	44	48
Percentage of mothers wanting their children to go to college		
Business class	93	90
Working class	23	83
Percentage of parents stressing independence in children		
Business class	46	82
Working class	17	68

SOURCE: Caplow and Chadwick, 1979. "Inequality and Life-Styles in Middletown, 1920–1978." Social Science Quarterly 60 (3). ©1979 by The University of Texas Press, 2100 Comal, Austin, TX 78722. Reprinted with permission of The University of Texas Press and authors.

Conclusion

Social class is important in U.S. society. It affects our attitudes, behaviors, values, health, and opportunities. Although some social-class differences have been reduced through federal programs and through mass culture, many important differences—including very substantial income inequality—remain. Next we look at the factors that explain this continuing inequality.

Explanations of Inequality

Steven Spielburg earned more than $300 million in 1993, Troy Aikman earned more than $10 million, Bill Clinton earned $200,000, and the average police officer and teacher earned about $30,000; some 10 percent of U.S. families have annual incomes of less than $10,000. How can we account for such vast differences in income? Why isn't somebody doing anything about it?

We begin our answers to these questions by examining the social structure of stratification—that is, instead of asking about Steven and Troy and Bill, we ask why some *statuses* routinely get more scarce resources than others. After we review these general theories of stratification, we will turn to explanations about how individuals are sorted into these various statuses.

Structural-Functional Theory

The structural-functional theory of stratification begins (as do all structural-functional theories) with the question, Does this social structure contribute to the maintenance of society? This theoretical position is represented by the work of Davis and Moore (1945), who conclude that stratification is necessary and justifiable because it contributes to the maintenance of society. Their argument begins with the premise that each society has essential tasks (functional prerequisites) that must be performed if it is to survive. The tasks associated with shelter, food, and reproduction are some of the most obvious examples. They argue that we may need to offer high rewards as an incentive to make sure that people are willing to perform these tasks. The size of the rewards must be proportional to three factors:

1. *The importance of the task.* When a task is very important, very high rewards may be necessary to guarantee that it is done.
2. *The pleasantness of the task.* When the task is relatively enjoyable, there will be no shortage of volunteers and high rewards need not be offered.
3. *The scarcity of the talent and ability necessary to perform the task.* When relatively few have the ability to perform an important task, high rewards are necessary to motivate this small minority to perform the necessary task.

Let us apply this reasoning to two tasks, health care and reproduction. The tasks of the physician require quite a bit of skill, intelligence beyond the average, long years of training, and long hours of work in sometimes unpleasant and stressful circumstances. To motivate people who have this relatively scarce talent to undertake such a demanding and important task, Davis and Moore would argue that we must hold out the incentive of very high rewards in prestige and income. Society is likely to determine, however, that little reward is necessary to motivate women to fill the even more vital task of reproducing and raising a new generation. Although the function is essential, the potential to fill the position is widespread (most women

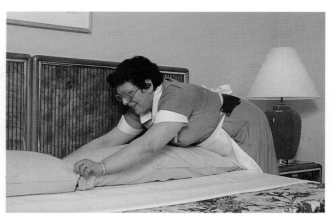

Which job is the hardest? Which one pays the most? Ironically, people who work the hardest often earn the least; they earn the least income and the least honor. A more critical problem for structural-functional theory is the issue of how these two jobs are assigned. Would you feel safe in concluding that the woman who cleans hotel rooms for a living has less talent and ability than the white-collar worker? Critics point out that social-class background and race are often more important than ability in determining who gets the good jobs and who gets the bad jobs.

between 15 and 40 can do it), and the job has sufficient non-cash attractions that no shortage of volunteers has arisen.

In many ways, this is a supply-and-demand argument that views inequality as a rational response to a social problem. This theoretical position is sometimes called consensus theory because it suggests that inequality is the result of societal agreement about the importance of social positions and the need to pay to have them filled.

Criticisms

This theory has generated a great deal of controversy. Among the major criticisms are these: (1) High demand (scarcity) can be artificially created by limiting access to good jobs. For example, keeping medical schools small and making admissions criteria unnecessarily stiff reduce supply and increase demand for physicians. (2) Social-class background, sex, and race or ethnicity probably have more to do with who gets highly rewarded statuses than do scarce talents and ability. (3) Many highly rewarded statuses (rock stars and professional athletes, but also plastic surgeons and speech writers) are hardly necessary to the maintenance of society.

The Conflict Perspective

A clear alternative to the Davis and Moore theory is given by scholars who adhere to conflict theory. They explain inequality as the result of class conflict rather than as a result of consensus about how to meet social needs. We review traditional Marxist thought first and then describe more recent applications of conflict theory to the study of inequality.

Marxist Theory

Marx argued that inequality was rooted in private ownership of the means of production. Those who own the means of production seek to maximize their own profit by minimizing the amount of return they must give to the proletarians, who have no choice but to sell their labor to the highest bidder. In this view, inequality is an outcome of private property, where the goods of society are owned by some and not by others.

In Marxist theory, stratification is neither necessary nor justifiable. Inequality does not benefit society; it benefits only the rich.

Although Marx did not see inequality as either necessary or justifiable, he did see that it might be nearly inevitable. The reason lies in the division of labor. Almost any complex task, from teaching school to building automobiles, requires some task specialization: Some people build fuel pumps and others install them, some teach algebra and others teach poetry. To make such a division of labor function effectively, somebody has to coordinate the efforts of all the specialists. Individuals who do this coordination are in a unique position to pursue their own self-interest—to hire their own children in preference to others', to give themselves more rewards than they give others, and generally to increase the gap between themselves and those they coordinate. Marx's patron and coauthor, Friedrich Engels, explained it this way:

> It is therefore the law of the division of labor that lies at the basis of the division into classes. But this does not prevent the division into classes from being carried out by means of violence, and robbery, trickery and fraud. It does not prevent the ruling class, once having the upper hand, from consolidating its power at the expense of the working class, from turning its social leadership into an intensified exploitation of the masses (Engels [1880] 1965, 79).

Modern Conflict Theory

During the early days of industrialization when Marx was writing, those who owned the factories also managed them. In the modern economy, however, control and ownership may be independent; ownership may be divided among many stockholders, while management is controlled by a handful of hired experts (Wright 1985).

Like the earlier Marxist theory, modern conflict theory recognizes that the powerful can oppress those who work for them by claiming the profits from their labor (Wright 1985). It goes beyond Marx's focus on ownership, however, by considering how control also may affect the struggle over scarce resources (Grimes 1989).

Criticisms

There seems to be little doubt that people who have control (through ownership or management) systematically use their power to extend and enhance their own advantage. Critics, however, question the conclusion that this means that inequality is necessarily undesirable and unfair. This is certainly a debatable assumption. First, people *are* unequal. Some people are harder working, smarter, and more talented than others. Unless forcibly held back, these people will pull ahead of the others—even without force, fraud, and trickery. Second, coordination and authority *are* functional. Organizations simply work better when those trying to do the coordinating have the power or authority to do so.

A Contemporary Synthesis

Structural-functional theory and conflict theory address important issues in the explanation of inequality; each also has a blind side. Structural-functional theory disregards how power may be used to create and enhance inequality; conflict theory generally ignores the functions of inequality.

In 1989, Beegley provided a general theory of contemporary stratification that pulls together some ideas from structural-functional and conflict theory. His synthesis rests on three major points:

1. *Power is the major determinant of the distribution of scarce resources.* People who have power, whether because of ownership or because of control, will use that power to enhance, protect, and extend their resources. For example, they will try to shape the labor market in ways that benefit themselves.
2. *The distribution of power (and hence of scarce resources) is socially structured.* The level of poverty in a society, the salaries of rock stars, and the opportunities to get ahead depend on public and private policies. For example, if professional schools require annual tuitions of $25,000 and more a year, then it is almost automatic that most of the professionals of the next generation will come from today's advantaged families. One implication of this idea is that, if we wanted to, we could challenge this policy and increase access to professional statuses.
3. *Individuals can make a difference.* As we saw when we looked at the development of the self-concept, individual behavior is not wholly determined by social structures. Individual characteristics such as talent and ambition play a role. Thus, scarce talents and abilities may allow some to rise to the top despite a disadvantaged position in the social structure.

The first two of Beegley's points are drawn from modern conflict theory, the third from structural functionalism. Although his theory does allow a role for scarce talents and abilities, it focuses on the social structure of power as the major determinant of inequality.

Conclusion

So, why *do* Steven Spielberg and Troy Aikman make so much money? Neither Spielburg nor Aikman appears to be engaged in force, fraud, or trickery or to be using his authority

CONCEPT SUMMARY	*A Comparison of Three Models of Stratification*		
BASIS OF COMPARISON	**STRUCTURAL-FUNCTIONAL THEORY**	**CONFLICT THEORY**	**BEEGLEY'S SYNTHESIS**
1. Society can best be understood as…	Groups *cooperating* to meet common needs	Groups *competing* for scarce resources	Groups competing for scarce resources
2. Social structures…	Solve problems and help society adapt	Maintain current patterns of inequality	Determine opportunities and their allocation
3. Causes of stratification are…	Importance of vital tasks, unequal ability, pleasantness of tasks	Unequal control of means of production maintained by force, fraud, and trickery	Inequalities in power
4. Conclusion about stratification…	Necessary and desirable	Difficult to eliminate, but unnecessary and undesirable	Inevitably built into social structure; no value judgment
5. Strengths…	Consideration of unequal skills and talents and necessity of motivating people to work	Consideration of conflict of interests and how those with control use the system to their advantage	Value-free; recognizes that structure is more important than individual talents
6. Weaknesses…	Ignores importance of power and inheritance in allocated rewards; functional importance overstated	Ignores the functions of inequality and importance of individual differences	Applies mainly to modern capitalist societies

to exploit others. Beegley's theory suggests an explanation for their earnings. First, of course, Spielburg and Aikman are very good at what they do (see Beegley's third point). At least as important, however, is the fact that entertainment is a huge industry in the United States, which supports Beegley's second point. Finally, and this is Beegley's first point, the people who *control* the entertainment industry (as opposed to those who merely work in it) have incomes that dwarf Spielburg's and Aikman's. One of the reasons, in fact, that Spielburg makes more than Aikman is that, over the years, he has acquired control of his own production; Troy Aikman still works for the owner of the Cowboys. Although it takes a stretch of the imagination to see Aikman as exploited, the professional football strike of the 1988 season illustrates the conflict of interest between management and labor even at this level. The fact that the players lost and the owners won also illustrates Beegley's first point: Those in control have the resources to protect and maintain their advantage.

The Determinants of Social-Class Position

With each generation, the social statuses in a given society must be allocated anew. Some people will get the good positions and some will get the bad ones; some will receive many scarce resources and some will not. In a class system, this allocation process depends on two things: the characteristics of the individuals (their education, aspirations, skills, and so on) and the characteristics of the labor market. We refer to these, respectively, as micro- and macro-level factors that affect achievement.

Microstructure: Status Attainment

If *Sports Illustrated* gave you the job of predicting the top 20 college football teams in the country next year, you could go to the trouble of finding out the average height, weight, and experience level of each team's members, the dollars allocated to the athletic department, the years of coaching experience, and the attendance at games. From this information you could devise some complex system of predicting the winners. You would probably do a better job for a lot less trouble, however, if you predicted that last year's winners will be this year's winners. The same thing is true in predicting winners and losers in the race for class, status, and power. The simplest and most accurate guess is based on social continuity.

U.S. culture values achievement rather than ascription, and occupations are not directly inherited. Yet people tend to have occupations of a status similar to that of their parents. How does this come about? The best way to describe the system is as an **indirect inheritance model.** Parents' occupations do not directly cause children's occupations, but the family's status and income determine children's aspirations and opportunities.

The **indirect inheritance model** argues that children have occupations of a status similar to that of their parents because the family's status and income determine children's aspirations and opportunities.

Inherited Characteristics: Help and Aspirations

The best predictor of your eventual social class is your education—and the best predictor of your education is your parents' education. A small part of this is directly financial. Better-educated parents are more often able to afford their children's college expenses. Most of the impact of parents' education, however, is less direct. If your parents graduated from college or have middle-class jobs, then you have probably always assumed that you, too, would go to college. You automatically signed up for algebra and chemistry in high school. If your parents didn't graduate from high school and tend to think that education is a necessary evil, then you probably bypassed algebra for a shop or sewing class.

The atmosphere of the home and the parents' support and encouragement may have important effects on the child's success. Bright and ambitious lower-class children may find it hard to do well in school if they have to study at a noisy kitchen table amidst a group of people who think that their studies are a waste of time; middle-class children with even modest ambitions and intelligence may find it hard to fail within their very supportive environment.

The Wild Cards: Achievement Motivation and Intelligence

The social-class environment in which a child grows up is the major determinant of educational attainment. There are, however, two wild cards that keep education from being directly inherited: achievement motivation and intelligence. Neither of these factors is strongly related to parents' social class, and both act as filters that allow people to rise above or fall below their parents' social class (Duncan, Featherman, & Duncan 1972).

Achievement motivation is the continual drive to match oneself against standards of excellence. Students who have this motivation are always striving for As, are never satisfied with taking easy courses, and have a real need to compete. Not surprisingly, students with high achievement motivation do better than others in school. Because achievement motivation is not strongly related to parents' social class, it is one way for people to move out of their inherited position.

Intelligence is another important factor in determining educational and occupational success (Duncan, Featherman, & Duncan 1972). Because intelligent people are born into all social classes, intelligence is a factor that allows for both upward and downward intergenerational mobility. (Chapter 12 looks at the issue of social class and intelligence in greater detail.)

> **Achievement motivation** is the continual drive to match oneself against standards of excellence.

Summary

The indirect inheritance model summarizes the processes of individual status attainment. It shows how some people come to be well prepared to step into good jobs, while others lack the necessary skills or credentials. By themselves, however, skills and credentials do not necessarily lead to class, status, or power. The other variable in the equation is the labor market.

Macrostructure: The Labor Market

If there is a major economic depression, you will not be able to get a good job no matter what your education, achievement motivation, or aspirations. The character of the labor market and the structure of occupations it provides have a significant effect on individual achievements.

As Figure 6.4 shows, the proportion of positions at the top of the U.S. occupational structure has increased dramatically over the last 90 years. Everyone, however, has not benefited equally from these new opportunities for upward mobility. Regardless of credentials or talent, women and minorities remain virtually excluded from some high-earning occupations. In 1960, the National Basketball Association was 80 percent white Americans; by 1980 it was composed of 80 percent African Americans. This shift did not occur because African American basketball players suddenly became a lot better. Instead, it was a result of changes in the structure of the labor market. Employers were finally willing to hire African American players.

The preparation that you receive for the labor market—your education, training, and aspirations—has an enormous impact on whether you get a good or bad job. Nevertheless, your fate ultimately depends on the job market. If there is a government hiring freeze, a well-trained economist may be unemployable; if the aerospace industry is in a slump, aeronautical engineers will have to look hard for work. Even in a booming economy, many labor markets implicitly discriminate against people on the basis of their age, sex, race, or ethnicity. These market factors are critical for understanding why some groups consistently have worse jobs and lower earnings than others.

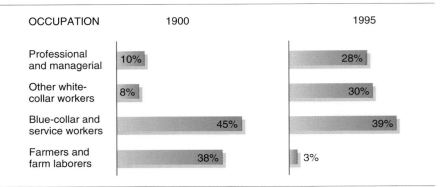

FIGURE 6.4
The Changing Occupational Structure, 1900–1995
Since the turn of the century, the occupational structure of the United States has shifted away from farm labor. Today there are many more white-collar, professional, and managerial jobs.
SOURCE: U.S. Bureau of the Census, 1996 Statistical Abstract, no. 637.

Labor market theorists suggest that the United States has a segmented labor market: one labor market for good jobs (usually in the big companies) and one labor market for bad jobs (usually in small companies). Women and minorities are disproportionately directed into companies with low wages, low benefits, low security, and short career ladders. (The contemporary economic structure is discussed in more detail in Chapter 13.)

Summary

The stratification structure of any society depends upon both macro- and micro-level processes. Some aspects of inequality are best explained on the macro level. For example, if we want to know what percentage of workers have good jobs or why some groups have bad jobs despite their credentials, then we look at the structure of the labor market and how it is changing. If we want to know which individuals are prepared to take the good jobs, then we need to look at micro-level processes that determine individual characteristics.

The American Dream: Ideology and Reality

A system of stratification is an organized way of ensuring that some categories of individuals get more social rewards than others. As we have seen, sometimes this means a great disparity not only in income but also in health, honor, and happiness. Yet, in most highly stratified systems, there are no revolutionary movements, the rich aren't always fearing attack, and the poor don't sit around stoking the fires of resentment. For the most part, inequality is accepted as fair and natural, even as God-given.

This consensus about inequality indicates the role of the normative structure in reinforcing and justifying a system of stratification. Each system furnishes an **ideology**— a set of norms and values that rationalizes the existing social structure (Mannheim 1929). The ideology is built into the dominant cultural values of the society—often into its religious values. For example, the Hindu religion maintains that a low caste in this life is a punishment for poor performance in a previous life. If you live well in this life, however, you can expect to be promoted to a higher caste in the next life. Thus, the Hindu religion offers mobility (extragenerational mobility, we might call it) and also an incentive to accept one's lot in life. To attack the caste system would be equivalent to saying that the gods are unfair or the religion is stupid.

In the United States, the major ideology that justifies inequality is the *American Dream,* which suggests that equality of opportunity exists in the United States and that

An **ideology** is a set of norms and values that rationalizes the existing social structure.

your position in the class structure is a fair reflection of what you deserve; that is, if you are worthy and if you work hard, you can succeed; because your position comes entirely from your own efforts, no one but you can be blamed for your failures. The upper class is the most likely to believe that the United States is a land of opportunity and that everybody receives a fair shake, but most others believe this, too. There are, to be sure, some grumblers, especially among minority groups, such as the disenchanted person who stated, "The rich stole, beat, and took. The poor didn't start stealing in time, and what they stole, it didn't value nothing, and they were caught with that!" (Huber & Form 1973).

The grumblers, however, are few, and most of them are less interested in changing the rules of the game than in being dealt into it. A survey of U.S. adults found that equality was a dirty word: Fewer than 20 percent agreed that "it would be a good thing if the president decided to distribute all the money in the United States equally among all the population" (Bell & Robinson 1978).

People in the United States believe in "fair shares" rather than in "equal shares" (Ryan 1981). They believe that people who work harder and people who are smarter deserve to get ahead. A typical attitude toward the wealthy is represented by the comments of one unemployed laborer: "If a person keeps his mind to it, and works and works, and he's banking it, hey, good luck to him!" (cited in Hochschild 1981, 116).

Variations on a Theme: The Rich, the Working Class, and the Poor

The United States is a middle-class nation. If given only three categories for self-identification, more than two-thirds of the population consider themselves middle class. U.S. norms and values are the norms and values of the middle class. Everybody else becomes a subculture. This section briefly reviews the special conditions of the nonmiddle class in the United States.

When was the last time you drank champagne and ate hors d'oeuvres at a football game? When was the last time you even saw anyone who looked like these New Orleans Saints fans at the stadium? Because they generally live in exclusive areas and belong to exclusive clubs, social scientists know relatively little about them. More important than their luxurious lifestyles, however, is the ability of the "rich and famous" to affect the lives and incomes of the rest of us. The ability of the wealthy to create or terminate jobs, affect elections, and subsidize education and the arts is the real focus of the sociological study of the upper class.

MAP 6.1
Where the Affluent Live
SOURCE: Atlas of Contemporary America ©1994, p. 124.

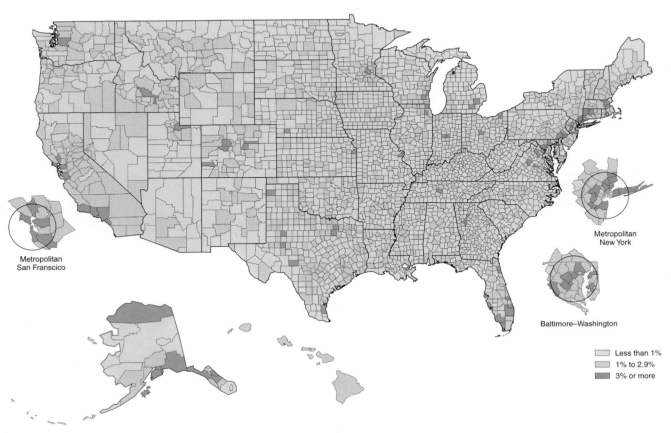

Metropolitan
New York

Metropolitan
San Franscico

Baltimore–Washington

Less than 1%

1% to 2.9%

3% or more

The Upper Class

In 1995, a family required an income of $113,000 to be in the richest 5 percent. Thus, a variety of more-or-less ordinary salespersons, doctors, lawyers, and managers in towns across the nation qualify as being very rich compared to the majority. Although their incomes are nothing to sneeze at, most of this upper 5 percent is still middle class. Like members of the working class, they would have a hard time making their mortgage payments if they—or their spouses—lost their jobs.

The true upper class, however, is the top 5 percent of the richest 5 percent—the very small minority who earn millions. There are nearly a half million millionaires in the United States—people whose assets total more than $1 million. At the very top of the heap are the nearly 5,000 whose earnings in a single year top $1 million. These people pursue a variety of occupations—most often making their fortunes in computers, the stock market, or media (see Table 6.3). Many inherit their way to the top; of the self-made, few went from rags to riches. Most have had at least middle-class parents with business experience, who sent their children to excellent schools.

TABLE 6.3
The Ten Richest People in the United States, 1996
Each year *Forbes* magazine publishes a list of the 400 richest people in the United States. About 40 percent of these fabulously wealthy individuals inherited their fortunes; two-thirds are largely individually responsible for generating their vast wealth. Their major avenues to riches were the stock market, manufacturing, and the media.

Name, Age	College	Estimated Net Worth	Major Wealth Source
William Henry Gates III, 40	Harvard (dropout)	18.5 billion	Microsoft, Inc.
Warren Buffett, 66	Columbia	15 billion	Stock Market
Paul G. Allen, 43	Harvard	7.5 billion	Microsoft, Inc.
John W. Kluge, 82	Columbia	7.2 billion	Metromedia
Lawrence Ellison, 52	U. of Illinois (dropout)	6 billion	Oracle Corp.
Philip H. Knight, 58	Stanford	5.3 billion	Nike, Inc.
Jim C. Walton, 48	none given	23.7 billion	Inheritance (Wal-Mart)
John T. Walton, 50	none given		
Alice L. Walton, 47	Trinity		
Helen Walton, 77	Oklahoma		
S. Robson, 52	Columbia		
Donald E. Newhouse, 66	none given	9.0 billion	Advance Publications
Samuel I. Newhouse, 68	none given		
Barbara Cox Anthony, 73	none given	8.0 billion	Inheritance (Cox)
Anne Cox Chambers, 76	none given		
Ronald O. Perelman	Wharton	4 billion	Leveraged Buyouts

SOURCE: "The Forbes Four Hundred," http://www.forbes.com/Richlist/richquer.htm

By controlling media resources, contributing money to political campaigns, building museums, or deciding to open and close manufacturing plants, the rich have the power to affect the lives of millions. The power they wield makes us very interested in their politics and attitudes. Yet, we know relatively little about the lives of the rich. Although their pictures may occasionally appear in the news, they do not participate in Gallup polls, nor do they hang out in the public places where social researchers can readily observe them. The only aspect of the rich that has been extensively studied is their attitude toward inequality. According to one social scientist who studies elites, ". . . [They] believe in equality of opportunity rather than absolute equality" (Dye 1983, 273). As conflict theory would suggest, they are interested in maintaining a system that has been good to them.

The Working Class

Who are the members of the working class? The answer is determined partly by occupation, partly by education, and partly by self-definition. Generally, the working class includes those who work in blue-collar industries and their families. They are the men and women who work in chemical, automobile, and other manufacturing plants; they load warehouses, drive trucks, and build houses. Although they sometimes receive excellent wages and benefits, it is the working class that suffers 10–15 percent unemployment during economic recessions and slumps.

To find out what life was like for the working class, sociologist David Halle (1984) spent seven years studying the blue-collar employees of a New Jersey chemical plant. His participant observation study involved hanging around in the plant itself and going to taverns, football games, and Christmas parties. His work suggests that the following dimensions make working-class life distinct from that of the middle class: education, job, and economic prospects, leisure, and gender differences.

Education

The working class are better educated than they used to be. A majority are probably high school graduates. Still, an eleventh-grade education is more common than a year of college. More important, many of these people did not do well in school, and they did not absorb the school's middle-class values.

> What was school like? It was horrible, horrible! . . . They (the teachers) were cuckoos. They gave you *Romeo and Juliet* to read, and I looked at it and I said, "What is this! What has this got to do with me?" I looked on the flyleaf. . . . And I saw Joe Smith's name from three years ago. I knew he was digging ditches now, so I said to myself, "This book didn't do anything for him. What's it going to do for me?" (Halle 1984, 49)

Although there are working-class women and men who read Shakespeare and discuss existentialism, much more common are the people who "ain't got no" interest in elite culture. Halle found that these men were painfully aware of the low prestige associated with their taste, but they were really much more interested in bowling than opera. Although sensitive about his lack of cultured interests, the typical worker excused himself as "just a working slob."

Economic Prospects

Quite a few members of the working class have incomes as good or better than the lower-middle class. Some make as much or more than public school teachers and people in retail sales. As a result, they live in the same neighborhoods as these members of the lower-middle class. Their economic prospects differ from their white-collar neighbors, however, in two ways. First, they have little or no chance of promotion. The barrier between manual labor and management is virtually impassable. The height of one's earning power may be reached at age 25. Second, layoffs and plant closings expose them to more economic uncertainty. As a result of low prospects and economic uncertainty, members of the working class tend to place a higher value on security than others. One aspect of this is home ownership. Nearly three-quarters of the working class own their own homes. For most, it is their major financial asset and their only effective savings plan. Their resulting interest in property taxes, property values, and neighborhood maintenance drives much of their political activity.

Leisure

Most working-class jobs are not enjoyable. Although middle-class workers may have a hard time separating their leisure time from their work time (going to the office on Saturdays, taking their work home with them, or having business dinners), members of the working class have no difficulty separating work from leisure—and they much prefer leisure. According to Halle, the working-class man does his work without joy and lives for his leisure. The high point of his life is going hunting and fishing with his friends, having a few drinks in the tavern after work, watching football on Sunday afternoons, and fooling around in his yard. When he dreams of the future, he dreams of his new boat or maybe a cabin in the mountains; he doesn't dream of a better job.

Gender Differences

Scholars of working-class life emphasize the sharp differences between the worlds of men and women. As the description of leisure activities suggests, men's leisure interests lie largely in sex-segregated activities: watching football, going hunting, drinking in

Social class affects all facets of our lives, from the number of years we can expect to live to the way we spend our leisure time. Symphony or ballet audiences are likely to come primarily from the upper or upper-middle classes. Research shows that wrestling fans such as these are more likely to be members of the working class. Even though the leisure activities of the working class are often segregated by gender, working-class women make up a sizable share of the wrestling audience (Ball 1990).

taverns. Even when social activity is based on couples, the women will talk in the kitchen and the men in the den. Although lack of common interests may damage some marriages, social class is not strongly related to marital happiness. One worker explains his recipe for a happy marriage:

> We [his wife and he] do a lot together. We both bowl, and we go out to eat together, and go to parties. I'm happy.
>
> My wife is very liberal. She doesn't bother if I'm out, and I don't bother if she's out. I know she's with her girlfriends or something. She doesn't mind if I come home late unless it's very late, maybe 2:00 or 3:00 in the morning. . . .
>
> There are some things you can't do. Like my wife doesn't mind if my friends come over to play cards so long as it's not every night. Once or twice a week is OK, but not every night. Then she'll say, "No way!" because she has to clean up afterward, sweep and clean away cigarette butts and all that. That's a lot of work. (Halle 1984, 57)

Gender segregation of leisure activities is not confined to the working class, but working-class women and men appear to have less in common than their middle-class counterparts. Perhaps more important, the working class are more likely to hold norms suggesting that men and women ought to be different and that they "naturally" will have different interests and roles.

Summary

Income is an inadequate criterion to distinguish the working from the middle class. The working class are often less well off than the middle class, but the primary distinctions are in occupation, education, and lifestyle rather than in income. This may change in the future. As Chapter 13 will document, the jobs that have provided relative affluence for the unionized working class are declining. In their place are minimum-wage service jobs that will push a growing sector of the working class to the poverty level.

The Poor in America

Each year, the U.S. government fixes a poverty level that is calculated to be the amount of money a family needs to meet the minimum requirements of a decent standard of living. The poverty level adjusts for family size, and in 1995, the poverty level for a family of four was $15,569 (U.S. Bureau of the Census, 1995). Under this definition, 36.4 million people, 13.8 percent of the population were classified as poor in 1995.

Who Are the Poor?

Poverty cuts across several dimensions of society. It is found among white Americans as well as nonwhites, in rural areas as much as in urban centers, in families as well as in single households. As Table 6.4 indicates, half (49.5 percent) of the poor in 1995 were too old or young to work. Of those in the working ages, a substantial proportion could not earn a wage that would lift them out of poverty.

A significant portion of the women and children who live in poverty do so simply because they have no husband or father in their house. Granted that having a man in the house is no guarantee of being out of poverty (3.3 million persons in male-headed families are below the poverty level), it does significantly decrease the likelihood of being in poverty: Six percent of individuals living in male-headed families are below the poverty level, whereas 35 percent of individuals living in female-headed families are below that level.

Although some categories of people, such as minorities or those in female-headed households, top the poverty charts year after year, a recent study shows that there is less continuity than one might suppose in the individual experience of poverty. A study of a large sample of the U.S. population between 1969 and 1978 found that one-quarter

TABLE 6.4
The Population Below the Poverty Level in 1995

	Millions of People	Percentage of Poverty Population	Percentage of Group in Poverty
Total	36.4	100	13.8
Race and Hispanic origin			
White	24.4	67.0	11.2
Black	9.9	27.2	29.3
Hispanic origin	8.6	23.6	30.3
Residence			
Central cities	16.3	44.8	20.6
Other urban	12.1	33.2	9.1
Nonmetropolitan	8.1	22.3	15.6
Living in families*			
Male headed	3.3	40.6**	6.1
Female headed	4.2	52.6**	34.6
Living alone*			
Male	3.3	8.6	17.8
Female	5.0	13.2	24.9
Children under 18	14.7	40.4	20.8
People over 65	3.3	9.1	10.5

SOURCE: U.S. Bureau of the Census, http://www.census.gov/hhes/poverty/pov95/povest1.html

*1994 data

**Percentage of Families in Poverty

were poor at least one year between 1969 and 1978, but only one-third of those who were poor in 1978 had been poor for 8 or more of those 10 years. The persistently poor fell into one of two categories: They were elderly or they lived in households headed by a black woman (Duncan 1984).

How Poor Are the Poor?

An important issue that arises in discussing U.S. poverty is how poor the poor actually are. Two concepts are important here: absolute poverty and relative poverty. **Absolute poverty** means the inability to provide the minimum requirements of life. **Relative poverty** means the inability to maintain what your society regards as a decent standard of living.

> **Absolute poverty** is the inability to provide the minimum requirements of life.
>
> **Relative poverty** is the inability to maintain what your society regards as a decent standard of living.

The poor in the United States come in both forms. Those who live at or close to the poverty level have a roof over their head and food to eat. On the other hand, their car is broken down and so is the television, the landlord is threatening to evict them, they are eating too much macaroni, and they cannot afford to take their children to the doctor or the dentist. Although they may not be absolutely poor, they are deprived in terms of what is regarded as a decent standard of living. There are also those who are absolutely poor: the homeless and the truly disadvantaged.

THE HOMELESS.   There are at least 300,000 homeless people in the United States. Many of them are single adults, but 23 percent are families with children. Not only do they lack a roof over their heads but 37 percent also eat one meal a day or less ("Another Winter" 1989). They are homeless, hungry, and often ill. By any standards, they are absolutely poor.

THINKING

CRITICALLY

What macro-level processes do you think are most responsible for the increasing numbers of homeless in the United States? What micro-level processes?

THE TRULY DISADVANTAGED.   At the very bottom of the social-class hierarchy is a group that has been called the **underclass,** a group that is unemployed and unemployable, a miserable substratum that is alienated from U.S. institutions (Myrdal 1962). This underclass is disproportionately black. It is a group characterized by high nonmarital birthrates, high drug use, high murder rates, and high unemployment rates (Wilson 1987). Children born in this environment start at the bottom and most will stay there.

> The **underclass** is the group that is unemployed and unemployable, a miserable substratum that is alienated from U.S. institutions.

Causes of Poverty

Earlier in this chapter, we said that both micro- and macro-level processes are at work in determining social-class position. The causes of poverty are simply a special case of these larger processes. At the micro level, poverty is explained by the hypothesis that there is a "culture of poverty;" at the macro level, poverty is explained by the lack of an adequate structure of opportunity.

THE CULTURE OF POVERTY: BLAMING THE VICTIM?   The indirect inheritance model suggests that people born into poverty are likely to stay there: They have poorer preparation for school, lower aspirations, and less help at every step of the way.

An additional mechanism that has been proposed to explain the inheritance of poverty is what anthropologist Oscar Lewis (1969) called the **culture of poverty.** Lewis argued that in rich societies, people who are poor develop a set of values that protects their self-esteem and maximizes their ability to extract enjoyment from dismal circumstances. This set of values—the culture of poverty—emphasizes living for the moment rather than thrift, investment in the future, or hard work. Recognizing that success is not within their reach, that no matter how hard they work or how thrifty they are, they will not make it, the poor come to value living for the moment.

> The **culture of poverty** is a set of values that emphasizes living for the moment rather than thrift, investment in the future, or hard work.

The culture-of-poverty hypothesis fits neatly into the U.S. ideology, and a substantial majority of the people of the United States agree that the poor are poor because of their values. In one survey, the two reasons most often endorsed as causes of poverty

MAP 6.2
Where the Poor Live
SOURCE: Atlas of Contemporary America ©1994, p. 120.

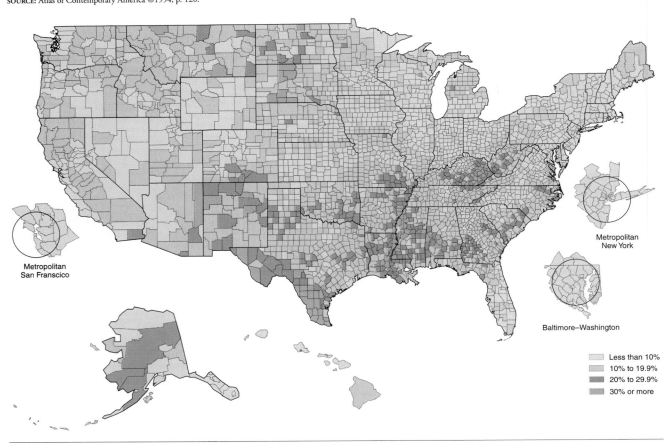

Metropolitan
New York

Metropolitan
San Franscico

Baltimore–Washington

Less than 10%
10% to 19.9%
20% to 29.9%
30% or more

were that the poor "are not motivated because of welfare" and that they show a "lack of drive and perseverance" (Smith & Stone 1989). Blaming failure—or success—on personal characteristics, however, overlooks the role of social structure in shaping both values and opportunities.

THE STRUCTURE OF OPPORTUNITY   The culture-of-poverty hypothesis implicitly blames the poor for perpetuating their condition. Critics of this hypothesis suggest that we cannot explain poverty by looking at micro-level processes. To understand poverty, they argue, we need to look at the structures of opportunity. If there are no jobs available, then we don't need to psychoanalyze people in order to figure out why they are poor.

The structural issues—the changing labor market and the growing link between education and wages—are particularly critical for understanding contemporary poverty. As we documented in Figure 6.4, the shift from an agricultural to an industrial society produced major structural pressure for upward mobility earlier in this century. Now, at the end of the century, the deindustrialization of the United States is squeezing the lower middle of the U.S. occupational structure and creating structural pressure for downward mobility among the traditional working class. Good jobs have virtually disappeared for the high school graduate who has no advanced training. Instead of the good union jobs that their parents held, today's high school graduates

Although we often think of poverty as a problem of racial minorities or of inner cities, the simple facts are that most poor people in the United States are white and almost equally likely to live in nonmetropolitan areas as in central cities. In parts of Appalachia and the rural South, there are many people for whom poverty is neither temporary nor relative. It is absolute. Without income, perhaps illiterate, and sometimes living in areas where there is virtually no economic growth, it is unlikely that these children will develop the skills or be given the job opportunities that will allow them to break out of their parents' social class.

often find themselves working at dead-end jobs for the minimum wage. A little arithmetic shows that the minimum wage means poverty.

This macrostructural approach to poverty suggests that a major cause of poverty is the absence of good jobs. The issue is a critical one, and we will look at the changing occupational structure in more detail in Chapter 13.

The Future of Inequality: Public Policy in the United States

Social Policies

If the competition is fair, inequality is acceptable to most people in the United States. The question is how to ensure that no one has an unfair advantage. Social policy has taken three different approaches to this: taxing inheritances, outlawing discrimination, and fostering educational opportunity.

An Application

SUBSIDIES, INCENTIVES, AND WELFARE: WHO BENEFITS?

The U.S. budget deficit has soared to more than $1 trillion, and despite their campaign promises, neither Presidents Reagan, Bush, nor Clinton have managed to reduce it. Middle-class families complain bitterly about their increasing tax burden and point their finger at the cost of antipoverty programs such as Medicaid, Aid to Families with Dependent Children, and Food Stamps. If lawmakers want to get control over the budget, they will need to more carefully consider the issue of just exactly who gets what.

Can you answer the following questions about "welfare" benefits?

_____ 1. Does the government spend more on (a) job training for the poor, (b) Head Start for low-income children, (c) Women, Infants and Children nutrition subsidies, or (d) health care for the richest 10 percent of elderly Medicare beneficiaries?

_____ 2. Government housing subsidies end up giving (a) 10 times as much to the poor as to the middle class, (b) five times as much to the poor, (c) more to the middle class than to the poor, or (d) about equally to both groups.

_____ 3. Taking spending programs and tax subsidies together, on the average which of the following gets the most money in federal benefits: a person with income of (a) less than $10,000, (b) from $10,000 to $40,000,

(c) from $40,000 to $100,000, or (d) more than $100,000? (Waldman 1992, 56)

The correct answer to question 1 is "d"; in 1989 the government spent more for medical care for well-off seniors than it did for all of the other programs combined. The answer to question 2 is "c"; taking tax credits for mortgage interest and other homeownership tax breaks into account, the federal government spends four times as much money to support middle- and upper-income families' housing than it does to house the poor. Finally, question 3 is correctly answered "d"; if the entire range of subsidies, incentives, and welfare programs are considered, an average upper-income individual will get considerably more than a typical poor person, even though it is precisely those middle- and upper-middle class individuals who are most likely to complain that the government is ignoring them.

> **If lawmakers want to get control over the budget, they will need to more carefully consider the issue of just exactly who gets what.**

According to a study by one conservative taxpayers' group, the average person with an income over $100,000 receives direct cash benefits—such as

social security—that are slightly higher than those received by persons with income of less than $10,000. If such tax breaks as mortgage interest deductions, health care reimbursement accounts, and the benefits of IRAs are also considered, the poor receive substantially less than families with incomes ranging from $30,000–$100,000 a year. The average benefit for the wealthiest group, $9,283, is highest of all.

In Oregon, angry taxpayers have insisted that welfare mothers be required to work for their assistance. In Nebraska, legislators have passed a maximum two-year period of eligibility for welfare benefits. Meanwhile, dealing with "the other half," Senator Hank Brown from Colorado was defeated on a proposal in the early 1990s that would have cut payments for holding land out of production if a farm family's income exceeded $120,000.

U.S. Census figures show that the portion of the benefit pie that goes to the poor has been shrinking for the past 25 years, while the portion going to the rest of the country has grown. As we grapple with the federal budget deficit, it would be wise to consider again the question of "Just who are those welfare 'kings and queens'?"

SOURCE: Steve Waldman, 1992. "Benefits 'R' Us," *Newsweek* (August 10); 56–58.

Estate Taxes

The policy regarding estate taxes is designed to reduce the direct inheritance of social position, to create greater equality at the start of the race. There is substantial consensus that although it is acceptable for ambitious, lucky, or clever persons to amass large fortunes, it is not fair that their children should start their race with such a large advantage. Thus, since 1931, the United States has had a progressive estate tax. The maximum tax has varied over the years from 50 to 90 percent of the estate.

In fact, however, inheritance taxes have not significantly reduced unequal advantage. If wealthy individuals die at 70, their children are already middle aged. The $90,000 the parents spent to give them the best private education money could buy, the new homes they bought for the children when they married, the businesses they set them up

Headstart programs and specially designed education programs alone cannot reasonably be expected to yield equality of opportunity. If the children in these two pictures went to the same school, their achievements would probably be very different. Equality of opportunity can only be achieved where there is already considerable equality in background. When children come to school from very unequal backgrounds, their achievements in school are likely to repeat the patterns of their parents.

in, the trusts they set up for their grandchildren—none of these are part of the estate. By the time the parents die, the children have already been established as rich themselves (Lebergott 1975). Unless we ban private schools, transfers of money to one's children, and giving one's children good jobs, this inheritance is outside the scope of public policy.

Outlawing Discrimination

Antidiscrimination and affirmative action laws are not aimed at reducing the inequalities that one starts the race with; rather they attempt to ensure that no unfair obstacles are thrown in the way during the race. Antidiscrimination laws have some effect. Able people have been and still are held back unfairly. If the race itself is not rigged, however, those who work very hard and are very able can overcome the handicaps they begin with. Because people start out with unequal backgrounds, however, they do not have an equal chance of success just by running the same course.

Education

Education is widely believed to be the key to reducing unfair disadvantages associated with poverty. Prekindergarten classes designed to provide intellectual stimulation for children from deprived backgrounds, special education courses for those who don't speak standard English, and loan and grant programs to enable the poor to go to school as long as their ability permits them to do so—all these are designed to increase the chances of students from lower-class backgrounds getting an education.

The programs have had some success. Certainly colleges and universities see many more students from disadvantaged backgrounds than they used to do. Because students spend only 35 hours a week at school, however, and another 130 hours a week with their families and neighbors, the school cannot reasonably overcome the entire deficit that exists for disadvantaged children. A study entitled *The Beginning School Study* documents the fact that poor children and better-off children perform at almost the same level in first- and second-grade mathematics while school is in session. For children in poverty, however, every summer means a loss in learning, whereas every summer means a gain for those out of poverty (Entwisle & Alexander 1992). The home environments of less-advantaged children do not often include trips to the library and other activities that encourage them to use and remember their schoolwork. Consequently, for every step they take at school, they slide back half a step at home during the summer.

Conclusion

This review of programs designed to reduce unfair advantages or disadvantages leads to several conclusions. First, the family is at the root of the inheritance of both advantage and disadvantage. As long as some people are born in tenements or shacks, as long as their parents are uneducated and have bad grammar and small vocabularies, and as long as they have no encyclopedias or intellectual stimulation—while others are born to educated parents with standard speech patterns who flood them with intellectual stimulation and opportunity—there can never be true equality of opportunity. To some extent, the pursuit of equal opportunity will come at the expense of the family; any attempt to reduce inheritance of status requires weakening the influence of parents on their children.

In any culture, individuals espouse values that conflict with one another and values that are so idealistic that few attempt to live by them. The United States's ambivalent feelings about inequality are no exception. Do we want equal opportunity badly enough to pay the costs, or will it remain, like premarital chastity or marital stability, an ideal but not a reality?

Summary

1. Stratification is distinguished from simple inequality in that (a) it is based on social roles or membership in social categories rather than on personal characteristics, and (b) it is supported by norms and values that justify unequal rewards.

2. Marx believed that there was only one important dimension of stratification: class. Weber added two further dimensions. Most sociologists now rely on Weber's three-dimensional view of stratification, which embraces class, status, and power.

3. Inequality in income and wealth is substantial in the United States and has changed little over the generations. The inequality has widespread consequences and affects every aspect of our lives. Although the negative consequences of being working class or lower class are less than they were 50 years ago, the lower class and working class continue to be disadvantaged in terms of health, happiness, and lifestyle.

4. Structural-functional theorists use a supply-and-demand argument to suggest that inequality is a functional way of sorting people into positions; inequality is necessary and justifiable. Conflict theorists believe that inequality arises from conflict over scarce resources, in which those with the most power manipulate the system to enhance and maintain their advantage.

5. Beegley provides a contemporary synthesis of conflict and structural-functional theories. His theory argues that power and social structure are critical factors, but it allows some role for individual characteristics.

6. Allocation of people into statuses includes macro and micro processes. At the macro level, the labor market sets the stage by creating demands for certain statuses. At the micro level, the status attainment process is largely governed by indirect inheritance.

7. There is a great deal of continuity in social class over the generations. In the United States, this inheritance of social class is indirect and works largely through education. Achievement motivation and intelligence, however, are factors that allow for upward and downward mobility.

8. In spite of high levels of inequality, most people in any society accept the structure of inequality as natural or just. This shared ideology is essential for stability. In the United States, this ideology is the American Dream, which suggests that success or failure is the individual's choice.

9. Approximately 15 percent of the U.S. population fall below the poverty level. Many of the poor are children or elderly. Although some part of poverty may be due to micro-level processes (the culture of poverty and indirect inheritance), the structure of opportunity determines how extensive poverty is in a society.

10. Because families pass their social class on to their children, any attempt to reduce inequality must take aim at the intergenerational bond between parents and children and must reduce the ability of parents to pass on their economic status and values.

Sociology on the Net

It is always fun to find out who is the richest of the rich, so let's go to the *List of Lists!*

```
http://hoovweb.hoovers.com
```

Once you are on the Hoover home page, open the selection called *Who's On Top?* and then click on the *List of Lists.* Take your pick, but be certain that you read the *Forbes 400 Richest Americans.* How many of these names do you recognize? Do you shop at their stores? Do you buy their gasoline? Do you root for any of their professional sports teams? If you were on this list, how would your life be changed? What would remain the same and what would you do differently?

We can dream about wealth, but the probability of ever being on this list is very slim. In fact there are far more people living in poverty than there are millionaires. *Poverty* is defined as a relative inability to subsist. What is the so-called "poverty level," and how does it vary with family size?

```
http://aspe.os.dhhs.gov/poverty/
poverty.htm
```

We have reached the United States Department of Health and Human Services. Browse around and check out the various offerings. When you have scrolled down near the bottom of the menu, click on the *1996 HHS Poverty Guidelines.* Read this brief summary. Exactly what is a *poverty guideline?* Why are there different guidelines for Alaska and Hawaii? How would you live if your family fell within these parameters? How do people that live in poverty differ from those on the *Forbes 400 Richest Americans* list? Who do you think lives longer and has better health?

In the past decade, there has been a significant shift in the distribution of wealth in the United States.

```
http://www.census.gov/population/www/
pop-profile/profile.html
```

We have reached a very rich source of data and information. It is the *United States Census Bureau Population Profiles.* You may wish to bookmark this address because we will return to it many times throughout these exercises. Scroll down to the section entitled *Money Income.* Open this document and read it thoroughly. Pay particular attention to the colored graph showing the changes in household income by quintile. What has been the trend in income since 1989 for the lowest 20 percent? For the middle 60 percent? For the top 20 percent? For the top 5 percent? Who are the winners and the losers? (Hint: Calculate who had the largest percentage increase and who had the largest percentage decrease in income from 1989 through 1993.) Do you support a tax break for the rich?

FIND IT ON INFOTRAC COLLEGE EDITION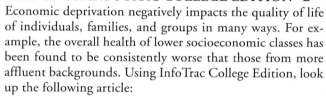

Economic deprivation negatively impacts the quality of life of individuals, families, and groups in many ways. For example, the overall health of lower socioeconomic classes has been found to be consistently worse that those from more affluent backgrounds. Using InfoTrac College Edition, look up the following article:

> "U.S. Socioeconomic and Racial Differences in Health: Patterns and Explanations." David R. Williams and Chiquita Collins. *American Sociological Review,* 1995, v21 p349.
>
> (Hint: Enter the search term *David R. Williams* using the Subject Guide.)

Explain the interactions between race and socioeconomic status in regards to health. How does socioeconomic stratification in the United States impact the relative health of African Americans? What solutions can you offer to this problem?

Suggested Readings

Ellis, Lee (ed.) 1994. *Social Stratification and Socioeconomic Inequality.* Westport, Conn.: Praeger. An up-to-date edited collection of empirical and theoretical writings on social stratification, primarily in the United States.

Hochschild, Jennifer L. 1995. *Facing up to the American Dream: Race, Class, and the Soul of the Nation.* Princeton, N.J.: Princeton University Press. A book that examines the extent to which the American Dream is inaccessible to many people in the United States and addresses the issue of why there is so little support for redistribution among the U.S.'s poor.

Lenski, Gerhard. 1966. *Power and Privilege: A Theory of Social Stratification.* New York: McGraw-Hill. A classic, this major work distinguishes the fundamental characteristics found in different types of societies, particularly in terms of socially structured inequality.

McFate, Katherine, Lawson, Roger, and Wilson, William Julius. 1995. *Poverty, Inequality, and the Future of Social Policy: Western States in the New World Order.* New York, N.Y.: Russell Sage Foundation. A contemporary discussion of the incidences of poverty and inequality throughout the world, with emphasis on policy decisions that will affect the future.

Patterson, James T. 1994. *America's Struggle against Poverty, 1900–1994.* Cambridge, Mass.: Harvard University Press. A history of poverty, poor people, and poverty legislation in the United States during this century.

Sharma, K.L. (ed.). 1995. *Social Inequality in India: Profiles of Caste, Class, Power, and Social Mobility.* Jaipur, India: Rawat Publications. A collection of theoretical, histori-cal, and empirical essays on the topics of caste and class in India.

Wilson, William Julius. 1987. *The Truly Disadvantaged: The Inner City, the Underclass, and Public Policy.* Chicago: University of Chicago Press. A strong statement about the growing U.S. underclass by one of the nation's most promi-nent experts. Wilson argues convincingly for new govern-ment programs to put a floor under all citizens.

CHAPTER 7

Global Inequality

Outline

Social Change: Inequality and Development

The central fact in the international political economy is the vast inequality that exists in today's world. In 1995, the average U.S. citizen produced $26,980 worth of goods and services; the average citizen of Mozambique produced only $80 worth. The average North American will live to age 76 compared to age 46 for individuals living in Mozambique (Population Reference Bureau 1997). The massive disparities that exist not only in wealth and health but also in security and justice are the driving mechanism of current international relationships.

Because massive inequality leads to political instability and unjustifiable disparities in health and happiness, nearly every nation—whether more or less developed—thinks that reductions in international inequality are desirable. The most accepted way to do this is through development—that is, by raising the standard of living of the less-developed nations.

What is development? First, development is *not* the same as Westernization. It does not necessarily entail monogamy, three-piece suits, or any other cultural practices associated with the Western world. **Development** refers to the process of increasing the productivity and raising the standard of living of a society, leading to longer life expectancies, better diets, more education, better housing, and more consumer goods.

> **Development** refers to the process of increasing the productivity and standard of living of a society—longer life expectancies, more adequate diets, better education, better housing, and more consumer goods.

Three Worlds: Most- to Least-Developed Nations

Almost all societies in the world have development as a major goal: They want more education, higher standards of living, better health, and more productivity. Just as social scientists often think of three social classes in the U.S. stratification system—upper, middle, and lower—nations of the world can also be stratified into roughly three levels. One familiar scheme for classifying countries based on their position in the international system is the Three Worlds typology. According to this model, the First World consists of rich, industrialized countries; the Second World includes the somewhat less-industrialized countries of the former Soviet bloc; and the Third World includes the more than 100 other nations, largely nonindustrialized, impoverished, and politically vulnerable. The end of the Cold War and the considerable variation found in levels of development across the Third World has led many social scientists to adopt the modified classification system we use here.

The **most-developed countries** include those rich nations that have relatively high degrees of economic and political autonomy: the United States, the Western European nations, Japan, Canada, Australia, and New Zealand. Taken together, these nations make up roughly 16 percent of the world's population, produce between 55 to 60 percent of the gross world product, and consume approximately 54 percent of the world's energy (Chirot 1986). Politically, economically, scientifically, and technologically, they dominate the international political economy.

> **Most-developed countries** include those rich nations that have relatively high degrees of economic and political autonomy: the United States, Western Europe, Canada, Japan, Australia, and New Zealand.

Less-developed countries include the former Soviet Union and the former Communist bloc nations of Eastern Europe, plus several nations in Southeast Asia and Central and South America that have seen substantial improvements in standard of living during the past few years. These countries hold an intermediate position

> **Less-developed countries** include the former Soviet Union and the nations of the former Communist bloc, plus several of the developing nations of Southeast Asia and Central and South America.

in the world political economy. They have far lower living standards than the most-developed nations but are substantially better off than the least-developed nations.

The remaining 75 percent of the world's population lives in the **least-developed countries** that are characterized by poverty and political weakness. Although these nations vary in their populations, political ideologies, and resources, they are considerably behind on every measure of development.

Least-developed countries include those nations that share a peripheral or marginal status in the world capitalist system.

The Human Development Index

The differences among the world's nations are obvious: In most-developed countries, people are healthier, more educated, and richer, but how important are these differences in the day-to-day quality of the average person's life?

One approach to answering this question is to develop an index that measures the average achievements of a country along the basic dimensions of human experience: longevity, knowledge, and a decent standard of living. Another approach focuses not only on these three aspects of development but also takes into account the unequal opportunities of men and women. Map 7.1 shows the location of the most- to least-developed countries worldwide. Table 7.1 compares several basic quality-of-life indicators for 15 nations, representing most-developed, less-developed, and least-developed countries. In addition to information about longevity and economic productivity, Table 7.1 also includes each country's overall ranking on the composite Human Development Index and the Gender-Related Development Index. These indexes are based on information about adult literacy rates and educational attainment, life expectancy, and per capita gross domestic product; the greater the disparity between men's and women's

TABLE 7.1
The Extent of International Inequality: Comparing 15 Countries

Development	Per Capita GNP (U.S. dollars)	Life Expectancy at Birth	Infant Mortality Rate	Human Development Ranking	Gender Related Development Ranking
Most-developed countries					
Canada	19,380	78	6.2	1	1
Japan	39,640	80	4.0	7	7
Norway	31,250	78	5.2	3	3
Singapore	26,730	76	4.0	26	26
United States	26,980	76	7.3	4	4
Less-developed countries					
Brazil	3,640	67	48.0	68	60
Bulgaria	1,330	71	14.8	69	49
Ecuador	1,390	69	40.0	72	73
Lithuania	1,900	70	13.0	76	55
Russian Federation	2,240	65	18.0	67	46
Least-developed countries					
Haiti	250	50	48.0	156	130
India	340	59	75.0	138	118
Kenya	280	54	62.0	134	112
Mozambique	80	46	118.0	166	139
Rwanda	180	40	85.0	174	—

SOURCE: United Nations Human Development Report 1997, Population Reference Bureau 1997.

MAP 7.1
Human Development Around the World
The human development index measures the average achievements in a country in three basic dimensions of human development—longevity, knowledge, and a decent standard of living.
Source. http://www.undp.org/undp/hdro/tableZ.htm.

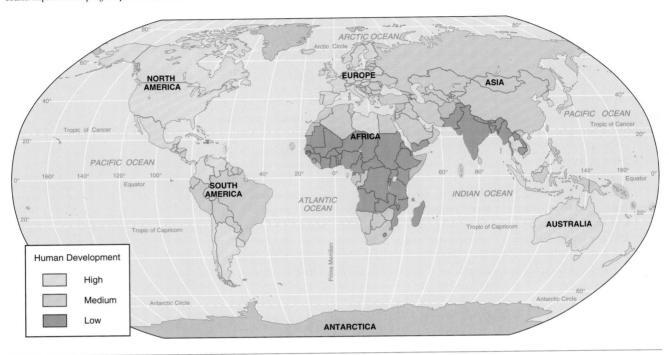

quality of life, the lower a country's Gender-Related Development ranking will be compared to its overall Human Development ranking.

In general, the more productive a nation is, the better its quality of life. Japan, with a per capita gross national product of $39,640, also has one of the lowest infant mortality rates in the world: Each year, for every 1,000 live births, 4 children who have not yet reached their first birthday will die. This compares to figures for the world's poorest nation—Mozambique—where per capita gross national product is only $80 and nearly 118 children per 1,000 births will die before their first birthday. Note, however, that the relationship between economic productivity and quality of life is not perfect. Canada, with a per capita gross national product of $19,380, nevertheless holds the number one ranking in the world on overall development and on gender-related development. In large part, Canada's number one ranking reflects the fact that access to health care, to education, and to adequate nutrition is more universally available there than in the more affluent Norway or United States.

The gap between quality of life for the world's most affluent and poorest continues to grow. The poorest 20 percent of the world's people share only 1.1 percent of global income. This figure is down from 1.4 percent in 1991 and 2.3 percent in 1960. Put another way, the ratio of the income of the top 20 percent to that of the poorest 20 percent rose from 30 to 1 in 1960, to 61 to 1 in 1991, to a new high of 78 to 1 in 1994. Women and children are particularly at risk in poor communities. A half-million women die each year in childbirth at rates 10–100 times those in the most-developed nations. Worldwide, more than 160 million children are moderately or severely malnourished (Human Development Report 1997). International inequality is indeed dramatic.

Although many of the differences between nations are matters of culture, there is general consensus that rich is better than poor, security is better than insecurity, and health is better than sickness. No nation wants to be poor and underdeveloped. What are the causes of current inequalities and what are the prospects for their relief? We examine two general theories of development—modernization and world system theory—and discuss their implications for resolving global inequalities.

Modernization Theory

Modernization theory sees development as the natural unfolding of an evolutionary process in which societies go from simple to complex institutional structures and from primary to secondary and tertiary production. This is a structural-functional theory based on the premise that adaptation is the chief determinant of social structures. According to this perspective, developed nations are merely ahead of the developing nations in a natural evolutionary process. Given time, the developing nations will catch up.

Modernization theory was very popular in the 1950s and 1960s. It implied that developing nations would follow pretty much the same path as the developed nations. Greater productivity through industrialization would lead to greater surpluses, which could be used to improve health and education and technology. Initial expansion of industrialization would lead to a spiral of ever-increasing productivity and a high standard of living. These theorists believed this process would occur more rapidly in the least-developed nations than it had in Europe because of the direct introduction of Western-style education, health care, and technology (Chodak 1973).

From today's vantage point, modernization theory seems naïve. The least-developed countries have not caught up with the developed world. In many cases, the poor have become poorer, while the rich have become richer.

Why haven't the less-developed nations followed in our footsteps to modernization? The primary reason is that they face an entirely different context: Population, environment, and world social organization are all vastly different now than they were in the seventeenth and eighteenth centuries. England, for example, was able to rely on surpluses extracted from her colonies to fuel industrialization. When improving conditions caused an overpopulation problem in the nineteenth century, the excess population was able to migrate to America and Australia.

The developing nations of the late twentieth century face many obstacles not faced by earlier developers: population pressures of much greater magnitude and with no escape valve, environments ravaged by early colonialists, plus the disadvantage of being latecomers to a world market that is already carved up. These formidable obstacles have given rise to an alternative view of world modernization—world system theory.

World System Theory

The entire world may be viewed as a single economic system that has been dominated by capitalism for the past 200 years. Nation-states and large transnational corporations are the chief actors in a free-market system in which goods, services, and labor are organized to maximize profits (Chirot 1986; Turner & Musick 1985). This organization includes an international division of labor where some nations extract raw materials and others fabricate raw materials into finished products.

Modernization theory sees development as the natural unfolding of an evolutionary process in which societies go from simple to complex institutional structures and from primary to secondary and tertiary production.

THINKING

CRITICALLY

Consider how the major themes of U.S. stratification covered in Chapter 6 apply to international inequality. *You are a hundred times better off than the average person in Mozambique. Is this a necessary and just reflection of your greater contribution to society? Is it surprising that so many Third World peoples ascribe to a conflict theory of international inequality?*

a global perspective

WOMEN AND DEVELOPMENT

*T*he Fourth World Conference on Women, held in Beijing, China, during the summer of 1995, pointed to a fact already known by the participants: Third World women are important. In fact, in many parts of the Third World, women are the main producers of food crops. Where women do not actually cultivate crops, they are almost always the main cooks and food processors. In several parts of sub-Saharan Africa, for instance, women are responsible for at least 70 percent of the total production, processing, and marketing of food. In addition to providing food and an almost exclusive responsibility for the rearing of children, women also have the added burden of collecting fuelwood and water; in areas where there are severe fuelwood deficits—Burkina Faso, Haiti, and Nepal, for example—women spend at least five to ten hours a week seeking wood for their cooking fires (United Nations Population Fund 1991). Although women and girls form only half of the world's population, their labor appears to account for at least two-thirds of the total work hours expended on any given day or in any given year. Tellingly, they receive only one-tenth of the world's income in return and own only about one-hundredth of the world's property (AID Horizons 1983). Despite the economic contributions of women to both their families and their nations, at least until the mid-1970s development agencies routinely directed educational programs about the advantages of using new types of seed, fertilizer, and cropping systems to almost exclusively male audiences. No wonder these programs failed to increase agricultural productivity when the people doing the actual farming were female!

Most specialists now recognize that sustainable development will only become possible when women are full partners in the process. Research clearly shows that overall economic growth has been fastest in those areas of the world where women have higher status and slowest where they face the greatest disadvantage (State of World Population 1992). Where women attain higher levels of education, their productivity in both the paid and unpaid labor market is enhanced. Furthermore, because women appear to be more likely than men to invest new resources in the nutrition, medical care, and education of their children (Schultz 1993), investments in women's education may be particularly crucial to development efforts.

As a result of the growing awareness of the critical role that women play in societal as well as domestic economies, new development strategies are more often aimed at women. These include programs designed to improve women's literacy, programs to help women market handicrafts and agricultural produce, changes in legal systems so that women can own and manage their own property, and giving women access to contraceptive knowledge and technology so that they can control their reproduction. Such programs cannot take the place of broader programs designed to reduce massive international and national disparities, but including women in development programs will help to ensure that their benefits reach down to the household level.

Although women and girls constitute only about half of the world's population, they do about two-thirds of the work. In addition to childcare, cooking, and, in many parts of the world, much of the agricultural production, collecting firewood and water is also typically the responsibility of women in the household. In return for their labor, women worldwide receive approximately one-tenth of the income earned and control about one-hundredth of the property.

Nation-states can pursue a variety of strategies to maximize their profits on the world market. They can capture markets forcibly through invasion, they can try to manipulate markets through treaties or other special arrangements, or they can simply do the international equivalent of building a better mousetrap. The Japanese auto industry (indeed, all of Japanese industry) is a successful example of the latter strategy.

On a global basis, capitalism operates with less restraint than it does within any single nation. There is no organized equivalent of welfare or Medicaid to take care of indigent nations. In the absence of a world political structure, economic activity is regulated only by market forces such as supply and demand (Turner & Musick 1985).

World system theory is a conflict analysis of the economic relationships between developed and developing countries. It looks at this economic system with a distinctly Marxist eye. Developed countries are the bourgeoisie of the world capitalist system, and underdeveloped and developing countries are the proletariat. The division of labor between them is supported by a prevailing ideology (capitalism) and kept in place by an exploitive ruling class (rich countries and transnationals) that seeks to maximize its benefits at the expense of the working class (underdeveloped and developing countries).

World system theory is a conflict perspective of the economic relationships between developed and developing countries, the core and peripheral societies.

Despite the rapidity with which new ideas and new technologies can be diffused throughout the world, a very large portion of the world's population does not share in the increased standard of living that development makes possible. In Vietnam, as well as throughout much of Asia, Africa, and Central and South America, rickshaws, bicycles, mules, and good old feet are more common modes of transportation than automobiles.

Core societies are rich, powerful nations that are economically diversified and relatively free of outside control.

Peripheral societies are poor and weak, with highly specialized economies over which they have relatively little control.

THINKING

CRITICALLY

Some people speculate that the most serious drain on world resources in the next decades will not be due to simple population growth but rather to the growing affluence of the Chinese—who can now afford more food, cars, and material goods. Consider both the advantages and disadvantages of growing Chinese affluence for people of the United States.

World system theory distinguishes two classes of nations: core societies and peripheral societies. **Core societies** are rich, powerful nations that are economically diversified and relatively free of outside control. They arrive at their position of dominance, in part, through exploiting the periphery. On an international level, they are very similar to the industrial core of the United State's dual economy (Chapter 13).

Peripheral societies, by contrast, are poor and weak, with highly specialized economies over which they have relatively little control (Chirot 1977). Like the small-business sector in the U.S. dual economy, peripheral societies are vulnerable to change and little able to control their environment. Some of the poorest countries rely heavily on a single cash crop for their export revenue. For example, 93 percent of Guinea-Bissau's export earnings come from cashews, coffee accounts for 62 percent of Uganda's total exports; and in Cambodia, one of the poorest countries in the world, rubber and timber together provide 94 percent of the export revenue (Europa Yearbook 1997). The economies of these and many other developing nations are vulnerable to conditions beyond their control: world demand, crop damage from infestation, flooding, drought, and so on.

A key element of world system theory is the connectedness between First World prosperity and Third World poverty. The message is clear: Our prosperity is their poverty. In other words, our inexpensive shoes, transistors, bananas, and so on depend on someone in a least-developed nation receiving low wages. Were their wages to rise, our prices would rise and our standard of living would drop. Furthermore, if their wages were to rise they also would want more consumer goods; the ensuing demand on resources would simply be more than the most-developed economies or the planet could sustain.

A Case Study: Economic Development, the Environment, and the Less-Developed World

Development refers to the process of raising the standard of living and improving the quality of life of a society. In theory, development programs should be as concerned with improving access to education and health care and with ensuring the rights of all citizens—regardless of age, gender, race, or ethnicity—as the programs are with increasing labor productivity. In practice, however, aid to least-developed nations has focused almost exclusively on fostering economic development, and its success is most often measured simply in terms of increases in gross national product or access to consumer goods. Under this model of development, affluent lifestyles become the goal to which poor nations aspire.

Interestingly, even in Europe and the United States, the experience of mass consumption by large numbers of people has occurred only recently. In the late 1940s, for instance, only one British household in 25 had a washing machine and only one in 50 had a refrigerator. However, by the mid-1980s in Kuala Lumpur, Malaysia, 13 percent of the very poorest households and 50 percent of those in the next poorest group owned a refrigerator. In the Chinese capital of Beijing, two-thirds of all families currently own a fridge. Given these trends in consumption and with an estimated world population of 10 billion people by the year 2050 it is estimated that then there will be only enough aluminum to last another 20 years, enough copper and oil to last another 4.5, and enough coal to last another 51 (State of World Population 1992).

Paradoxically, threats to the so-called "renewable" resources may be even more severe. Water and trees, for example, are renewable resources. With wise use, they are self-replenishing. Overuse, however, can destroy such resources within a generation. In the late-1980s, 11 nations—primarily in the Middle East—were already using water at a rate much faster than it can be replaced. In addition, although only 6 African nations faced water stress or scarcity in 1982, it is estimated that by the year 2025,

21 countries will be affected. These nations will be home to 1,100 million people, or approximately two-thirds of the population of the African continent.

Irrigated agriculture consumes most of the water used by people—nearly 70 percent worldwide. Industry accounts for about 23 percent of worldwide water use; and households use only about 8 percent. Although irrigation and industry have improved nutrition and access to goods in the short run, in the long-term both may have serious detrimental consequences. Agriculture, which accounts for the largest share of water consumption, is the least efficient. It is not uncommon for 70 to 80 percent of the water diverted to irrigation systems to be lost before it ever reaches the fields (Falkenmark & Widstrand 1992); this runoff leaches important nutrients from the soil and contributes to increased soil salinity. In the long run, then, irrigation may reduce agricultural productivity. Industry is the second largest user of water; moreover, the kinds of industry often held up as models for Third World development—food processing, petroleum refining, and pulp and paper refineries—are particularly water intensive. Thus, in many of the least-developed countries, economic development is likely to place extraordinary demands on already fragile water supplies. Water is used by factories not only for cooling, generating steam to run equipment, and as a method of transport but also as a dumping ground for industrial waste. Industrial contamination of rivers and lakes has reached such epidemic proportions in Eastern Europe and the former Soviet Union that it is estimated that 50 to 80 percent of the drinking water in many of these nations is unfit to drink (French 1991).

Industrialization and the Environment: A Case Study of Eastern Europe

Eastern Europe and the former Soviet Union are good examples of what can happen to the environment when industrialization and economic growth and development are pursued at any cost.

- According to a 1979 report by the local Communist youth league, a lake near one Soviet chemical plant was so polluted that citizens of the town disposed of stray dogs by throwing them into the water. The report added that the lake was so contaminated by phenols that the carcasses dissolved within days.
- The Vistula River, which runs through Poland, is so laden with poisons and corrosive chemicals that it is considered unusable, even for factory coolant systems (Jensen 1990).
- Near the industrial center of the eastern part of Germany, life expectancy is six years less than in other parts of the region and four out of five children in the region will develop chronic bronchitis or heart problems by the age of seven (Painton 1990).
- When 300 children attending two kindergartens in Estonia began to lose their hair, residents were horrified. Eventually, the former director of a local factory revealed that his company had dumped radioactive waste where the schools were later built (French 1991).

A report prepared by the Russian Environmental Ministry for the 1992 Earth Summit in Rio de Janeiro blamed the ecological woes of the nation on the "growth-at-any-cost" mentality that drove Soviet economic development at least since the late 1920s. Coupled with concerns about national security and the rapid rise of a huge military-industrial complex, Soviet bloc emphasis on production over efficiency created both a badly outdated industrial sector and environmental hazards of unparalleled magnitude (Stanglin 1992). Of course, this creates a real quandry: How can the countries of Eastern Europe and the former Soviet Union revitalize their economies *and* clean up

THINKING CRITICALLY

If affluence means more environmental destruction, it seems like we're caught on the horns of a dilemma. Is eliminating poverty more or less important than preserving the environment?

the environment when the costs of repairing the ecological damage alone are estimated at $100 billion each, just for Poland and the former Czechoslovakia (Marshall 1991)?

Conclusion

With a population of 5.8 billion and growing, it is clear that planet Earth simply cannot sustain the kind of energy-hungry, wasteful, pollution-riddled development that has driven the economies of First and Second World nations. In 1987, the important Brundtland Commission defined **sustainable development** as "development that meets the needs of the present without compromising the ability of future generations to meet their own needs." There is broad agreement within the international community that two major changes will be necessary to achieve this type of development.

Sustainable development is development that meets the needs of the present without compromising the ability of future generations to meet their own needs.

First and foremost, sustainable development will require the elimination of absolute poverty so that the 1.1 billion poorest of the world's people can produce or buy the food, clothing, and housing necessary to ensure health and self-respect (State of World Population 1992). Ending absolute poverty will depend, then, on improving access to education, health care, clean water, and sanitation. It will depend, too, upon in-

Throughout Eastern Europe and the former Soviet Union, the result of government policies that encouraged industrialization and economic growth at any cost has been pollution of almost unprecedented magnitude. Here, President Václav Havel of the Czech Republic responds to an industrial waste dump in his country. For those who must live near the site, the health consequences are likely to be far more serious than olfactory discomfort. Western lifestyles and Western models of development are simply too costly—in both economic and ecological terms—to be sustainable in either the First, Second, or Third World.

creasing the status of women and reducing the buildup of military arms that already costs least-developed nations an enormous share of their scarce resources (Sen & Grown 1987). Ending poverty is a goal in itself, but because poverty and rapid population are often found together, it will also enhance development by limiting population size.

In addition to eliminating destitution, sustainable development will require meeting the legitimate aspirations of the 3 billion people who are neither very poor nor rich. In today's world, the most-affluent 1 billion satisfy their aspirations with little regard for sustainability, but the world will not be able to satisfy the aspirations of billions more people in this way. Consequently, in order to meet the needs of the middle third of the world's population, the economic inequalities that exist within and between countries will have to be reduced and the benefits of development more fairly distributed.

Second, sustainable development will require slower population growth. Smaller increases in world population, coupled with an evolution to lower-consumption lifestyles and more efficient production, will reduce the environmental impact of development and help to ensure equal access to "the good life" across generations (State of World Population 1992).

Global Poverty and Dependence

Most contemporary scholars use some form of world system theory to understand development. Although many would reject the Marxist implications of the theory, there is general agreement that international inequality is the result of competition for scarce natural resources within an international capitalistic economic framework that rewards immediate profitability over longer-term sustainable development.

Forms of Economic Dependency

The primary characteristics of the periphery is that these nations have relatively little control over their economies. They are dependent. In many cases, these dependent relationships represent a triple alliance among three actors: transnational corporations (generally based in the most-developed nations), the governments of the peripheral societies, and local elites. Scholars who study the world system identify three types of dependency (Bradshaw 1988): the classic "banana republic," industrial dependency, and foreign-capital dependency.

The Banana Republic
The classic case of dependency occurs in a least-developed nation whose economy is dependent on the export of raw materials—bananas, fruit, or minerals. African nations, the Indian subcontinent, and Latin American countries have all fallen into this situation; many of the world's peripheral nations are still in it. Because the cost of manufactured imports is much higher than the prices paid for raw material exports, these countries accumulate huge debts. In order to pay the debts, they must produce and export more crops such as coffee, sugar, or even broccoli; this emphasis on export crops reduces the nation's ability to feed itself and retards the development of economic diversity.

Industrial Dependency
Increasingly, transnational corporations are making use of one of the other major assets of less-developed countries: cheap labor. Third World assembly plants have reduced labor costs dramatically for transnationals (say, from $10 an hour to $3 a

technology

INFORMATION TECHNOLOGY AND GLOBAL CULTURE

*E*ven the most creative society discoverers or invents only a small portion of all its innovations. Much of the content of U.S. culture, for instance, was acquired from successive waves of immigrants, each bringing their own language, values, and customs to American shores. In the late twentieth century, one of the primary vehicles for cultural diffusion is the new information technology. The information revolution, which began in the United States, is rapidly spreading over the world. Almost every nation on every continent now has some access to the Internet and E-mail technology (Watson 1995). One reason for this growth is that computer hardware that has become outdated in the United States is bought by international recyclers, who sell or donate it to countries such as Russia and Ethiopia (Bryant 1994). Worldwide annual sales of personal computers exceed 50 million units; to gain perspective on that number, consider that about 35 million cars are sold per year. The Internet has more than 20 million E-mail users worldwide, with the number growing 20–30% per quarter. Created in 1989 at a European physics lab, the World Wide Web allows Internet users to browse efficiently through enormous quantities of information; World Wide Web usage is increasing twice as fast as the Internet itself (Talero & Gaudette 1996).

Not only has the use of information technology spread throughout the world, the fact that it has done so means that other aspects of cultural diffusion also take place much more rapidly. Ideas can sweep the world within days and be introduced into the most remote villages within weeks and months. A fervor for democracy, for example, swept the world in 1989. The year began with pro-democracy student protests in Tienanmen Square in Beijing, China and ended with the fall of the Berlin Wall and the toppling of communist governments in Eastern Europe. Many of those seeking freedom and democracy relied on the ideals and symbols of the American revolutions; the Statue of Liberty lent symbolic support to the demonstration at Tienanmen Square.

The spread of information technology not only means more rapid change in all areas of life, it also means growing international similarity. In Moscow, Beijing, Nairobi, and Boston, business leaders are wearing the same kinds of suits. In the Western United States and in Africa, plans to develop virtual universities are well underway. Throughout the world, young people are listening to the same kinds of music. An article published in a Soviet sociology journal not long before the dissolution of the USSR, for example, analyzed the ill effects of "khard-roka and khevi-metallu" on Soviet youth (Sarkitov 1987). On a more serious level, the speed of diffusion has also meant that nuclear weapons and terrorist technologies are also widespread.

The "Goddess of Liberty," which symbolized the pro-democracy demonstrations in Tienanmen Square in Beijing in 1989, bears a marked resemblance to the U.S. Statue of Liberty. The ideals of democracy, its symbols, and even the constitutional forms that have been developed to embody it have been diffused throughout the world. The rapidity of global diffusion means that ideas, fashions, and technologies are spread rapidly. Nevertheless, economic and political realities may distort or repulse ideas that do not fit with established patterns.

day) and have also saved them money on the transportation costs of goods that will be sold in these countries.

Some observers have hoped that investment in Third World industrial plants would improve local health and transportation facilities, provide jobs, spur development of indigenous subsidiary industries, and generally galvanize local economies. Empirical studies demonstrate that few of these good effects are actually realized.

Foriegn-Capital Dependency

When there is a relatively strong indigenous elite, dependency may take the form of dependency on loans and investment (Wimberly 1990). In this case, the firms will be owned and operated by locals, rather than by foreign corporations. In theory, this should encourage development by providing additional capital.

Instead, the extension of loans to Third World nations has usually resulted in a dramatic debt crisis. High inflation and worldwide recession in the early 1980s decreased the demand for Third World products at the same time they increased the interest rates debtor nations were paying on existing loans. To meet debt payments, countries had to reduce investment in their own economies and squeeze their own people even harder (Wimberly 1990).

The Consequences of Dependency

By definition, dependency is a bad thing. It means that you do not have control over your economy and, perhaps, over your government. Although ties to richer, stronger economies might be expected to benefit less-developed nations, research does not bear this expectation out. Scholars have identified at least three unfortunate consequences of foreign-capital penetration.

First, efforts to attract foreign capital usually result in excessive investment in urban areas and almost no investment in the rural areas, where most of the nation's population is likely to live. One long-term consequence of this investment pattern is a substantial decrease in the overall standard of living: per capita food production actually decreases as people migrate to huge urban areas that are ill-equipped to receive them (Stokes & Anderson 1990).

This worker is assembling garments for a well-known U.S.-owned clothing design firm. By locating assembly plants in the least-developed nations, transnational firms realize huge savings in labor costs and U.S. consumers get cheaper products. Among the undesirable side effects, however, are loss of jobs for U.S. workers, and the growth of inequality in Third World nations.

Second, only the urban elites and the small urban middle class actually benefit from foreign capital (Bradshaw 1988). After all, low wages are what attract foreign investment in the first place. Thus, transnational corporate penetration and foreign aid dependence increase income inequality. Partially as a result of that inequality, infant mortality rates rise (Wimberly 1990).

A final result of dependence on foreign capital is that governments often become more authoritarian as they ally themselves with local cities in an attempt to ensure continued investment. Because this alliance occurs at the expense of the majority of the people, anger and conflict increase; transnational penetration has been shown to increase the risk of political violence (London & Robinson 1989).

Some scholars have questioned whether or not foreign capital penetration actually has all of the negative effects outlined above. Asking the question "Does foreign capital really do more harm than good?" virtually all studies have at least agreed that domestic investment in poor nations is far more beneficial than is foreign investment (Dixon and Boswell 1996; Firebaugh 1996).

Competition, Change, and International Relationships

Haiti: A Case Study

Haiti, a Third World nation in the Caribbean, has a predominantly agricultural economy. Approximately 61 percent of the labor force is involved in agriculture, most on small family plots worn away by severe soil erosion. Coffee and cocoa are the major cash crops, but light manufacturing—cosmetics, textiles, toys, and baseballs—provides the major source of export revenues (Europa Yearbook 1997). With a population of about 6.7 million people and an annual growth rate of 1.8 percent, Haiti is one of the fastest growing and poorest countries in the Western hemisphere. Per capita productivity is $250 per person. Yet even this figure does not accurately reflect Haitian poverty, given the wide earnings gap between local elites and the remaining population who make less than $100 a year (Nelan 1993). Almost 55 percent of all Haitians are illiterate (Europa Yearbook 1997). Life expectancy is only 50 years, and the infant mortality rate is almost ten times higher than it is in many of the most developed nations.

Background

From the beginning Haiti's role in the world system has been unique. Originally a French colony, the Haitian Republic gained its independence in 1804 as the result of a slave uprising. As the first black republic in a world system still dominated by slavery, Haiti was diplomatically isolated for more than three decades by both the United States and European powers, who feared that the Haitian revolution would set a dangerous example for blacks in their own countries and colonies (Plummer 1985). This concern did not, however, stop foreign companies from trading with Haiti, and Haitian migrant labor played a major role in the Cuban and Dominican Republican sugar economies.

Because political unrest threatened U.S. economic interests and because the United States wanted to use Haiti as a base to protect its routes to the Panama Canal, the U.S. Marines occupied Haiti from 1915 to 1934. Working with local elites, the Marines helped found the modern Haitian military and ran the customs and banking business of the country (Wilentz 1993). Overt racism also played a role in the U.S. decision to occupy Haiti. U.S. Assistant Secretary of State William Phillips explained at the time,

"These facts all point to the failure of an inferior people to maintain the degree of civilization left them by the French, or to develop any capacity of self-government entitling them to international respect and confidence" (Schmidt 1971, 63).

The Duvalier Dictatorship

Between 1957 and 1986, Haiti was governed by the Duvalier family. In cooperation with foreign investors, the Duvaliers and a small economic and military elite amassed fortunes. The Duvaliers robbed the country of somewhere between $20–800 million before a popular uprising led to their ouster. Although the United States issued strong verbal protests about human rights violations during the Duvaliers' reign, U.S. foreign policy provided tacit support for the dictators and even helped François Duvalier (Papa Doc) stay in power in exchange for supporting sanctions against Cuba and its "communist threat."

Aristide and Modern Haiti

In December 1990, the Haitian people elected the socialist Jean-Bertrand Aristide as president by a 67 percent majority, in the first free and fair election since the Duvaliers took power. In February 1991 Aristide was sworn in, but in September he was overthrown by the military in a coup led by two U.S.-trained Haitian generals. Although the United States had previously expressed some concern about Aristide's

Although U.S. military intervention and the restoration of the Aristides presidency have helped to resolve some of the political problems in Haiti, the economic problems are likely to be far more intractable. In order to improve the lives of the vast majority of Haitians who are very poor, economic development will require a healthy dose of foreign aid.

The distribution of U.S. grain in India, Haiti, Somalia, and in Rwandan refugee camps serves many purposes: helping those in need, supporting U.S. farm prices, and furthering U.S. foreign policy objectives. The interdependence of the global food market is a potent symbol of the world political economy.

economic agenda, they have played the key role in negotiating and militarily enforcing his return to power, perhaps in hope that his return will stem the tide of Haitians seeking refuge in the United States. Ironically, the international embargo spearheaded by the United States in the period preceding the reinstatement of Aristides had little impact on the Haitian elite; it did, however, substantially increase the suffering of Haiti's poor (Post 1993).

Conclusion

Haiti remains very much a least-developed country, one that may, in fact, be more vulnerable and more dependent than most. Haiti has no valuable natural resources; the land has been all but stripped bare of its once rich forests, and its soil is damaged almost beyond repair. Although Haiti can offer cheap labor (the average worker earns the equivalent of 14 cents an hour to assemble toys, electronics, and clothes) (Wilentz 1993), better-educated labor in Mexico and the newly developing nations of Southeast Asia make Haitian labor less attractive. Finally, with the end of the Cold War, the United States no longer needs Haiti even as a strategic foothold against communism. In short, Haiti has virtually nothing to trade on the world market. The primary reason for U.S. intervention seems to be its desire to stem the flood of refugees. If it is to survive as a nation, let alone develop, Haiti will need resources from abroad. Research suggests, however, that those resources must take the form of foreign aid (gifts) if they are to help Haiti on the road to development (Wimberly 1990). Foreign investment and loans are only likely to help a very determined elite to maintain its power.

Challenges to the Status Quo

Serious income and wealth inequalities can make for conflict among nations. During the last four decades, there has been a high and increasing level of political violence in poor nations. There have been nationalist revolutions or violent class struggles in Vietnam, Angola, Cuba, Chile, Algeria, Ethiopia, Nicaragua, El Salvador, Zimbabwe, and Afghanistan, to name just some (Kerbo 1991, 512). Now that the Cold War is over, some observers believe that the least-developed nations may be a more real source of peril. Nearly a half-dozen peripheral and semiperipheral nations—Argentina, Brazil, India, Iraq, Pakistan, and South Africa—now have nuclear capacities. Information technology, media, and increasing international travel all contribute to a rising sense of deprivation and frustration as citizens of peripheral nations are exposed to the lifestyles of more affluent countries. When relative deprivation leads to organized nationalist movements, it can result in violent, revolutionary challenges to the status quo (Kerbo 1991, 513).

The Developed World's Response

Responses to the threats and opportunities posed by the least-developed nations can be placed in three broad categories: military, humanitarian, and market-based strategies.

Military Responses

One response to possibly revolutionary challenges from peripheral nations has been military action to quiet unrest. Such action can be *overt* or *covert*. U.S. actions in places such as Vietnam, Panama, and Iraq, for example, illustrate overt intervention. Covert actions include propaganda campaigns, rigged elections, and help in staging coups. CIA involvement in the Chilean coup that ousted President Allende and its consideration of a plan to damage Fidel Castro's image by giving him LSD before a major public address (Kerbo 1991, 517) are examples of covert action.

Humanitarian Responses

In recent years, the United States has begun to use its military primarily for humanitarian interventions. The 1992–1993 food relief effort in Somalia and the more recent efforts at keeping peace in Bosnia are examples. Meanwhile, citizens in core nations provide both tax dollars and private donations to aid people in poor nations, and many people in the United States volunteer in agencies such as Peace Corps to make conditions better in less-fortunate parts of the world.

The Least-Developed World as a Market

Why do developed nations interfere in least-developed countries? Why have countries such as Nicaragua and Cuba become battlegrounds for more developed nations? Why, as in the recent war with Iraq, do First and Second World nations find themselves at war with the Third World nations they have armed? Although ideology plays a part, the importance of the least-developed nations as a market provides a more compelling answer.

Nearly one-third of all world trade is with the least-developed world. These nations are both an important market for manufactured goods and an important source of cheap labor and raw materials. Any closure of trade with the least-developed world would strike a major blow to international trade and world capitalism. Although the avowed aim of U.S. foreign policy has been to ensure each nation's right to self-determination, in practice U.S. policy has been more concerned with keeping nations open to capitalist interests than with protecting political liberty. In an effort to keep Third World markets open to U.S.-owned transnationals, the United States has provided economic and military support to repressive, right-wing, totalitarian regimes in nations such as South Korea, Vietnam, Chile, Nicaragua, and Iran. Despite evidence of continuing human rights violations, the Clinton administration has once again renewed China's most-favored-nation trade status, presumably because a market for U.S. goods that consists of 1 billion people is simply too lucrative to penalize or ignore. Likewise, it is hard to assess the extent to which the recent presence of European and U.S. peace-keeping forces in Somalia was a result of humanitarian concerns and how much was a

THINKING

CRITICALLY

One reason nations use coercion to force their will on others is that there is no authority to compel them to get along. Consider how relations among U.S. states might differ if federal power was as weak as the United Nations is in international affairs.

result of pressure from the European and U.S. petroleum firms who hold large oil exploration contracts in the country.

In the Cold War era, as a matter of official policy, the Soviet Union and Eastern European nations espoused the right of Third World peoples to employ force as a means to liberation. In Soviet foreign policy, however, liberation frequently meant liberation from capitalism and an alliance against Western capitalist states; it did not mean independence for least-developed nations (McFarlane 1985). Pressured by a stagnating domestic economy and by heavy military expenses in nations such as Afghanistan, Soviet foreign policy under Gorbachev became much less ideological; under Yeltsin, Russian diplomacy is increasingly driven by its own need for foreign currency and by the need to ensure a supply of raw materials and a market for its manufactured goods. In that respect, it is like U.S. policy.

Nowhere are the political and economic interdependencies of most-to-least developed nations more clear than in the recent Iraqi conflict. Although the Middle East has a long history of religious and ethnic strife, the conflicts have changed dramatically over the last two decades. Ever since oil prices quadrupled in 1973, First and Second World nations have exchanged arms for oil, with little attention to either their own diplomatic objectives or those of their clients' states. The United States, one of the world's largest arms exporters, enamored with cheap energy and unwilling to pay a higher price for oil and gasoline, faced the grand irony of waging the Desert Storm campaign against an Iraqi army equipped with weapons provided by U.S., British, French, German, and Italian arms suppliers. As diplomats continue to struggle with the political consequences of the most recent war in the Middle East and a possible new one, defense industries in the developed world have already begun to rearm the region.

Whatever its ultimate resolution, the issue reminds us of a lesson we learned in Chapter 6: Power is one of the most important determinants of the distribution of scarce resources, internationally as well as locally.

With the end of the Cold War, arms manufacturers in the former Soviet Union, Western Europe, and the United States wonder what to do with their excess production capacity. Although some manufacturers will successfully shift their attention to the production of peacetime products—passenger airplanes, computer technology, nuclear power and safety equipment, for example—others will continue to supply arms to volatile areas of the world. Countries such as Saudi Arabia and Kuwait still shop for advanced military technology at markets such as this Paris air show, and less sophisticated weaponry still finds its way into the hands of terrorists and Third World armies, alike.

Summary

1. Inequality is the key fact in the international political economy. Reducing this disparity through the development of less-developed and least-developed countries is a common international goal. Development is not the same as Westernization; it means increasing productivity and raising the standard of living.

2. The world's nations can be divided into the rich, diversified, independent most-developed nations; the less-developed nations include not only the former Soviet Union and former Communist-bloc countries of Eastern Europe but also countries in Southeast Asia and South America that are experiencing economic growth; and the least-developed nations of the periphery. Least-developed nations can be further subdivided into banana republic, industrial, and foreign-capital dependencies.

3. The Human Development Index and the Gender-Related Development Index use literacy and educational attainment, life expectancy, and economic productivity to assess quality of life overall, as well as quality of life adjusted for the effects of gender inequality.

4. In many least-developed countries, women are the principal food producers, but at least until the mid-1970s they were largely ignored by development programs. New development strategies are more often aimed at women.

5. Modernization theory, a functionalist perspective of social change, rests on the assumption that less-developed countries will evolve toward industrialization by adopting the technologies and social institutions used by the developed countries.

6. World system theory, a conflict perspective, views the world as a single economic system in which the already industrialized countries control world resources and wealth at the expense of less-developed countries. Empirical studies document that foreign-capital penetration has adverse consequences for Third World nations.

7. Where economic development has been pursued without attention to resource management, pollution, or wasteful consumption, the environmental consequences have been devastating. In order to improve the quality of life in least-developed nations, more attention will need to be paid to these issues.

8. Sustainable development is the ability to meet the needs of the present without compromising the ability of future generations to meet their own needs. Sustainable development will require (a) the elimination of absolute poverty and the reduction of inequality within and between nations, and (b) a reduction in population growth, coupled with more efficient production and less wasteful lifestyles.

9. As the international political economy undergoes competition, conflict, and change, the most-developed nations' responses can be classified as military, humanitarian, and marketplace strategies.

Sociology on the Net

The wealth and riches of the United States can blind us to the extent of international inequality throughout the world and even in our own hemisphere. One of the more impoverished nations is Haiti.

```
http://www.odci.gov/cia/
```

Surprise: You are now at the *Central Intelligence Agency Home Page!* Click on the *Publications* section. Now open the *1995 World Fact Book.* Scroll down the table of contents to *Haiti.* Open the section on *Haiti* and carefully review this information. Some of the more telling statistics are the infant mortality rate, life expectancy, and the literacy rate. Return to the table of contents and open the document on the *United States.* How does the United States compare to Haiti? Try a few more countries such as Japan, India, and Germany. These statistics tell us that inequality refers to not only the quality of life but also the length of life. Now try to come up with a picture of everyday life in Haiti. How would you be living if you were born there? There are other nations around the world that are poorer than Haiti. Can you find some of them?

The United Nations has a great deal of information on international inequality. One of the more dramatic indicators is the World Poverty Clock.

```
http://www.undp.org
```

Click on the highlighted phrase entitled *Poverty Clock.* Think back to the previous chapter, and remember the definitions of *absolute poverty* and *relative poverty.* What kind of poverty is the United Nations measuring? How does the United Nations definition of *poverty* differ from the one used by the United States government? How many people in the world are currently living on less than a dollar a day?

FIND IT ON INFOTRAC COLLEGE EDITION
In a recent article in *The Ecologist,* Edward Goldsmith critiques the development of third-world countries. Using InfoTrac College Edition, look up Goldsmith's article:

"Development as Colonialism." Edward Goldsmith. *The Ecologist,* March–April 1997, v27 n2 p69.

(Hint: Enter the search term *Development as Colonialism* using the Key Words Guide.)

Compare the strategies used in pre-1900s colonization and the post-World War II development of third-world nations. Do you agree with Goldsmith's point that formal colonialism ended because the economic control of less-developed countries could be achieved by what are considered politically acceptable and more-effective methods?

Suggested Readings

Chirot, Daniel. 1986. *Social Change in the Modern Era.* New York: Harcourt Brace Jovanovich. A historical and comparative approach to understanding social changes in the world. Excellent introduction to a complex area.

Eckstein, Susan (ed.). 1988. *Power and Popular Protest Latin American Social Movements.* Berkeley, Calif.: University of California Press. A collection of essays outlining the causes and consequences of Latin American protest and resistance movements. Factors unique to the Latin American experience are discussed, and features common to all protest movements are identified.

George, Susan. 1984. *Ill Fares the Land: Essays on Food, Hunger, and Power.* Great Britain: Writers and Readers Publ. A very readable collection of essays by a well-known science and policy writer who employs a conflict and world system perspective in her analysis of international inequalities.

George, Susan. 1988. *A Fate Worse Than Debt.* New York: Grove Press. A readable analysis of the results of international debt in peripheral nations by a well-known and respected world system analyst.

Gradolph, Rebecca Sue. 1994. *Social Stratification in Soviet Russia, 1991: Inequality of Opportunity and Outcome.* A contemporary study of social inequality in Russia after the demise of the Communist regime there in 1989.

Kurtz, Lester, with Dillard, John, and Benford, Robert. 1988. *The Nuclear Cage: A Sociology of the Arms Race.* Englewood Cliffs, N.J.: Prentice-Hall. A primer on the nuclear arms race and a recipe for its resolution. This explicitly sociological account gives special attention to the symbolic meanings that fuel conflict and impede resolution.

Robinson, Kathryn. 1986. *Stepchildren of Progress: The Political Economy of Development in an Indonesian Mining Town.* Albany: SUNY Press. A monograph based on two years of fieldwork in an Indonesian village that is an indictment of multinational investment. It is an enormously readable book, more like a horror story than a scholarly compendium.

Racial and Ethnic Inequality

Outline

Race and ethnicity, are ascribed characteristics that define categories of people. Each has been used in some times and places as bases of stratification; that is, cultures have thought it right and proper that some people receive more scarce resources than others simply because they belong in one category rather than another. *Racism* is the ideology used to justify these patterns of inequality. In the following section, we provide a basic framework for looking at categorical inequality; then we turn to a separate analysis of the patterns of inequality that exist in the United States along each of these dimensions. We examine cross-cutting statuses, such as race and class, and conclude the chapter by looking at a case study of ethnic antagonisms in Bosnia.

A Framework for Studying Group Inequalities

How is it possible for groups to interact on a daily basis within the same society and yet remain separate and unequal? In this section, we introduce sociological concepts that help explain how societies maintain and reinforce group differences.

The Semicaste Model

Most contemporary scholars use some form of conflict theory to explain how racial and ethnic inequalities are developed and maintained. In conflict over scarce resources, this theory suggests that historical circumstances such as slavery and technological advantage gave some groups an edge over others. Early stratification theorists confidently expected that industrialization would be followed by the virtual elimination of caste-like statuses in favor of achieved status. However, ascribed characteristics such as race and sex continue to be important in allocating scarce resources in our society. Thus, many scholars of race relations now believe that at least in the United States there are two stratification systems, class and caste. This semicaste structure is a hierarchical ordering of social classes within racial categories that are also hierarchically ordered (see Figure 8.1). Data from 1994 (Table 8.1) show that, although the median income of white families is more than one and a half times that of black families ($38,787 versus $23,482), the races display very similar patterns of internal inequality. In both black and white populations, the wealthiest 20 percent of families receive about half of all income. We can also see another implication of the semicaste model in this table; wealthy blacks have less net worth than wealthy whites, and poor blacks are much poorer than poor whites. As Oliver and Shapiro (1995) point out, it is these racial differences in wealth, not racial differences in income, that are primarily responsible for the continuing U.S. racial divide.

Majority and Minority Groups

A **majority group** is a group that is culturally, economically, and politically dominant.

Rather than talk about white and black or Jew and Arab, sociological theories of race and ethnic relationships usually refer to majority and minority groups. A **majority group** is

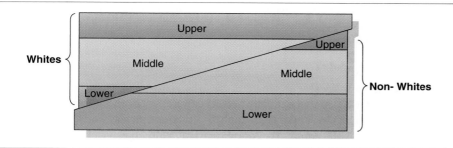

FIGURE 8.1
The Semicaste Model
Race and ethnicity are factors in the stratification system of the United States. There is a social class hierarchy within each race, but there is also a castelike barrier between races. Upper-class nonwhites are not as upper class as upper-class whites, and the lowest positions in the social class hierarchy are reserved for nonwhites.

TABLE 8.1
Income and Wealth Distributions Among Families by Race and Ethnicity, 1994
United States income data support the semicaste model. The wealthiest 20 percent of black and Hispanic families receive 51 and 49 percent of all family income, a figure very comparable to the 48 percent received by the wealthiest 20 percent of the white families. Nevertheless, wealthy white families are much better off than wealthy black and Hispanic families and poor black and Hispanic families have only a small fraction as much wealth as poor white families do.

| | PERCENTAGE OF INCOME RECEIVED | | | MEDIAN NET WORTH | | |
	Black	Hispanic	White	Black	Hispanic	White
Poorest Fifth	3%	4%	5%	$250	$499	$7,605
Second fifth	8	9	10	3,406	2,900	27,057
Third fifth	15	15	16	8,480	6,313	36,341
Fourth fifth	25	24	23	20,745	20,100	54,040
Richest fifth	49	48	46	45,023	55,923	123,298
Median net worth				$4,418	$4,656	$45,740

SOURCE: U.S. Bureau of the Census, 1996a and http://www.census.gov/hhes/wealth/wlth93f.html

one that is culturally, economically, and politically dominant. A **minority group** is one that is culturally, economically, and politically subordinate. Although minority groups are often smaller than majority groups, that is not always the case. In the Republic of South Africa, for example, whites until recently were the majority group, although they made up only 15 percent of the population, because they controlled all major political and social institutions. Some scholars regard women as a minority group because they have been economically, politically, and culturally subordinate to men.

A **minority group** is a group that is culturally, economically, and politically subordinate.

Patterns of Interaction

Relations between majority and minority groups may take one of four general forms: conflict, accommodation, acculturation, or assimilation.

Conflict

Conflict is a struggle over scarce resources that is not regulated by shared rules; it may include attempts to neutralize, injure, or destroy one's rivals. Although some intergroup conflicts are expressed in violence (for example, genocide), conflict may also be expressed in laws forbidding social, political, or economic participation by the minority group.

Conflict is a struggle over scarce resources that is not regulated by shared rules; it may include attempts to destroy, injure, or neutralize one's rivals.

When blacks or Latinos are categorically excluded from voting or from some forms of political or social participation, this is a form of conflict. It excludes them from the competition for scarce resources.

Accommodation

Accommodation occurs when two groups coexist as separate cultures in the same society.

When two groups coexist as separate cultures in the same society, we speak of **accommodation.** They are essentially parallel cultures, each with its own institutions. The term was originally developed to apply to situations such as Canada's French and English provinces and Switzerland's German, Italian, and French cantons. As these examples imply, separate is seldom truly equal. Nevertheless, accommodation gives at least outward support to the norm of equality. Sometimes referred to as pluralism, these systems are often difficult to maintain.

Acculturation

Acculturation occurs when the minority group adopts the culture of the majority group.

Another possible outcome of intergroup contact is for the minority group to adopt the culture of the majority group. This process is called **acculturation.** It includes learning the language, history, and manners of the majority group; it also involves accepting the loyalties and values of the majority group as one's own. As middle-class African Americans have discovered, however, full acculturation does not necessarily mean full acceptance.

Assimilation

Assimilation is the full integration of the minority group into the institutions of society and the end of its identity as a subordinate group.

When full acceptance comes—when the minority group is fully integrated into the institutions of society and ceases to be a subordinate group—we speak of **assimilation.** Assimilation means that category membership no longer affects social relationships. In the case of racial and ethnic groups, it includes going to the same schools, living in the same neighborhoods, belonging to the same social groups, and being willing to marry one another. In the case of sex categories, assimilation would mean that sex was no longer relevant for social behavior such as occupational choices or leisure activities.

Although the groups involved may differ, nations are remarkably similar in both the causes and the consequences of prejudice and discrimination. As Germany copes with the costs of reunification and of slowing economic growth, racial and ethnic antagonisms have increased. So, too, have the personal losses and pain. Here, Turkish victims of neo-Nazi violence mourn the victims of an arson attack that left three children injured and five adults dead.

The Maintenance of Inequality

To understand the persistence of racial and ethnic inequality, we need to look at the processes that promote what sociologists call **social distance.** Social distance is operationally defined by questions such as, "Would you be willing to have members of this group as good friends?" It is a measure of the degree of intimacy and equality in the relationship between two groups.

> **Social distance** is the degree of intimacy in relationships between two groups.

Prejudice and discrimination are processes that allow social distance to be maintained even when physical distance is absent. Most societies also use segregation, or physical distance, as an aid to maintaining social distance.

Prejudice

The foundation of prejudice is stereotyping, a belief that people who belong to the same category share common characteristics—for example, that athletes are dumb or that African Americans are naturally good dancers. Stereotyping is not always bad. If your seatmate on a bus is an elderly woman, you might start a conversation about the high prices of groceries; if your seatmate is a 30-year-old man, it might be about the prospects of next year's football team. These conversational topics are based on stereotyped notions of what these people are likely to be interested in. The woman might be able to give you details about middle guards, and the man might be completely uninterested in sports, but stereotypes often prove useful at least as a starting point in secondary relationships.

Prejudice moves beyond stereotyping in that it is always a negative image and it is irrational. It exists in spite of the facts rather than because of them (Pettigrew 1982). A person who believes that all Italian Americans are associated with the Mafia will ignore all instances of the law-abiding behavior of Italian Americans. If confronted with an exceptionally honest man of Italian descent, the bigot will rationalize him as the exception that proves the rule.

> **Prejudice** is irrationally based negative attitudes towards categories of people.

A startling example of prejudice was the decision by the United States to inter its West Coast Japanese-American citizens during World War II. The decision to go ahead with the internment in spite of a lack of evidence of treason is a fascinating study in the irrationality of prejudice. Said General John DeWitt (1943): "The very fact that no sabotage has taken place to date is a disturbing and confirming indication that such action will be taken."

Prejudice is a powerful barrier to the kinds of interaction that might reduce social distance. What causes prejudice? We review three factors: cultural norms, institutional patterns, and personal factors.

Cultural Norms

We learn to hate and fear through the same processes by which we learn to love and admire. Prejudice is a shared meaning that we develop through our interactions with others. Most prejudiced people learn prejudice when very young at the same time they are internalizing other social norms. This prejudice may then grow or diminish depending on whether groups and institutions encountered during adulthood reinforce these early learnings (Wilson 1986).

Institutional Patterns

Prejudice arises from and is reinforced by institutionalized patterns of inequality. In a stratified society, we tend to rate ourselves and others in terms of economic worth. If we observe that no one pays highly for a group's labor, we are likely to conclude that the members of the group are not worth much. Through this learning process, members of the minority as well as the majority group learn to devalue the minority group (J.Q. Wilson 1992).

Prejudice is also powerfully reinforced by patterns of segregation and discrimination. A child growing up in a society where separation by race or sex is well established is very likely to learn prejudice. "They are not as good as us" is only one short step from "They are not like us."

Of course, competition over scarce resources, admission to prestige universities, good jobs, and nice homes, for example, also increases prejudice. For all racial and ethnic groups, prejudicial attitudes are closely associated with the belief that gains for other racial and ethnic groups necessarily spell losses for one's own (Bobo & Hutchings 1996).

Personal Factors

Personality factors cannot explain widespread prejudice (Stone 1985). Nevertheless, some people are more prone to prejudice than others. Three factors that dispose people to prejudice are authoritarianism, frustration, and beliefs about stratification.

Authoritarianism is a tendency to be submissive to those in authority and aggressive and negative toward those lower in status (Pettigrew 1982). Regardless of their own race or ethnic group, authoritarians in the United States tend to be strongly anti-African American and anti-Semitic.

Frustration is another characteristic associated with prejudice. People or groups who are blocked in their own goal attainment are likely to blame others for their problems. This practice, called **scapegoating,** has appeared time and again. From the anti-Chinese riots in nineteenth-century California to anti-Jewish pogroms in Nazi Germany, setbacks for majority-group members often result in attacks against the minority group.

Finally, prejudice is more likely to exist among individuals who believe strongly in the American Dream. People who subscribe to the view that we can all get ahead if we work hard and that poor people have only themselves to blame are substantially more likely to attribute poverty or disadvantage to personal deficiencies. In the case of disadvantaged minorities, the American Dream ideology supports the belief that the disadvantage is the fault of undesirable traits within the minority group (Kluegel & Smith 1983; Pettigrew 1985).

Racism

Racism is the belief that inherited physical characteristics determine the presence or absence of socially relevant abilities and characteristics and that such differences provide a legitimate basis for unequal treatment.

Racism is an ideology. As defined in Chapter 6, this means that it provides a justification for inequality. People who are racists believe that African Americans are biologically less fit than whites and that it is therefore only fair that they are worse off than whites.

Conflict theorists explain this prejudiced ideology as part of the general strategy of stratification. Whenever any group has access to scarce resources, its first step is to try to exclude others from them. If it can exclude others categorically (that is, on the basis of category membership such as ethnicity and race), the need to compete individually is reduced. Racism then is a means of restricting access to scarce resources.

The Self-Fulfilling Prophecy

An important mechanism for maintaining prejudice is the **self-fulfilling prophecy**—where acting on the belief that a situation exists causes it to become real. A classic

---

Authoritarianism is the tendency to be submissive to those in authority, coupled with an aggressive and negative attitude toward those lower in status.

Scapegoating occurs when people or groups who are blocked in their own goal attainment blame others for their failures.

Racism is the belief that inherited physical characteristics determine the presence or absence of socially relevant abilities and characteristics and that such differences provide a legitimate basis for unequal treatment.

The **self-fulfilling prophecy** occurs when acting on the belief that a situation exists causes it to become real.

example is the situation of women in feudal Japan (or in more recent Western cultures). Because women were considered to be inferior and capable of only a narrow range of social roles, they were given limited education and barred from participation in the institutions of the larger society. The fact that they subsequently knew nothing of science, government, or economics was then taken as proof that they were indeed inferior and suited only for a role at home. In fact, most women were unsuited for any other role: Being treated as inferiors had made them ignorant and unworldly. The same process reinforces boundaries between racial and ethnic groups. If we assume that Jews are clannish, then we don't invite them to our homes. When we subsequently observe that they associate only with one another, we take this as confirmation of our belief that they are clannish.

Discrimination

Treating people unequally because of the categories they belong to is **discrimination.** Prejudice is an attitude; discrimination is behavior. Often, discrimination follows from prejudice, but it need not. Figure 8.2 shows the possible combinations of prejudice and discrimination. Individuals may fit into the two consistent cells: They are prejudiced, so they discriminate (bigots); or they aren't prejudiced, so they don't discriminate (friends). Some people, however, are inconsistent, usually because their own values are different from those of the dominant culture. Fair-weather friends do not personally believe in racist or sexist ideologies; nevertheless, they discriminate because of what their customers, neighbors, or parents would say. They do not wish to rock the boat by acting out values not shared by others. The fourth category, the timid bigots, have the opposite characteristics: Although they themselves are prejudiced, they hesitate to act on their feelings for fear of what others would think (Merton 1949).

Public policy directed at racism is aimed almost entirely at reducing discrimination—allowing fair-weather friends to act on their fraternal impulses and putting some timidity into the bigot. As Martin Luther King, Jr. remarked, "The law may not make a man love me, but it can restrain him from lynching me, and I think that's pretty important" (cited in Rose 1981, 90).

Discrimination is the unequal treatment of individuals on the basis of their membership in categories.

THINKING

CRITICALLY

In thinking about the relationship between prejudice and discrimination, we generally assume that prejudice is the cause of discrimination. Can you think of a time or a situation when the reverse might be true, that is, that prejudice would follow from discrimination?

Segregation

Prejudice and discrimination may occur between groups in close, even intimate, contact; they create social distance between groups. Differences between groups are easier to maintain, however, if social distance is accompanied by **segregation**—the physical separation of minority- and majority-group members. Thus, most societies with strong divisions between racial or ethnic groups have ghettos, barrios, Chinatowns, and Little Italies, where, by law or custom, members of the minority group live apart.

Historical studies suggest that high levels of residential segregation of Hispanic, Asian, and African Americans are not new; they have existed since at least 1940 and have changed relatively little. Such segregation is no longer established in law, but it is no historical accident. It occurs partly as a result of social class segregation of neighborhoods. Blacks, for instance, are substantially less likely than whites to leave poor neighborhoods and substantially more likely to move into them. However, even the most educated blacks remain substantially less likely than the least-educated whites to escape "distressed" neighborhoods (South & Crowder 1997). Similarly, socioeconomic status is related to living in the suburbs and to better-quality housing for all racial and ethnic groups. Yet, the suburbs in which middle-class African Americans and Hispanics, especially Puerto Ricans, live are considerably poorer and more problem ridden than those in which middle-class Asian and European Americans reside (Rosenbaum 1996).

Segregation refers to the physical separation of minority- and majority-group members.

FIGURE 8.2
The Relationships Between Prejudice and Discrimination

Prejudice is an attitude; discrimination is behavior. They do not always go hand-in-hand. Some people act on their attitudes, whereas others suppress their own attitudes to conform to community standards. Fair-weather friends are unprejudiced people who will discriminate anyway; timid bigots are prejudiced people who are deterred from discrimination by community standards.

PREJUDICED?

	No	Yes
DISCRIMINATE? No	Friend	Timid bigot
DISCRIMINATE? Yes	Fair-weather friend	Bigot

A **race** is a category of people treated as distinct on account of physical characteristics to which *social* importance has been assigned.

An **ethnic group** is a category whose members are thought to share a common origin and to share important elements of a common culture.

Racial segregation remains a fact of life in the United States. Studies show that African Americans, even those from the middle class, are substantially less likely than European Americans to leave poor neighborhoods and are substantially more likely to move into them.

As the authors of one study conclude, "Blacks experience a consistent, powerful, and highly significant penalty in the process of spatial assimilation" (Massey & Denton 1988, 621). This penalty is not just the result of economic differences between racial and ethnic groups but also reflects the variety of political and social pressures that whites have used to keep minorities "in their place" (Saltman 1991).

Race and Ethnic Inequalities in the United States

Race and Ethnicity

A **race** is a category of people treated as distinct on account of physical characteristics to which *social* importance has been assigned. An **ethnic group** is a category whose members are thought to share a common origin and to share important elements of a common culture—for example, a common language or religion (Marger 1994). Both race and ethnicity are handed down to us from our parents, but the first refers to the genetic transmission of physical characteristics whereas the second refers to socialization into cultural characteristics.

Although race is based loosely on physiological characteristics, such as skin color, both race and ethnicity are socially constructed categories. Both individual self-identity, and institutional forces play a role in the creation and maintenance of racial and ethnic statuses. For example, in 1930, the U.S. Bureau of the Census declared that those with Mexican background should be classified as nonwhite. The Mexican government complained, and the bureau reversed itself. Now the census bureau defines Hispanic Americans as an ethnic group, declaring that Hispanics can be of any race (U.S. Bureau of the Census 1994, 4). The 1980 census revealed that there were 6.7 million people who claimed Native American (American Indian) as their ethnic group but only 1.4 million who claimed it as their race (Lieberson & Waters 1988). More recently, the shift from "black" to "African American" is an example of chang-

ing from a racial to an ethnic group identification. Then, too, the growing number of multiracial births in the United States helps to blur the very concept of race (Kalish 1995; Morganthau 1995). Elsewhere in the world, new ethnic identities are forged by changing national borders. Swedish-speaking Finnish intellectuals embarked on a campaign to reconstruct Finnish folklore and music during the mobilization for Finnish independence; likewise, after decades of Russian influence, individuals in Lithuania are now on a journey to uncover the real cultural and historical meaning of Lithuanianness (as reported in Nagel 1994). As these examples illustrate, racial and ethnic statuses are not fixed. Over time, individuals may change their racial and ethnic identification, and society, too, may change the statuses it recognizes and uses.

Race and ethnicity *could* be of primary importance to sociologists as the basis of subcultures. Sociologists could (and some do) focus on racial and ethnic differences in musical preferences, language use, and values. Overshadowing these subcultural differences among racial and ethnic groups, however, is the issue of inequality. In the United States today, African Americans, Hispanics, and Native Americans do not simply comprise subcultures; they comprise *disadvantaged* subcultures. In the rest of this chapter, we analyze the types and degree of disadvantage involved.

White Ethnics

The earliest immigrants to North America were English, Dutch, French, and Spanish. By 1700, however, English culture was dominant on the entire Eastern seaboard. The English became the majority group, and everybody who came after that time became a minority group in North America.

As the dissolution of the former Soviet Union has clearly demonstrated, changes in national boundaries are often accompanied by changes in ethnic identity. By reclaiming traditional customs and dress, these young women are helping to redefine what it has meant and what it now means to call oneself a Lithuanian.

The Melting Pot

The extent of interaction and assimilation among white ethnic groups led some idealistic observers to hope that a new race would emerge in North America, where "individuals of all nations are melted into a great race of men" (Crévecouer [1782] 1974).

In fact, careful observers suggest that the melting pot never existed. Certainly, our language is peppered with words borrowed from other languages (*frankfurter, ombudsman, hors d'oeuvre, chutzpah*), and some of us are such mixtures of nationalities that we would be hard pressed to identify our national heritage. Instead of a blending of all cultures, however, what has occurred is a specific form of acculturation—**Anglo-conformity,** the adoption of English customs and English language. To gain admission into U.S. society, to be eligible for social mobility, one has to learn correct English, become restrained in public behavior, work on Saturday and worship on Sunday, and, in general, act like the American version of the English prototype.

The Future of White Ethnicity

Despite generations of acculturation, many white people in the United States still identify with their ethnic heritage. They are proud to be Italians, Greeks, Norwegians, or Poles. Nevertheless, high rates of intermarriage are blurring these identities. As a result of intermarriage, a growing segment of the population cannot identify themselves with a single ethnic group; they are simply "unhyphenated whites" (Lieberson & Waters 1993). Although ethnicity for this group may be largely symbolic and largely a matter of choice, this should not obscure the fact that it has important consequences. In many ways, its very invisibility means that the ethnicity of the unhyphenated white becomes the national "American" identity and the standard against which all other cultural and ethnic identities are judged (Doane 1997).

Nevertheless, ethnicity has ceased to be a basis for stratification among white, non-Hispanic people in the United States. Although ethnic differences do exist, these differences are not related to structured inequality. The integration of 80 percent of the population from diverse backgrounds and conditions is a remarkable achievement. Yet it leaves out a significant portion of these people. Here we consider the situation of the other 20 percent: the nonwhites and the Hispanics.

African Americans

African Americans are the largest racial minority in the United States, representing one-ninth (or 12 percent) of the entire population. Their importance goes beyond their numbers. Next to Native Americans, African Americans have been the greatest challenge to the United States' view of itself as a moral and principled nation.

The history of African Americans has two essential elements that distinguish it from the history of other ethnic groups. First, black migration was involuntary; blacks came as slaves, not as settlers. Second, African Americans are almost uniformly descendants of people who have been here since the founding of the nation; they have roots in this country deeper than those of the Swedish, Norwegian, Italian, Irish, and German settlers who followed them.

In many ways, World War II was a benchmark for African Americans. The move from the rural South to the industrial North and Midwest that had begun during World War I was greatly accelerated. The defense effort sharply increased the demand for labor and made possible some real gains in income for blacks relative to whites. In addition, the Nazi slaughter of six million Jews in the name of racial purity deeply shocked the Western world, causing a renewed soul-searching about racism in the United States.

Compared with the century before, the years following World War II have seen rapid social change: segregation banned in the armed forces (1948), school segregation outlawed (1954), the Civil Rights Act passed (1964), affirmative-action laws passed

Anglo-conformity is the process of acculturation in which new immigrant groups adopt the English language and English customs.

(1968). For the first time, African Americans appeared on television, on baseball diamonds, in ballet companies, and on the Supreme Court. In the following sections, we review some evidence about the differences in life chances for blacks and whites. In some cases, comparisons over time show that differences have been significantly reduced.

Political Change

African Americans have been entitled to vote and hold office since the passage of the Fourteenth Amendment immediately after the Civil War. It took the Civil Rights Act of 1964, subsequent voter registration laws, and the Civil Rights activism of the late 1960s to make these political rights effective, however. The results have been dramatic, and African American voters are now an active and influential political force, especially in the Democratic party. African American political leadership is also growing. The number of African Americans holding elected positions increased more than 500 percent between 1970 and 1991 (Joint Center for Political Studies 1991); cities without African American majorities have elected African American mayors (Seattle, New York, and Los Angeles, for example), and 1989 saw the election of the first African American governor (L. Douglas Wilder in Virginia). Nevertheless, African Americans remain significantly underrepresented at all levels of government.

Education

African Americans have made significant progress in education too. In 1940, young white adults were nearly three times more likely to have graduated from high school than blacks of the same age (39 versus 11 percent). By 1995, the racial difference in high school education was only eleven percentage points (85 and 74 percent, respectively). Unfortunately, the educational gap remains wide at higher levels. Among young adults, blacks are only half as likely as whites to have graduated from college. Despite considerable improvement, then, significant educational differences remain between white and black Americans.

Economic Disadvantage

Black income continues to lag behind white income. In 1964, the median income for black families was only 54 percent of the income for white families. By 1996, this figure had increased but only to 64 percent. This striking economic disadvantage is due to two factors: black workers earn less than white workers, and black families are less likely to have two earners.

FEMALE-HEADED FAMILIES.   About half of the gap between black and white family incomes is due to the fact that African American families are less likely to include an adult male. Because women earn less than men and because a one-earner family is obviously disadvantaged relative to a two-earner family, these female-headed households have income far below those of husband–wife families. The fact that so many more black than white families are headed by females—48 percent compared to 13 percent—has led some commentators to conclude that poverty is the result of bad decisions by African American men and women. This type of argument is an example of "blaming the victim," and empirical evidence suggests that it simply isn't true. Rather than causing poverty, research indicates that female headship results from the severe lack of adequately paying employment opportunities for African American males (Lichter, LeClere, & McLaughlin 1991).

Low Earnings   One of the reasons individual blacks earn less than whites is that they are twice as likely to be unemployed. In 1995, black unemployment was 10 percent compared to 6 percent for whites; among teenagers, the figures were 36 and 14 percent (U.S. Bureau of the Census 1996a). Even when employed, however, blacks earn less than white workers.

Blacks are less well educated than whites (this is especially true of older blacks), and a relatively high proportion live in the South, where wages are low; the average black worker is also somewhat less experienced than the average white worker. However, these differences account for only part of the earnings gap between black and white people in the United States (Cancio, Evans, & Maume 1996). The other part is the result of a pervasive pattern of discrimination that produces a very different occupational distribution, a very different pattern of mobility, and a very different earnings picture for black and white people in the United States. Thanks largely to government employment opportunities, there is a growing African American middle class (Hout 1986). However, even among African American professionals there is evidence of considerable inequality. For instance, many African American executives and managers have been assigned to corporate positions that were created explicitly to deal with African American demands for civil rights. The existence of these jobs depends, then, on the continuing sensitivity of employers to racial pressures (Collins 1993, 1997). The positions, often in public relations or personnel, are in full public view. Yet, the people who hold these positions usually remain outside the true corporate power structure. Having less authority at work, African Americans employed in these positions also receive a lower economic return for that authority than do whites employed in comparable positions (Smith 1997; Wilson 1997). Thus, even though they enjoy heightened visibility, the African American middle class remains economically vulnerable. In addition, the African American middle class may pay a psychological price for their success. Employed in predominantly white settings, the black elite must demonstrate more competence than peers; they may also experience multiple demands of being black, a loss of black identity and a sense of isolation. All of these factors can contribute to higher levels of anxiety and depression (Jackson, Thoits, & Taylor 1995). In the end, despite growth in the middle class, African Americans continue to make up 27 percent of the nation's child-care workers and 42 percent of all household maids (U.S. Bureau of the Census 1996a).

Continued Concerns

Since World War II, important improvements have been made in many areas of African American life. However, some areas have seen virtually no change. Black unemployment remains almost twice as high as white, and the black-white earnings gap has not closed. Neighborhood segregation has declined very little; it remains so high and has such damaging consequences that one scholar has referred to it as "American apartheid" (Massey 1990). Black infants are still almost two-and-a-half times as likely as white infants to die before their first birthday (U.S. Bureau of the Census 1996a), and African American men's life expectancy is still eight-and-a-half years less than white men's.

How do we reconcile this troubling picture with the improvement noted earlier? Many perceive a fissure in the African American population: on the one hand, a working- and middle-class population that is increasingly integrated into U.S. society; on the other, an African American underclass that has not been included in the overall improvement. In fact, for them, the situation has deteriorated. This African American underclass has been left behind by the more prosperous members of all races (W.J. Wilson 1978).

Hispanics

Hispanics or Latinos are an ethnic group rather than a racial category, and a Hispanic may be white, black, or some other race. This ethnic group includes immigrants and their descendants from Puerto Rico, Mexico, Cuba, and other Central or South American countries. Hispanics constitute about 10 percent of the U.S. population. The largest group of Hispanics is of Mexican origin (65 percent), with 10 percent from Puerto Rico and the remaining 25 percent from Cuba and elsewhere.

It is almost impossible to speak of Hispanics as if they were a single group. Their experiences in the United States have been and continue to be very different. Recent evidence suggests that their economic and political experiences may be growing more divergent rather than more similar (Portes & Truelove 1987). Although Cubans are becoming increasingly assimilated, as signaled by a very high rate of U.S. citizenship, the Mexican American, or Chicano, population is not.

Figure 8.3 compares the various Hispanic groups to one another and to the overall white and black populations on three measures: education, poverty, and family structure. (Because race and Hispanic origin are overlapping categories, Hispanics are included twice in this table—once under their ethnic group and again under their racial identification.) On two of these measures, a Hispanic group comes out at the very bottom; Mexican Americans are the most poorly educated racial or ethnic group, and Puerto Ricans are the most likely to live in poverty. In addition, Puerto Ricans are almost as likely as African Americans to live in female-headed households.

Special Concerns

If current immigration patterns continue, Hispanics are expected to constitute 19 percent of the U.S. population by 2030 (U.S. Bureau of the Census 1996a). This rapid growth raises three concerns for the status of Hispanics: First, because the new immigrants are young and poorly educated, the socioeconomic position of the Hispanic population is falling. Between 1975 and 1993, Hispanic poverty rates grew considerably faster than those for blacks (U.S. Bureau of the Census 1996a). Second, the growing Hispanic population has raised concerns among non-Hispanics about competition for jobs and cultural change, thus spurring greater prejudice and discrimination. Finally, rapid growth is associated with increasing residential segregation (Massey & Denton 1988). Segregation, in turn, may retard the rate at which new immigrants learn English and become integrated into U.S. society.

Despite these concerns, studies show that the castelike barrier separating races operates much less dramatically in the case of Hispanics. For white Hispanics, the problem is largely one of class. As a result, by the second and third generation, Hispanics are largely able to translate education into occupational prestige and leave the segregated barrios (Massey & Mullan 1984). The exception is the 7 percent of Hispanics (mostly Puerto Ricans and other Caribbean islanders) who suffer the triple disadvantage of being Hispanic, poor, and black (Massey & Bitterman 1985).

Asians

The Asian population of the United States (Japanese, Chinese, Filipinos, Koreans, Laotians, and Vietnamese) almost doubled between 1980 and 1990; yet it still constitutes only 2.9 percent of the total population. The Asian population can be broken into three segments: the nineteenth-century immigrants (Chinese and Japanese), the post- World War II immigrants (Filipinos, Asian Indians, and Koreans), and the recent refugees from Southeast Asia (Cambodians, Laotians, and Vietnamese).

A century ago, Asian immigrants were met with sharp and occasionally violent racism. Today, incidents of racial violence directed at Vietnamese and other Asians continue to make headlines. Despite these handicaps, Asian Americans have experienced high levels of social mobility. People in the United States of Japanese and Chinese descent have surpassed the educational attainment of white people there, and it appears that many of the more recent streams of Asian immigrants will follow the same path.

Hispanics or Latinos are an ethnic group rather than a racial category. The largest group of Hispanics is of Mexican descent, the second largest group is Puerto Rican, and the third largest group is from Cuba. The experiences of each of these groups in the United States has been quite different. Some groups are doing quite well. Although they retain much of their language and many of their customs, Cuban Americans such as these have high school graduation rates only slightly below and poverty rates only slightly above those of European Americans. For the most recent waves of Cuban immigrants, however, the experience may be quite different. Like Puerto Ricans, many bear the double burden of being both Hispanic and black; the high poverty rates of both groups reflect the fact that problems associated with learning a new culture are being compounded by the problems produced by racism.

FIGURE 8.3
Education, Poverty, and Family Structure by Race and Hispanic Origin, 1994
On two out of three measures of disadvantage, a Hispanic ethnic group comes out at the bottom. A significant portion of this disadvantage, however, can be traced to recent immigration and poor English skills.
SOURCE: U.S. Bureau of the Census http:www.census.gov/population/socdemo/hispanic/ed94text.
[a]Each racial category includes people of all ethnic backgrounds, including Hispanics.

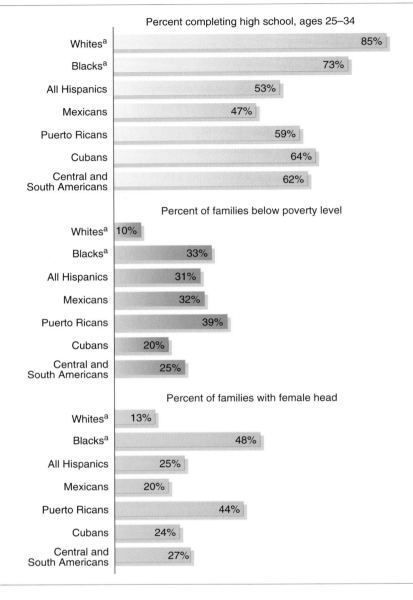

Percent completing high school, ages 25–34

Whites[a]	85%
Blacks[a]	73%
All Hispanics	53%
Mexicans	47%
Puerto Ricans	59%
Cubans	64%
Central and South Americans	62%

Percent of families below poverty level

Whites[a]	10%
Blacks[a]	33%
All Hispanics	31%
Mexicans	32%
Puerto Ricans	39%
Cubans	20%
Central and South Americans	25%

Percent of families with female head

Whites[a]	13%
Blacks[a]	48%
All Hispanics	25%
Mexicans	20%
Puerto Ricans	44%
Cubans	24%
Central and South Americans	27%

For example, although many of the Southeast Asian refugees who came to the United States between 1975 and 1984 began their lives there on welfare, almost twice as many Vietnamese youths aged 20–24 are enrolled in school as white youths of the same age.

The high level of education is a major step in opening doors to high-status occupations; the 1990 census showed that Asian American families had the highest average incomes of any major racial or ethnic group in the United States. Yet, discrimination is not all in the past. Asian American applicants are less likely to be accepted at elite colleges and universities than whites with the same credentials (Takagi 1990). Recent studies have concluded that highly educated, native-born Asian males earn substantially less than similarly qualified white men (Suzuki 1989, 16) and that they are promoted more slowly (Tang 1993). Discrimination can be even more direct. In 1982 two unemployed autoworkers beat a Vietnamese man to death because they thought he was Japanese and they blamed the Japanese for the loss of their jobs. The startling thing in this case was that the judge found their stress understandable and sentenced them to only three years of probation (Saigo 1989).

Native Americans

Native Americans (American Indians) are one of the smallest minority groups in the United States (about 0.5 percent of the entire population), and nearly half of their members live in just four states: Oklahoma, Arizona, California, and New Mexico. Native Americans are widely regarded as our most disadvantaged minority group. Incomes of husband–wife families are lower for American Indians than for African Americans, and the former tend to live in more crowded conditions (U.S. Department of Commerce 1994). Native Americans suffer the highest rates of alcoholism and premature death of any U.S. racial or ethnic group. This situation exists despite hopeful new signs of economic vitality on some Indian reservations over the past 15 years—for example, development of mineral reserves on the Navajo reservation in the Southwest (Schaefer 1990, 199–200) and (to a lesser degree) the advent of gambling casinos on other reservations (Johnson 1994).

As shown in Table 8.2, even within this general picture of low social and economic status, there is enormous variability. Native Americans represent more than 200 tribal groupings, with different cultures and languages. Some have been successful: fish farmers in the Northwest, ranchers in Wyoming, bridge builders in Maine. In urban areas, Native Americans experience much less racial segregation than do other nonwhite groups (Bohland 1982), and some have entered the professions and other occupations of modern industrial society.

More than any other minority, Native Americans have remained both unacculturated and unassimilated. It is clear that Native American culture is not simply different from the dominant culture; in many ways it is a shattered culture. Those who cling to earlier values face a severe case of anomie because the means to achieve the old goals are gone forever. In addition, prejudice and discrimination stand in the way of achieving the new goals. The story of the Ojibwa described in Chapter 2 represents an extreme case of what has happened to Native American culture.

Summary

In the United States today, racial and ethnic inequalities remain important. Being African American is particularly critical. Despite all the changes that have occurred in the last decades, being African American is a better predictor of poverty and unemployment today than it was 25 years ago (Lichter 1988).

The major difference in racial and ethnic inequalities over the last decades is that these differences are no longer normatively supported. Traditional racism has decreased; few people believe that blacks earn less money than whites because of their innate inferiority, and few believe that people should be discriminated against on account of their race and ethnicity. Nevertheless, subtle forms of racism persist. Widespread belief in the ideology of the American Dream means that many people continue to blame the poor for their poverty; the predominant white explanation for poverty stresses the lack of motivation among blacks. As a result of such beliefs, whites are reluctant to support governmental policies designed to promote economic equality between blacks and whites (Bobo & Kluegel 1993; Quillian 1996).

Another factor leading to the persistence of racial inequalities is simply the indirect inheritance model described in Chapter 6. A major implication of this model is that

During World War II, many U.S. citizens were held against their wills in large detention camps—not in Germany or Eastern Europe but at sites in the western United States. These individuals had committed no crimes. Nevertheless, they were deprived of both liberty and property for no other reason than that they were of Japanese ancestry. Although Congress has passed legislation permitting limited financial compensation for the land and businesses Japanese Americans lost more than 50 years ago, sufficient funds have not yet been appropriated to make good on even this modest reparation.

TABLE 8.2
Social and Economic Characteristics of the American Indian Population: 1990

Characteristic	American Indian, total	Cherokee	Navajo	Sioux	Chippewa	Choctaw	Pueblo	Apache	Iroquois	Lumbee
Percent 65 years old and over	5.9	7.2	4.6	4.4	4.7	8.0	5.8	3.4	6.7	5.6
Percent high school graduates or higher, persons 25 years and older	65.6	68.2	51.0	69.7	69.7	70.3	71.5	63.8	71.9	51.6
Percent families with female heads	26.2	20.8	28.6	36.0	33.1	20.0	29.2	24.7	25.5	23.9
Percent families below poverty	27.2	19.4	47.3	39.4	31.2	19.9	31.2	31.8	17.3	20.2
Median family income, 1989	21,619	24,907	13,940	16,525	20,249	24,467	19,845	19,690	27,025	23,934

SOURCE: U.S. Bureau of the Census 1996a.

Institutionalized racism occurs when the normal operation of apparently neutral processes systematically produces unequal results for majority and minority groups.

patterns of inequality established in past generations will persist. **Institutionalized racism** is the term used to describe the persistence. It means that apparently color-blind forces such as educational attainment produce systematically unequal results for members of majority and minority groups. This more subtle form of racism is an important dimension of contemporary race and ethnic inequalities.

Cross-Cutting Statuses: Class and Race

In this chapter and Chapter 6, we have dealt with unequal life chances by social class and race and ethnicity. For each of these characteristics, we have been able to demonstrate that there is a hierarchy of access to the good things in life and that some groups are substantially disadvantaged.

When a person has a lower status on more than one of these dimensions, we speak of **double or triple jeopardy.** This means that disadvantages snowball. Black teenagers are twice as likely as white teenagers to be unemployed; old women are more likely than old men to be poor. In this section, we review briefly the special problems of the cross-cutting statuses of class and race.

Double or triple jeopardy means having low status on two or three different dimensions of stratification.

Class and Race

U.S. scholars have hotly debated whether race or class is more important for understanding the structure of inequality in the United States today. The question most often asked is, "Is the status of lower-class African Americans due to the color-blind forces of stratification, or is it due to racism?" A 1978 book titled *The Declining Significance of Race* set the tone for much of this debate. In it, black sociologist W.J. Wilson argued that the status of African Americans had less to do with racism than with the simple inheritance of poverty and the changing nature of the U.S. economy. As well-paying factory jobs have disappeared and as other forms of employment have shifted from the inner cities to the suburbs, the position of the poorest third of the African American population has clearly disintegrated. Joblessness is up, the number of female-headed households is up, rates of drug use are up, and so on. For this reason, Wilson has argued that the solutions to racial inequality are primarily economic. By developing strategies that create full employment and better jobs for *all* Americans, Wilson (1987) believes that the serious problems faced by African Americans will largely be eliminated.

An Application

focus on

ENVIRONMENTAL RACISM

A growing body of research shows that minority communities are exposed to a disproportionately large number of health and environmental risks in their neighborhoods and on their jobs. Nationally, three of the five largest commercial hazardous-waste landfills are located in areas where blacks and Hispanics make up the majority of the population (Bullard 1990). Farm workers (the vast majority of whom are members of minority groups) and their children are exposed to the poisons of pesticides sprayed on crops. Indian reservations are increasingly asked to serve as high-level radioactive waste disposal sites for public utilities that are running out of space to store used nuclear fuel rods. The United States and other Western industrial powers export hazardous waste and household garbage to Third World nations in Africa, Asia, and Central and South America. American manufacturers currently export about a third of their total domestic production of pesticides; about one-fourth of these exports are products which cannot be sold in the United States for any use whatsoever (Bright 1990).

In December 1991, Lawrence Summers, the chief economist of the World Bank, sent a note to his colleagues that may well summarize the view of many movers and shakers about the environment. An excerpt follows:

> Just between you and me, shouldn't the World Bank be encouraging more migration of the dirty industries to the LDCs [Less-Developed Countries]? I can think of three reasons:
>
> (1) The measurement of the costs of health-impairing pollution

depends on the forgone earnings from increased morbidity and mortality. From this point of view a given amount of health-impairing pollution should be done in the country with the lowest cost which will be the country of the lowest wages....

(2) The costs of pollution are likely to be non-linear as the initial increments of pollution will probably have very low cost. I've always thought that under-populated countries in Africa are vastly under-polluted; their air quality is probably vastly inefficiently low [sic] compared to Los Angeles or Mexico City. Only the lamentable facts that so much pollution is generated by non-tradeable industries (transport, electrical generation) and that the unit transport costs of solid waste are so high prevent world-welfare-enhancing trade in air pollution and waste.

(3) The demand for a clean environment for aesthetic and health reasons is likely to have very high income-elasticity. The concern over an agent that causes a one-in-a-million change in the odds of prostate cancer is obviously going to be much higher in a country where people survive to get prostate cancer than in a country where under-five mortality is 200 per thousand...(as quoted in Foster 1993).

Because poor and minority communities in the United States and abroad are least able to resist, they are most likely to be on the receiving end of pollution. Institutionalized racism is responsible, then, for the pattern that shows communities

Third World countries in Africa and South America, like minority communities and Indian reservations in the United States, are often on the receiving end of industrial contaminants. In the short run, nations like Argentina may find storing toxic waste a viable way of earning the hard currency necessary to pay off large foreign debts. In the long run, however, the communities that host hazardous-waste disposal sites bear all of the health costs and receive far fewer economic benefits (jobs) than the communities that produce it.

hosting hazardous-waste disposal sites receive fewer economic benefits (jobs) than the communities that generate the waste. The people who benefit the most bear the least burden (Bullard 1993).

Most sociologists disagree. They doubt that policies based on social class alone will be enough to resolve the problem of racial inequality in the United States. True, there is an African American middle class (even an African American upper class) and it is a serious mistake to assume that racism keeps all racial minorities poor and powerless. Nevertheless, race continues to be a fundamental cleavage in U.S. society. Not only does being African American prove to be a handicap in social-class attainment, it

Because of the various prejudices that characterize our society, being old, poor, physically challenged, female, or a member of a racial or ethnic minority often entails significant social and economic disadvantage; disadvantages not only accumulate but actually compound each other when the same person occupies more than one devalued status, a condition that sociologists refer to as one of double or even triple jeopardy.

also continues to be important in social relationships. The finding that African American physicians are as apt as unemployed African Americans to live in segregated neighborhoods suggests that the issue goes beyond class.

Ethnic Relations in Comparative Perspectives

THINKING

CRITICALLY

Some scholars contend that the major cause of racial/ethnic inequality in the United States today is institutionalized, not individual, racism. If this is so, what recommendations would you offer to policy makers interested in reducing racial or ethnic differences in quality of life?

In 1991 in Brooklyn, New York, a black child was accidentally struck and killed by a car driven by a Hasidic Jewish man. Charges that the slightly injured driver was treated while the child lay trapped beneath the car led to four days of violence in the Crown Heights neighborhood. In 1992, following the acquittal of four police officers charged with the videotaped beating of black motorist Rodney King, Los Angeles erupted in riots that would leave 51 people dead. Dramatic events such as these tend to obscure the more subtle nature of most ethnic conflict in the United States; they also may obscure the fact that processes that operate to deny U.S. minority groups their rights and opportunities are at work in other nations as well.

Germany has witnessed a rising wave of neo-Nazi attacks on Turkish immigrants. In Rwanda, civil war between the Tutsi and Hutu has decimated the population. In Bosnia, tens of thousands of people have been killed and millions displaced by the strife between Serbians and Croatians.

Many social scientists believed that industrialization, the emergence of mass communications and education, and the development of impersonal bureaucratic forms of government would reduce interethnic antagonisms by creating new loyalties, based on a unified nation state. This has not happened and the question is, of course, "Why not?" Ethnic conflict in the former Yugoslavia illustrates both how majority/minority relations vary across societies and how they are similar.

Bosnia: A Case Study

Called the Kingdom of the Serbs, Croats, and Slovenes, Yugoslavia came into existence in 1918. It was created out of five territories. Serbia, the largest of the republics, and Bosnia/Herzegovina had been under Turkish rule for centuries and were predominantly peasant economies. Slovenia and Croatia had been under the rule of Austria or Hungary for about the same period of time. As a result, these two republics were Catholic, relatively "westernized," and considerably more economically developed than the other regions of the country, especially the fifth territory, the mountainous Montenegro (Lydall 1989). In addition to the three major ethnic groups residing in each of these territories, the new Yugoslavia included a significant number of ethnic Albanians. Moslems—Serbs or Croats who adopted the Islamic faith during the period of Turkish rule but who self-identify as a unique nationality—constituted another major group.

Despite Serbian political dominance between the two world wars, Croatia and Slovenia maintained a significant economic advantage over the other regions. Perhaps as a result of these political and economic tensions, Croatian Nazi collaborators, known as the Ustashi, killed hundreds of thousands of Serbs, Jews, and Gypsies during World War II. When the Ustashi surrendered in 1945, many thousands of Croatians were put to death by Serbian partisans (Bell-Fialkoff 1993).

In the United States in recent years, prejudice and discrimination have most often been aimed at individuals who differ physically from the majority group; in other words, people who belong to a racial minority. In some societies—Northern Ireland and Bosnia, for instance—group conflict originates in ethnic, not racial, differences. Although it is hard for many Americans to understand how people who look so similar can hate each other so much, the results of both types of prejudice—racial and ethnic—are similar and unmistakeable. The devastation produced by racial tensions and riots in cities such as Los Angeles and Miami differ only in degree from this scene of interethnic conflict and destruction in Sarajevo.

In the wake of this conflict, Field Marshall Tito (Josip Broz) came to power determined to turn Yugoslavia into a modern, unified nation state. The system of government and of workplace management that Tito instituted was consistent with the accommodation model of ethnic relations discussed earlier in this chapter. The cultural integrity and equality of each ethnic group was to be maintained at the same time that industrialization and education were expected to forge a collective Yugoslav identity.

Although there were periods of ethnic unrest, the postwar period did see a gradual increase in the number of people who identified as Yugoslavians rather than Serbs, Croatians, Moslems, Albanians, or some other ethnic group. Why did Yugoslavia disintegrate, then, erupting into enormous ethnic strife? Two factors seem critical. First, when Tito died, political power shifted from the central government to republic-level Communist parties, and local politicians had a great deal to gain from increasing the influence and power of their own region. Second, during the economic crisis of the mid-1980s, living standards in Yugoslavia declined by at least one-fourth, with inflation reaching more than 2,500 percent in 1989 (Sekulic, Massey, & Hodson 1994). While the political process that underlies the current ethnic strife in Bosnia may be unique to this society, the economic processes appear to be typical of those accompanying ethnic conflict in societies throughout history: Racial and ethnic hostilities are most pronounced when economic resources are scarce and the majority group advantage is most threatened.

Summary

1. The concepts of majority and minority groups provide a generic framework for examining structured inequalities by ascribed status. Interaction between majority- and minority-group members may take the form of conflict, assimilation, accommodation, or acculturation.

2. Prejudice, discrimination, the self-fulfilling prophecy, and the ideology of racism help create and maintain social distance. Segregation also helps reinforce differences and inequalities.

3. On many fronts, African Americans have improved their position in U.S. society. Nevertheless, black families continue to have a median income that is only 64 percent of that of white families. Major areas of continued concern are high rates of female-headed households, unemployment, and housing segregation. Being African American is now a better predictor of poverty than it was 25 years ago.

4. Hispanics are one of the fastest growing minority groups in the United States; within 50 years, they are likely to replace African Americans as the largest minority group.

Because many Hispanics are recent immigrants from less-developed countries, they generally have poor educations and low earnings. Studies show, however, that they face fewer barriers to achievement than African Americans.

5. Native Americans are the least prosperous and least assimilated group in the United States.

6. Asian Americans have used education as the road to social mobility. Even the newest immigrant groups far outstrip white Americans in their pursuit of higher education. Despite some discrimination, Asian Americans have higher family incomes than white Americans and very low levels of residential segregation.

7. Both race and class are important for understanding the experience of African Americans. Although the color-blind forces of the indirect inheritance model make it hard enough to erase economic disadvantage, prejudice and discrimination continue to pose additional handicaps.

8. From a comparative perspective, the key factor increasing ethnic and racial antagonisms appears to be an increase in competition for scarce resources.

Sociology on the Net

The dynamics of racial change can be seen throughout the contemporary United States. As the racial numbers change, so will the political landscape on which the racial wars of the past have been won and lost. The United States Census Bureau keeps close tabs on these changing figures.

http://www.census.gov/population/www/pop
-profile/profile.html

We are back to the current Population Profile that we first discovered in Chapter 6. This time let's look at the characteristics of the different minority groups in the United States. Begin by browsing through the selections on *The Black Population, The Hispanic Population, The Asian and Pacific Islander Population,* and *The American Indian, Eskimo, and Aleut Population.* Can you figure out one or two unique findings from each of these reports?

Now open and read the sections entitled *National Population Trends* and *National Population Projections.* What is the overall trend of our population growth? What is meant by the term *natural increase?* What are the differences in the rate of natural increase for the various racial categories? How will the relative proportions of the minority populations change by the year 2050? How will this changing racial composition touch you as you grow older?

Statistics tell one story of possible change. If we are to understand how minorities have experienced America, we must heed Weber and exercise some *verstehen* and try to understand what is happening from the standpoint of the participants. Let's take a look at the American Indian Movement.

```
http://dickshovel.netgate.net/McCloud.html
```

You have just entered a network of *Native American Internet* home pages. Browse around. Pay special attention to *Leonard Peltier's Pre-Sentencing Statement and Russel Means'* speech, "For America to live, Europe Must Die." How do you react to this statement? Does your status as a minority or a majority group member influence you as you consider what is being said? Can you identify with the situation of the Native Americans now that you have come in contact with some of their views?

For a further look into one of the more infamous Indian massacres, click on the *Wounded Knee Home Page.*

FIND IT ON INFOTRAC COLLEGE EDITION

For an example of how racism is expressed in everyday life, look up the following article in InfoTrac College Edition:

"The New Racism." Hans J. Massaquoi. *Ebony,* August 1996, v51 n10 p56. (Hint: Enter the search term *new racism* using the Subject Guide.)

What is your perception of the current racial climate in the United States? Do you believe that the racial climate has improved, stayed the same, or become worse in the past 20 years? What evidence can you cite to support your position?

Suggested Readings

Benjamin, Lois. (1991). *The Black Elite Facing the Color Line in the Twilight of the Twentieth Century.* Chicago: Nelson-Hall. An absorbing account of the everyday lives of today's most successful African Americans. Benjamin documents that racism continues to impact the existence of the black middle class, raising troubling questions about just how much progress has been made in U.S. race relations.

Ezorsky, Gertrude. (1991). *Racism and Justice.* Ithaca, N.Y.: Cornell University Press. A timely, readable analysis of affirmative action. Ezorsky addresses questions such as "Does affirmative action really counteract racism?" and "Is affirmative action morally or legally justifiable?" A useful starting point for a discussion of this controversial policy.

Haizlip, Shirlee Taylor. 1994. *The Sweeter the Juice: A Family Memoir in Black and White.* New York: Simon & Schuster/Touchstone. A wonderfully readable personal story of the African American author's search to find her mother's sister and other relatives, who had left her when she was a child because they passed for whites. The book illustrates that racial categories are not accurate—and that our own racial genes may not be exactly what we think they are.

Mills, Nicholaus, ed. 1994. *Arguing Immigration: Are New Immigrants a Wealth of Diversity . . . or a Crushing Burden?* New York: Simon & Schuster/Touchstone. A collection of reprinted essays from policy makers on both sides of this issue.

Oliver, Melvin L., and Thomas M. Shapiro. 1995. *Black Wealth/White Wealth.* New York: Routledge. Winner of the 1995 C. Wright Mills Award, this book demonstrates the important effects of wealth inequality in maintaining racial stratification systems.

Takaki, Ronald. (1993). *A Different Mirror.* Boston: Little, Brown and Company. A powerful retelling of America's distant and very recent past, told from the standpoint and often in the words of the many different ethnic groups usually left out of U.S. history books.

Terkel, Studs. 1992. *Race: How Blacks and Whites Think and Feel about the American Obsession.* New York: Doubleday/Anchor Books. Honest interviews in the style and tradition of Terkel's classic book *Working,* this collection offers insights into the U.S.'s race conflict as interviewees from both races "tell it like it is."

Wilson, William J. 1987. *The Truly Disadvantaged: The Inner City, the Underclass, and Public Policy.* Chicago: University of Chicago Press. A controversial book by one of the U.S.'s leading African American scholars.

Sex and Gender

Sexual Differentiation

Men and women are different. Biology differentiates their physical structures, and cultural norms in every society differentiate their roles. In this chapter, we describe some of the major differences in men's and women's lives as they are socially structured in the United States. We will be particularly interested in the extent to which the ascribed characteristic of sex has been the basis for structured inequality.

Sex Versus Gender

In understanding the social roles of men and women, it is helpful to make a distinction between gender and sex. **Sex** refers to the two biologically differentiated categories, male and female. It also refers to the sexual act that is closely related to this biological differentiation. **Gender,** on the other hand, refers to the normative dispositions and behaviors that cultures assign to each sex. Although biology provides two distinct and universal sexes, cultures provide almost infinitely varied **gender roles.** Each man is pretty much like every other man in terms of sex—whether he is upper class or lower class, black or white, Chinese or Apache.

Gender, however, is a different matter. The rights and obligations, the dispositions and activities, of the male gender are very different for a Chinese man than for an Apache man. Even within a given culture, gender roles vary by class, race, and subculture. In addition, of course, individuals differ in the way they act out their expected roles. Some males play an exaggerated version of the "manly man," whereas others display few of the expected characteristics.

Just how much of the difference between men and women in a particular culture is normative and how much is biological is a question of considerable interest to social scientists (Udry 1994). For the most part, however, social scientists are more interested in gender than in sex. They want to know about the variety of roles that have been assigned to women and men and, more particularly, about what accounts for the variation. Under what circumstances do women have more or less power, prestige, and income? What accounts for the recent changes that have occurred in gender roles in our society?

Sex is a biological characteristic, male or female.

Gender refers to the expected dispositions and behaviors that cultures assign to each sex.

Gender roles refer to the rights and obligations that are normative for men and women in a particular culture.

Cross-Cultural Evidence

A glance through *National Geographic* confirms that there is wide variability in gender roles across cultures. The behaviors we normally associate with being female and male are by no means universal. Despite the wide variety across human cultures, there are two important universals: In all cultures, child care is primarily a female responsibility, and in all cultures that we know about, women have less power than men.

In spite of the fact that women do substantial amounts of work in all societies, often supplying more than half of the food as well as taking care of stock, children, and households, women universally have less power and less value. A simple piece of evidence is parents' almost universal preference for male children (Sohoni 1994). This preference can be life threatening to girls. Demographers have determined that worldwide there are 100 million fewer women than there would be if boys and girls were equally valued. A 1992 Bombay study found that of 8,000 abortions performed after parents had used amniocentesis to determine the sex of the fetus, only one aborted fetus was male. Beyond selective abortion and female infanticide (killing of infant girls), boys often receive preferential treatment—for example, more food and medical care—while girls are more likely to be neglected. The preference for boys is less strong in modern industrial nations, but parents in the United States prefer their first child to be a boy by a 2-to-1 margin (Holloway 1994; Pebley & Westoff 1982; Sohoni 1994).

Determinants of Women's Power

There are no known societies in which women have more power than men, but important variations exist from society to society in the amount of power and prestige women have. In some societies, women's power is very low, whereas in others it approaches equality with men's. Four key factors determine women's power in any society: (1) the degree to which women are tied to the home by bearing, nursing, and rearing children; (2) the degree to which economic activities in a society are compatible with staying close to home and caring for children; (3) the degree of physical strength necessary to carry on the subsistence activities of the society; and (4) the degree to which the society is militaristic, valuing waging war and weapons.

Until the sharp fertility declines of the last 200 years, the first factor showed relatively little variability: Most women in most societies were more or less continually tied close to home by pregnancy and subsequent responsibility for nursing and rearing children. The degree to which women could participate in the economic life of their society and contribute to subsistence depended substantially on the second and third conditions. When economic activities required little physical strength and could be carried on while caring for children, women made major contributions to providing subsistence for their families and communities (Chafetz 1984).

As a result of these factors, women have the highest power and prestige in gathering and simple horticultural societies (Quinn 1977). In these societies, the major subsistence activities (gathering and simple hoe agriculture) are compatible with women's child-related roles, and women may be responsible for 60–80 percent of a society's subsistence (Blumberg 1978). Moreover, their economic activities make them an active part of their community and increase the likelihood of their being involved in community and group decisions. Partly because war traditionally required physical strength, the power of women in warring societies tends to be lower.

These four factors help explain most of the important differences in women's power across societies, and they allow us to understand why women's power was low during industrialization but now shows signs of rising. Industrialization moved work away from home and made it difficult for women to be economically productive while bearing and rearing children. Reduced fertility, however, has allowed women to leave the household and increase their participation in society's economic and public life. As a result, the status of women is improving. Nevertheless, men remain substantially advantaged in prestige and power. A recent study sponsored by the United Nations concludes that, at the current rate of change, women will not be equal to men economically until the year 2490 (Wright 1995).

One result of female power disadvantage is widespread violence toward women. "Domestic violence is a leading cause of female injury in almost every country in the world," concluded an international study by the Human Rights Watch organization in 1995 (Wright 1995). For example:

- In the United States, in 1992, more than a half-million women were murdered, raped, assaulted, or robbed by an intimate (spouse, ex-spouse, or boyfriend). That number is more than 10 times the number of male victims of intimate violence for the same year, 49,000 (U.S. Department of Justice 1994, 3). Violence between intimates, or domestic violence, is further discussed in Chapter 11.

- In Peru, the beating of women by their husbands makes up 70 percent of all reported crime, according to a United Nations study (Wright 1995).
- In Russia, where an estimated 15,000 women were killed by their mates in 1994, there are no laws that effectively protect women from domestic violence, although a man beating his wife can be charged with the minor offense of hooliganism (Edwards 1995).
- An estimated 110 million women, mostly in African countries but also in Asia, South America, and Europe, have undergone the ritual of genital mutilation—removal of some or all of the clitoris and surrounding genitalia in order to control sexual desire and behavior (Daly 1991; Holloway, 1994; Woods & Clouse 1994).
- In India, since 1990, more than 20,000 brides have been murdered in "dowry deaths," the consequence of a new wife's bringing an inadequate dowry (sum of money and goods) to her husband's family (Wright 1995).

At home and abroad, violence against women derives essentially from the lower status accorded to women in the family (Wright 1995). In growing numbers, women around the world are demanding equal rights. The 1995 World Conference on Women in Beijing, China may well have marked a turning point for the international women's movement. In developing nations, of course, the primary question arising from the conference will be how to pay for the programs that are necessary to improve the quality of women's lives significantly. In the United States and the rest of the developed world, the questions may be somewhat more subtle: How are gendered identities developed, and what are the institutional forces that can and do maintain inequality, sometimes even in the absence of overt violence and discrimination? In the rest of the chapter, we focus on gender in the United States.

Gender in the United States

Developing a Gendered Identity

Gender differences begin with pink or blue blankets in the hospital and extend throughout life. One of the first elements that youngsters distinguish as part of their self-concept is their gender identity. By the age of 24 to 30 months, they can correctly identify themselves and those with whom they come into contact by sex, and they have some ideas about what this means for appropriate behavior (Cahill 1983).

Young children's ideas about what it means to be a boy or a girl tend to be quite rigid. They develop strong stereotypes for two reasons. One is that the world they see is highly divided by sex: In their experience, women usually don't build bridges and men usually don't crochet. The other important determinant of stereotyping is how they themselves have been treated. Substantial research shows that parents treat boys and girls differently. They give their children "gender-appropriate" toys; they respond negatively when their children play with cross-gender toys; they allow boys to be active and aggressive; and they encourage their daughters to play quietly and visit with adults (Orenstein 1994). If parents do not exhibit gender-stereotypic behavior and if they do not punish their children for cross-gender behavior, however, the children will be less rigid in their gender stereotypes (Berk 1989).

As a result of this learning process, boys and girls develop fairly strong ideas about what is appropriate for girls and what is appropriate for boys. Because males and male behaviors have higher status than female behaviors, boys are punished more than girls

MAP 9.1
Women in Public Life, 1994

SOURCE: United Nations, 1995. The World's Women 1995: Trends and Statistics. Social Statistics and Indicators, Series K, No. 12. New York: United Nations.

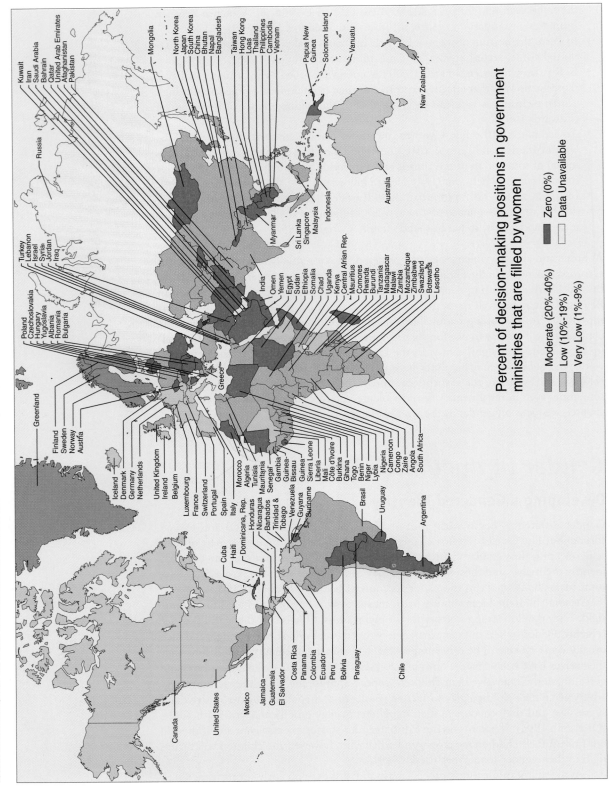

Percent of decision-making positions in government ministries that are filled by women

Moderate (20%-40%)

Low (10%-19%)

Very Low (1%-9%)

Zero (0%)

Data Unavailable

for exhibiting cross-gender behavior. Thus, little boys are especially rigid in their ideas of what girls and boys ought to do. Girls are freer to engage in cross-gender behavior, and by the time they enter school, many girls are experimenting with boyish behaviors.

Differences in Life Chances By Sex

In terms of race and social class, women and men start out equal. The nurseries of the rich as well as the poor contain about 50 percent girls. After birth, however, different expectations for females and males result in their having very different life chances. This section examines some of the structural social inequalities that exist between women and men.

Education

In the late 1990s, women and men are equally represented among high school graduates and among those receiving bachelor's and master's degrees. It is not until the level of the Ph.D. or advanced professional degrees (law and medicine) that women are disadvantaged in quantity of education.

Probably more important than the differences in level of education are the differences in types of education. From about the fifth grade on, sex differences emerge in academic aptitudes and interests: Boys take more science and math, whereas girls more often excel in verbal skills and consequently tend to focus their efforts on language and literature. As a result of these early interests, women college graduates are overrepresented in the fields of education and the humanities, and men are overrepresented in the fields of physical sciences and engineering.

Table 9.1 shows the proportion of bachelor's degrees earned by women in various fields of study in 1971 and in 1993, the latest statistics available. You can see from the table that there were changes over this period. Women were far more likely in 1993 than in 1971 to major in pre-law. Furthermore, women and men graduates in 1993 were about equally likely to have majored in business or management—a significant change since 1971, when only 9.1 percent of such graduates were women. Women and men were also about equally divided in 1993 in the fields of mathematics and social sciences.

The most striking differences between men and women were in the fields of education, home economics, library sciences, and engineering. In 1993, only 14 percent of graduates in engineering were women. Meanwhile, 78 percent of graduates in education, 89 percent in home economics, 83 percent in the health sciences, and 89 percent in library sciences were women (U.S. Bureau of the Census 1996, Table 302). Because engineers make a great deal more money than teachers, librarians, and most health science workers, these differences in educational aspirations have implications for future economic well-being. This situation is an example of institutionalized discrimination against women—a persistence of disadvantage through presumably neutral social processes, such as educational attainment, that produces systematically unequal results for members of a designated group or category. (Recall that Chapter 8 discussed institutionalized discrimination in the form of racism.)

Children

Although childbearing itself is a biologically determined capacity, many aspects of childbearing are socially determined. How many children to have, when, and who will care for them are issues that are determined by social structure rather than biology.

Bearing children has enormous consequences for women's status. Although the last few years have seen increased emphasis on fathers' involvement with their children, the

THINKING CRITICALLY
Suppose you would really like your daughter to grow up to be a mathematician. Short of coercion, how would you go about encouraging her to make this career choice?

TABLE 9.1
Bachelor's Degrees Earned, by Field, 1971 and 1993
Between 1971 and 1992, there was a substantial narrowing of the sex gap in educational focus. Nevertheless, engineering continues to be largely a male preserve, while education attracts a disproportionate number of women undergraduates. Because engineers earn roughly three times what teachers earn, this difference in educational direction is one reason why, on average, women earn less than men.

	PERCENT FEMALE	
Field of Study	1971	1993
Business and management	9.1	47.2
Computer and information sciences	13.6	28.1
Education	74.5	78.4
Engineering	0.8	14.4
Health sciences	77.1	83.1
Home economics	97.3	89.2
Library and archival sciences	92.0	89.2
Pre-law	5.0	67.6
Mathematics	37.9	47.2
Social sciences	36.8	45.8

SOURCE: U.S. Bureau of the Census, 1996a, Table 302.

major responsibility for child care in most U.S. households rests with the mother. Perhaps as a result, women are more likely than men to be employed in part-time or part-year work. Some economists have suggested that mothers also invest less effort in their work because family obligations consume so much time and energy. However, research shows that, on average, women allocate more effort to work than men with comparable family situations (Bielby & Bielby 1988). If childbearing and child care account for the female disadvantage in status and power, then, they appear to do so more because employers *expect* women to devote less effort to their careers than because they actually do so.

Nevertheless, the social and economic hardship that childbearing and childrearing can impose on women probably explains the very low fertility of U.S. women. The average U.S. woman has only one or two children, and there is growing tolerance for deliberate childlessness. Even so, 90 percent of U.S. women expect to have children (U.S. Bureau of the Census 1989j).

Labor-Force Participation

In 1995, 93 percent of men compared with 75 percent of women aged 25–54 were in the labor force (U.S. Department of Labor 1993). This gap is far smaller than it used to be but probably larger than it will be in the future (Figure 9.1). Recent studies show no sex differences at all in the proportion of male and female college students who expect to be employed at age 25 or age 50 (Affleck, Morgan, & Hayes 1989). Although most young women expect to be mothers, they also expect to be full-time permanent members of the labor force.

Despite the growing equality in labor-force involvement, major inequalities in the rewards of paid employment persist. In 1997 women who were full-time, full-year workers earned 75 percent as much as men, an all-time high. Overall, this percentage has not changed much since 1950. Why do women earn less than men? The answers fall into two categories: differences in the types of jobs men and women have and differences in earnings of men and women in the same types of jobs.

THINKING

CRITICALLY

Chapter 8 discussed institutionalized discrimination. Can you think of some specific examples of how institutionalized discrimination works against women in the workplace? Against men? How specifically might affirmative action programs (also discussed in Chapter 8) help to alleviate this discrimination?

FIGURE 9.1
Labor-Force Participation Rates of Men and Women Aged 16 and Over, 1980 to 2000 (est.)
In the 20 years between 1980 and 2000, two major changes occur in labor-force participation: Men's and women's rates become very similar, and fewer of either sex work past age 60.
SOURCE: U.S. Bureau of the Census 1995, Table 627; U.S. Department of Labor.

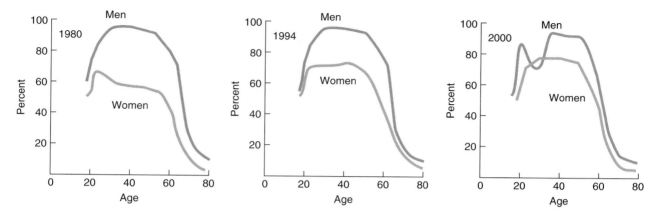

DIFFERENT OCCUPATIONS: DIFFERENT EARNINGS.   Women are employed in different jobs than men (see Figure 9.2) and the jobs women hold pay less than the jobs men hold. The major sex difference in Figure 9.2 is that women dominate clerical jobs, whereas men dominate blue-collar jobs. The proportion of men and women in professional and managerial jobs is nearly equal. Generally, though, men are doctors and women nurses; men manage steel plants and women manage dry cleaning outlets.

There are three major reasons why men and women have different jobs: gendered jobs, different qualifications, and discrimination.

1. *Gendered jobs.* Because of historical circumstances, many jobs in today's labor market are regarded as either "women's work" or "men's work." Manual labor is almost exclusively men's work; nursing and keyboarding are largely women's work. These occupations are so sex segregated that many men and women would feel uncomfortable working in a job where they were so clearly the "wrong" sex. Women in male-dominated occupations report being made to feel uncomfortable by such subtle forms of discrimination as exclusion from informal leadership and decision-making networks, sexual harassment, and other forms of hostility from male co-workers (Jacobs 1989). These forms of informal discrimination serve as "glass ceilings"—invisible barriers to women's promotions in traditionally male careers (Freeman 1990). In contrast, men working in female-identified occupations report that negative stereotypes about men who do "women's work" form the major barrier to more men entering occupations such as nurse, elementary school teacher, librarian, or social worker. However, when men do work in nontraditional fields, they are likely to encounter "glass escalators"—rapid promotions "up" to more masculine administrative positions (Williams 1992).

Although there has been virtually no change in the number of men employed in female-dominated occupations, a growing number of jobs that used to be reserved for men—such as insurance adjusters, police officers, and bus drivers—have opened up to women in the last decade. Research shows, however, that this is not because of women's increased access to good jobs. Rather, women are moving into jobs that men are abandoning because of deteriorating wages and working conditions (Reskin 1989).

Despite many changes in U.S. gender roles, boys and girls still tend to experience large doses of traditional gender socialization. Boys are more likely to want to go hunting or target shooting, and girls are more likely to want to play dress-up. Although some of these differences may be due to our genetic heritage, studies show that parents' and teachers' expectations have a tremendous influence on the degree to which children ascribe to and act out traditional roles. Dad and mom, after all, provided this lipstick, not nature.

FIGURE 9.2
Differences in Occupation by Sex, 1995
Men and women continue to be employed in different occupations in the United States. The most striking differences are in technical, sales, and clerical work, where women predominate, and in various blue-collar jobs, held primarily by men. Within categories, men typically hold higher-status positions than women.
SOURCE: U.S. Bureau of the Census, 1996a, Table 637.

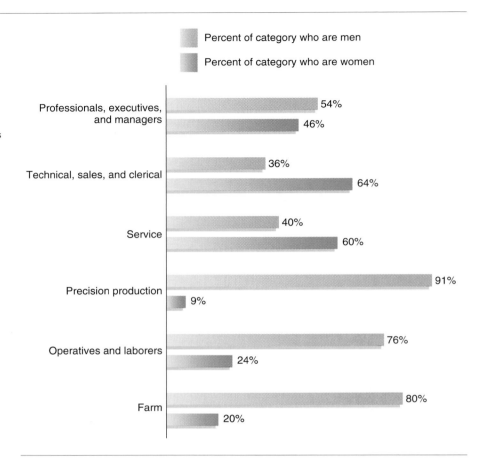

2. *Different qualifications.* Although the differences are smaller than they used to be, women continue to major in fields of study that prepare them to work in relatively low-paying fields, such as education, while men are more likely to choose more lucrative fields. More important than these differences in educational qualifications, however, are the disparities in experience and on-the-job training. Because many women have taken time out of the labor force for childrearing, they have less job experience than men their own age (Marini 1989). Perceiving that women are more likely to be short-term employees than men, employers have invested less in training women. As a result, women are less likely to be promoted to management positions.

3. *Discrimination.* Although men and women have somewhat different occupational preparations, a large share of occupational differences is due to discrimination by employers (Bielby & Baron 1986). Employers reserve some jobs for men and some for women based on their own gender-role stereotypes. This covers not only major occupational differences (men drive forklifts and women type) but also very minor distinctions in job titles. Within the same objective task, men and women will be given different titles—women will be executive assistants, and men doing the same work will be assistant executives, often at a much higher pay scale.

Why do employers discriminate? An important reason is that those who do are sexists. They believe that men are competent and analytical and that women are emotional and timid. It is not surprising, then, that they prefer to hire men for managerial or other responsible positions.

Several studies document that another important reason that men and women have different jobs is that women are less likely to be promoted. One study of women engineers, for example, found that 5 to 10 years after finishing school, women were already significantly behind their male peers, even though they were identical in terms of education and experience (Robinson & McIlwee 1989).

SEXUAL HARASSMENT.   A special form of discrimination that is especially problematic for female workers, military personnel, and students is **sexual harassment**—unwelcome sexual advances, requests for sexual favors, and other unwanted verbal or physical conduct of a sexual nature. Harassment may be just an annoyance. It may, however, turn into real discrimination. The courts use the following guidelines to determine when unwelcome sexual advances constitute unlawful discrimination (Schapiro 1994).

- *Quid pro quo harassment.* This occurs "when an employee's submission to unwelcome sexual conduct becomes an explicit or implicit condition of employment, or when personnel actions such as promotion, transfer, compensation, or discipline are determined on the basis of an employee's response to such conduct."
- *Hostile environment.* A hostile environment develops when there is "unwelcome sexual conduct which unreasonably interferes with an individual's job performance or creates an intimidating, hostile, or offensive work environment."

Sexual harassment ranges from subtle hints about the rewards of being more friendly with the boss or teacher to rape. In the less-severe instances (and sometimes even in the more severe instances), the subordinate may be reluctant to make, literally, a federal case of it. This pattern of reluctance to report the matter became an issue several years ago when Anita Hill, a law professor, accused Supreme Court nominee Clarence Thomas of sexual harassment that allegedly had occurred years earlier.

SAME JOB: DIFFERENT EARNINGS.   Not all jobs are substantially sex segregated. Some, such as flight attendant, teacher, research analyst, contain considerable proportions of both men and women. Generally, however, men earn substantially more than women even within the same occupation (see Table 9.2), and these same occupation wage differences are a major source of sex inequities (Brown et al. 1980; M. H. Stevenson 1975).

Within integrated occupations, such as hotel managers or law, men tend to be in high-paying firms and women tend to be in low-paying firms (Blau 1977). Men tend to be in large firms and women in small. These differences in type of firm reflect what is called a segmented labor market, discussed in more detail in Chapter 13. The gist of this concept is that *where* you work may be as important as what you do in determining your income and opportunity.

Another important reason that women are less successful than men within the same occupation is that women's ability to maximize their careers is sharply curtailed by their family responsibilities. More often than men, women have to choose jobs close to home so that it is easy to drop junior off at the baby-sitter or take the child to the doctor during the lunch hour. Similarly, women are much less likely than men to be able to uproot their families and move to advance their careers.

Recognizing these family claims on women employees, employers are less likely to hire them for jobs requiring long career tracks or geographical moves.

Sexual harassment consists of unwelcome sexual advances, requests for sexual favors, or other verbal or physical conduct of a sexual nature.

In the contemporary labor force, few jobs require sex-specific abilities, such as being able to lift 100 pounds or do 100 push-ups. As a result, there are a growing number of jobs that include women and men. Just having the same job, however, does not produce equal earnings. Even among full-time, full-year employees with the same occupational title, women earn substantially less than men. As women's employment becomes a lifelong role and fewer women take sustained periods off for child rearing, some of this earnings gap may be reduced.

As we have seen, they are also less likely to invest in training for them and less likely to promote them. This pattern affects all women, even those who intend to work for 40 years or who are the sole support for themselves or their families (Marini 1989). However, studies show that discrimination is strongest against women who are married and have children. Partly as a result of the greater opportunities that are open to them, women with fewer family ties earn more than women with husbands and children (Shellenbarger 1992).

Power Inequalities

As Max Weber pointed out, differences in prestige and power are as important as differences in economic reward. When we turn to these rewards, we again find that women are systematically disadvantaged. Whether we are talking about the family, business, or church, we find that women are less likely to be given positions of authority.

UNEQUAL POWER IN SOCIAL INSTITUTIONS.   Women's subordinate position is built into most social institutions. From the church to the family, we find that norms specify that women's roles are subordinate to men's. The Bible's New Testament, for example, urges, "Wives, submit yourselves unto your own husbands" (Ephesians 5:22); the traditional marriage vows require women to promise to obey their husbands; until 1919, women were not allowed to vote in the United States.

In politics, prejudice against women leaders is declining—but it is still quite strong. Whereas women comprise 53 percent of the voters, they represent only 21 percent of all state legislators and only 11 percent of the U.S. Congress. In every institution, traditional norms have specified that men should be the leaders and women the followers. Social scientists call this situation patriarchy.

UNEQUAL POWER IN INTERACTION.   As we noted in Chapter 4, even the informal exchanges of everyday life are governed by norms; that is, they are patterned regularities, occurring in similar ways again and again. Careful attention to the roles men and women play in these informal interactions shows some rather clear differences—all of them associated with women's lower prestige and power in U.S. society.

An easy-to-see example is that women smile more (Goffman 1974a). They smile to offer social support to others, and they smile to express humility. Studies of informal conversations also show that men regularly dominate women in verbal interaction.

TABLE 9.2
Sex Differences in Earnings from the Same Occupation
Even when women have the same occupation as men, they tend to earn substantially less money. In part because of their family responsibilities, women tend to be employed in smaller and lower-paying firms. They also experience substantial discrimination in both employment and promotional opportunities.

	MEDIAN EARNINGS		
Occupation	Males	Females	Male/Female Ratio
Accountants	$40,469	$27,750	1.46
Engineers	46,512	40,341	1.15
Natural scientists and mathematicians	42,308	34,208	1.24
Computer equipment operators	29,131	21,595	1.35
Lawyers and judges	71,530	50,296	1.42

SOURCE: U.S. Bureau of the Census 1992a, pp.153, 156.

Men take up more of the speaking time, they interrupt women more often, and most important, they are more often successful interrupters. Finally, women are more placating and less assertive in conversation than men, and they are more likely to state their opinions as questions ("Don't you think the red one is nicer than the blue one?" (Tannen 1990). This pattern is also apparent in committee and business meetings—one reason women employees are less likely than men to get credit for their ideas (Tannen 1994).

Laboratory and other studies show that this male/female conversational division of labor is largely a result of differential status (Kollock, Blumstein, & Schwartz 1985; Tannen 1990; Wagner, Ford, & Ford 1986). When clear-cut situational factors, such as a student/teacher situation, give women more status in a conversation, they cease to exhibit low status interaction styles. Nevertheless, a study of conversations between physicians and patients showed that patients are much more likely to interrupt when the doctor is a woman (West 1984). It would appear that the lower status accorded to women in U.S. society cannot be overturned merely by changes in occupational or political roles.

The Other Side: Male Disadvantage

Women are at a substantial disadvantage in most areas of conventional achievement; in informal as well as formal interactions, they have less power than men. They pay for their disadvantage in higher levels of mental illness and poverty. Men, too, face some disadvantages from their traditional gender roles.

MORTALITY. Perhaps the most important difference in life chances involves life itself. In 1995, men in the United States could expect to live 72.5 years and women 79.3 years (U.S. Bureau of Census 1996a). On the average, then, women live seven years longer than men. Although part of this difference is probably biological (Waldron 1983), the social contribution to differential life expectancy begins at a very early age and continues through the life course. At high school and college ages, males are nearly three times more likely than females to die. To a significant extent, this difference has to do with the more dangerous lifestyle (particularly drinking and driving) associated with masculinity in our culture.

The sex differential in life expectancy is more complex than the greater risk taking of young men, however. Even at the earlier ages just mentioned, men are more likely than women to die of cancer or heart disease. Consider men's disadvantage in heart disease, for example. Evidence suggests that men have this disadvantage not because they experience more stress than women but because, under the same levels of stress, they are more vulnerable to heart disease than women. Current thinking attributes men's greater risk of heart disease at least partly to the male gender role's low emphasis on nurturance and emotional relationships. Where women's personalities and relationship characteristics seem to protect them from this consequence of stress, lack of social support appears to leave men especially vulnerable to stress-related diseases (Nardi 1992). This suggests that, despite increasing female participation in politics and the labor force, women's life expectancy will continue to remain substantially higher than men's. The sex differential in mortality is likely to decline only when differences in personality and dispositions are reduced.

SOCIAL INTEGRATION. The low emphasis on nurturance and expressiveness in the male role appears to reduce men's interest in and ability to form close relationships with their children, other kin, and friends. Maintaining family relationships is usually

Women sometimes point and men sometimes hug one another, but the genders do differ in characteristic styles of nonverbal communication. Consistent with their generally higher social status and power, men typically occupy more physical space while communicating and make use of more assertive gestures than women do. Although this picture does not necessarily make it clear, women are less likely to fling an arm across the back of someone else's chair or stare than men are, but they are also more likely to smile.

THINKING

CRITICALLY

If men are the more powerful gender, why is it that they die earlier and have higher rates of heart disease, suicide, and alcoholism? As women gain power, should we expect them to have similar health problems? Why or why not?

viewed as women's work, and when men end up without women to do this work for them (never married, divorced, or widowed), they also frequently end up alone (Kessler & McLeod 1984; Stroebe & Stroebe 1983; Wallerstein & Kelly 1980). Ultimately, this leaves them substantially disadvantaged in health.

HIGHER STRESS.   In addition, the focus of the male gender role on achievement and success can prove stressful. Even men who are successful by any reasonable standard may feel pressured by the constant striving, and those who fail often compensate by excessive aggressiveness in other spheres. Then, too, some men today experience anxiety and stress over losing some of their higher prestige and power in the workplace (Galen 1994) and at home. This stress may be reflected in higher rates of heart disease, a suicide rate that is four times higher than women's, and an alcohol-related death rate that is almost four times higher (U.S. Bureau of the Census 1996a).

Summary

Although there are individual exceptions, men as a category have more prestige, more income, and more power than women. One consequence of this is that five times more female-headed than male-headed households are living in poverty. Men pay a price, however, in terms of foregone intimacy and higher mortality. Many changes have occurred in the last decades, and most of them have reduced the differences between women and men. However, in two areas—life expectancy and earnings—the gap between men and women is as large as or larger than it was 20 years ago.

Perspectives on Gender Inequality

The fact that women bear children is due to physical differences between men and women. Most of the differences in men and women's life chances, however, are socially structured. Different sociological theories offer different explanations for the persistence of gender inequality.

Structural-Functional Theory: Division of Labor

The structural-functional explanation of gender inequality is based on the premise that a division of labor is often the most efficient way to get a job done. In the traditional sex-based division of labor, the man does the work outside the family and the woman does the work at home. According to this argument, a gendered division of labor is functional because specialization will (1) increase the expertise of each sex in its own tasks, (2) prevent competition between men and women that might damage the family, and (3) strengthen family bonds by making men dependent on women and vice versa.

Of course, as Marx and Engels noted, any division of labor has the potential for domination and control. In this case, the division of labor has a built-in disadvantage for women because by specializing in the family, women have fewer contacts, less information, and fewer independent resources. Because this division of labor contributes to family continuity, however, structural functionalists have seen it as necessary and desirable.

Conflict Theory: Segmented Labor Markets

According to conflict theorists, women's disadvantage is not an historical accident. It is designed to benefit men. In addition, contemporary sex differences are designed to benefit the capitalist class.

American Diversity

WHO IS DIANE NASH?

During the first half of the nineteenth century in the United States, many civic-minded Northern women were active in the movement to abolish slavery. The discrimination they experienced in trying to speak out against slavery radicalized many of these women, who soon realized that they couldn't free anyone else while they themselves were not free. As a result, some women split away from the abolitionist movement to promote women's rights. Many who stayed within the ranks of the abolitionist movement thought that once rights were extended to black men, they would surely be extended to white and other women. Of course, many activists continued to work hard both for the end of slavery and for women's rights. Among the most eloquent of these was Sojourner Truth. In response to male hecklers who claimed that women were weak and needed the protection, she responded:

> The man over there says women need to be helped into carriages and lifted over ditches, and to have the best place everywhere. Nobody ever helps me into carriages or over puddles, or gives me the best place— and ain't I a woman?
>
> Look at my arm. I have ploughed and planted and gathered into barns, and no man could head me—and ain't I a woman? I could work as much and eat as much as a man—when I could get it—and bear the lash as well! And ain't I a woman? I have born thirteen children, and seen most of 'em sold into slavery, and when I cried out with my mother's grief, none but Jesus heard me—and ain't I a woman? (Cited in Flexner 1972, 90–91).

Despite Sojourner Truth's effective advocacy of women's rights, many women of color have found the struggle for women's rights and that for minority rights difficult to balance. Feeling that issues of race were too often overlooked in the women's movement, many also found their concerns as women were ignored within the minority rights organizations. Most students of the 1960s Civil Rights movement will recognize the names of Stokely Carmichael, Martin Luther King, Jr., Ralph Abernathy, and Julian Bond but will wonder where the women were. "Who," they might ask "is Diane Nash?"

Analysts have, in fact, noted that despite their seeming progressiveness, the 1960s antiwar and Civil Rights movements were male-dominated. Although women made up a significant portion of the members of these groups, men filled virtually all of the leadership positions. Newspaper accounts, photos, and interviews from the time period show men "manning" the barricades while women busily support them by cooking, typing, or running errands. But is the conventional wisdom about this time period true?

Using the archives of the Martin Luther King, Jr., Center for Nonviolence and the Civil Rights Documentation Project at Howard University, Belinda Robnett (1996) has recently demonstrated the pivotal leadership role that African American women played in the modern Civil Rights movement. Although they received little of the media attention, Robnett illustrates the crucial leadership role that women played as bridges between African American communities in the rural south and the formal, male civil rights leadership. Serving as fundraisers and coordinators of voter education projects, these women not only were responsible for increasing membership and resources for the movement, they did so with a considerable amount of authority and control over day-to-day activities. Furthermore, by keeping in touch with the desires of the community, they were often able to influence the strategies adopted by formal movement leaders.

So, who is Diane Nash? Here is the story that Robnett (1996) tells: The Congress of Racial Equality (CORE) developed the Freedom Rides Project in order to spur the desegregation of public transportation. On May 4, 1961, young African American and white riders left Washington, D.C., on the first ride south. They encountered few difficulties until reaching Alabama, where riders were attacked and severely beaten, many near to death. Though the CORE riders were unable to continue, Diane Nash, who participated both in the Student Nonviolent Coordinating Committee (SNCC) and in the Southern Christian Leadership Conference (SCLC) phoned Fred Shuttlesworth, a minister and formal SCLC leader, in Birmingham to insist that the rides continue. She told him, "The students have decided that we can't let violence overcome. We are going to come to Birmingham to continue the Freedom Ride." Shuttlesworth responded "Young lady, do you know that the Freedom Riders were almost killed?" "Yes," she replied, "that's exactly why the rides must not be stopped. If they stop us with violence, the movement is dead. We're coming; we just want to know if you can meet us" (Branch 1988, 430; as quoted in Robnett 1996, 1685–1686). Clearly, Diane Nash was a civil rights leader. In bridging the desires of the participants with those of the formal leaders, her actions earned her considerable recognition and power. The Freedom Rides, of course, became a powerful instrument in eroding racism in the South.

It is relatively easy to see how women's lower status can benefit men, but how can it benefit capitalism? The answer lies in the segmented labor market (discussed in Chapter 13). The segmented labor market is two-tiered and sharply gendered. It creates one set of jobs for women and another, better set for men. In this way, capitalists divide the working class, coaxing working-class men into a coalition against working-class women. Further, female labor can be used to provide a cushion against economic cycles. Women can be laid off in slack times and hired back when employment demands are up (Bonacich 1972).

One of the most important mechanisms for keeping women in their place is **sexism**—the belief that women and men have biologically different capacities and that these differences form a legitimate basis for unequal treatment. Conflict theorists explain sexism as an ideology that is part of the general strategy of stratification. If others can be categorically excluded, the need to compete individually is reduced. Sexism, then, is a means of reducing access to scarce resources.

Sexism is a belief that men and women have biologically different capacities and that these form a legitimate basis for unequal treatment.

Summary

Structural functionalism stresses how society benefits from a gendered division of labor, while conflict theory assumes sex stratification should be minimized. Borrowing many elements of conflict theory, the new feminist theories all emphasize that gender inequality is socially structured and culturally condoned. Whether liberal, socialist, or radical, feminist theories offer blueprints for reducing inequality based not only in gender but at the intersections of class, race, and gender stratification systems.

Cross-Cutting Statuses: Sex and Race

Women of color face a two-pronged dilemma. First, they have not benefited from the sheltered position of traditional women's roles. Women of color have always worked outside the home: For example, married African American women were six times more likely to be employed at the turn of the century than married white women (Golden 1977). Although they worked, they still had to face the economic and civic penalties of being women. Consequently, minority women traditionally have had less to lose and more to gain from abandoning conventional gender roles. On the other hand, women of color face a potential conflict of interest: Is racism or sexism their chief oppressor? Should they work for an end to racism or an end to sexism? If they choose to work for women's rights, they may be seen as working against men of their own racial and ethnic group.

Current income figures indicate that sex is more important than race in determining women's earnings: The difference between Hispanic, African American, and European American women is relatively small compared to the difference between women and men. This suggests that fighting sex discrimination should be more important than fighting racial discrimination. But this conclusion overlooks the fact that the total income of women and children depends to a significant extent on the earnings of their husbands and fathers. For example, black women and children are three times more likely than white women and children to live below the poverty level. The reason, of course, is the low earnings and employment opportunities of African American and Hispanic men. From this perspective, women of color could advance their cause most effectively by fighting racial and ethnic discrimination.

The dilemma remains a real one. The woman's rights movement is often seen as a middle-class white social movement; racial and ethnic movements have been seen as men's movements. Nevertheless, minority women have a long history of resistance to both forms of discrimination, racism and sexism; increasingly, women of color strive for gender equality within traditional racial and ethnic organizations (Collins 1991).

technology

HOW IS TECHNOLOGY AN EQUITY ISSUE?

"Technology is an equity issue," writes Corlann Bush of Montana State University. It has "everything to do with who benefits and who suffers, whose opportunities increase and whose decrease, who creates and who accommodates" (Bush 1993, 206). To a significant extent, who benefits and who suffers depends on the social conditions in place when technological innovations appear (McGinn 1991). We can examine a few ways in which technology has affected African Americans and women in our society. In some cases, technology has apparently benefitted African Americans or women, while in other cases it assuredly has not.

Technology and African Americans

○ Eli Whitney's 1793 invention of the cotton gin made large-scale plantations profitable and created the need for a large supply of cheap labor— a problem solved by importing many more African slaves (McGinn 1991, 118).

○ Automobiles made it easier for people to live in neighborhoods distant from their workplaces, but the resultant suburbs have been primarily white, with blacks concentrated in central cities. New expressways destroyed cohesive city neighborhoods and parks, further alienating racial minorities (Johnson 1993).

○ In the 1950s and 1960s, television gave impetus to the civil rights movement as millions of Americans watched Dr. Martin Luther King's demands for racial justice and also saw police dogs and fire hoses unleashed on demonstrators (McGinn 1991, 120). Yet, contemporary television tends to depict African Americans either relatively infrequently or in negatively stereotyped or lesser roles (Johnson 1993, 272).

○ Good jobs in today's information society are disproportionately unlikely to go to African Americans, who are less inclined to show interest in computer skills (Wessells 1990). Largely because of economic disparities, African American children are less likely to attend computer camps and are only one third as likely to have access to a computer (http://www.census.gov/population/socdemo/computer/compl.txt).

○ Current biomedical technologies concern some African American observers, who worry, for instance, that "dead or dying Black bodies" will be seen as little more than resources for human organs going to whites (Johnson 1993, 280).

Technology and Women

○ Beginning with the development of the factory during the Industrial Revolution, women have increasingly engaged in paid employment— a situation that gradually liberated them from sole reliance on family roles as their only means of economic support.

○ A proliferation in recent decades of new technological jobs in many sectors from dentistry to machine maintenance has meant expanding job options for women. According to some research findings, working-class women see new work technologies as generally improving their circumstances (Walshok 1993).

○ Nevertheless, the debilitating physical condition RSI (repetitive motion syndrome), which can result from keyboarding on word processors, has increasingly afflicted women; many have difficulty getting compensation from their employers (Elmer-Dewitt 1994).

○ Despite electronic "labor-saving" devices, women spend just about as much time today in domestic work as they did 50 years ago (Walshok 1993). Partly, this situation is due to more rigorous cultural standards for cleanliness. Also, technology helped change housekeeping tasks like clothes washing from communal or neighborhood activities to individual, isolated ones (Cowan 1993).

○ Software development and programming as well as other high-paying computer jobs are currently less available to women than to (white) men. Reasons include a lack of female role models in computer programming and an emphasis in video games and computer-based learning programs on activities of interest primarily to boys (Rosenberg 1994; Wessells 1990).

○ Many women find much discussion on the Internet rude and insensitive (Rosenberg 1994). Furthermore, snuff porn (depicting women being mutilated, raped, and murdered) has infiltrated this technology (Elmer-Dewitt 1995) a situation of dire concern to many feminists and others.

Because technology has two faces and social institutions ultimately shape just how technology affects social change, women and African American spokespersons alike warn that we must all become knowledgeable about technology today and participate in forming technology-related social policies (Bush 1993; Johnson 1993).

focus on

The feminist movement today is very diverse. On the one hand, many business and professional women fight for equal rights in legislatures and courtrooms. On the other hand, radical feminists want to dismantle the entire set of institutions, which they see as embodying principles of male domination. One issue that unites nearly all feminists is defense of legal abortion, a woman's right to choose.

THINKING CRITICALLY

What specific issues separate white non-Hispanic women from women of color? What specific issues unite them? Do you think that the problems and dilemmas faced by African American women are mainly the same as or different from those faced by Hispanic American women? Why? What about the issues faced by Asian American women? Does it make sense to talk about "women" as one single minority group? Why, or why not?

An understanding of women's status can gain much from an analogy to race relations. Women can be viewed as a minority group and men as the majority (dominant) group. The same mechanisms that help maintain boundaries between racial groups also maintain boundaries between the sexes: prejudice, discrimination, self-fulfilling prophecies, segregation, and physical violence. The analogy extends to the use of labels to keep minority groups in their place. Just as African American men were called *boys* to remind them of their inferior status, middle-aged women continue to be called *girls*. Is this just a friendly term that doesn't have any political meaning, or is it a subtle suggestion that women are not full grown-ups with responsibilities equal to men's? Although people who use the term *girls* usually deny any negative intent, the fact that this usage is most common toward women in lower-status occupations (for example, keyboard operators and housewives) suggests that it is a mechanism for reinforcing status.

Gender roles have changed over the past 30 years. With few exceptions, those changes have affected us deeply. Changing gender roles have brought stress to many people—to men who have had to give up rights and power and also to women. Nevertheless, in many respects, continuity with the past is much more important than change.

Summary

1. Although there is a universal biological base for gender roles, a great deal of variability exists in the roles and personalities assigned to men and women across societies. Universally, however, women have had less power than men.

2. From earliest childhood, females and males integrate ideas about sex-appropriate behavior into their self-identities. Nevertheless, differences in aptitudes and personality are surprisingly small and are declining.

3. Women and men are growing more similar in their educational aspirations and attainments and in the percentage of their lives that they will spend in the work force. Family roles remain sharply gendered. Parenting and household production remain largely female responsibilities.

4. Women who are full-time, full-year workers earn 75 percent as much as men. This is because they have different (poorer-paying) jobs and because they earn less when they hold the same jobs. Causes include different aspirations

and educational preparation, discrimination, and women's greater family obligations.

5. Women's subordinate position is built into family, religious, and political institutions. Although some of this has changed, men disproportionately occupy leadership positions in social institutions. They also dominate women in conversation.

6. Men also face disadvantages due to their gender roles. These include higher mortality, fewer intimate relationships, and higher stress.

7. Structural-functional theorists argue that a division of labor between the sexes builds a stronger family and reduces competition. Conflict theorists stress that men and capitalists benefit from a segmented labor market that relegates women to lower-status positions.

8. Sex stratification is maintained through socialization and learned expectations and by the same types of mechanisms used to maintain boundaries between racial groups: prejudice, discrimination, and sexism.

9. Women of color are affected by both sexism and racism; they are economically disadvantaged by their own low earnings as well as by the low earnings of their husbands. To combat this double economic jeopardy, women of color increasingly pursue gender equality within traditional minority organizations.

Sociology on the Net

The United States has experienced a Civil Rights movement that has improved the status of women and minorities. Although true equality is still a hope and not a reality, much of the rest of the world still lags far behind the United States. To more fully grasp the worldwide status of women, let's turn to the organization called *Human Rights Watch*.

```
gopher://gopher.humanrights.org:5000/11
/int/hrw
```

Feel free to browse through the various reports starting with "About Human Rights Watch." Then click on the *Human Rights Watch Women's Rights Project*. Scroll down to the *Global Report on Women*. How many kinds of rights violations can you find listed on this report? What are the goals set forth by Human Rights Watch for the women of the world? How would you rate the United States by these standards?

Westerners are particularly offended by the practice of female genital mutilation. We are offended by the acts as well as repulsed by the pain and suffering that these women and children must endure. Feminists see the practice as one more form of male domination over women. The practice is so widespread that the World Health Organization and the United Nations have both condemned it.

```
http://www.hollyfeld.org/fgm/
```

This is the *FGM* home page. It is an excellent source for information on genital mutilation and circumcision. Browse through the extensive set of topics, and make certain to read the first three selections entitled *What is FGM? What Population Groups Practice FGM?* and *How Widely Practiced is FGM?* What exactly is female genital mutilation and where in the world is it practiced? What kinds of beliefs sustain this practice? Why would people of the United States and others from the industrialized nations not engage in this behavior? For a more detailed look at the subject, scroll down to and open the *WWW FGM Resources*. You might care to look at the statement by the World Health Organization listed as *For the Elimination of FGM (WHO)*.

FIND IT ON INFOTRAC COLLEGE EDITION

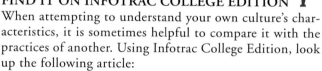

When attempting to understand your own culture's characteristics, it is sometimes helpful to compare it with the practices of another. Using Infotrac College Edition, look up the following article:

"Hindu Men Talk About Controlling Women: Cultural Ideas as a Tool of the Powerful." Steve Derne. *Sociological Perspectives,* Summer 1994, v37 n2 p203.

(Hint: Enter the search term *Hindu men* using the Subject Guide.)

What comparisons can be made between the treatment of women in the United States and Hindu women in India? Do men in the United States benefit from the limitations of womens' power? Defend your position.

Suggested Readings

Bly, Robert. 1990. Iron John: *A Book about Men*. Reading, Mass.: Addison-Wesley. The much publicized classic "bible" of one branch in the 1990s men's movement.

Kimmel, Michael. 1995. *Manhood in America: A Cultural History.* New York: Free Press. An authoritative, entertaining, and wide-ranging history of men in the United States by a respected social scientist in the field of men, manhood, and masculinity.

Orenstein, Peggy. 1994. *School Girls: Young Women, Self-Esteem, and the Confidence Gap.* New York: Doubleday. A very good book, written in association with the American Association of University Women, on young women in today's middle schools. Based on participant observation in two California schools.

Ruth, Sheila. 1995. *Issues in Feminism,* 3rd ed. Mountain View, Calif.: Mayfield. A textbook collection of essays by and about feminists and feminism.

Sapiro, Virginia. 1994. *Women in American Society,* 3rd ed. Mountain View, Calif.: Mayfield. A textbook in women's studies designed for undergraduates. A thorough review of U.S. gender inequalities.

Tucker, Judith E. 1993. *Arab Women: Old Boundaries, New Frontiers.* Bloomington: Indiana University Press. A collection of essays addressing women's issues in traditional and changing Islamic thought and culture.

Zavella, Patricia. 1987. *Women's Work and Chicano Families.* Ithaca, N.Y.: Cornell University Press. An ethnographic account of the Chicanas who work in northern California's fruit and vegetable canneries. Zavella documents the linkages between Chicano family life and gender inequality in the labor market, concluding that the rigidity of work in the canneries tends to reinforce traditional family roles.

Age Inequalities and Health

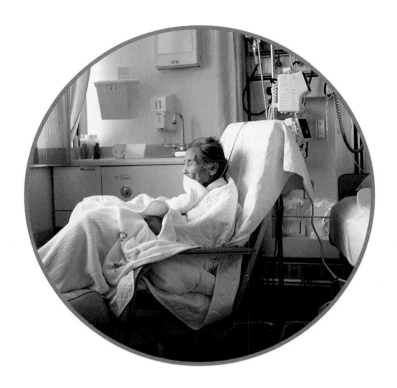

Outline

In this chapter, we consider two related topics—aging and health. Because children are seldom able to perform difficult tasks as well as adults and because the elderly experience declining strength and endurance as they age, all societies are differentiated by age. However, the physical and psychological changes associated with aging are not of primary interest to sociologists. Rather, we are concerned with the ways that age norms and roles structure the behavior and opportunities of people in different age categories. We want to consider whether the rights and privileges that are associated with age roles represent another system of structured inequality.

Although the vast majority of people over the age of 65 live independently and are in good health, after the age of 85 health concerns and limitations become a major factor in daily life. However, it is not just the elderly who are interested in health and who are impacted by changes in the health care system. More than 15 percent of the U.S. population has no health insurance and health-related debt is a major cause of personal bankruptcy. Changes in medical technology have altered the way we think about the viability and the quality of life at both ends of the age structure—for both the infant born at six months gestation and for the very old. Furthermore, health is the single most-important factor that influences overall quality of life. Thus, we want to consider why the U.S. health-care system has taken the particular form it has and who will pay the bill.

Age Differentiation and Inequality

Society is run mainly by adults between the ages of 30 and 65. These adults control jobs, industry, education, and wealth. The various benefits of this control are reflected in age-related suicide rates (see Figure 10.1). Among white males, suicide rates have increased since 1970 for those age 15 through 44 and 75 and older; suicide rates have declined for white men between 45 and 75 years old. One worrisome trend is the dramatic increase in teen suicides: The rate more than doubled (from 9.4 to 19.3 per 1,000) between 1970 and 1990. Moreover, the rate for white males in their early twenties is higher than for the middle aged. The very high rates for elderly white males in part reflect the fact that they are less well integrated into society than either younger men or women. A general conclusion from Figure 10.1 is that today the young and the old have relatively high suicide rates compared with the middle aged. In the following sections, we examine the extent to which young and elderly persons can be considered structurally disadvantaged relative to this middle-aged group.

The Status of Young People

Legal Restrictions

In law, people under 18 are called infants (Sloan 1981). They are not responsible for their contracts (thus, they usually are not allowed to make any), and they are considered less responsible than adults for their bad deeds. Just as legal statutes once declared women and minority members incapable of self-government, the law still declares the young to be incapable. As you may be painfully aware from personal experience, young

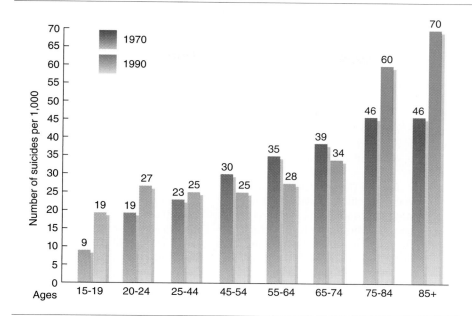

FIGURE 10.1
Suicide Rates (Number of Suicides per 1,000 Individuals) of White Males for Selected Age Groups, 1970 and 1990
SOURCE: U.S. Bureau of the Census 1993, Table 137.

people have few legal rights. Their rights to drink, drive, work, own property, and marry are abridged. In some cities, curfew laws require them to be off the streets at night. They have to pay adult prices but cannot see adult movies. In church, school, and industry, they find they must pass age barriers before they are allowed full participation. Unlike the case of women and minorities, there is no movement to reduce inequality for teenagers. The voting age was reduced from 21 to 18 in 1970, but the drinking age has been raised to 21. Once again, we face the irony of staffing our armed forces with young people we will not trust with a beer. Although raising the drinking age springs from the praiseworthy objective of reducing deaths from drinking and driving, it is nevertheless a reminder that many people believe the young cannot make responsible decisions and need to be protected.

For the most part, these age-related rules reflect society's belief that age is a reasonable indicator of competency; moreover, it is one that can be administered easily and efficiently. Thus, instead of devising some test of maturity that one must pass before obtaining voting privileges, society uses a simple rule of thumb: Persons under 18 are not mature, but persons 18 and over are mature enough for this purpose.

Although there is growing sentiment that neither ethnicity nor sex is a good guide to competency, no such change of sentiment has occurred about the utility of age for making relevant judgments about youth. In fact, discrimination against the young may be the last bastion of approved inequality before the law. The Supreme Court, in siding with parents who wished to prohibit a teenage daughter's abortion, concluded that "the rights of children cannot be equated with those of adults" (Justice Powell in *Bellotti v. Baird* [1979], cited in Eglit 1985, 532). These age rules receive very high consensus approval from society; little, if any, pressure exists to repeal them.

Economic Status

Owing to underdeveloped work skills, limited experience, and desire for part-time or flexible schedules, and also owing to simple discrimination, a very large share of youth in the United States earn minimum wage. Even this wage is under a two-pronged assault. First, the law that raised the minimum wage in 1990 provides for a

THINKING

CRITICALLY

The Focus on American Diversity section introduces the question of whether changes over time are due to age effects or cohort effects. Think of some age effects that might explain age-related suicide rates. Think of some cohort effects that could help explain changes in age-related suicide rates between 1970 and 1990.

THINKING

CRITICALLY

Should U.S. society treat all people, regardless of age, as individuals? Do age requirements for voting, driving, and so on assume that, for example, *all* 18-year-olds are more responsible and better-informed citizens than any 17-year-old? Isn't this stereotyping? If so, is there some justification for it?

focus on

American Diversity

DO YOU GET BETTER WITH AGE?

As we move from age 20 to 40 to 60, our lives change in more or less predictable ways. We leave home, form families, work, and retire. In early adulthood, the process is largely one of role acquisition; in middle and old age, it is a process of role loss. According to some symbolic interactionists, the addition of roles in adulthood should improve self-concept, whereas the loss of roles in later life should decrease it. Others suggest that adulthood is a period of growth and maturation so that older individuals become more, not less, satisfied with themselves and their roles.

A study by Walter Gove, Suzanne Ortega, and Carolyn Style (1989) addressed this issue. They asked a national random sample to read a list of adjectives and check the ones that described them. The adjectives measured four dimensions of self-concept:

○ *Competent:* hardworking, well-organized, self-confident, strong, logical, and intelligent
○ *Supportive:* helpful, flexible, considerate, content

○ *Calm:* not emotional, not nervous, not frustrated
○ *Cooperative:* not lazy, not disorganized, not stubborn.

Because the researchers reasoned that these characteristics might be affected by social class, race, and sex, they adjusted for these factors. The adjusted results, presented in Figure 10.2, show clearly that the older people surveyed felt better about themselves than the younger people. Young people rated themselves as relatively uncooperative, incompetent, uncalm, and unsupportive. In contrast, the older respondents felt very positive about their identities.

In fact, getting older seems to be associated with getting better—with being a nicer and more capable person.

These results suggest that, despite the high esteem in which youth is held in U.S. society, youth is not all wonderful, and growing old is not altogether a bad thing. In fact, getting older seems to be associated with getting better—with being a nicer and more capable

person. These results led Gove, Ortega, and Style to conclude that maturation is more important than role loss in affecting self-concept.

In interpreting these data, however, the authors ran into a serious problem. These are *cross-sectional data*. That means that the respondents who were 65 to 74 and those who were 18 to 24 were of entirely different generations. This raises the very realistic possibility that the older people did not become more cooperative, competent, and so forth with age but that their generation was always more competent and cooperative.

Whenever we look at age differences from a cross-sectional sample, we face two possibilities: that the differences result from *age effects* or that they result from *cohort effects*. The age effect takes the data at face value and hypothesizes that, as today's young people age, they will become more competent, cooperative, supportive, and calm. The cohort effect suggests that these are permanent generational differences and will not change with age—that even at 65, today's young people will feel

subminimum training wage; this training wage is targeted largely at young people. The second assault on youthful earnings is a very high rate of unemployment: In 1995 when 6 percent of the civilian labor force was unemployed, the unemployment figure was 17 percent for those aged 16–19 and 9 percent for those 20–24 (U.S. Bureau of the Census 1996a). For African American youth, these figures are 36 and 18 percent, respectively. As a consequence, families headed by people under 25 are more than three times as likely to be in poverty than the average family.

Crime

Failure to integrate youth fully into major social structures is reflected in several indices of social disorganization. Although youth are underrepresented among voters and workers, they are overrepresented in accident and crime statistics. In 1994, people between the ages of 10 and 24 represented only 22 percent of the population, but they

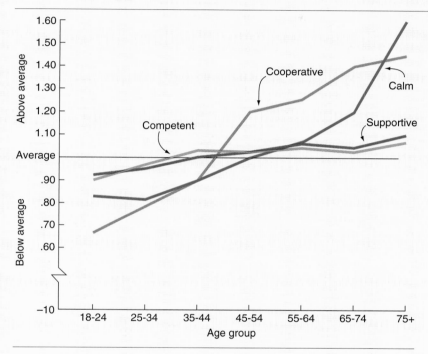

FIGURE 10.2
Differences in Self-Concept by Age
In a large national study, older people reported substantially better self-concepts than younger people. Older people reported being more competent, supportive, calm, and cooperative. It is possible that today's older generation always had these positive traits, but the study's authors concluded that people tend to develop these characteristics as they grow older.

Source: Adapted from Gove, Ortega, and Style 1989.

uncooperative, unsupportive, uncalm, and incompetent.

Gove, Ortega, and Style chose to believe that their results showed an age effect. They concluded that maturation does bring about positive changes. They supported their conclusion by citing survey research from as far back as 1957 showing the same pattern. In the final analysis, however, the choice of a cohort or an age interpretation rests on theory. Which interpretation makes the most sense to you? Do you expect to get better with age?

accounted for 42 percent of the rape arrests, 42 percent of the arrests for aggravated assault, 56 percent of murder arrests, and 72 percent of arrests for motor-vehicle thefts (U.S. Bureau of the Census 1996).

Summary

Young people are systematically excluded from participation in many of our society's institutions. In some cases, they are excluded by formal rules, such as restrictions on drinking, driving, or renting a car. In other cases, they are excluded because there is no room for them, as in the labor force. In terms of class, status, and power, youth rank significantly behind adults. To a large extent, this inequality represents stratification: The lower status of youth is justified by institutionalized norms applicable to an entire category of individuals. For most of us, this lower status is temporary. We are hired for jobs, become married, finish school, have children, and generally become plugged into society. For the minority who never become integrated into society's economic and so-cial structures—and these are disproportionately racial and ethnic minorities—the lower status of youth becomes permanent.

The Status of the Over 65

Although they may have less power than they used to, the elderly in our own society do not seem to be particularly disadvantaged—as long as they keep their health. Here we generally refer to the condition of the "young old," those roughly between the ages of 65 and 75 whose status can be regarded as more socially than physically determined.

Health, Life Expectancy, and Changing Definitions of Old

THINKING

CRITICALLY

Society has greater tolerance for the inequalities of the young than for those of the old. We more often think that the disadvantages of old age are "unfair." Why do you suppose society is so tolerant of the disadvantages of young people?

Perhaps the most important change in the social structure of the population over 65 is that being over 65 is now a common stage in life—and often a long one. The average person, on reaching 65, can now expect to live 17 more years—19 more if a white woman and 15 more if a white man; for African Americans, the figures are 17 and 13 years, respectively (U.S. Bureau of the Census 1996a, Table 20). Because more and more people in the United States are living into their 80s and beyond, demographers now tend to divide the over-65 population into two categories: the "old old" (age 75 and above) and the relatively "young old" (ages 65–74) (Treas 1995).

For the majority of us, most of the "young old" and even many of the "old old" years will be healthy years. Certainly most people at 65 or 70 experience some loss of energy and physical stamina, and most older people report having at least one chronic health problem, such as arthritis, hypertension, or hearing loss (Treas 1995, 32–35). But a remarkably large proportion experience no major health limitations. Demographers estimate that three-fourths of the years of life remaining after age 70 will be spent in good enough health to permit independent living in the community (Crimmins, Hayward, & Saito 1994).

Income

The economic condition of the elderly has improved sharply over the last 35 years. In 1959, 35 percent of the population over 65 had incomes below the poverty level; in 1995, this figure was only 10 percent. This improvement is directly related to increases

As more Americans live into old age—and actively!—you can expect to see a few more run-ners and surfers like these "old old" athletes.

in Social Security, expanded coverage of private pension plans, and a more comprehensive system of benefits for the elderly. As a result of this system, the proportion of the aged below the poverty level is *less* than the proportion of the entire population that is poor.

Although only 10 percent of the population over 65 are below the poverty level, the other 90 percent are not all that wealthy. The average person experiences only a minor income loss in the years immediately following retirement, but there are two dark spots in this picture. First, more than one-quarter of the population experiences an income drop of more than 50 percent following retirement. Those people who were living close to the economic margin during middle age—who accumulated no assets, who didn't pay off a home mortgage, and who contributed relatively little to either Social Security or a pension fund—find that retirement brings poverty or near poverty. These people are disproportionately women and minorities, people whose work-life earnings are low. Second, widowhood produces a severe economic blow for most older women, many of whom find that a substantial portion of their retirement income dies along with their husband (Holden, Burkhauser, & Feaster 1988).

Discrimination

In the past, mandatory retirement regulations forced some people out of the labor force regardless of their physical ability, economic need, or desire to work. In 1986, however, federal legislation outlawed mandatory retirement except for a few cases (such as police officers) where age-related abilities are considered legitimate bases for discrimination.

A far more important problem than mandatory retirement is age discrimination that begins during middle age, when women and men over 40 are considered too old to learn new skills or take new jobs. Age-based discrimination is particularly hard on people who need a new job after 40. This is why affirmative-action legislation uses age 40 to define when age discrimination begins. One recent study examined what happened to men who were laid off after plant closings. The results showed that workers over 55 had to wait twice as long to find a new job as workers under 45 and that the new jobs of older workers were substantially worse than the new jobs of younger workers (Love & Torrence 1989).

Political Power

Age and even infirmity in no way abridge one's legal rights. The elderly are more likely to vote than other age groups and this makes them a potentially powerful group. There is no evidence, however, that the older population acts together as a cohesive voting bloc. Cleavages of race, class, and sex work against such unity. Major political victories for the aged, most recently the defeat of threatened decreases in Social Security, owe more to broadly based interest groups such as organized labor than they do to senior-based political groups (Hudson & Strate 1985).

Honor and Esteem

Societal rewards distributed by stratification systems go beyond income and power. They also include prestige, esteem, and social honor. It is in this regard that the status of older people has been most seriously at issue. Our idealized version of the past holds that elderly people used to be highly regarded and respected. Careful examinations of cross-cultural and historical data, however, suggest that older people seldom have as much prestige as younger adults (Stearn 1976) and that, historically, our attitude toward them has been ambivalent at best (Achenbaum

1985). Hags, crones, dirty old men, and other negative stereotypes of older people crowd our literature. Whether scheming hags or dear old things, older people are seldom regarded with complete respect or admiration (Cool & McCabe 1983; Levin 1988).

Stereotypes about older people—as inflexible or less competent, for example—form the basis of **ageism**—the belief that chronological age determines the presence or absence of socially relevant characteristics and that age therefore legitimizes unequal treatment. In U.S. society, many circumstances work to reduce the esteem in which we hold older people. Age is associated with reductions in vigor and physical beauty, both of which are highly prized in U.S. society (Thorson 1995). Moreover, because of rapid expansion of the educational system within the last several decades, older people are substantially less well educated than younger adults. In 1995 only 59 percent of those over 75 had completed high school, compared with 82 percent of all people in the United States over age 25 (U.S. Bureau of the Census 1996a, Table 243). In a society where knowledge and education are taken as indications of intelligence and worth, older people are considered less worthy than others. Then, too, the life experience of older people is considered less valuable in a society that is changing rapidly, as U.S. society is today. Finally, because people over 65 (or 70) are generally outside the mainstream of economic competition, they seldom receive the prestige of active earners (Thorson 1995). Together, these factors mean that, although individual older people may have good educations, incomes, and health and be politically active, they often find themselves being patronized and condescended to by younger individuals.

> **Ageism** is the belief that chronological age determines the presence or absence of socially relevant characteristics and that age therefore legitimates unequal treatment.

Social Integration

We have defined adulthood as a period of maximum integration into social institutions. This involvement is gradually reduced during old age. Active parenting is the first role to drop off, followed by the work role. Many, especially women, also experience the loss of the marital role through the death of their spouses. The world of church, family and friends, and community, however, often remains active, at least until the advent of old, old age, when health limitations may restrict activities.

Studies of older people demonstrate that one of the most important predictors of life satisfaction is relationships with close friends. These are usually age peers and often, at the end of the life course, brothers and sisters. Interestingly, close ties with their children are not a uniform blessing. Especially when age starts to become a serious handicap, ties with children become tinged with dependency and ambivalence (Thorson 1995). Because ties with age peers usually lack this element, they tend to be more gratifying.

Summary

An assessment of the status of the elderly during the last 35 years would show very substantial economic improvements. With this improvement in economic circumstance has come much greater independence from their families, better health, and greater community involvement. Nevertheless, among the "old old," a substantial group are "ill fed, ill housed, ill clothed, and just plain ill" (Hess 1985, 329). This group is disproportionately female and minority.

Youth Versus Age: The Battle Over the Public Purse

The ratio of children to old people was 29 to 1 in 1820; in 1990, it was only 2.2 to 1. In numerical terms, then, the relative power of young people has declined substantially. Nevertheless, if numbers were the only criterion, we would still expect youth to retain advantages over age. In point of fact, however, older people vote and children do not. Children have various representatives to act in their interests (parents, social workers, and so on), but they are basically disfranchised. One result is that, increasingly, children lose out in the battle for scarce public resources (Preston 1984).

Federal expenditure on child-related programs (AFDC, Head Start, food stamps, child nutrition, child health, and aid to education) is only one-quarter the size of federal expenditures on programs for the elderly. When figured on a per capita basis, the federal government spends 9 percent as much on the average child as it does on the average person over 65 (Preston 1984). This situation has been gradually worsening. The major programs hit by federal cutbacks on social expenditures during the Reagan-Bush years were child-related programs. Programs for the elderly, especially Social Security, have become too politically risky to cut, and they have remained largely immune to major decreases in social spending.

A consequence of these changes is that the condition of children in the United States has deteriorated, whereas the condition of the elderly has improved. Table 10.1 compares changes in poverty rates between 1970 and 1992. These data show that poverty has increased among children and decreased among the elderly. Within 22 years, we have gone from a society in which the elderly were much *more* likely than children to be poor to a society in which the elderly are much *less* likely to be poor.

Why have children become less advantaged than the elderly? Several explanations have been suggested. An obvious one is that, although everybody has parents, not everybody has children. Some childless people have little sympathy with the problems of providing for children. A related explanation is that of family norms: In the United States, we generally believe that parents are obligated to support their children but that children are not obligated to support their parents. We have given government the responsibility for allocating resources to the elderly, and it does so relatively evenhandedly. Distribution of resources to children, however, still rests largely with their parents and varies widely, depending on the earnings of their parents and whether they live with one or two parents. Some observers have also wondered if the decreasing benefits to children might not reflect the fact that 24 percent of the population under 15 are Hispanic or African American, greater than the percentage in the entire population (Leach 1994).

TABLE 10.1
Changing Levels of Poverty Among Youth and the Elderly, 1970–1995

| | PERCENTAGE BELOW THE POVERTY LEVEL | |
	Under 18	Over 65
1970	14.9%	24.5%
1975	16.8	15.3
1980	17.9	15.7
1985	20.7	12.6
1990	21.8	11.7
1992	21.9	12.9
1995	20.8	10.5

SOURCE: Data from 1970, 1975, and 1980 are from U.S. Bureau of the Census 1982. Current Population Reports, Series P-60, No. 133, p. 11. Data for 1985 are from U.S. Bureau of the Census 1993. Current Population Reports, Series P-60, No. 185, p. 4. Data for 1990 and 1995 are from U.S. Bureau of the Census, http://www.census.gov/hhes/poverty/pov95/povest1.html.

Explanations for Age Stratification

In everyday usage, most people rely on a physiological explanation of age stratification. The young and the old have less status because they are less competent and less productive. To some extent, this explanation is correct. It does not, however, explain cross-cultural or historical variations in the status of age groups. Conflict and structural-functional perspectives, as well as modernization theory, furnish insightful explanations about why age stratification exists and varies across societies.

Structural-Functional Perspective

The structural-functional perspective focuses on the ways in which age stratification helps fulfill societal functions. From this perspective, the restricted status of youth is beneficial because it frees young people from responsibility for their own support and gives them time to learn the complex skills necessary for operating in society.

At the other end of the age scale, the functionalist perspective is the basis for **disengagement theory.** The central argument of this theory is that older people voluntarily disengage themselves from active social participation, gradually dropping roles in production, family, church, and community even before actual disability connected with age requires it. This disengagement is functional for society because it allows younger people, with new ideas and skills appropriate to a fast-changing society, to take their places. Disengagement also makes possible an orderly transition from one generation to the next, avoiding the dislocation caused by people dropping dead in their tracks or dragging down the entire organization by decreased performance. It is functional for the individual because it reduces the shame of declining ability and provides a rest for the weary. Disengagement theory is a perfect example of why functional theory is sometimes called consensus theory: The lack of participation of older people is agreed on by older people and others and benefits all (Hendricks & Hendricks 1981).

Disengagement theory, a functionalist theory of aging, argues that older people voluntarily disengage themselves from active social participation.

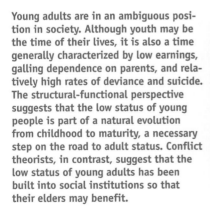

Young adults are in an ambiguous position in society. Although youth may be the time of their lives, it is also a time generally characterized by low earnings, galling dependence on parents, and relatively high rates of deviance and suicide. The structural-functional perspective suggests that the low status of young people is part of a natural evolution from childhood to maturity, a necessary step on the road to adult status. Conflict theorists, in contrast, suggest that the low status of young adults has been built into social institutions so that their elders may benefit.

Conflict Perspective

The conflict view of age stratification produces a picture of disengagement that is far less benign—a picture of simple rejection and discrimination springing from competition over scarce resources. These resources are primarily jobs but also include power within the family. Conflict theorists suggest that barring young and older people from the labor market is a means of categorically eliminating some groups from competition and improving the prospects of workers between 25 and 65.

There is empirical evidence to support this view. Early in U.S. history, few people retired. Not only could they not afford to, but their labor was still needed. As immigration provided cheap and plentiful labor, the need for the elderly worker decreased. In addition, as unions established seniority as a criterion for higher wages, older workers became more expensive than younger ones. In response to these trends, management instituted compulsory retirement to get rid of older workers. The mandatory retirement rules occurred long before Social Security, in an era when few employees had regular pension plans. Thus, compulsory retirement usually meant poverty for the older worker (Atchley 1982).

It was not until 1965 that social structures such as Social Security and private pensions began to make retirement a desirable personal alternative. Between 1959 and 1992, the proportion of older people who were poor declined dramatically, from 35 percent to about 13 percent. At present, retirement suits both the aging worker and the economic system. As the growing elderly population becomes an increasing burden on a shrinking working-age population, however, conflict may once again emerge between the generations over economic interests.

Modernization Theory of Aging

According to the **modernization theory of aging,** older people have low status in modern societies because the value of their traditional resources has eroded. This is due to three simultaneous events: the decline in importance of land (disproportionately owned by older people) as a means of production, the increasing productivity of society, and a more rapid rate of social change (Cowgill 1974). Next, we look at each of these in turn.

First, when land is the most vital means of production, those who own it have high status and power. In many traditional societies, land ownership is passed from father to son. This gives fathers a great deal of power even if they live to an age when they are physically much less able than their sons. This explanation, of course, applies only in a society where wealth resides in transferable property, either land or animals.

Second, in societies with low levels of productivity, it is not feasible to exclude either young or elderly persons from productive activity. Everyone's labor is needed. In industrial societies, however, productivity is so high that many people can be freed from direct production. They can study, they can do research, they can write novels, or they can do nothing at all. In such a society, the labor of young and elderly persons becomes expendable: Society doesn't need it anymore.

Third, technological knowledge has grown at an ever-accelerating pace. Because most of us learn the bulk of our technological skills when we are young, this rapid change produces an increasing disadvantage for older workers. Their technical skills become outdated. Thus, rapid social change works to the disadvantage of older people (Thorson 1995).

According to the modernization theory of aging, the elderly in modern societies have low status because the value of their traditional resources has eroded. In preindustrial societies, the aged are revered because they are responsible for teaching traditional beliefs and important survival skills to the young. When technology is changing rapidly, however, as in industrial and postindustrial societies, an individual's knowledge and skills are likely to be obsolete by the time he or she reaches old age. A loss of honor and esteem for older persons may be one of the negative consequences of "modernizing" traditional societies (Cowgill 1986), such as this one in Papua New Guinea.

The **modernization theory of aging** argues that older people have low status in modern societies because the value of their traditional resources has eroded.

THINKING

CRITICALLY

Chapter 2 discussed basic values in the U.S.'s dominant culture (for example, achievement, efficiency, progress, and individual freedom). Based on what you know about modernization theory, how might these basic values affect the prestige of older people in U.S. culture and society?

CONCEPT SUMMARY	*A Comparison of Three Explanations of Age Stratification*			
	MAJOR ASSUMPTIONS	CONCLUSIONS ABOUT YOUTH	CONCLUSIONS ABOUT OLD AGE	OVERALL EVALUATION
Structural-Functional Theory	Age groups cooperate for common good	Young people's exclusion from full social participation is for good of self and society	Older people disengage voluntarily; good for self and society	Currently adequate to explain status of older people but not compelling to explain status of young people
Conflict Theory	Age groups compete for scarce resources	Young people are excluded so that others may benefit	Older people are excluded so that senior positions open up for younger adults	Useful to explain status of young people but not to explain today's older people
Modernization Theory	Changes in institutions alter the value of special resources that age groups hold (land, labor, and knowledge)	Unspecified; by implication, status goes down because labor not necessary	Status of older people has decreased because traditional bases of power have eroded	Useful to explain low social honor of older people and young people

Evaluation

Although both young people and old people have lower status than adults in midlife, their experiences are rather different. Thus, theories that explain the status of young people may not be as effective in explaining the status of older people (see the Concept Summary).

Overall, it would appear that all three theories can contribute to understanding the status of the young. Certainly, as functional theorists suggest, when people are protected from full responsibilities while very young, both the young people and society benefit. Nevertheless, the continued disadvantage of young adults and their subsequent poverty, lack of social integration, and deviance are hardly functional for them or society. In this case, it seems appropriate to attribute their low status to the systematic disadvantages they face in competing for scarce resources controlled by an older generation. Part of this disadvantage stems, as modernization theory suggests, from devaluation of their traditional resource: the capacity for low-skill, physically demanding work.

The status of older people is harder to account for. Modernization theory is not really applicable to their rising economic status, though it may explain their low social honor. Nor does conflict theory seem entirely adequate to explain the economic status of older people: Rather, at this point in history, the disengagement of the older worker seems mutually attractive to younger and older people. As noted in Chapters 8 and 9, however, systems of inequality often overlap. Thus, the status of both young and old varies for men and women and for members of majority and minority groups.

Cross-Cutting Statuses: Age, Sex, and Race

Aging and Gender

Aging poses special problems for women. First is the problem of the double standard of aging: The signs of age—wrinkles, loose skin, gray hair—are considered more damaging for women than for men. Thus, age is associated with greater decreases in prestige and esteem for women than for men.

The life expectancy gap between men and women also makes the experience of old age very different for women than for men. On average, women live about seven years longer than men. Taken together with the fact that women are usually two years or so younger than their husbands, this works out to a nine-year gap between when a woman's husband dies and when she dies. This mortality difference has enormous consequences for the quality of life. First, it means that most men will spend their old age married, have the care of a spouse during illness, and be able to spend their last years at home being cared for by a spouse. The average woman, on the other hand, will spend the last years of her life unmarried, will spend most of these years living alone with no one to care for her, and is more likely to be cared for in a nursing home during her last illness.

Sex differences in mortality mean that old age is disproportionately a society of women. Above age 75, there are nearly twice as many women as men. The decreasing availability of men in older age groups has consequences for social roles. It means that, as people get older, heterosexual contacts, whether through marriage or in bridge groups, are less important for structuring social life. It also means that fewer of the elderly are married and more of them live alone (U.S. Bureau of the Census 1994, Table 63). The feminization of old age is an increasing component of the poverty of old age. Female-headed households are poorer than male-headed households at all ages, and older people are no exception. As the elderly population rose out of poverty over the last three decades, female-headed households were disproportionately left behind.

Aging Among Minority Groups

Ethnic minorities earn substantially less during their peak years than do non-Hispanic whites, and this means that they are less likely to have accumulated assets such as home ownership to cushion income loss during retirement. Further, there is evidence that the economic status of racial/ethnic minorities becomes worse, relative to that of non-Hispanic whites, in old age (Taeuber 1992). That is, the disparity in wealth and income between non-Hispanic whites and others increases from adulthood into old age.

But the most significant link between minority status and aging is that minorities are less likely to live to experience old age! Whereas 75 percent of white males can expect to survive until they are 65, only 58 percent of black males will live until retirement age. For women, the figures are 85 and 75 percent, respectively.

An issue that applies to minority aging in today's increasingly diverse society concerns how elderly immigrants experience U.S. culture. Many immigrants, such as those from Asian or Latino countries, have internalized their native culture's belief that an old person deserves respect and prestige simply because she or he has lived a long time. However, U.S. culture places high value on activity, productivity, and individual achievement—values that are inconsistent with high prestige for old people. The assimilation of elderly immigrants and their children and grandchildren requires mutual adjustments in age roles and expectations (Lin & Liu 1993; Paz 1993).

Health and Health Care

Health and health care are basic concerns for most Americans. Over the last decade, Americans have increasingly tried to quit smoking, cut down on cholesterol, and increase their exercise as medical scientists and the media have

The quality of life one experiences in old age probably has less to do with age than with social class, sex, and race ethnicity. Those who are most disadvantaged in old age are those who were close to the poverty line during their earning years. Even among relatively affluent retirees, widowhood often means sharp reductions in income.

proclaimed it a moral duty to take care of one's health. Of course, taking these precautions makes good sense because health status is the single most important factor influencing overall quality of life. For many Americans, however, worry over medical expenses is the companion to concern about living a healthy lifestyle. These worries are realistic; medical costs have become one of the leading causes of personal bankruptcy in the United States.

The sociological study of health, then, is central to understanding almost every facet of life; it is relevant to sociologists who study fertility and mortality, socialization and self-identity, to those concerned with the family, and to those interested in studying inequalities based in class, race, gender, and age. It is also of concern to those who study the changing nature of work. Nowhere are the themes of professionalization and an expanding tertiary sector more visible than in the arena of health care. Nowhere is the impact of technology more vivid or the political and economic stakes higher. In this section, we want to consider health and the political economy of health care: How do systems of inequality impact health, why has the U.S. health-care system taken the particular form it has, and who will pay the bill?

Good health is not simply a matter of taking care of yourself and having good genes. Although both elements play important parts in health, we also find that good health is related to such social statuses as gender, social class, and race or ethnicity. The study of how social statuses relate to the distribution of illness and mortality in a population is called **social epidemiology** (Rockett 1994). In this section, we provide an overview of social epidemiology in the United States, and then briefly examine how changes in social structure have influenced life expectancy in Russia and Eastern Europe.

Social epidemiology is the study of how social statuses relate to the distribution of illness and mortality.

Social Epidemiology

In the United States, the average newborn can look forward to 75.5 years of life (U.S. Bureau of the Census 1996a Table 120). Although some will die young, the average

Medicine is one of our most familiar institutions. Nearly all children have enough experience with the medical institution to add playing doctor and playing nurse to their role-playing repertoires. Although playing doctor has some comical overtones, the popularity of doctor and nurse play sets and the large number of children who aspire to medical professions are indicators of how important medicine is to contemporary society.

person in the United States now lives to be a senior citizen (see Table 10.2). This is a remarkable achievement given that life expectancy was less than 50 years at the beginning of this century. Not everyone benefited equally, however, and men and African Americans are significantly disadvantaged in terms of years of life, as are people of lower socioeconomic status.

There is a great deal more to health, of course, than just avoiding death. The incidence of nonfatal conditions is at least as important as the distribution of mortality in evaluating a population's overall well-being (Verbrugge 1989a). Although only one out of every 100 people dies each year in the United States, well over half the people experience some sort of long-term or serious illness that affects the quality of their lives and their ability to hold jobs or maintain social relationships. In the following sections, we consider why gender, social class, and race are related to ill health and mortality.

Gender

On the average, U.S. women live almost seven years longer than U.S. men (Table 10.2). Some of this difference appears to be the result of a lifelong biological advantage: From conception to old age, females experience lower death rates than males. Ironically, however, women report significantly worse health than men: more high blood pressure, arthritis, asthma, diabetes, cataracts, corns, and hemorrhoids (U.S. Bureau of the Census 1995, Table 215).

TABLE 10.2
Health and Life Expectancy by Sex, Race, and Family Income, United States, 1992–1993
Generally, people of higher status report better health: Men report better health than women, white Americans report better health than black Americans, and those with higher incomes report better health than those with lower incomes. For the most part, differentials in life expectancy parallel these differentials in health. The exception is sex: Despite their better health, men have lower life expectancy than women.

	Life Expectancy At Birth, 1993	Percentage Reporting Excellent Health
Total	75.5	38%
Sex		
Male	72.1	41
Female	78.9	35
Race		
White	76.3	39
Black	69.3	30
Family Income		
Under $10,000	NA	25
$10,000–$19,000	NA	29
$20,000–$34,999	NA	36
$35,000 and over	NA	49

SOURCE: U.S. Bureau of the Census 1993a; U.S. Bureau of the Census 1995, Table 114; U.S. National Center for Health Statistics 1994.

Why do men have higher mortality rates despite apparently better health? Some of the answer may lie in biology, but social factors also play a very important and perhaps dominant role. Two aspects of the male gender role in the United States appear to put men at a disadvantage in terms of mortality.

First, contemporary gender roles encourage males—particularly young males—to be rowdy, aggressive, and risk-taking. There is *normative* approval for higher rates of drinking, fighting, fast driving, and dangerous behavior for males. As a result, young men are two and one-half times as likely to die in motor vehicle accidents and six times more likely to be homicide victims. Second, men appear to cope less well with stress than women do. Generally, studies find that men report lower levels of stress than women (Ulbrich, Warheit, & Zimmerman 1989) and that men are less likely than women to have high blood pressure. Nevertheless, men are more likely than women to die of stress-related diseases such as heart attack and stroke. Why is stress more deadly for men than women? On the average, men do not take care of themselves as well as women do—men are less likely to go to the doctor when ill, to follow doctors' recommendations, and to watch their diets. Men are also less likely to have a network of intimates in whom they can confide and from whom they can seek support (Nathanson 1984). Thus, men's stress is more likely than women's stress to develop into life-threatening proportions.

Social Class

The higher one's social class, the longer one's life expectancy and the better one's health (see Table 10.2). The effects of social class are complex. They are partially attributable to the fact that poorer people cannot afford expensive medical care. However, they are also related to the fact that lower-income people are more likely to live in unhealthy conditions, near an air-polluting factory, or in substandard housing. Low-income people are also less likely to have control over the world around them than those who are better off. As a result they are likely to experience higher levels of stress and to develop poor coping strategies, like drinking, smoking, and risky behavior.

Race

Because a higher proportion of African Americans, Hispanics, and Native Americans than non-Hispanic whites are poor, the impacts of low socioeconomic status on health disproportionately affect these minorities. Because of lower incomes, African Americans and Hispanic Americans are twice as likely as non-Hispanic whites to be without any health insurance (U.S. Bureau of the Census 1994, Table 165). Because they are poor, racial and ethnic minorities are also 47 percent more likely than others to live near a hazardous waste facility, which may emit toxins into the surrounding ground, air, or water (Ember 1994).

However, even after we control for income, racial and ethnic minorities face obstacles to good health *because they are minorities.* A language barrier, for instance, often separates Hispanic patients from health-care professionals (Vega & Amero 1994). Furthermore, regardless of income, minority group members experience prejudice and discrimination that raise their risk of physical and psychological distress (Ulbrich, Warheit, & Zimmerman 1989).

A Case Study: Social Structure and Declining Life Expectancy in Russia and Eastern Europe

The single most important social factor affecting mortality is the standard of living—access to good nutrition, safe drinking water, protective housing that is free from exposure to environmental hazards, and decent medical care. Differences in living standards help to explain why black infants in the United States are more than twice as likely as white infants to die in their first year of life (U.S. Bureau of the Census 1995) and why the average life expectancy of African American men is seven years less than that of the average European American male. Differences in living standards also help to explain why the average U.S. person can expect to be healthier and live 22 years longer than the average Nigerian. Throughout the world, improvements in living standards have been accompanied by increased life expectancy. Consequently, the 20+ year decline in life expectancy in Russia and Eastern Europe is one of the most surprising current developments in world health. Although there is some question about the reliability of Soviet health statistics, it is quite clear that Russian men today can expect to live almost eight years less than their Soviet counterparts did in 1960 and Russian women can expect to live one-and-a-half years less (Cockerham 1997). Why? The answer to this question probably rests on four basic aspects of social structure.

First, as was discussed in Chapter 7, Eastern Europe is plagued by extensive environmental pollution. Pollution is strongly implicated in the onset of life-threatening illnesses, such as cancer and respiratory diseases. Research conducted in the United States demonstrates that it is the lower social classes who are most likely to be exposed to environmental hazards (Austin & Schill 1994). Correspondingly, the decline in Russian life expectancy appears to be greatest among individuals with the least education living in the most developed, industrialized regions (Cockerham 1997). According to some estimates, pollution-related diseases were responsible for about 12 percent of the 1992–1993 increase in deaths in Russia (Haub 1994).

Second, some analysts have blamed the decrease in life expectancy in republics of the former Soviet Union on problems within the Soviet health-care system. Funded by the money left over after high priority defense and industrial needs were addressed and relying heavily on physician assistants and rather poorly equipped hospitals, the Soviet health-care system was designed to prevent the spread of infectious disease more than it was to deal with the treatment of chronic health problems. So long as the major causes of death were related to epidemics and contagious disease, the Soviet health-care system made substantial progress in improving life expectancy. Once these problems were controlled, however, the Soviet system has not proven very effective in dealing with chronic illnesses, such as heart disease, that now constitute the major causes of death in industrial societies. Most important, even though Soviet-style medicine did guarantee a universal and equal right to health protection, it did not end differences in the quality of care provided to political elites and the less-privileged rural classes (Cockerham 1997). A similar pattern of unequal access to quality health care also appears to be a major contributor to health differences between poor and more affluent people in the United States (Waldholz 1991).

Third, some medical researchers have suggested that increased stress is the single most important variable in the health of large populations. Thus, as Japan moved up in affluence and standing in the world community, Japanese life expectancy increased; likewise, as Russia and Eastern Europe moved down in status and standard of living, the stress caused by these changes produced a decrease in life expectancy (Hertzman, Frank, & Evans 1994).

Finally, Cockerham (1997) has provided evidence that unhealthy lifestyles may be the most important social cause of increasing mortality in Russia and Eastern Europe. Eastern Europeans are among the world's heaviest drinkers and smokers and their diet is loaded with fat. Rather than being a problem of "bad attitudes" or poor individual choices, however, Cockerham makes the following argument: Social structural constraints seriously limit the life chances of the least privileged segments of society, sometimes leaving people with little or no choice in exposing themselves to unhealthy conditions and practices. In Eastern Europe, where fresh fruits and vegetables are often absent from store shelves and heavy alcohol consumption is encouraged by the workplace norms, external constraints place sharp limits on the range of health behaviors from which one might choose. Both at home and abroad, then, unequal access to education, employment, and other social rewards have life-**and**-death consequences.

The U.S. Health-Care System

Medicine may be regarded as a social institution. It has a complex and enduring status network, and the relationships among actors are guided by shared norms and roles. Most of us occupy the status of patient in this institution. However, there are dozens of other statuses. Approximately 10 million U.S. people are employed in health institutions. They include phlebotomists and X-ray technicians, aides, pharmacists, and hospital administrators. We will focus on just two of these statuses: physicians and nurses.

Physicians

Less than five percent of the medical workforce consists of physicians. Yet they are central to understanding the medical institution. Physicians are responsible both for defining ill health and for treating it. They define what is appropriate for those with the status of patient, and they play a crucial role in setting hospital standards and in directing the behavior of the nurses, technicians, and auxiliary personnel who provide direct care.

Modern medical technology has enhanced our ability to extend and save lives. It is, however, extraordinarily expensive. In China, a largely rural society with a population in excess of 1 billion, Western-style medicine has been rejected in favor of prevention; community organization for health-care delivery; improved sanitation, housing, and food; and traditional healing practices, the benefits of which are only now being recognized by U.S. and European physicians. China's medical practices have effectively reduced the death rate at a relatively low cost. Throughout the less-developed nations, the Chinese model of health care would probably be more effective in reducing mortality than building high-tech hospitals; at home, critics argue that U.S. tax dollars also would be more wisely spent on nutrition and preventive health-care programs than on expensive new technology.

As will be described in Chapter 13, a *profession* is a special kind of occupation that demands specialized skills and permits creative freedom. No occupation better fits this definition than that of physician. Until about 100 years ago, however, almost anyone could claim the title of physician; training and procedures were highly variable and mostly bad (Starr 1982). About the only professional characteristic early practitioners shared was that their product was highly unstandard! With the establishment of the American Medical Association in 1848, however, the process of professionalization began; the process was virtually complete by 1910, at which point strict medical training and licensing standards were adopted.

Learning the Physician Role

Most people who enter the medical profession have high ideals about helping people. Studies of the medical school experience, however, suggest that the strenuous training schedule of student physicians deals a temporary blow to these ideals. On crowded days, patients come to be defined as enemies who create unnecessary work. Thus, interns and residents learn to GROP (get rid of patients) by referring patients elsewhere, giving them the minimum amount of time, and discharging them as soon as possible (Mizrahi 1986). A patient who escapes GROP and actually corners the busy intern's attention usually has a really interesting, curable disease, is intelligent, cooperative, and appreciative, and is not ill because of his or her own destructive lifestyle.

Medical education is a grueling experience. There is more and more technical information to learn, and the biggest rewards in medicine go to those who have the most technical and least personalized types of skills and practices. Once they survive the rigors of medical school, many physicians renew their commitment to helping people, but nowhere in the medical profession does the structure of rewards encourage personal care (Bloom 1988).

Understanding Physicians' Income and Prestige

The medical profession provides a controversial case study of stratification theories. Why are physicians predominantly male and nurses predominantly female? Why are physicians among the highest-paid and highest-status professionals in the United States?

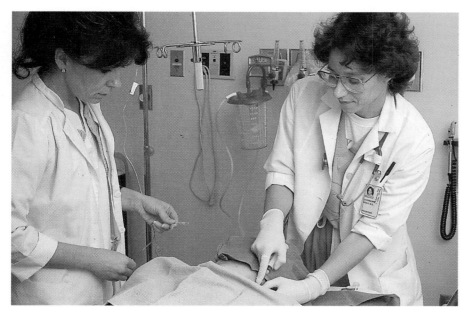

One of the most striking changes in U.S. medicine is the growth in the number of women physicians. Nearly half of the students currently enrolled in medical school are women, and it seems likely that sex will no longer be a good guide to which green-suited specialists are nurses and which are physicians. One consequence, however, is the growing shortage of women entering nursing.

In 1990, the average physician earned $164,500 after paying for deductible professional expenses. The lowest average was $102,700 for general practitioners, and the highest average was $236,400 for surgeons (U.S. Bureau of the Census 1993a).

According to the structural-functional explanation, there is a short supply of persons who have the talent and ability to become physicians and an even shorter supply of those who can be surgeons. Moreover, physicians must undergo long and arduous periods of training. Consequently, high rewards must be offered to motivate the few who can do this work to devote themselves to it. The conflict perspective, on the other hand, argues that the high income and prestige accorded physicians have more to do with physicians' use of power to promote their self-interest than with what is best for society.

Central to the debate on whether physicians' privileges are deserved or are the result of calculated pursuit of self-interest is the role of the American Medical Association (AMA). The AMA sets the standards for admitting physicians to practice, punishes physicians who violate the standards, and lobbies to protect physicians' interests in policy decisions. Although only about half of all physicians belong to the AMA, it has enormous power. One of its major objectives is to ensure the continuance of the capitalist model of medical care, where the physician remains an independent provider of medical care on a fee-for-service basis. In pursuit of this objective, the AMA has consistently opposed all legislation designed to create national health insurance, including Medicare and Medicaid. It has also tried to ban or control a variety of alternative medical practices such as midwifery, osteopathy, and acupuncture. In 1987, the U.S. Supreme Court found that the AMA was unfairly restraining trade by trying to drive osteopaths out of business. As a result of these apparent attempts to protect physicians' profits and independence rather than improve the nation's health care, the AMA has lost credibility among the public (Cockerham 1989).

The Changing Status of Physicians

Thirty or forty years ago, the physician was an independent provider who had substantial freedom to determine the conditions of work and who was regarded as a nearly godlike source of knowledge and help by patients. Much of this is changing. The many signs of changes include the following (Light 1988):

1. A growing proportion of physicians work in incorporated group practices, where fees, procedures, and working hours are determined by others. As a result, physicians have lost a significant amount of their independence. These bureaucratized structures are also more likely to have profit rather than service as a dominant goal.
2. The public has grown increasingly critical of physicians. Getting a second opinion is now general practice, and malpractice suits are about as common as unquestioning admiration. Patients are critical consumers of health care rather than passive recipients.
3. Fees and treatments are increasingly regulated by insurance companies and the government. The vast number of patients whose bills are paid by private or government insurance agencies allows these agencies' pay-out structures to determine what treatments will be given at what fee.

Being a physician is still a very good job, associated with high income and high prestige. It is also part of an increasingly regulated industry that is receiving more critical scrutiny than ever before.

Nurses

Of the nearly 10 million people employed in health care, the largest category includes the 1.8 million who are registered nurses. Nurses play a critical role in health care, but they have relatively little independence. Although the nurse usually has much more contact with the patient than the physician, the nurse has no authority over patient care. Nurses are subordinate to physicians both in their day-to-day work and in their training. Physicians determine the training standards that nurses must meet, and they enforce these standards through licensing boards. On the job, physicians give instructions and supervise. Because the majority of physicians are male and the majority of nurses are female, the income and power differences between doctors and nurses parallel the gender differences in other institutions (see Table 10.3). This makes the hospital a major arena in the battle for gender equality.

In part, women have fought the battle by joining them rather than beating them. Many women who previously would have become nurses are now raising their aspirations: Between 1975 and 1986, the number of female physicians increased twice as fast as the number of male physicians. Although women constitute only 24 percent of practicing physicians, they are nearly one-half of current medical students.

Within the field of nursing, training standards have risen and so have salaries. A decade ago, the standard credential in the field was the RN (registered nurse), which represented three years of classroom and practicum experience in a hospital training program. As nurses have attempted to raise their status in medical care, two new positions have developed. At the top of the nursing hierarchy is a relatively new status, the nurse practitioner, who may provide direct patient care (for example, prescribing birth-control pills) with only very general supervision from a physician. Below her are the nurses with a BSN (bachelor of science in nursing) degree, who have the training of an RN plus a full bachelor's of science college degree. The BSN is becoming the new standard in nursing, and greater education is a lever nurses are using to demand higher wages and a greater role in health-care management.

Despite the higher wages, a serious nursing shortage is developing. Like teaching, nursing is a traditional women's occupation that is having difficulty attracting members of the quality and quantity that it used to. Now that women have wider occupational opportunities, fewer choose a job that requires weekend and midnight shifts, makes severe emotional demands, includes relatively little independence, and has a short career ladder.

Hospitals

The hospital was once idealized as the "temple of healing." Today it is more often part of a complex bureaucracy whose major concern is the bottom line—that is, money.

THINKING CRITICALLY

Nurses earn less than physicians, but they earn more than many other jobs with equal training requirements. Why do so few males enter nursing? What could change this gender gap?

TABLE 10.3
Physicians and Registered Nurses: Income, Sex, and Race, 1995
Nurses earn less than a quarter of what physicians earn. Critics wonder whether this reflects real differences in training and responsibility or whether it is another instance of traditional women's jobs being evaluated as less worthy than traditional men's jobs.

	Physicians	Registered Nurses
Mean income	$164,500	$33,488
Percentage female	24.4%	93.1%
Percentage black	4.9%	8.4%

SOURCE: U.S. Bureau of the Census, 1996 Statistical Abstract, no. 637.

focus on

technology

TECHNOLOGY AND THE MEANING OF LIFE

*I*n 1997, scientists announced a very important new arrival—Dolly. Dolly, the lamb had been created as the result of the first cloning of an adult mammal. Of course, Dolly is just the most recent and perhaps the most vivid example of the way in which technological change can provide the impetus for a society to rethink its norms and values. Through advances in medical technology, babies born at 5½ to 6 months gestation and weighing less than a pound are now surviving. Developments in human genetics make it possible to determine in advance whether adults and fetuses carry the genetic markers that predispose them to mental and physical abnormalities or life-threatening diseases, such as cancer. Medical interventions can prolong the lives of the terminally ill, and medical technologies can sustain the human body even after the heart stops beating and brain activity has stopped.

Developments in medical technology pose challenges that no society or culture is prepared to address. They raise questions, for instance, about what constitutes a good life and about how much we ought to be willing to pay for it.

In 1908, physician William Osler reported on a study of 486 deaths at Johns Hopkins Hospital in Baltimore. He found that only about 20 percent of the deceased seemed to be suffering in their final days and that for the great majority "death was a sleep and forgetting." (Horgan 1997). Recent research, on the other hand, has found that most of those who are conscious while dying feel distress (Horgan 1997) and that 30 percent of those with life-threatening illnesses would rather die than live in a nursing home (http://www.lastacts .org/7-30-97.htm). In this context, it is not surprising, perhaps, that euthanasia is winning so many adherents and that Jack Kevorkian, alias Dr. Death, was acquitted of charges related to helping two women kill themselves. Many doctors, however, feel that physicians have absolutely no role to play in euthanasia, but instead should focus on easing pain and addressing the psychological and spiritual needs of the dying.

Computerized axial tomography (CAT) and magnetic resonance imaging (MRI) have enormous potential for diagnosing tumors, cancer, heart mur-

murs, and other conditions. Yet, the minimum installation fee for each is $1 million. Should all hospitals have this equipment? If not, will the cardiac patient in a small rural hospital or the indigent patient in a central city emergency room receive the same quality of services as the more affluent patient receives at a major teaching hospital? One billion dollars will buy 500 people a liver transplant or fund the Arkansas public school system for a year; can we have it all and, if not, which should we choose? Could the millions of dollars spent on saving the lives of extremely premature infants be better spent on inoculations and appropriate nutrition for the children of the poor in the United States? The drug regimen of azidothymidine (AZT) and protease inhibitors used to prolong the lives of AIDS patients can cost $18,000 annually for one person. Who should pay the cost?

In the wake of medical advances, societies find themselves struggling not only with questions about medical costs and what constitutes quality of life; they also find themselves struggling with questions regarding the very meaning and nature of life. If human beings can create life in a laboratory and can alter the very forms that life will take through ge-

---

Hospitals are run by professional administrators who hire nurses and other staff and provide rooms, food, and equipment. Yet, what goes on in the hospital is significantly determined by physicians, who often are not hospital employees. Nurses are the group most often caught in the middle: They get their orders from the physicians and their paychecks from the hospital (Cockerham 1989).

The High Cost of Medical Care: Who Pays?

Medical care is the fastest-rising part of the cost of living. Although the overall inflation rate between 1983 and 1992 was 40 percent, it was 90 percent for medical care. Doctors' services and hospital treatments that cost $340 in 1970 cost $1,901 in 1992 (U.S. Bureau of the Census 1993a).

netic interventions, where does God fit in? Who will have the power and authority to decide who or what gets cloned? Is it ethical to "harvest" organs from beings created through cloning?

Sociologist William T. Ogburn (1922) pointed out that societies can hardly adapt to new technologies before they are introduced. Thus, he proposed the concept of **cultural lag** as the time interval between the arrival of a technological change in society and the completion of the structural and cultural adaptations to it. Structural and cultural adaptations to medical technology are only beginning. In Oregon, legislators have made an explicit attempt to control the costs of medicine and medical technology by prioritizing the types of services that will be available to Medicaid recipients (Ginzberg 1994). The Nebraska Commission on Human Genetic Technologies has just been formed and is charged with answering questions about who has access to a person's genetic information, the appropriateness of DNA sampling of people released from prison, and cloning technology. What other structural and cultural adaptations will have to be made? Who do you think should be involved in making these decisions? Who do you think will be?

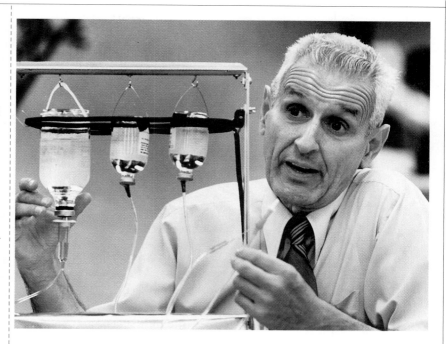

Pictured here with his "suicide machine" is Jack Kevorkian, a Detroit physician. Kevorkian has been called Dr. Death by the media since 1990, when the first of the suicides with which he assisted became public. On trial in Michigan for refusing to obey the state's assisted-suicide law, Kevorkian argued that a higher authority guides him to help relieve suffering in those who find their lives intolerable because of incurable disease or pain. Kevorkian said that many of his fellow physicians agree with his position, and a Oregon survey of physicians found that 60 percent supported legalizing doctor-assisted suicide ("Surveys Find" 1996). From a sociological standpoint, the controversy sparked by Kevorkian is essentially about defining—or redefining—the doctor's role in health care. Do you think assisting suicides should be part of the physician's role? Why or why not?

Who Pays the Bills?

Underlying many of the analyses of health care is one question: "Who pays?" There are three primary modes of financing health care in the United States: private payments, insurance, and government. The cost of health care is so high that only the very rich can afford private payments. The bulk of the population must rely on private insurance or government programs.

PRIVATE INSURANCE. Approximately 70 percent of people in the United States are covered by a private health insurance plan. Almost all of these plans are available through the place of employment, and insurance tends to be limited to those employed adults (and their families) who have a job in the corporate core. Many jobs in the periphery and most minimum-wage jobs do not include insurance benefits. Less than

Cultural lag is the time interval between the arrival of a change in society and the completion of the adaptations that this change prompts.

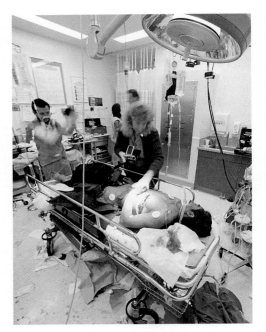

Access to medical care is a serious problem for the 37 million Americans who lack health insurance. The uninsured often have no family physician or other regular source of medical care. They tend to postpone seeking care until their health problems have become extremely serious. Furthermore, because many are also very poor, they live in neighborhoods and communities that make them particularly prone to accidents and injuries. For these reasons, the uninsured are particularly likely to receive their medical care in emergency-room settings where the cost of care is very high and the prospects for good follow-up treatment are poor.

one-third of jobs in retail sales or food services (the fastest growing sector of our economy) are covered by health insurance (Renner & Navarro 1989).

GOVERNMENT PROGRAMS.   The government has several programs that support medical care. The federal government provides some health care through its Veterans Administration hospitals, but its two largest programs are Medicaid and Medicare. In addition, local governments provide medical care through public health agencies and public hospitals.

Medicare is a government-sponsored health insurance policy for citizens over 65; premiums are deducted from Social Security checks. The enactment of this program in 1965 did a great deal to improve the quality of health care for the elderly. Over 95 percent of the elderly are now covered by health insurance. This is not a cheap program, however, and in 1994, the government paid more than $129 billion in Medicare benefits (U.S. Bureau of the Census 1996a).

Technically a welfare program, *Medicaid* is a federal cost-sharing program that provides federal matching funds to states that provide medical services to the poor. Although the program was originally limited to people who were on welfare, in many states it is now available to poor children and pregnant women who are low income but not actually on welfare. The eligibility of individuals, however, and the services available are determined by states. As a result, some states offer much more generous medical care than others.

THE UNINSURED.   A significant portion of the U.S. population—15 percent—has no medical coverage (Table 10.4). Thanks to Medicare, nearly 100 percent of the elderly are insured; those who fall through the cracks are adults who are unemployed or the working poor. Ironically, those with poor health are nearly twice as likely to be uninsured as those with good health (19 percent versus 12 percent) (U.S. National Center for Health Statistics 1987). Map 10.1 shows the proportion of uninsured living in each state.

The uninsured are not entirely without health care. Every county in the United States makes some provision for the so-called medically indigent. The care provided for these people is largely emergency treatment rather than prevention and diagnosis, however, and patients are often treated as unworthy, kept standing in long lines, and sometimes given second-rate treatment by overworked and underpaid staff members who aren't crazy about associating with the poor.

Why Doesn't The United States Have National Health Insurance?

Health care in the United States is available on a fee-for-service basis. Like dry cleaning, you get what you can afford. If you cannot afford it, you might not get any. The United States is "the only industrialized nation in the world that does not provide financial coverage for medical care of a majority of its citizens" (Cockerham, Kunz, & Lueschen 1988, 114). In the rest of the industrialized world, medical care is like education—regarded as a good that should be available to all regardless of ability to pay.

Why is the United States alone among industrialized nations in having no national health insurance? Certainly, the AMA has strongly opposed national health insurance, but that cannot be the whole reason. It opposed Medicaid and Medicare too, and those programs have been in place for 30 years. Nor is public opposition the reason. A recent poll showed that 71 percent of the U.S. public believes that adequate medical care is a right to which a person is entitled as a citizen rather than a privilege that must be earned (Public Opinion 1988), and more than half are willing to

TABLE 10.4
Americans' Health Insurance Coverage, 1995
Approximately 15 percent of the U.S. population has no health insurance. These people, generally the unemployed and the working poor, are concentrated among those who are under 65 and have family incomes of less than $20,000. Ironically, those without health insurance have worse health than those with insurance.

(Numbers in thousands)	Total Number	Total Number	Covered Percent	Not Covered	
				Number	Percent
All persons	264,315	223,733	84.6	40,582	15.4
Sex					
Male	129,144	107,496	83.2	21,648	16.8
Female	135,171	116,237	86.0	18,934	14.0
Age					
Under 18 years	71,148	61,353	86.2	9,795	13.8
18 to 24 years	24,843	17,847	71.8	6,996	28.2
25 to 34 years	40,919	31,561	77.1	9,358	22.9
35 to 44 years	43,078	35,946	83.4	7,132	16.6
45 to 64 years	52,668	45,668	86.7	7,000	13.3
65 years and over	31,658	31,358	99.1	300	0.9
Race and Hispanic Origin					
White	218,443	187,338	85.8	31,105	14.2
Black	33,889	26,782	79.0	7,107	21.0
Hispanic origin[1]	28,438	18,964	66.7	9,474	33.3
Household Income					
Less than $25,000	78,435	59,722	76.1	18,713	23.9
$25,000–$49,999	84,459	70,762	83.8	13,697	16.2
$50,000–$74,999	53,453	48,479	90.7	4,974	9.3
$75,000 or more	47,967	44,770	93.3	3,197	6.7

[1]Persons of Hispanic origin may be of any race.

SOURCE: U.S. Bureau of the Census, March 1996 Current Population Survey. http://www.census.gov/hhes/hlth.ns/cover95/c95taba.html.

pay higher taxes to provide adequate health care to all who need it (Gallup Poll 1993). The reason that we have no national health insurance is suggested by a recent study showing that the quality of national health insurance across countries and the rapidity with which it was implemented vary directly with the political strength of the working class (Navarro 1989). If this is true, then the absence of national health insurance is linked directly to the absence of a working-class or socialist party in the United States.

Summary

Sociological analysis suggests that health and illness are socially structured. To paraphrase C. Wright Mills again, when one person dies too young from stress or bad habits or inadequate health care, that is a personal trouble, and for its remedy we properly look to the character of the individual. When whole classes, races, or sexes consistently suffer significant disadvantage in health and health care, then this is a social problem. The correct statement of the problem and the search for solutions require us to look beyond individuals to consider how social structures and institutions have fostered these patterns.

MAP 10.1
Percent of Persons Without Health-Care Coverage, 1995

Approximately 15 percent of the U.S. population has no health insurance. These people, generally the unemployed and the working poor, are concentrated among those who are under 65 and have family incomes of less than $20,000. Interestingly, those without health insurance have worse health than those with insurance.

SOURCE: http://www.census.gov/hhes/hlthins/cover95/c95taba.html

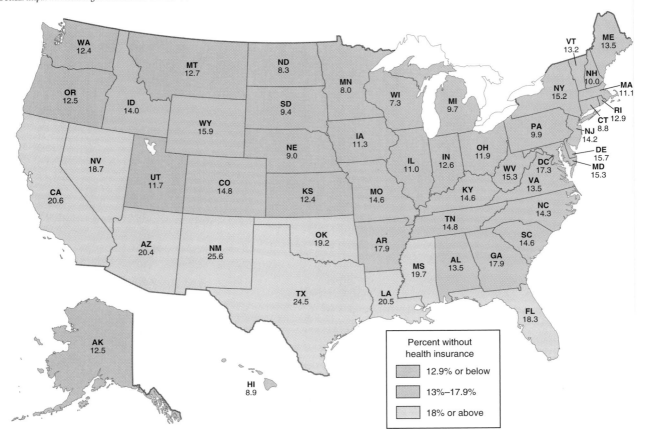

Percent without health insurance

- 12.9% or below
- 13%–17.9%
- 18% or above

The sociological imagination suggests that significant improvements in the nation's health will require changes in social institutions. Many of these changes will need to take place outside of the medical institution itself.

Summary

1. Young people suffer many structured inequalities and are not well integrated into society's institutions. Among the consequences of this are a high crime rate and a high level of poverty among families headed by young adults.

2. The population over 65 is perhaps better off now than it ever has been. Although the older population may suffer from low esteem, there have been sharp improvements in economic conditions. The disadvantages of aging are most pronounced for women and those who were living on the economic margin during their working years.

3. As indexed by the poverty rate, children are more disadvantaged than older people. This is due to two factors: the greater willingness to support older people out of the public purse and the increasing proportion of children whose fathers do not help support them.

4. Structural-functional theory (disengagement theory), conflict theory, and modernization theory provide competing explanations of why young and old people tend to have lower status in U.S. society. Right now, conflict theory seems more appropriate to explain the disadvantages of young people and structural-functional theory more appropriate to explain the improved status of older people.

5. Three statuses are especially relevant to the social epidemiology of health in the United States: gender, social class, and race/ethnicity. Men, racial and ethnic minorities, and those with lower socioeconomic status have higher mortality rates.

6. The health disadvantage associated with lower socioeconomic status goes beyond a simple inability to afford health care. Lower social class is associated with lower standards of living, more stress, lower education, and poorer coping strategies, all of which increase the likelihood that individuals will need health care.

7. Physicians are professionals; they have a high degree of control not only over their own work but also over others in the medical institution. Structural functionalists argue that physicians earn so much because of scarce talents and abilities, but conflict theorists argue that high salaries are due to an effective union (the AMA). Physicians' independence is lower now than it used to be.

8. Nurses comprise the largest single occupation in the health-care industry. Nurses earn much less than physicians, have less prestige, take orders instead of give them, and are predominantly female. The hospital is a major arena in the battle for gender equality.

9. Most people in the United States (70 percent) are covered by private insurance. Medicare has covered most senior citizens successfully, but the poor are seriously underinsured: Fifteen percent have no insurance. The uninsured are less healthy than the insured.

10. The United States does not have national health insurance because we do not have a strong workers' party. We are one of the few nations in the Western world that does not make medical care available regardless of the patient's ability to pay.

Sociology on the Net

One of the most important changes in the United States is the aging of our population. This has recently led to political fights over Social Security and medical care for the elderly and has even fostered attacks on organizations that represent the interests of the elderly. Let us return once more to the *Census Bureau's Population Profiles.*

    http://www.census.gov/population/www/pop
    -profile/profile.html

Return to the selection on *National Population Projections* and reread the part called "The U.S. population will be older than it is now." In what year will the median age peak? Who are the baby boomers and why are they an important part of the aging population? When were they born?

Now go back to the main menu and open the selection entitled *The Elderly Population.* How has the age structure of the United States changed in the last 90 years? What is the fastest growing segment of the aging population? What kinds of problems might this create?

A lot of the controversy centers on the future of the social security program. What does the Social Security Administration have to say about this? Go to:

    http://www.ssa.gov/coss_speech.html

Browse through the commissioners speech on social security. What is social security, and why do some people feel that it is in trouble? What impact will the baby boomers have on the social security system?

Older people get out and vote in much larger proportion than any other age group. This makes them a formidable voting block. One of the strongest organizations representing the elderly is the American Association of Retired Persons.

    http://www.aarp.org

This is the home page of AARP. Start your browsing by clicking on the section called *Who We Are.* Take a look around and then open the section on *Advocating For Our Members.* When you have read to the end of this selection, open the section entitled *Where We Stand.* Click on the highlighted phrase *Where We Stand* in this section. What issues are listed here that are currently hot topics in Congress? How do the interests of this group coincide with your own interests? On what issue do you differ? If the AARP achieves all of these goals, who will pay for them?

FIND IT ON INFOTRAC COLLEGE EDITION

During the 1992 presidential elections, health-care reform hit center stage. The debate continues. Using InfoTrac College Edition, look up the following article:

"Universal Health Care Coverage in the United States: Barriers, Prospects, and Implications." Stephen Gorin. *Health and Social Work,* August

1997, v22 n3 p223. (Hint: Enter the search term *Stephen Gorin* using the Subject Guide.)

Describe the problems facing the current health-care system in the United States. What are the barriers facing the prospect of universal coverage? Do you believe the United States should provide medical coverage for all its citizens? Why or why not?

Suggested Readings

Blank, Robert H., and Bonnicksen, Andrea (eds.). 1994. *Medicine Unbound: The Human Body and the Limits of Medical Intervention.* New York: Columbia University Press. Essays and research reports on the fascinating technological advances in medicine in recent years, along with warnings that technology cannot solve all our health challenges.

Chapman, Audrey R. (ed.). 1994. *Health Care Reform: A Human Rights Approach.* Washington, D.C.: Georgetown University Press. Research reports and essays that argue for health care reform of various types and point out ways to achieve it.

Cockerham, William. 1995. *Medical Sociology,* 6th ed. Englewood Cliffs, N.J.: Prentice-Hall. A textbook on medical sociology by an active researcher in the field. The volume provides a balanced presentation of theoretical views along with a detailed analysis of how medical institutions and professions operate.

Friedan, Betty. 1993. *The Fountain of Age.* New York: Simon and Schuster. Most known for her book, *The Feminine Mystique,* published in 1963, Friedan now has tackled the issue of aging in our society, particularly as aging affects women in the United States.

Kertzer, David, and Schaie, K. Warner (Eds.). 1989. *Age Structuring in Comparative Perspective.* Hillside, N.J.: Erlbaum. A collection of articles showing how the experience of age from childhood to old age is socially structured around the world.

Leach, Penelope. 1994. *Children First.* New York: Knopf. A policy book by an author who has become a television personality. The book argues that the industrialized world is forgetting its children and that we need to put children first in our family and national political decisions.

Navarro, Vincente. 1993. *Dangerous to Your Health: Capitalism in Health Care.* New York: Monthly Review Press. A conflict analysis of how inequalities in medical care and in health are produced by an unequal society, which, in turn, is produced by capitalism.

Treas, Judith. 1995. Older Americans in the 1990s and Beyond. *Population Bulletin 50* (2) (May). Washington, D.C.: Population Reference Bureau, Inc. Short, readable booklet with up-to-date information on many issues regarding the over-65 population in the United States today.

Family and the Life Course

There have been many changes in American family life in recent decades. Birthrates have declined sharply, divorce rates have reached record levels, the proportion of single-parent families has increased, and, for the first time, women with small children have entered the labor force in large numbers. In addition to these statistical trends, major shifts in attitudes and values have occurred. Sexual activity outside marriage has become increasingly acceptable; unmarried couples living together and homosexuals have become more open. Related to many of these changes are the dramatic changes in the roles of women in our society.

These changes in family life have been felt, either directly or indirectly, by all of us. Is the family a dying institution, or is it simply a changing one? In this chapter, we examine the question from the perspective of sociology. We begin with a broad description of the family as a basic social institution.

Around the world, the family is a central social institution. Although families have different tasks in some cultures than in others, in all cultures family members are charged with the responsibility for taking care of one another. Parents must take care of children and husbands and wives of each other. As this photograph of a family in Papua New Guinea shows, the family is also often a primary source of personal satisfaction and warmth.

Marriage and Family: Basic Institutions of Society

To place the changes in the U.S. family into perspective, it is useful to look at the variety of family forms across the world. What is it that is really essential about the family?

In every culture, the family has been assigned major responsibilities, typically including the following (Murdock 1949; Pitts 1964):

1. Replacement through reproduction.
2. Regulation of sexual behavior.
3. Economic responsibility for dependents—children, the elderly, the ill, and the handicapped.
4. Socialization of the young.
5. Ascription of status.
6. Provision of intimacy, belongingness, and emotional support.

Because these activities are important for individual development and the continuity of society, every society provides some institutionalized pattern for meeting them. No society leaves them to individual initiative. Although it is possible to imagine a society in which these responsibilities are handled by religious or educational institutions, most societies have found it convenient to assign them to the family.

The importance of these tasks varies across societies. Status ascription is a greater responsibility in societies where social position is largely inherited; regulation of sexual behavior is more important in cultures without contraception. In our own society, we have seen the priorities assigned to these family responsibilities change substantially over time. In colonial America, economic responsibility and replacement through reproduction were the family's primary functions; the provision of emotional support was a secondary consideration. More recently, however, some of the responsibility for socializing the young has been transferred to schools and day-care centers; financial responsibility for the dependent elderly has been shifted to the government. At the same time, intimacy has taken on increased importance as a dimension of marital relationships.

Unlike most social structures, the family is a biological as well as a social group. The **family** is a relatively permanent group of persons linked together in social roles by ties of blood, marriage, or adoption who live together and cooperate economically and in the rearing of children. This definition is very broad; it would include a mother living alone with her child as well as a man living with several wives. The important criteria for families are that their members are bound together—if not by blood, then by some cultural ceremony such as marriage or adoption that ties them to each other relatively permanently—and that they assume responsibility for each other.

Marriage is an institutionalized social structure that provides an enduring framework for regulating sexual behavior and childbearing. Many cultures tolerate other kinds of sexual encounters—premarital, extramarital, or homosexual—but all cultures discourage childbearing outside marriage. In some cultures, the sanctions are severe, and almost all sexual relationships are confined to marriage; in others, marriage is an ideal that can be bypassed with relatively little punishment.

Marriage is important for childbearing because it imposes socially sanctioned roles on parents and the kin group. When a child is born, parents, grandparents, and aunts and uncles are automatically assigned certain normative obligations to the child.

This network represents a ready-made social structure designed to organize and stabilize the responsibility for children. Children born outside marriage, by contrast, are more vulnerable. The number of people normatively responsible for their care is smaller, and, even in the case of the mother, the norms are less well enforced. One consequence is higher infant mortality for children born outside of marriage in almost all societies, including our own.

Marriage and family are among the most basic and enduring patterns of social relationships. Although blood ties are important, the family is best understood as a social structure defined and enforced by cultural norms.

The **family** is a relatively permanent group of persons linked together in social roles by ties of blood, marriage, or adoption who live together and cooperate economically and in the rearing of children.

Marriage is an institutionalized social structure that provides an enduring framework for regulating sexual behavior and childbearing.

The U.S. Family Over the Life Course

Family relationships play an important role in every stage of our lives. As we consider our lives from birth to death, we tend to think of ourselves in family roles. Being a youngster means growing up in a family; being an adult often means having our own family; being elderly means being a grandparent. Family ties and family roles are an important part of the developmental process from birth to death (Juster & Vinovskis 1987).

Because of the close tie between family roles and individual development, we have organized this description of the U.S. family into a life course perspective. This means that we will approach the family by looking at age-related transitions in family roles.

Childhood

U.S. norms specify that childhood should be a sheltered time. Children's only responsibilities are to accomplish developmental tasks such as learning independence and self-control and mastering the school curriculum. Norms also specify that children should be protected from labor, physical abuse, and the cruder, more unpleasant aspects of life.

Childhood, however, is seldom the oasis that our ideal norms specify. A sizable number of children are physically or emotionally abused by their parents; current estimates suggest that one out of four girls experiences sexual abuse during childhood (Finkelhor 1986). Nearly one-fifth of all children grow up in poverty, and evidence suggests that the gap between rich and poor children is growing (Lichter & Eggebeen 1993).

An important change in the social structure of the child's world is the sharp increase in the proportion of children who grow up in single-parent households: Twenty-five percent are born to single mothers. Many more experience the divorce of their parents. Of those children whose parents divorce and remarry, nearly half will experience the breakup of the second marriage too (Bumpass 1984). It is estimated that fully 59 percent of all children born in the 1980s will spend some time in a single-parent household before they are 18 (Glick 1984). Perhaps because single parents cannot provide as much money or time as two parents, studies show that, on the average, children raised in single-parent families have lower self-esteem and academic performance and poorer quality social relations (Amato & Keith 1991a; McLanahan & Booth 1989). It is important to note, however, that children of divorce do no worse on most of these measures than children raised in conflict-ridden families (Amato & Keith 1991b).

The increasing participation of women in the labor force has added another social structure to the experience of young children: the day-care center. In 1992, 54 percent of mothers of babies under one year of age were employed. Approximately half of all preschoolers are cared for by relatives while their mother is at work; about one out of five children of employed married mothers is cared for by his or her father (Presser 1989). However, about one-quarter of all preschool children with an employed mother are enrolled in a day-care center. Research shows that children experience no ill effects of early enrollment in high-quality day-care programs and may actually outperform other children in tests of persistence, independence, school achievement, vocabulary, and low anxiety (Andersson 1989). In the United States, however, high-quality programs are hard to find; few day-care centers have a large stable staff, well trained in early childhood development. Children placed in low-quality day-care centers may

As increasing numbers of U.S. women—including those with infants under the age of one—have entered the labor force, day-care centers have become much more important aspects of early childhood socialization. In contrast to other nations where most women work, for instance, the countries of Eastern and Western Europe and China, when it comes to locating dependable, high-quality care, U.S. mothers and fathers are pretty much on their own. Unfortunately, in day care as in most other things, these parents get exactly what they pay for. Research shows that children in day-care facilities with large, stable, well-educated staffs may actually outperform children who stay at home with their mothers on many cognitive tasks. However, day care of this kind is both very scarce and very expensive. Thus, children from low-income families are least likely to benefit from the advantages that good day care can provide.

have more difficulty attaching themselves to their parents and may show more problem behaviors (Berk 1989, 456). Because low-quality day-care centers are also less expensive, children from low-income families are more likely to experience these disadvantages.

Adolescence

Contemporary social structures make adolescence a difficult period. Because society has little need for the contributions of youth, it encourages young people to become preoccupied with trivialities—such as concern over personal appearance or the latest music. Yet, because adolescence is a temporary state, the adolescent is under constant pressure. Questions such as "What are you going to do when you finish school?", "What are you going to major in?", "How serious are you about that boy [girl]?", and "What went wrong in Friday night's game?" have an urgent reality that creates strain. Adolescents are supposed to become independent from their parents, acquiring adult skills and their own values. They are supposed to shift from the family to peer groups as a source of self-esteem. They must learn how to impress new people and, last but not least, they are supposed to have fun (Campbell 1969). The average person begins to date at about age 14, and an adolescent who is far behind may find that parents and friends are concerned. Thus, despite the fact that society does not appear to expect much from them, adolescents experience a great deal of role strain. Survey data show that adults are twice as likely to pick adolescence as the worst rather than the best time of their lives (Harris & Associates 1975).

Rites of passage mark the end of one status and the beginning of another. In our society, rituals such as graduations and weddings continue to have symbolic significance, but the transitions they mark are less clear than they used to be. When as many as half of all couples cohabit before their weddings, when many people receive advanced degrees years after they have married and borne children, the transition to adulthood becomes somewhat fuzzy.

The Transition to Adulthood

Some societies have **rites of passage,** formal rituals that mark the end of one age status and the beginning of another. In our own society, there is no clear point at which we can say a person has become an adult.

Although expectations about adulthood vary greatly by sex, in the United States adulthood usually means that a person adopts at least some of the following roles: being employed and supporting oneself and one's dependents, being out of school, voting, being a church member, marrying, and having children. Some of these social roles are optional, and people may be considered adults who never vote, marry, or, in the case of women, hold a paid job (Hogan & Astone 1986). Nevertheless, the exit from adolescence always entails "escaping" from dependence on parents and family.

The normative and most common transition sequence is to finish school, get a job, marry, and have children in that order, but major changes have taken place in this sequence over the last two decades. In the United States there is a great deal of fluidity and reversibility in late adolescence and early adult years (Rindfuss, Swicogood & Rosenfeld 1987). Youth may leave home and return several times before becoming independent (White 1994).

Rites of passage are formal rituals that mark the end of one status and the beginning of another.

Dating and Mate Selection

Nearly all Americans marry. In fact, the United States is the "marryingest" of industrialized nations. By the time they reach 30, a very high proportion of people in the

Romantic love is the ideal that people in the United States use in judging their dating relationships. Although love and physical attraction can cross many barriers, most people fall in love with people who are similar to themselves in terms of race, age, interests, and social-class background.

Propinquity is spatial nearness.

Homogamy is the tendency to choose a mate similar to oneself.

THINKING

CRITICALLY

Analyze the mate selection processes that you (or someone close to you) have undergone. Show how propinquity, homogamy, and appearance were or were not involved. What role did parents play?

United States have been married at least once. At first glance, it appears as if all persons are on their own in the search for a suitable spouse; we do not have matchmakers or formally arranged marriages in our society. On further reflection, however, it is clear that parents, schools, and churches are all engaged in the process of helping young people find suitable partners. Schools and churches hold dances designed to encourage heterosexual relationships; parents and friends introduce somebody "we'd like you to meet." Although dating may be fun, it is also an obligatory form of social behavior—it is normative.

Recent Trends

In the 1950s, teenagers dated in order to find a spouse. Many did so very quickly, and more than 50 percent of U.S. women were married before their 21st birthday. Times have changed. Teenagers no longer date with the expectation of settling down early. Because people are marrying later and more people are marrying for a second or even third time, dating is no longer an activity restricted to the teen years. Nearly one-third of women and one-half of men are unmarried at ages 25–29. Although not all of these people are looking for a spouse, most are looking for at least a temporary partner. Courtship and dating are activities of 28 or 35 year olds as well as teens.

Narrowing Down the Field

Over the course of one's single life, one probably meets thousands of potential marriage partners. How do we narrow down the field?

Obviously, you are unlikely to meet, much less marry, someone who lives in another community or another state. In the initial stage of attraction, **propinquity,** or spatial nearness, operates in this and a much more subtle fashion, by increasing the opportunity for continued interaction. It is no accident that so many people end up marrying co-workers or fellow students. The more you interact with others, the more positive your attitudes toward them become—and positive attitudes may ripen into love (Homans 1950).

Spatial closeness is also often a sign of similarity. People with common interests and values tend to find themselves in similar places, and research indicates that we are drawn to others like ourselves. Of course, there are exceptions, but faced with a wide range of choices, most people choose a mate of similar class, race, religion, age, and interests (Rawlings 1978). This tendency is called **homogamy.**

Physical attractiveness may not be as important as advertisers have made it out to be, but studies do show that appearance is important in gaining initial attention (Saks & Krupat 1988). Its importance normally recedes after the first meeting.

Dating is likely to progress toward a serious consideration of marriage (or cohabitation) if the couple discover similar interests, aspirations, anxieties, and values (Reiss 1980). When dating starts to get serious, couples begin sharing such expectations as the desire for children and the division of household labor. If he wants her to do all the housework and she thinks that idea went out with the hula hoop, they will probably back away from marriage.

LOCAL MARRIAGE MARKETS.   One of the factors that influences both who will date but also who and how many couples will ultimately marry is the local supply of "economically attractive" men. As early as 1987, William Julius Wilson had noted that one of the reasons why African American women were much less likely to marry than white women was the shrinking pool of well-educated black men with good jobs and earnings. The "male marriagable pool index" is the tool that researchers have developed to assess this hypothesis. Results are clear; local marriage markets do matter. A shortage of males employed in good jobs with adequate earnings sharply reduces the likelihood of both women's first marriage and first cohabitation rates (Lichter et al. 1992; Raley 1996). In fact, differences in the availability of marriageable men accounts for at least 40 percent of the race difference in overall marriage rates. In an interesting sidebar, researchers have found that "economically attractive" women are also more likely to marry. Their greater attractiveness to potential male partners apparently more than makes up for the fact that women with full-time employment and higher earnings tend to be choosier about the men they date and marry (Lichter et al. 1992).

The Sexual Side of Courtship

Some of the more important norms surrounding dating behavior are concerned with the amount of acceptable sexual contact. In the United States, we have seen two revolutions in premarital sexual norms and behavior. The first occurred in the 1920s, when there was a major increase in the proportion of both women and men who engaged in premarital sexual intercourse (Kinsey 1948, 1953). The second began in the late 1960s. Studies of adolescents and college students indicate that this second revolution had two components: an increase in permissiveness and a decline of the double standard.

All major surveys have found increases in permissiveness in recent years; more people engage in sex before marriage, and fewer see anything wrong with it. Increasingly, both men and women believe that a strong commitment is unnecessary for a sexual relationship to be acceptable. Moreover, these changes in the last decade have been more pronounced for women than for men, with the result that men and women are now much more alike in both attitudes and levels of experience (Laumann et al. 1994).

It is possible that widespread concern over AIDS is leading to a third sexual revolution in which people are becoming more conservative in their sexual behavior. It is too early to have empirical evidence of such a trend, but the necessity of a more prudent approach to the selection of sexual partners is so widely discussed that it seems likely to have had some effect on sexual behavior and sexual standards. Such concerns have led some participants in the dating game to ascribe to a standard sometimes described as "secondary virginity."

Marriage and Divorce

Getting Married

Despite the seeming disorganization of the dating process and the later average age at marriage, 95 percent of people in the United States end up being married at least once by the time they reach 45. A sizable proportion also get divorced at least once, but most people who divorce eventually remarry. In 1994, however, approximately 9 percent of the population aged 40–44 had never married, and many of these people never will. Some of them would have married if their health or choices had been better, but perhaps half chose the independence of remaining single (Austrom & Hanel 1985).

Getting Divorced

In the United States, more than two million adults and approximately one million children are affected annually by divorce. The **divorce rate,** calculated as the number of divorces each year per 1,000 married women has risen steadily in the post World War II period. In 1988, it stood at 21—that is, 21 out of 1,000, or 2.1 percent—of all married women in the United States divorce annually. Another way of looking at divorce is to calculate the probability that a marriage will *ever* end in divorce—the **lifetime divorce probability.** Of marriages begun in 1890, for example,

The **divorce rate** is calculated as the number of divorces each year per 1,000 married women.

Lifetime divorce probability is the estimated probability that a marriage will ever end in divorce.

the proportion eventually ending in divorce was approximately 10 percent (Cherlin 1981). Today, however, experts estimate that about half of first marriages contracted in the last decade will end in divorce (see Figure 11.1) (Bumpass, Raley, & Sweet 1995).

Remarrying

Although a large number of people end their first marriage with divorce, most remarry. In recent years, 80 percent of men and 75 percent of women have remarried following their divorces (Schoen et al. 1985). Remarriage is more common for young people, and aging is particularly hard on women's chances of remarriage. As a result, it takes women about twice as long as men to find a new spouse.

Parenthood

In the past, the majority of couples entered marriage with the expectation of becoming parents—often immediately. As discussed below in the section on contemporary family choices, this link between marriage and parenthood is being broken in two ways: nonmarital births and delayed childbearing.

Middle Age

The busiest part of most adult lives is the time between the ages of 20 and 45. There are often children in the home and marriages and careers to be established. This period of life is frequently marked by role overload simply because so much is going on at one time. Middle age, that period roughly between 45 and 65, is by contrast a quieter and often more prosperous period. Studies show that both men and women tend to greet the empty nest and then retirement with relief rather than regret (White & Edwards 1990; Goudy et al. 1980).

The Sandwich Generation

Many middle-aged couples, however, look forward to the empty nest period only to find that they are sandwiched between the demands of their own parents and their

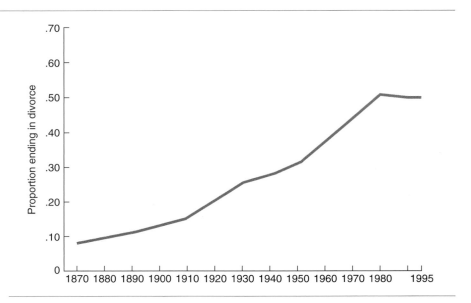

FIGURE 11.1
Changing Probability of Divorce, 1870–1995

There has been a dramatic increase in the likelihood that marriages will end in divorce. Half of recent first marriages are expected to end in divorce.

SOURCE: Adapted from Cherlin 1981 (reprinted by permission, Harvard University Press); Martin and Bumpass 1989; Bumpass, Raley, and Sweet 1995.

children. This generation squeeze has been fostered by two distinct trends. The first is the rising life expectancy of their own parents. In 1940, only 37 percent of 50-year-old women had a mother who was still alive; by 1980, that figure was up to 65 percent (Menken 1985). Thus, middle-aged couples are nearly twice as likely to face caretaking responsibilities for elderly parents. At the same time, rising ages of marriage and increasing rates of separation and divorce have increased the likelihood that their own children will be hanging on. In 1992, 51 percent of all men ages 20 to 24 were still living at home; 29 percent of the divorced and separated men that age were living with mom and dad. Even at ages 30–34, 19 percent of all divorced men return to their parents' homes (U.S. Bureau of the Census 1993f). As a result of these pressures from both sides, many middle-aged people do not find the relief from family responsibility that they hoped for.

Age 65 and Beyond

One of the most important changes in the social structure of old age is that it is now a common stage in the life course—and often a long one. Almost all of us can count on living to 65. Furthermore, those who do will live an average of 16 more years—an average of 18 more years if a woman and 14 if a man. Most of these years will be healthy ones. Although most people at 65 and 70 experience some loss of stamina, a remarkably large proportion experience no major health limitations. In fact, only one-third of those over 85 report that their health places major limitations on their activities (Neugarten & Neugarten 1986, 35).

Family roles continue to be critical in old age. Being married, having children, grandchildren, and brothers and sisters all make substantial contributions to well-being. Marriage is an especially important relationship, one that provides higher income, live-in help, and companionship. Because of men's shorter life expectancy, however, marriage is unequally available: 79 percent of men aged 65–74 are still married compared to only 51 percent of the women that age.

Whether married or not, relationships with children and grandchildren are an important factor in most older people's lives.

However, both older people and their children are happiest when these relationships are free of dependency. The elderly much prefer to live alone rather than with their children, and they don't want to baby-sit.

Intergenerational Bonds

Relations between age groups, like relationships between the sexes, are qualitatively different from relationships between races or classes. Generations are intimately tied to one another through family. Some people have worried that geographical mobility and the increased role of the state in providing support for dependents might weaken these intergenerational ties. In this respect, as in many others, the death of the family has been much exaggerated.

Youth and Parents: The Generation Gap

The relationship between teens and their parents is not as bad as is often suggested. Although many families go through a prolonged period of conflict, with both parents and children being relieved when they are able to live apart, they maintain a strong bond and many common values (Dornbusch 1989). Dozens of studies of college students demonstrate that apparent differences are more often over style than substance and that young people and parents agree about basic values. As Mark Twain is alleged

For most people, the grandparent–grandchild relationship is positive and significant. For many grandparents, involvement with grandchildren is an important source of personal satisfaction. Grandparents are a particularly important source of support for single parents and mothers of high-risk infants (Werner & Smith 1982). For grandchildren, too, the relationship is significant. In a recent study, 82 percent of college age students ranked their relationship with their grandparents as very or moderately important (Vernon 1988).

to have said, "When I was a boy of 14, my father was so ignorant I could hardly stand to have the old man around. But when I got to be 21, I was astonished at how much the old man had learned in seven years."

Adults and Their Parents

At the other end of the age scale, sociological concern has focused on the nature of the relationship between older people and their children and grandchildren. Empirical studies document a reassuring level of family commitment and involvement across the generations.

The nature of intergenerational relationships between adult children and their older parents depends very substantially on the ages of the generations. When the older generation falls into the "young old" category, the older generation is, on the average, still providing more help for their children than children are for their parents (Hogan, Eggebeen, & Clogg 1993). They are helping with down payments and grandchildren's college educations or providing temporary living space for children who have divorced or lost their jobs.

As the senior generation moves into the "old old" category, however, relationships must be renegotiated (Mutran & Reitzes 1984). Even in the "old old" category, most

THINKING
CRITICALLY
Do you know anyone who is taking care of an elderly parent or grandparent? Why do you think that person rather than some other family member has assumed that responsibility? What personal characteristics and what relational characteristics are involved?

people continue to be largely self-sufficient, but eventually most need help of some kind—shopping, home repairs, and social support. Although these services are available from community agencies, most older people rely heavily on their families, especially their daughters.

Roles and Relationships in Marriage

Marriage is one of the major role transitions to adulthood, and most people do indeed marry at least once. In fiction, the story ends with the wedding, and we are told that the couple lived happily ever after. In real life, though, the work has just begun. Marriage means the acquisition of a whole new set of duties and responsibilities, as well as a few rights. What are they and what is marriage like?

Gender Roles in Marriage

Marriage is a sharply gendered relationship. Both normatively and in actual practice, husbands and wives, mothers and fathers have different responsibilities. Although many things have changed, U.S. norms specify that the husband *ought* to work; it is still considered his responsibility to be the primary provider for his family. Similarly, norms specify that the responsibility for housework falls on the wife, not the husband. Although attitude surveys indicate this norm is changing, the actual division of labor remains virtually unchanged whether the wife works or not (Ferree 1991; Potuchek 1992). As a result, wives who work, and especially mothers who work, often end up with a severe case of role overload, or role strain. One adaptation families make to this overload is to lower their standards for cleanliness, meals, and other domestic services. They eat at McDonald's and let the iron gather dust.

Sexual Roles in Marriage: A Changing Script

In few areas of our lives are we free to improvise. Instead, we learn social scripts that direct us toward appropriate behaviors and away from inappropriate ones. Sex is no exception. Unfortunately, we know relatively little about the sexual script for marriage partners and about how it has changed. All the attention in the sexual revolution has gone to young people. Did the revolution pass married folks by, or have sexual roles changed within marriage as well as outside it?

The serious studies that have been done find that frequency of sexual activity seems to have changed very little among married people in the last six years (Call, Sprecher, & Schwartz 1995; Laumann et al. 1994). There have, however, been two notable trends. One is an increase in oral sex, a practice that was limited largely to unmarried sexual partners and the highly educated in earlier decades. The second is that women have reached parity with men in their probability of having an affair. The double standard has disappeared in adultery, and recent studies suggest that as much as 50 percent of both men and women have had an extramarital sexual relationship (Laumann et al. 1995; Thompson 1983).

One of the most consistent findings about sexuality in marriage is that the frequency of intercourse declines steadily with the length of the marriage (see Table 11.1). The decline appears to be nearly universal and to occur regardless of the couple's age, education, or situation. After the first year, almost everything that happens—children, jobs, commuting, housework, finances—reduces the frequency of marital intercourse (Call, Sprecher, & Schwartz 1995).

TABLE 11.1

Frequency of Sexual Intercourse Per Month in the Early Years of Marriage

The frequency with which married couples engage in sexual intercourse declines steadily after the first year of marriage for most couples. Couples attribute this decline to such things as work, childrearing, fatigue, and familiarity.

Year of Marriage	Average	Range
First	14.8	4–45
Second	12.2	3–20
Third	11.9	2–18
Fourth	9.0	4–23
Fifth	9.7	5–18
Sixth	6.3	2–15

SOURCE: Greenblat 1983, 292.

Approximately 50 percent of U.S. mothers are back in the labor force before their child's first birthday. This trend has exerted pressure on fathers to increase their role in child care, and children of married parents now spend about equal time with mom and dad. Nevertheless, research shows that parenting remains a sharply gendered behavior, with fathers being more likely to play with their children than to either bathe or diaper them.

Oh, it's getting worse all the time. Maybe it's three or four times a month now instead of three or four times a week. But I guess it's natural—it's like "I'm tired, you're tired, let's forget it" (cited in Greenblat 1983, 296).

Sex has become less important now—in the beginning there was a feeling that newly-weds screw a lot; therefore, we ought to. It was great and I loved it, but now I think that other things have become more important as we found other things that are satisfying to do besides sex (cited in Greenblat 1983, 297).

The overall conclusion drawn from research is that, after the first year of marriage, sex is of decreasing importance to most people. Nevertheless, satisfaction with both the quantity and the quality of one's sex life is essential to a good marriage (Blumstein & Schwartz 1983; Laumann et al. 1994).

The Parental Role: A Leap of Faith

The decision to become a parent is a momentous one. Children are extremely costly, both financially and in terms of emotional wear and tear. Recent estimates suggest that it may cost as much as $133,000 to raise a middle-class child to adulthood (Kalish 1994). Parenthood, however, is one of life's biggest adventures. Few other undertakings require such a large commitment of time and money on so uncertain a return. The list of disadvantages is long and certain: It costs a lot of money, takes an enormous amount of time, disrupts usual activities, and causes at least occasional stress and worry. Also, once you've started, there is no backing out; it is a lifetime commitment. What are the returns? You hope for love and a sense of family, but you know all around you are parents whose children cause them heartaches and headaches. Parenthood is really the biggest gamble most people will ever take. In spite of this, or maybe because of it, the majority of people want and have children.

Mothering Versus Fathering

Despite some major changes, the parenting roles assumed by men and women still differ considerably. Mothers are the ones most likely to drop out of the

technology

KINKEEPING, GENDER ROLES, AND THE INTERNET

One of the most frequently cited impacts of the industrial revolution was the shift in family structure from one characterized by geographic proximity and multigenerational extended family ties to one based on the isolated nuclear family. From the 1940s well into the 1960s, most sociologists believed that the nuclear family form, consisting of a husband, wife, and dependent children, had at least two serious negative consequences. First, scholars argued that in isolating couples from their broader kinship group, the nuclear family placed an enormous obligation on husbands and wives to provide each other with all of the love, acceptance, emotional support, and companionship the other might need. The pressure of this highly unrealistic expectation, in turn, was held to be a major cause of the rising divorce rates in industrial societies. Second, because of the high levels of residential mobility that accompany the nuclear family form, many sociologists feared that the net result would be a weakening of family ties in a way that jeopardized the well-being of older family members at precisely the time in their lives when they were most likely to need help from adult children and other family members.

Although the verdict may still be out on the first issue, more than 25 years of research convincingly demonstrates that older people in the United States have not been abandoned by their children. Instead, more than two-thirds of U.S. adults talk to their parents at least once a week, and one-fifth talk to a parent every day (Gallup Report 1989c). Grandchildren keep in touch with their grandparents and siblings maintain contact with each other. The fact that family members remain strongly connected throughout the life course, even though they often live many miles apart, has led some sociologists to propose that the typical family form in modern industrial societies is not the nuclear family but one that might more appropriately be called the modified extended family (Litwak 1960).

Technology has been a major factor in the creation of this new family form; rather than relying on close physical proximity to ensure continuing contact and support, the modified extended family uses planes, trains, automobiles, **and** the telephone to link nuclear family members to their more extended kin. Unless or until a major catastrophe occurs—the onset of a serious chronic illness or disability, for example, or the stress of poverty or marital disruption—families can rely on modern modes of transportation and telecommunication to provide support to and maintain strong ties with their more distant relatives (Litwak & Kulis 1988).

The worldwide web has added yet another technological avenue for maintaining ties among far-flung family members. Across the internet, family members now plan celebrations, share information, solve problems, and generally keep track of one anothers' "comings and goings" via the family homepage. How might this use of the web create important changes in the family roles of men and women?

Although anyone can use a phone or drive a car, studies show that, at least up until now, women are overwhelmingly the ones to play the role of *kinkeeper*. They send the birthday cards, organize family reunions, provide most of the emotional support, and generally keep the family in touch with one another. One result of this gender-based division of labor is that female relatives are usually closer to each other than are male relatives. Because men are more likely than women to have access to the worldwide web and to make frequent use of it (see the Focus on Technology box in Chapter 9), the question is simply this: Will men take on a larger share of the kinkeeping role as families come to use the web more frequently to communicate and connect with one another? If so, will the bonds between male relatives begin to take on the closeness and intimacy now associated with the relationships between female family members? What do you think?

---

labor force to care for infants and young children; they are the ones most likely to care for sick children and to go to school conferences. Fathers, on the other hand, are the ones likely to carry the major burden of providing for their families. Generally, parenthood increases men's attachment to the work force at the same time as it decreases that of women (Thompson & Walker 1989); child rearing also reduces the size of women's social support networks and temporarily increases the number of kin included in men's (Munch, McPherson, & Smith Lovin 1997).

The growing proportion of mothers who are employed—about two-thirds in the early 1990s—has exerted pressure for fathers to increase their role in child care. Although research still finds that fathers "help" rather than "take responsibility" and that they are more likely to play with children than change diapers, fathers

THINKING

CRITICALLY

How many children do you plan to have? What do you think the advantages and disadvantages will be? How do you think you and your significant other will decide whose child-care responsibilities are whose?

have increased their role in child care. A growing proportion of fathers, however, do not live with their children. Among nonresidential fathers, contact tends to be low and child care virtually nonexistent.

Stepparenting

The latest estimates show that between 25 and 30 percent of children in the United States will live in a stepfamily before they are age 18—most often with a mother and a stepfather (Bumpass, Raley, & Sweet 1994). If parenting is difficult, stepparenting is more so. Research shows that the average stepparent enters the family with good intentions, but his or her efforts are often rejected by the children and even by the spouse (Hetherington & Jodl 1993). Research confirms the fairy tales: stepmother families are more problematic than stepfather families. The reasons include that stepmothers are more actively involved in their stepchildren's lives than stepfathers and also that the children's noncustodial, biological mother is more likely to be a frequent visitor than a noncustodial father.

The Launching Stage of Parenting

For the first 18 years or so of their lives, children usually live under the same roof as at least one of their parents. A generation ago, young adults married early and moved directly from their parents' homes into their own. Today, many young adults will live at home longer than they or their parents expected. After they move out the first time, nearly half will move back home at least once (Goldscheider & Goldscheider 1994). Some never leave. Nearly 15 percent of parents over 65 still have unmarried children living at home (Speare & Avery 1993).

The average parent copes quite well with children's continued coresidence in the parental home. When a large national sample of parents was asked to rate how well it worked for them to have an adult child in their home, nearly 70 percent gave a positive reply. Nevertheless, a full quarter of parents reported having disagreements once a week or more (Aquilino & Supple 1991). As a result, it is not surprising that research shows parents experience an increase in feelings of well-being when their children leave home (White & Edwards 1990). This feeling of well-being, however, depends on their

Most studies show that marital happiness peaks during the honeymoon, declines in middle age, and increases again in later life. Although the reasons why older couples report higher marital quality are not fully understood, several factors seem likely. First, for most married middle-aged and older persons, spouses provide an important source of companionship and support; shared experiences—"for better or worse"—give couples an important basis for friendship and personal well-being. Second, because retirement frees people from work-related demands, older couples may enjoy increased opportunities to focus on one another. Finally, it is possible that older couples are more satisfied with their marriages simply because those couples who were most unhappy have already divorced by the time they reach the retirement years.

children's having established themselves successfully in an alternative living arrangement and continued positive association between parents and children.

Contemporary Family Choices

As discussed in Chapter 2, U.S. norms have changed over time to permit much wider variation in the way that people achieve core values. Although a happy family life remains at the center of things people in the United States value, the ways that individuals choose to meet this value have changed considerably. An increasing number of people will find themselves making three choices about marriage and the family: to marry or cohabit; to have children, either within or outside of marriage, or remain childless; to juggle work demands to accommodate those of family or vice versa.

Marriage or Cohabitation

During the last 30 years, the chances that an individual will *ever* live in a cohabiting relationship has increased more than 400 percent for men and 1,200 percent for women. Cohabitation is also an increasingly common stage in the courtship process, with approximately half of all recently married couples cohabiting before they were married (Bumpass & Sweet 1995). Cohabitation is even more common among the divorced and separated than it is among the young and never married: 60 percent of persons recently remarrying lived with a partner before marriage—46% only with the person they then married and 14% with someone else (Bumpass & Sweet 1991).

Although cohabitation is often a prelude to marriage, there is good evidence that much of the decline in marriage and remarriage rates in the United States is linked to the increasing numbers of individuals who prefer to live with one another outside of legal marriage. Even if they plan to marry, most cohabitors simply do not believe that anything would change if they did so. Probably the most important exception to this rule is that nearly a third of men, but only about a sixth of women, report that their "freedom to do what they want" would be worse if they were married (Bumpass & Sweet 1991).

It is clear that normative pressures to marry are not as high as they once were. About one-fifth of cohabiting persons do not expect ever to marry or to marry again. Among all unmarried persons, only about one-third agree that it is better to marry than to go through life single, and one-third say it would be alright for them to have a child without ever marrying (Bumpass & Sweet 1991). Because of cohabitation, however, being unmarried does not necessarily mean being single or living alone.

Having Children

Although most people in the United States plan to have children, increasingly they choose to do so outside of marriage and many will not only choose to postpone parenthood, but they will also choose to remain childless.

Nonmarital Births

One-quarter (25 percent) of all births in the United States today are to unmarried women. Most of these births (about two-thirds) are to women 20 years of age and older. However, the rate of unmarried child bearing among teenagers has risen sharply; in just 20 years the rate has increased by more than 80 percent. Because teenage mothers are less likely to complete both high school and college, they are

focus on

American Diversity

IS THE GAY FAMILY A CONTRADICTION IN TERMS?

In a landmark 1986 decision upholding a Georgia law that criminalized same-sex sexual activity, Supreme Court Justice White argued that laws granting privacy rights to families did not apply to homosexuals. There was, he said, "No connection between family, marriage, or procreation on the one hand and homosexual activity on the other..." (Cited in Weston, 1991, p. 208). Is Justice White right? Are heterosexuals the only ones with families?

Because homosexuals have parents, siblings, and cousins, the definition of family at issue in this decision is obviously a narrow one that centers around marriage and children. On the question of marriage, the justice is correct by definition. In a few jurisdictions, gay or lesbian partners may gain some legal protection by registering their union as a "domestic partnership," but no state allows them to marry. Same-sex lovers obviously cannot produce children from their union, but they can have children in a variety of other ways: Many lesbians and gay men can (and do) have children from previous heterosexual relationships, in some states they can adopt children, and lesbian women can have children through artificial insemination. As a result, somewhere between 4 and 14 million children have homosexual parents (Bozett, 1987).

The issue of gay families has become a public issue in both the gay and straight communities (Editors, Harvard Law Review, 1989). Should gays be allowed to adopt children? Should being gay be sufficient by itself to make a parent unfit for child custody or visitation rights? Should lesbians be able to use artificial insemination? Should gay and lesbian partners be allowed to marry legally so that their health insurance and social security benefits could be shared by their partners?

Stepparenting and single parenting, whether because of divorce, nonmarital childbearing, or single-parent adoption, are only two of the increasingly common living arrangements that challenge the conventional U.S. wisdom that families are composed of a mother, a father, and their biological offspring. In addition, a sizable number of children in the United States are being raised and nurtured by gay or lesbian parents and their same-sex partners. In the long run, the definition of a family is likely to change as new methods of procreation (artificial insemination, for instance) and alternative family forms (more grandparents raising their grandchildren, for example) increase in prevalence.

These are questions that go to the heart of the family. The traditional view is that homosexual unions are both unnatural and sinful. Others define the family by long-term commitment, and they are willing to tolerate and encourage a variety of family forms—including gay and lesbian families—as long as they contribute to stable and nurturing environments for adults and children.

It is not only heterosexuals who question the desirability of a gay family (Weston 1991). Many gay men and lesbians reject the idea of family as an oppressive straight-jacket of conformity and patriarchy. They argue that family is an oppressive institution that traps women and men into rigid gender roles and ties us forever to people we don't

like. One of the good things about the gay community, they contend, is that people can form "chosen families" based on shared interests and affection.

Others within the gay community find that family is a good thing, and they would like to stretch the legal as well as the functional bonds of family to include same-sex marriage, adoptions, and protection for child custody. This is the group that is testing the court's protection and participating in what has been called "the lesbian baby boom."

Contrary to Justice White's opinion, homosexuals obviously do have families. The question society must address is whether such families should receive the same legal recognition and protection as other families.

also likely to suffer economic hardship. Consequently, many view the rise in child bearing among U.S. teenagers with alarm. Interestingly, increased rates of teen births have also been observed in Europe during the past several years (see Table 11.2). In contrast to the United States, however, the actual level of teen births is much lower. Studies indicate that teen child bearing is more frequent in the United States than in other industrialized countries, in large part because information about sexuality and contraception is less readily available to teenagers in the United States than in Europe. As a result, fewer U.S. women use any form of contraception and fewer still use the most effective forms (Ahlberg & De Vita 1992).

Delayed Childbearing

Many married women are choosing to postpone child bearing until five or even ten years after their first marriages. Today, 26 percent of U.S. women ages 30 to 34 are still childless (U.S. Bureau of the Census 1993e). Although most still intend to have children eventually, child bearing is no longer seen as an inevitable consequence of marriage

Currently, the average number of children born to U.S. women is approximately two. This small family size is due largely to changes in the role of women and changes in the security of family roles. Although the average woman does not yet place career roles over family roles, women today desire economic security and more personal freedom, both of which are adversely affected by taking time out for having children (McLaughlin & Associates 1988). If the divorce rate remains high and if women's labor-force participation rises—both of which appear likely—then having children will grow even less attractive.

Work Versus Family

Several developments in the labor force have squeezed the time that people have available to spend at home. First, more than three-quarters of women 25–34 are now in the labor force (U.S. Bureau of the Census, 1995). Second, for many of the nation's workers, workdays and work weeks are growing longer. Hourly employees are often coerced or bribed into working overtime hours—even double shifts. Professionals have to work early, late, and on weekends and take work home in order to demonstrate that they are serious players. As a result, parents experience a time bind at home. Family meals are increasingly rare, and time at home becomes rigidly scheduled as parents try to get themselves to work, their laundry done and cleaning picked up, and their children to school and to lessons.

This time bind is often explained as the inevitable result of decreasing real wages, global competitiveness in the workplace, and the growing taste for expensive consumer goods. In a recent study, however, sociologist Arlie Hochschild (1997) argues that many parents are choosing to spend more time at work because they find work more rewarding than being at home with their children. The more hectic it gets at home, the nicer the job looks. Bosses and co-workers hardly ever spill their juice, dirty their diapers, cry, or slam out of the house because they cannot use the car. Compared to home, the workplace tends to be relatively quiet, orderly, and rewarding. For many, work rather than home is the place where you can put your feet up and drink a quiet cup of coffee, work is the place where you can get advice on your meddlesome mother-in-law or crumbling marriage, and work is the place where they notice that you're under a lot of stress and provide free professional counseling—plus, of course, at work there are paychecks, promotion opportunities, and recognition ceremonies.

The link between marriage and parenthood is less direct and less inevitable than it once was. Although the average woman in the United States still desires to have children and places family values over those of career, her participation in the labor force increasingly gives her both the economic and the psychological resources to tackle the tough job of parenting on her own. Many women will postpone childbearing until their thirties. For some, this will come 5 to 10 years into their career and their first marriage; others will take advantage of changing adoption laws and their own occupational stability to start a family on their own.

TABLE 11.2
Percentage Births to Unmarried Women, Selected Countries

	1970	1980	1990	1992
United States	11	18	28	30
Canada	10	13	24	29*
Denmark	11	33	46	46
France	7	11	30	33
Germany	6	8	11	15
Italy	2	4	6	7
Japan	1	1	1	1
Netherlands	2	4	11	12
Sweden	18	40	47	50
United Kingdom	8	12	28	31

SOURCE: U.S. Bureau of the Census, http://www.census.gov/population/socdemo/fertility/htab08.dat
*Data from 1991

Much of the horror expressed over parents finding work less work than family can be traced to anxiety over changing gender roles. It has always been supposed that men would find work satisfying in itself and preferable to being tied down at home all day by the kids. Now that women report the same preferences, it is defined as a problem. One way that people handle this dilemma is simply by having fewer children, and the time bind is surely one reason why women in many European countries are only having one child (see Chapter 14). Another solution is to turn the care of children—day care, after-school care, swimming lessons, birthday parties, psychological counseling, summer vacations—over to paid professionals. These solutions, however, are simply means to help parents work ever longer hours with less family interference. Real solutions would require a reduction in overtime work, a reduction of the work week, and a cultural shift that valued children as much as careers. For the time being, it seems most likely that the family will continue to lose the time war, with men, women, and children spending increasingly long hours away from home and away from each other.

Problems in the American Family

There are couples who swear that they never have an argument and never disagree. These people must certainly be in the minority, however; most intimate relations involve some stress and strain. We become concerned when these stresses and strains affect the mental and physical health of the individuals and when they affect the stability of society. In this section, we cover two problems in the U.S. family: violence and divorce.

Violence in Families

Child abuse is nothing new, nor is wife battering. These forms of family violence, however, didn't receive much attention until recent years. In a celebrated court case in 1871, a social worker had to invoke laws against cruelty to animals in order to remove a child from a violent home. There were laws specifying how to treat your animals, but no restrictions on how wives and children were to be treated. In recent years, however, we have become much more aware of and less tolerant of abuse and violence.

**THINKING
CRITICALLY**
How would you test the hypothesis that "Although husband abuse is slightly more frequent than wife abuse, it is less serious"? What ethical issues would your research design raise?

TABLE 11.3
Violence in the American Family
Nobody likes to admit that they have behaved violently and tried to hurt members of their families. As a result of strong social desirability bias, these figures probably seriously underestimate the amount of violence that actually occurs in U.S. families.

Percentage Reporting Severe Violence

Parent to child	10.7%
Wife to husband	4.4
Husband to wife	3.0

SOURCE: Adapted from Straus and Gelles 1986 and Cornell and Gelles 1982.

The incidence of abuse is hard to measure. The social desirability bias on this question is enormously high. Nevertheless, a series of studies by scholars at the University of New Hampshire provides a relatively reliable set of figures on abuse. Table 11.3 shows the percentage of the population reporting severe violence in the year prior to being interviewed. The operational definition of *severe violence* is if one or more of the following actions is reported: kicked, bit, or hit with fist; hit or tried to hit with some object; beat up; threatened with a gun or a knife; used a gun or a knife. Using this definition, 10.7 percent reported severe violence from a parent to a child, 4.4 percent reported severe violence from a wife to a husband, and 3 percent reported severe violence from a husband to a wife (Straus & Gelles 1986). Although husband abuse is slightly more frequent than wife abuse, researchers believe it is less serious: A blow from a woman is much less likely to cause physical damage than is a blow from a man. A previous study by this team suggests that child-to-parent abuse is on approximately the same level as these others: about 3 percent per year (Cornell & Gelles 1982).

Family violence is not restricted to any class or race. It occurs in the homes of lawyers as well as the homes of welfare mothers. Studies suggest that violence is most typical in families with multiple problems: unemployment, alcohol and drug abuse, money worries, stepchildren, physically or mentally handicapped members, or members who were abused themselves as children (Gelles & Straus 1988).

Solutions to family violence are complex. The first step, however, is to make it clear that violence is inappropriate and illegal. New laws against spousal rape and other forms of family violence may clarify what used to be rather fuzzy norms about whether family violence was appropriate (Straus & Gelles 1986).

Divorce

About half of recent first marriages are expected to eventually end in divorce, but one-half will last. For second marriages, the odds of failure are a little bit higher. What are the factors that make a marriage more likely to fail? Table 11.4 displays some of the predictors of marital failure within the first five years of marriage. A review of empirical results over the last decades suggests that six factors are especially important (White 1990):

- *Age at marriage.* Probably the best predictor of divorce is a youthful age at marriage. Marrying as a teenager or even in one's early twenties *doubles* chances for one's divorce relative to those who marry later. Not surprisingly, if you are already on a second marriage before your twentieth birthday, your chances of failure are very high (see Table 11.4).

TABLE 11.4
Probability of Marriage Breaking up Within the First Five Years
The probability that a first marriage will ever end in divorce is about 50 percent. If we limit our focus on the first five years of marriage, about 23 percent of all first marriages and 27 percent of second marriages end in this period. Divorce is more likely for those who marry young, those with low levels of educational attainment, blacks, and those who had a child before the marriage.

	First Marriages	Second Marriages
Total	23%	27%
Age at marriage		
14–19	31	40
20–22	26	26
23–29	15	27
30+	14	14
Education		
Less than 12 years	33	36
12 years	26	26
13 years or more	16	22
Children before marriage		
No	21	24
Yes	36	28
Race		
White	22	26
Black	36	43
Hispanic	24	28

SOURCE: Martin and Bumpass 1989.

- *Parental divorce.* People who were raised in single-parent families because their parents divorced are more likely to divorce themselves.
- *Premarital child bearing.* Having a child before marriage reduces the stability of subsequent marriages. Premarital conception followed by a postmarital birth, however, does not seem to increase the likelihood of divorce.
- *Education.* The higher one's education, the less likely one's marriage is to end in divorce. Part of this is because people with higher educations are likely to come from two-parent families, avoid premarital child bearing, and marry later. Independent of these other factors, however, higher education does reduce the chances of divorce.
- *Race.* Black Americans are substantially more likely than white Americans (whether Hispanic or non-Hispanic) to end their marriages in divorce. Even if we restrict the comparison to women who marry late, go to college, and have no premarital births, black women are twice as likely as white women to divorce (Martin & Bumpass 1989).
- *Bad behavior.* As you might expect, alcohol and drug abuse, adultery, and abusive behavior are all predictors of divorce. Surveys that ask newly divorced people what happened in their marriages find that these bad behaviors crop up frequently. One woman said, "He was running around and the first time we had sex after the baby's birth, he gave me VD." Another said, "He's a liar and a cheater and a gambler" (Booth & Associates 1984). Although many people just drift apart and cite irreconcilable differences, nearly one-third of the people who seek divorce have a specific and important grievance (Kitson & Sussman 1982).

Societal-Level Factors

Age at marriage, premarital childbearing, education, and bad behavior affect whether a particular marriage succeeds or fails. These personal characteristics, however, cannot account for slightly more than 50 percent of all marriages to fail. The shift from a lifetime divorce probability of 10 percent to one of 50 percent within the last century is a social problem, not a personal trouble, and to explain it we need to look at social structure.

The change in marital relationships is probably most clearly associated with changes in economic institutions. Rising divorce rates are not unique to the United States (see Table 11.5). Although divorce has always been more prevalent in the United States than elsewhere, most industrialized nations have experienced substantial increases in divorce. In Sweden and Germany, for instance, the divorce rate doubled between 1960 and 1988 (Ahlberg & De Vita 1992). In these countries and in the United States, the shift from an industrial and agricultural to a service economy, a change detailed in Chapter 13, has revolutionized the technologies and relationships essential to production. One result of this revolution is that an earner's chief economic asset is education and experience. You can walk away from a marriage and take these assets along; the same is not true with land, which is often tied up in family relationships. Another result is the increased opportunity for women to support themselves outside marriage.

Thus women and men are less and less impelled to marry or to stay married by economic necessity. Because no new incentive for marriage has proven to be more effective than economic need, marriages have less institutional support than before.

How Serious Are Problems in the American Family? A Theoretical Approach

In Chapter 2, we noted that some people view institutions as constraints that force people into uncomfortable and perhaps oppressive relationships; others see institutions as providing the stability and comfort frequently associated with old shoes. This difference of views is nowhere more present than in the case of the family.

Theoretical viewpoints sharply influence perceptions of the health of the modern family (Adams 1985). If the family is an oppressive institution, then divorce is a form of liberation; if the family is the source of individual and community strength, then divorce undermines society. In this section we briefly review some of the major criticisms of the contemporary family and conclude with a perspective on the future.

TABLE 11.5
Divorce Rates for Selected Countries, 1960–1992

Country	DIVORCES PER 1,000 MARRIED WOMEN				
	1960	1970	1980	1990	1992
United States	9	15	23	21	21
Canada	2	6	11	11	11
France	3	3	6	8	9
Germany*	4	5	6	8	7
Japan	4	4	5	5	6
Sweden	5	7	11	12	12
United Kingdom	2	5	12	13	12

*Prior to 1991, data are for former West Germany.
SOURCE: U.S. Bureau of the Census 1995, Table 1366.

Loss of Commitment

Some critics argue that a major problem with the U.S. family—and they would further argue, with U.S. culture—is an accent on individual growth at the expense of commitment. This criticism is most likely to come from structural functionalists, who traditionally stress the subordination of individual to community needs, but it also comes from symbolic interactionists concerned with stable self-identity.

The most prominent symptom of the alleged emphasis on individual happiness and growth at the expense of long-term commitment is the rapid rise in the number of women who are raising children on their own. Because men don't want to be tied down by wives and children and because wives don't want to be tied down by husbands, fathers have walked away and mothers have let them, even encouraged them. The result is that the basic family unit, in the sense of long-term commitment to sharing and support, has become the mother–child pair. Husband–wife and father–child relations are increasingly viewed as temporary and even optional. To many critics, this increasingly voluntary nature of family ties is dysfunctional, reducing the stability of the family and decreasing its ability to perform one of its major tasks: caring for children. Among the ill consequences they note are the increasing proportion of women and children in poverty.

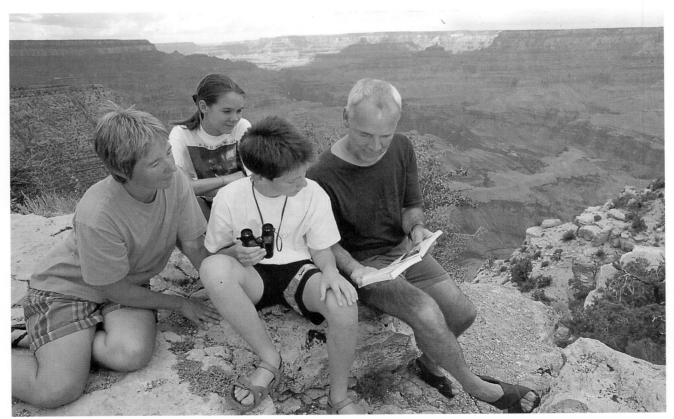

The contemporary family in the United States is experiencing many problems; high levels of divorce, teen pregnancies, and child abuse, among others. The publicity accorded to these problems has led some commentators to conclude that the U.S. family is dying. Despite these very real problems, however, family relationships continue to be a major source of satisfaction for most people in the United States across the entire life course.

Oppression

Critics from a conflict perspective are more apt to criticize the family for its oppression of women and children. They point to the lower power of women and children relative to men and to the number of women and children who suffer abuse in the family. For example, Denzin (1984, 487–88) declares, "A patriarchical, capitalist society which promotes the ownership of firearms, women, and children; which makes homes men's castles; and which sanctions societal and interpersonal violence in the forms of wars, athletic contests, and mass media fiction (and news) should not be surprised to find violence in its homes. . . . Violence directed toward children and women is a pervasive feature of sexual divisions of labor which place females and children in subordinate positions to adult males."

From this perspective, the family has not really changed for the worse: It has traditionally been an oppressive institution maintained by force and fraud. These critics do note the development of greater egalitarianism in the contemporary family, but generally find these changes too slow.

The major developing problem from this perspective is that women's and children's independence from husbands-fathers in the family is coming before their independence in the marketplace. The solution they recommend is equal opportunity and equal pay for women and state support for children in the form of family allowances.

Perspectives on the Future

Some of the functions performed by the traditional family, such as care of dependents, are not as important today as they used to be. Nevertheless, the family continues to perform vital functions for society and individuals. Despite social welfare programs, families remain the major source of economic support for children; families remain the major source of social support for all ages.

The family is also an important arena in which we develop our self-concept, learn to interact with others, and internalize society's norms. Without the strong bonds of love and affection that characterize family ties, these developmental tasks are difficult if not impossible. Thus, the family is essential for the production of socialized members, people who can fit in and play a productive part in society.

The family and the larger kin group are important not just in childhood but also throughout the life course. To cite just a few examples, people with close ties to their kin report greater satisfaction with their life, their marriages, and their health; they are less likely to abuse their children or get divorced; they are also more likely to be able to ride out personal and family crisis (Lavee et al. 1985; McGhee 1985; Rosenthal 1985).

Given these benefits that the family gives to both the individual and society, it would seem to be a reasonable goal to keep and support the family, while simultaneously reducing some of its more oppressive features. This goal is not impossible. Despite current rates of divorce, illegitimacy, childlessness, and domestic abuse, there are signs of health in the family: the durability of the mother–child bond, the frequency of remarriage, the frequency with which stepfathers are willing to step in and support other men's children, and the frequency with which elderly persons rely on *and get* help from their children.

There is no doubt that the family is changing. When you ask a young man what his father did when he was growing up, you are increasingly likely to hear, "What father?" or "Which father?" These recent changes must be viewed as at least potentially troublesome. At present we have no institutionalized mechanisms comparable to the family for giving individuals social support or for caring for children. The importance of these tasks suggests that the family and especially their children need to be moved closer to the top of the national agenda.

Summary

1. Marriage and family are the most basic institutions found in society. In all societies, these institutions meet such universal needs as regulation of sexual behavior, replacement through production, child care, and socialization.

2. Individual development over the life course is closely paralleled by changing family roles. The transition to adulthood is closely linked to adoption of marital and parental roles.

3. High rates of divorce and increases in the labor-force participation of wives and mothers have led to major changes in the social structure of childhood. It is estimated that 59 percent of U.S. children will spend some time in a single-parent household before they are 18. About one quarter of preschoolers with employed mothers attend day-care centers. If day-care programs are of a high quality children may actually benefit from the experience. In the United States, quality day care is expensive and hard to find, and low-quality day care may have detrimental effects.

4. Because of postponement of marriage and high rates of divorce, dating is no longer just a teen activity. Mate selection depends on love, but also on propinquity, homogamy, and value consensus. Sexual relations are a common aspect of contemporary courtship.

5. Intergenerational bonds are important to individuals. Studies show that parent-child bonds remain strong over the life course and that sibling bonds remain important even in old age.

6. The increasing participation of wives in the bread-winning role is a major change in family roles. The change has not been accompanied by a significant increase in the husband's housekeeping role.

7. One option an increasing number of people in the United States choose is cohabitation. Although it is often a prelude to marriage, a substantial percentage of cohabitors never intend to marry even though they do not believe marriage would have any effects on their relationships. Among male cohabitors, a sizeable minority believe that marriage would limit their ability and freedom to do the things they want.

8. Family violence is relatively commonplace in U.S. homes. It is more characteristic of parent–child relationships than of husband–wife relationships. It is strongly related to multiple family problems.

9. It is estimated that 50 percent of first marriages will end in divorce. Factors associated with divorce include age at marriage, parental divorce, premarital child bearing, education, race, and bad behavior. Reduced economic dependence on marriage underlies many of these trends.

10. Perceptions of the health of the family depend on the theoretical orientation of the viewer. Two problems are loss of commitment and inequality within the family.

Sociology on the Net

The changing U.S. family is cause for a great deal of concern and hope. Many are uneasy about changing what they see as a good situation, while others see the changes as bringing new freedoms and fairness to the family. Let's see how the family has changed in recent years by returning to our old friend, the U.S. Census Bureau.

> http://www.census.gov/population/www/
> pop-profile/profile.html

Start by reading the selection on *Households and Families*. What is a family and how does the government define the term *family?* Who is included and who is left out? What percent of families include children? What percent of families are headed by single parents, and how do blacks and whites differ?

Go back to the table of contents and open the selection entitled *Marital Status and Living Arrangements*. What is happening to the median age at first marriage? Why has it changed? What do you feel is a good age for young people to marry? Why do you feel that way?

With both parents working, finding decent child care has become an important part of family life. Return to the table of contents and click on *Child Care Arrangements*. Who provides child care in most families? How are these statistics changing? What are the reasons for the changes?

There are many different segments of our society that see the changing family as an important social issue. Many conservative religious organizations stress what they call "family values." One such organization is the American Family Association.

> http://www.afa.net/

Browse through this home page. Read the section on *Who is AFA?* Note their *Current Boycott Information*. Scroll down to the section entitled *AFA Resources*. Open the *AFA Journal* and check out the topics. What kind of family values are stressed? What actions are being taken by this organization to realize their goals? Have you encountered other organizations that share these goals? Hint: Remember the Internet exercises from Chapter 7.

As families change, there is mounting pressure to change the older definitions to accommodate the diverse family arrangements that are now becoming more evident. One segment of the population that supports broadening definitions of the family are gays and lesbians who want the same legal benefits of marriage that others enjoy.

```
http://www.eskimo.com/~demian/index.html
```

You have reached the home page of the *Partners Task Force for Gay and Lesbian Couples.* Begin your browsing by opening the *Table of Contents.* Click on the topic box under the heading of *Legal Marriage Essays.* Scroll down to and read the selection called *Everything Possible.* Now go back and read the *Most Compelling Reasons for Legal Marriage.* Finally, let's check out what two politicians have to say about a recent bill in Congress. Return to the table of contents and click on *A Mean Bill* and *I Have Anguished.* (Note: The bill under discussion passed the House of Representatives with an overwhelming majority.) What marriage benefits do gays and lesbians seek?

What effects would allowing gay and lesbian marriages have on society? Where do you stand?

FIND IT ON INFOTRAC COLLEGE EDITION

Among the professionals who study and/or deal with domestic violence, there is a debate surrounding methods effective in reducing the problem. In InfoTrac College Edition, look up the following article:

"Constraints Against Family Violence: How Well Do They Work?" Richard J. Gelles. *American Behavioral Scientist,* May–June 1993, v36 n5 p575.

(Hint: Enter the search term *Richard J. Gelles* using the Key Words Guide.)

What is Gelles's opinion regarding mandatory arrest for domestic violence offenders? What approach(es) do you feel would be most effective in reducing the rates of domestic violence?

Suggested Readings

Clinton, Hillary Rodham. 1996. New York: Simon and Schuster. *It Takes a Village.* A call for commitment and policy that supports America's children.

Fineman, Martha A., and Mykitiuk, Roxanne (eds.). 1994. *The Public Nature of Private Violence.* New York: Routledge. A collection covering all forms of domestic violence with essays and research articles by well-known researchers and writers in the field.

Gerson, Kathleen. 1993. *No Man's Land: Men's Changing Commitments to Family and Work.* New York: HarperCollins, Basic Books. Readable and convincing analysis of the changing options and expectations for men in today's constricting economy, along with the various choices men make as a result.

Rubin, Lillian B. 1994. *Families on the Fault Line: America's Working Class Speaks about the Family, the Economy, Race, and Ethnicity.* New York: HarperCollins. A book based on interviews conducted with 162 working-class and lower-middle-class families of various ethnic and racial backgrounds during a period in which the economy had contracted and shifted. The author examines the impact of this change on family life and the stress it caused.

Taylor, Ronald E. 1994. *Minority Families in the United States.* Englewood Cliffs, N.J.: Prentice Hall. A volume containing a chapter on each of the major racial minorities in the United States. Leading scholars discuss the historical and socioeconomic factors that have influenced minority family structure and processes. Attention is given to the diversity within ethnic communities as well as between them.

Thorne, Barrie, and Yalom, Marilyn (eds.). 1992. *Rethinking the Family: Some Feminist Questions.* Rev. ed. Boston: Northeastern University Press. A book that raises and addresses thorny questions, such as why women in industrialized societies will continue to choose to have children when their parenting is seldom or meagerly supported by advanced capitalistic societies.

Weston, Kath. 1991. *Families We Choose: Lesbians, Gays, Kinship.* New York: Columbia University Press. A book based on the premise that we create our own families; families are those persons whom we actively choose to include as members. This theme is specifically applied to lesbians and gay men as they create families through involved decision making.

CHAPTER 12

Education and Religion

Outline

This chapter examines two institutions, education and religion. Both are central components of our cultural heritage and have profound effects on us as individuals and on the norms and values of our society. Most Americans are directly and personally affected by these institutions: Ninety-nine percent of people in the United States have been to school, and 71 percent belong to a church or synagogue. Even those who do not go to school or church live in a society that is shaped by the norms and values of these two institutions.

Education in the United States: Theoretical Views

The **educational institution** is the social structure concerned with the formal transmission of knowledge. It is one of our most enduring and familiar institutions. Nearly 3 of every 10 people in the United States are involved in education on a daily basis as students or staff. As former students, parents, or taxpayers, all of us are involved in education in one way or another.

What purposes are being served by this institution? Who benefits? Structural-functional and conflict theories offer two different perspectives on these questions.

Structural-Functional Model

A structural-functional analysis of education is concerned with the consequences of educational institutions for the maintenance of society. It points out both how education contributes to the maintenance of society and how educational systems can be forces for change and conflict.

The Functions of Education

The educational system has been designed to meet multiple needs. Major manifest (intended) functions of education include cultural reproduction, social control, assimilation, training and development, selection and allocation, and promotion of change:

- *Cultural reproduction.* Schools transmit society's culture from one generation to the next by teaching the ideas, customs, and standards of the culture. We learn to read and write our language, we learn the pledge of allegiance, and we learn history. In this sense, education builds on the past and conserves traditions.
- *Social control.* Second only to the family, schools are responsible for socializing the young into patterns of conformity. By emphasizing a common culture and instilling habits of discipline and obedience, the schools are an important agent for encouraging conformity.
- *Assimilation.* Schools function to assimilate persons from diverse backgrounds. By exposing students from all ethnic backgrounds, all regions of the country, and all social backgrounds to a common curriculum, schools help create and maintain a common cultural base.

The **educational institution** is the social structure concerned with the formal transmission of knowledge.

In all societies, education is an important means of reproducing culture. In addition to neutral skills such as reading and writing, children learn many of the dominant cultural values. In pluralistic societies, the cultural heritage children learn at home may not be consistent with the values and history they learn at school. Native American children, taught by their parents never to embarrass a peer, are expected to correct one another publicly in the classroom; U.S. history texts may describe Indian-Anglo violence but fail to examine the factors that provoked the tribes to warfare. Value conflicts such as these are among the factors contributing to a high school dropout rate among Native Americans that is almost three times higher than it is among whites.

- *Training and development.* Schools teach specific skills—not only technical skills such as reading, writing, and arithmetic, but also habits of cooperation, punctuality, and obedience.
- *Selection and allocation.* Schools are like gardeners; they sift, weed, sort, and cultivate their products, determining which students will be allowed to go on and which will not. Standards of achievement are used as criteria to channel students into different programs on the basis of their measured abilities. Ideally, the school system ensures the best use of the best minds. The public school system is a vital element of our commitment to equal opportunity.
- *Promotion of change.* Schools also act as change agents. Although we do not stop learning after we leave school, new knowledge and technology are usually aimed at schoolchildren rather than at the adult population. In addition, the schools promote change by encouraging critical and analytic skills and skepticism. Schools, particularly colleges and universities, are also expected to produce new knowledge.

Latent Functions and Dysfunctions

In spite of its many positive outcomes, a system as large and all-encompassing as education is bound to have consequences that are either unintended or actually negative. They include generation gaps, custodial care, youth subculture, rationalization of inequality, and perpetuation of inequality:

- *Generation gap.* As schools impart new knowledge, they may drive a wedge between generations. Courses in sociology, English, history, and even biology expose students to ideas different from those of their parents. What students learn in school about evolution, cultural relativity, or the merit of socialism may contradict values held by their parents or their religion.
- *Custodial care.* Compulsory education has transformed schools into settings where children are cooped up seven to eight hours a day, five days a week, for nine months of the year (Bowles 1972). Young people are kept off the streets, out of the

labor force, and, presumably, out of trouble in small groups dispersed throughout communities in special buildings designed for close supervision. This enables their elders to command higher wages in the labor market and relieves their parents of the responsibility of supervising them.

- *Youth subculture.* By isolating young people from the larger society and confining them to the company of others their own age, educational institutions have contributed to the development of a unique youth subculture.
- *Rationalization of inequality.* One of the chief consequences of life in the schools is that young people learn to expect unequal rewards on the basis of differential achievement. Schools prepare young people for inequality. Some consider this preparation undesirable in that it leads young people to believe that inequality is a fair response to unequal abilities.
- *Perpetuation of inequality.* The most significant criticism launched against education is that schooling benefits some groups more than others. Abundant evidence exists that ascriptive characteristics of students (race, sex, and social class) have an impact on how students are treated in school. The evidence supports the conclusion that schools perpetuate inequality and function to maintain and reinforce the existing social-class hierarchy. This criticism is at the heart of the conflict theory of education.

Summary

A structural-functional analysis begins with the premise that any ongoing institution of society must be contributing to the maintenance of society. The enumeration of the functions of education clarifies what some of these contributions are. Although there are unanticipated side effects, both positive and negative, functionalists tend to concentrate on how education benefits society and individuals.

Conflict Model

Conflict theories of education look much like structural-functional theories, except in their value judgments on the final product. Conflict theorists agree that education reproduces culture, socializes young people into patterns of conformity, sifts and sorts, and rationalizes inequality. Because conflict theorists see the social structure as a system of inequality designed to benefit the rich at the expense of the rest of us, however, naturally they see any institution that reproduces this culture in a negative light. Three of the major conflict arguments are summarized here.

Education as a Capitalist Tool

Some conflict theorists argue that mass education developed because it benefited the interests of the capitalist class (Bowles & Gintis 1976). To support this argument, these theorists point to the schools' **hidden curriculum,** which socializes young people into obedience and conformity. This curriculum—learning to wait your turn, follow the rules, be punctual, and show respect—prepares young people for life in the industrial working class (Dale 1977).

The **hidden curriculum** socializes young people into obedience and conformity.

This argument has generated a great deal of controversy. Certainly, capitalists and industry have tried to affect the content of schooling in a variety of ways supportive of their interests. Most recently this has taken the form of demanding more basic skills—better reading, writing, and arithmetic skills—rather than more obedience. Nevertheless, the school system does prepare young people to accept inequality and hierarchy, and this does make it easier for their employers to control them.

Credentialism

One supposed outcome of free public education is that merit will triumph over origins, that hard work and ability will be allowed to rise to the top. Conflict theorists, however, argue that the shift to educational credentials as the mechanism for allocating high-status positions has had little impact on equalizing economic opportunity. Instead, a subtle shift has taken place. Instead of inquiring who your parents are, the prospective employer asks what kind of education you have and where you got it. Because these educational credentials are highly correlated with social-class background, they serve to keep undesirables out. Conflict theorists argue that educational credentials are mere window dressing; apparently based on merit and achievement, credentials are often a surrogate for social-class background. The use of educational credentials to measure social origins and social status is called **credentialism** (Collins 1979).

Credentialism is the use of educational credentials to measure social origins and social status.

Reproducing Inequality

In the modern world, the elite cannot directly ensure that their children stay members of the elite. To pass their status attainments on to their children, they must provide their children with appropriate educational credentials (Robinson 1984). To an impressive extent, they are able to do so: Students' educational achievements are very closely related to their parents' social status.

How does this happen? Do schools discriminate? Or, in a fair and open competition, do students with the most resources win? To answer these questions, we need to look at the processes that take place within the schools.

Inequality and the Schools

The central concern of the sociology of education in the last 40 years has been the link between education and stratification (Hallinan 1988). This concern exists on both micro and macro levels. On the micro level, we want to know what happens to individual children—how the school experiences of working-class and middle-class children differ. On a macro level, we want to understand whether the structure of the schools to which we send our children affects their learning.

Race, Ethnicity, and Schooling

In all societies, education is an important means of reproducing culture. In addition to neutral skills such as reading and writing, children learn many of the dominant cultural values. In the United States, this means two things: First, school curricula have typically focused on the history, art, literature, and scientific contributions of Western civilization: second, the way that students are taught has generally been designed to accommodate middle-class European American learning styles. In a pluralistic society such as ours, the cultural heritage children learn at home may not be consistent with the values and history they learn at school. For example, Native American children, taught by their parents never to embarrass a peer, are suddenly expected to correct one another publicly in the classroom. U.S. history texts may describe Indian-Anglo violence but fail to examine the factors that provoked Native American nations to warfare; they fail to mention the waves of anti-Chinese violence in the United States toward the turn of the twentieth century, the removal of Japanese Americans to relocation camps during World War II, or the rich cultural

Studies show that by third grade, one can predict with reasonable accuracy which students will successfully complete high school and which will not. Studies also show that dropouts will come disproportionately from the lower classes and from those racial and ethnic groups whose values, history, and experiences have been systematically omitted or distorted in the curricula of our schools. For this reason and because no account of U.S. history, art, or literature is complete without it, state and local school boards are increasingly requiring that K–12 students take classes that give appropriate attention to multiculturalism.

and economic presence of Mexicans who lived in the Southwest long before settlers from the United States arrived. Likewise, the contributions of African civilizations to modern science, mathematics, and engineering are ignored. Not surprisingly, minority students and parents increasingly ask the question "Education, for whom?" In response, more and more school districts and universities are adopting multicultural education programs. Educators hope that such programs will provide a potential solution to the joint problems of racism and the disproportionately high dropout rates characteristic of U.S. minority students.

THINKING CRITICALLY

Do you think colleges and universities should require all students to take at least one course dealing with multiculturalism or diversity? Why or why not?

Social Class and Schooling

Children from disadvantaged backgrounds are likely to experience economic hardships that work against them in all their daily experiences in school. They are unlikely to have a set of encyclopedias and a home computer. Furthermore, it is likely that their parents will be too caught up in the struggles of day-to-day living to have the time or energy to help them with their studies. Paying the bills may be more important than trying to improve their children's SAT scores.

The differences are far more subtle than simple economics. Poorly educated parents, for example, are likely to give more attention to conformity than to independent thinking when socializing their preschoolers (Alexander et al. 1987). Further, they are less likely to attend parent/teacher conferences, are less comfortable talking to their children's teachers, and are less able and often less willing to help their children with their schoolwork. In addition, children of the middle and upper-middle classes have more of what has been called **cultural capital**—social assets, such as familiarity and identification with elite culture (Bourdieu 1973). They are more likely to have been introduced to art, music, and books at home and to define themselves as cultured people. This doesn't mean they all prefer Beethoven to Pearl Jam, but it does mean that they accept books and reading as a natural and important part of life. It also means that they will dress and behave in a socially approved manner. The cultural capital will help

Cultural capital refers to social assets such as familiarity and identification with elite culture.

them to do well in school (DiMaggio & Mohr 1985; Farkas et al. 1990; Teachman 1987), and it will influence how effectively they are able to translate their education into occupational success.

Social Class and Experiences in School

To understand how social-class background translates into educational experiences, we need to look at two processes in the schools: cognitive development and tracking. Each is critical to eventual educational attainment, and each has been shown to be affected by social-class background.

Cognitive Development

One of the major processes that takes place in schools is, of course, that students learn. When they graduate from high school, many can type, write essays with three-part theses, and even differentiate equations. In addition to learning specific skills, they also undergo a process of cognitive development in which their mental skills grow and expand. They learn to think critically, to weigh evidence, to develop independent judgment. The extent to which this development takes place is related to both school and home environments.

An impressive set of studies demonstrates that cognitive development during the school years is enhanced by complex and demanding work without close supervision and by high teacher expectations. Teachers and curriculums that furnish this setting produce students who have greater intellectual flexibility and higher achievement scores. They are also more open to new ideas, less authoritarian, and less prone to blind conformity (Miller, Kohn, & Schooler 1985, 1986).

Unfortunately, the availability of these ideal learning conditions varies by students' social class. Studies show that teachers are most demanding when they are of the same social class as their students: The greater the difference between their own social class and that of their pupils, the more rigidly they structure their classrooms and the fewer the demands they place on their students (Alexander, Entwisle, & Thompson 1987).

Because students who are assigned to lower-track classes receive few rewards for their academic efforts from either parents or teachers, many will quit school and seek their rewards from delinquent peers. In the absence of smaller schools, more individualized instruction for both good and poor students, and enough flexibility in the curriculum to meet the needs of students who for one reason or another can no longer live at home, campaigns to improve the "quality" of education are likely to increase the number of disadvantaged students who drop out of both high school and society.

As a result, students learn less when they are from a lower social class than their teacher. Because the social-class gap tends to be largest when youngsters are the most disadvantaged, this process helps to keep them disadvantaged.

Tracking

When students enter first grade, they are sorted into reading groups on the basis of ability. This is just the beginning of a pervasive pattern of stratification in the schools. By the time they are out of elementary school, some students will be directed into college preparatory tracks, others into general education (sometimes called vocational education), and still others into remedial classes.

Tracking is the use of early evaluations to determine the educational programs a child will be encouraged to follow. Ideally, tracking is supposed to benefit both the gifted and the slow learners. By having classes that are geared to their levels, both should learn faster and should benefit from increased teacher attention. Instead, one of the most consistent findings from education research is that assignment to a high-ability group has positive effects whereas assignment to a low-ability group is more likely to have negative effects (Hallinan & Sorenson 1986; Shavit 1984).

An important reason students assigned to low-ability groups learn less is because they are taught less. They are exposed to less material, asked to do less homework, and, in general, not given the same opportunities to learn. One recent study found that the average student in high-track English classes was asked to do an average of 42 minutes of homework a night; in low-track English classes, the average was 13 minutes (Oakes 1985).

Less-formal processes also operate. Students who are assigned to high-ability groups, for instance, receive strong affirmation of their academic identity; they find school rewarding, have better attendance records, cooperate better with teachers, and develop higher aspirations. The opposite occurs with students placed in low-ability tracks. They receive fewer rewards from their efforts, their parents and teachers have low expectations for them, and there is little incentive to work hard. Many will cut their losses and look for self-esteem through other avenues such as athletics or delinquency (Rosenberg, Schooler, & Schoenbach 1989).

Recent studies show, however, that the effects of tracking vary across schools (Hallinan 1994). Between-track inequality is less pronounced, for instance, in Catholic schools and schools that permit more mobility between tracks. Where teachers are more optimistic about the possibility of remediation and schools continue to place academic demands on students in noncollege tracks, evidence suggests that between-track differences in academic performance are reduced. Importantly, the narrowing of the gap seems to occur because low-track students are brought up and not because high-track students are held down (Gamoran 1992).

SUMMARY.   There is no doubt that tracking benefits those in the high-ability groups and generally lowers the achievements of those in low-ability groups, nor is there any doubt that track assignment is related to social class. The result is that tracking plays a critical role in reproducing social class. If this is true, why do we keep doing it? The answer from a conflict perspective is obvious: It perpetuates the current system of inequality. From a structural-functional viewpoint, the answer is that it facilitates administration. A first-grade teacher cannot teach reading to 35 students simultaneously, so she must divide them into groups; homogeneous ability groups are easier to work with. Tracking (stratification) is thus necessary and justifiable, according to this view.

> **Tracking** occurs when evaluations made relatively early in a child's career determine the educational programs the child will be encouraged to follow.

focus on

American Diversity

WHAT DO IQ TESTS MEASURE?

- How many legs does a Kaffir have?
- Who wrote *Great Expectations*?
- Which word is out of place?
  sanctuary—nave—altar—attic—apse
- If you throw the dice and 7 is showing on top, what is facing down? 7—snake eyes—boxcars—little joe's—11

If you have answered two, Dickens, attic, and 7, then you get the highest possible score on this test. What does that mean? Does it mean that you have genetically superior mental ability, that you read a lot, that you shoot craps? What could you safely conclude about a person who got only two questions right?

The standardized test is one of the most familiar aspects of life in the schools. Whether it is the California or the Iowa Achievement Test, the SAT or the ACT, students are constantly being evaluated. Most of these tests are truly achievement tests; they measure what has been learned and make no pretense of measuring the capacity to learn. IQ tests, however, are supposed to measure the innate capacity to learn—mental ability. On these tests, African American, Hispanic, and Native American students consistently score below Anglo students, and working-class students score substantially below middle-class students. The obvious question is whether these tests are fair measures. Are African American, Hispanic, Native American, and working-class youths lower in mental ability than middle-class or Anglo youths?

Before we can answer this question, we must first ask another: What is mental ability? Most scholars recognize that it is a combination of genetic potential and prior social experiences. It is an aspect of personality, "the capacity of the individual to act purposefully, to think rationally and to deal effectively with his environment" (Wechsler 1958, 7).

Thus, reasoning capacity is not culture-free; it is determined by the opportunities to develop it.

Do questions such as those that opened this section measure any of these things? No. We can all imagine people who act purposefully, think rationally, and deal effectively with the environment but do not know who wrote *Great Expectations* and are ignorant about dice or church architecture. These people may be foreigners, they may have lacked the opportunity to go to school, or they may have come from a subculture where dice, churches, and nineteenth-century English literature are not important.

For this reason, good IQ tests try to measure reasoning ability as well as knowledge. These nonverbal tests are supposed to measure the ability to think and reason independently of formal education. Examples of items from such a nonverbal test are reproduced in Figure 12.1. Do these nonverbal tests achieve their intention? Do they measure the ability to reason independently of years in school, subcultural background, or language difficulties? Again, the answer seems to be no.

There are two ways in which these tests are not culture-free. The first is that they reflect not only reasoning and knowledge but also competitiveness, familiarity with and acceptance of timed tests, rapport with the examiner, and achievement aspiration. Students who lack these characteristics may do poorly even though their ability to reason is well developed.

The more serious fault with such nonverbal tests is their underlying assumption. Reasoning ability is not independent of learning opportunities. How we reason, as well as what we know, depends on our prior experiences. The deprivation studies of infant monkeys and hospitalized orphans (see Chapter 3) demonstrate that mental and social retardation occur as a result of sensory deprivation. Just as the body does not develop fully without exercise, neither does the mind. Thus, reasoning capacity is not culture-free; it is determined by the opportunities to develop it. For this reason, there will probably never be an IQ test that will not reflect the prior cultural experiences of the test taker.

1. From the array of four, choose that form that is identical to the target form.

Target form

 (a)    (b)

 (c)    (d)

FIGURE 12.1
Culture-Free Intelligence Tests?
What can we conclude about your intelligence from your score on this simple test? Does a high score mean that you are naturally intelligent or have some of your experiences in life and in school prepared you for these kinds of problems? Increasingly, scholars believe that it is impossible to make an intelligence test that is free of cultural influences.

SOURCE: From *Frames of Mind: The Theory of Multiple Intelligences* by Howard Gardner © 1983. Reprinted by permission of Basic Books, Inc., Publishers, a division of HarperCollins Publishers, Inc.

2. From the array of four, choose that form which is a rotation of the target form.

Target form

 (a)   (b)

 (c)   (d)

3. (For a, b, c,) indicate whether the second form in each pair is a rotation of the first or is a different form.

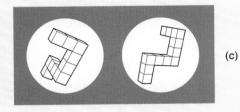 (a)

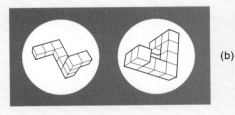 (b)   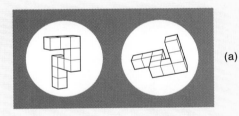 (c)

Answers: 1. d   2. d   3(a) same   3(b) same   3(c) different

THINKING

CRITICALLY

Given the same levels of funding now available for public education, how would you organize elementary and secondary classrooms to best meet the needs of all students? What would be the manifest functions of your system? The latent functions? The dysfunctions?

The Excellence Campaigns and Inequality

While sociologists criticize the educational institution for its failure to equalize opportunity for all social classes, the general public is most likely to criticize the education institution for its lack of quality. Excellence rather than equity is the major public concern. This concern is not unwarranted.

In 1983, an 18-member National Commission on Excellence in Education issued a report that was extremely critical of U.S. education. The report indicated that 13 percent of all 17-year-olds and as many as 40 percent of minority youths are functionally illiterate. In a comparison of U.S. students with students from 21 other nations, Americans scored the worst on 7 of 19 achievement tests and never came in either first or second. The commission argued that the problem was caused not by factors beyond our control but simply by lack of insight and will. The solutions recommended include (1) a more demanding sequence of basic courses; (2) longer school days and school years, and (3) higher standards for school achievement.

Will more basics, more time, and higher standards improve the quality of education? The answer is, on the average, yes. As with almost everything else in the school system, however, the greatest benefits seem to go to those who already are achieving.

The implementation of higher standards, especially the implementation of standard competency tests for graduation or promotion, is bound to have disproportionate effects on disadvantaged students. If these recommendations are directly implemented, the already high dropout rate (14 percent) is expected to soar. The result will be to tighten rather than loosen the strong chain of interconnected problems that reproduces poverty and disadvantage (McDill, Natriello, & Pallas 1986).

The most effective single way to raise the educational performance of children from every socioeconomic and cultural background is probably also the least expensive: raise teacher expectations. Teachers and curricula that couple complex, demanding schoolwork with opportunities for independent learning, extracurricular activities, and ample formal and informal rewards produce students with greater intellectual flexibility and higher achievement test scores.

Recent research suggests several strategies for increasing the scores and achievement of all of our students without losing half along the way. First, raise teacher expectations. This inexpensive mechanism has been found to raise attendance and performance levels of both good and poor students. This change should especially benefit poor students, many of whose teachers have such low expectations for them that they are given a passing grade for attendance or for not causing trouble. Second, reduce the size of schools. In smaller schools, teachers have greater ability to influence students and, perhaps as important, students have greater ability to influence their schools. They have more opportunities for extracurricular activities that will enhance their attachment to school. Third, build some flexibility into the system so that youth who work, marry, or bear children can participate. Fourth, use grants and loans to encourage college students from disadvantaged backgrounds to enter the teaching profession.

THINKING CRITICALLY

Given growing concerns about guns and violence in school, do you think the strategies listed here would help? Why or why not? If not, what would you do?

College: Who Goes and What Does It Do for Them?

Going to College

If your grandparents attended college, they were part of the educational elite. Even in 1950, only half of the 17-year-olds in the United States had graduated from high school. However, the period following World War II saw a tremendous growth in high school and college education, and today nearly 90 percent of young adults have a high school diploma.

As Figure 12.2 shows, all segments of the population have been affected by this expansion in education, but significant differences still remain. Because the data reported in Figure 12.2 are for the population over 25, they reflect not only the experiences of today's young people, but also those of people who went to school 20, 40, and even 60 years ago. Has the educational gap been eliminated among the current generation? Minority children are still disproportionately likely to attend schools that are poorly funded and study from textbooks that malign or ignore their people, their culture, and their history. Coupled with the inadequate attention usually given to English as a Second Language programs, the end product is a 55 to 70 percent high school completion rate among black, Hispanic, and Native American adults over the age of 25.

Today more than 40 percent of all high school graduates between the ages of 18 and 24 are enrolled in colleges or universities. This means that slightly more than half are not enrolled. Twenty percent of high school graduates aged 25–29 have graduated from college; this means that 80 percent have not. Thus, although you and many of the people you know have gone to college, you are in the minority.

Until just a few years ago, non-Hispanic white males were the group most likely to be enrolled in college, but this has changed (see Figure 12.3). Today, non-Hispanic white females are the group most likely to be in college. Except for professional and doctoral degrees, which still go mainly to white men, white female collegiates are also the most likely to graduate (U.S. Bureau of the Census 1996a). At least among women, racial/ethnic differences in college attendance are smaller today than they used to be (see Figure 12.3). Nevertheless, non-Hispanic whites—both women and men—are far more likely to attend and to graduate from college than are African Americans or Hispanics (U.S. Bureau of the Census 1996a).

FIGURE 12.2
Educational Achievement of Persons 25 and Older by Race and Ethnicity, 1940–1994
Among whites, the proportion of adults graduating from high school has tripled in the last 50 years; among blacks, the increase in education has been even more dramatic. Nevertheless, blacks and Hispanics continue to be disadvantaged in terms of quantity of education.

SOURCE: U.S. Bureau of the Census 1975, 380; U.S. Bureau of the Census 1989a, 1989b, and 1989c; U.S. Bureau of the Census 1993a and 1993g.

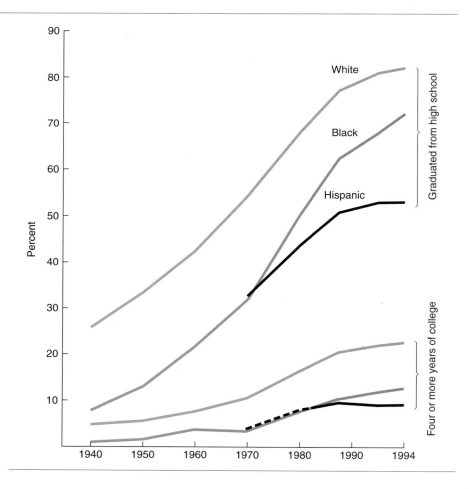

Impact of Education on Earnings and Occupation

For many people, the chief objective of college education is the attainment of a good job and higher earnings. This objective has been and continues to be realistic. In 1995, the average college graduate, male and female, earned nearly double the earnings of the average high school graduate. Some of this additional income is related to background characteristics rather than years in school; that is, people who complete college generally have higher high school test scores and higher-status parents. These factors would have raised their income even if they had not completed college. Nevertheless, however, college continues to pay a handsome profit. As Table 12.1 shows, this profit varies by sex. At every educational level, male earnings exceed female earnings.

What's the Bottom Line? What Will College Do for Me?

A major incentive for college attendance is the belief that it will pay off economically. The evidence clearly supports this belief. As Table 12.2 shows, people who graduate from college get better jobs, experience lower unemployment, and have higher incomes than those with less education. In this sense, your investment in a college education will pay off.

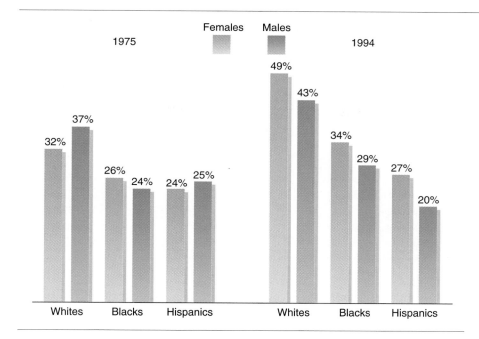

FIGURE 12.3
Percentage of High School Graduates Ages 18 to 21 Enrolled in College, by Race, Ethnicity, and Sex, 1975 and 1984
Comparisons by sex, race, and ethnicity show increasing similarity in the likelihood that high school graduates from each category will attend college.
SOURCE: U.S. Bureau of the Census 1996, Statistical Abstract of the United States.

It can pay off in other ways too. It is a value judgment to say that a college education will make you a better person, but it is a value judgment that the majority of college graduates are willing to make. Survey after survey demonstrates that people feel very positively about their college education, believing that it has made them better and more tolerant people (Bowen 1977).

Whether it makes you a better person or not, a college education is likely to have a lasting effect on your knowledge and values. If you finish college, you will sit through 30 to 45 different courses. Even the least dedicated students are bound

TABLE 12.1
Economic Returns of Additional Education, 1995
For all groups, there is a substantial economic payoff for educational attainment after high school. Full-time, full-year workers who completed college annually earned nearly twice as much as high school graduates. The impact of postgraduate education on earnings is especially strong for African Americans.

| | MEAN MONTHLY INCOME | | | | |
	4 Years High School	4 Years College	Master's Degree	% Income Increase from High School to Bachelor's	% Income Increase from Bachelor's to Master's
All	$1,380	$2,625	$3,411	90%	30%
Male	1,812	3,430	4,298	89	25
Female	1,008	1,809	2,505	79	38
White	1,422	2,682	3,478	89	30
Black	1,071	2,333	2,834	118	21
Hispanic	1,106	2,186	2,605	98	19

SOURCE: U.S. Bureau of the Census, 1996 Statistical Abstract of the United States.

TABLE 12.2
Socioeconomic Consequences of Education, 1994–1995
Education pays off in terms of good jobs and good income. The differences shown in this table, however, are not all directly related to additional years of schooling; that is, people who graduated from college have, on the average, higher high school grades and more background advantages than those who graduated from high school only. These circumstances may have as much to do with their achievements as do their additional years of schooling.

Education	Percentage with Managerial or Professional Occupation	Percentage Unemployed	Median Income of Full-Time Full-Year Workers (Male)
Less than 4 years of high school	1.6	10.0	$24,377
Exactly 4 years of high school	12.9	5.2	31,081
1–3 years of college	22.9	4.5	35,639
4 or more years of college	62.5	2.5	61,008

SOURCE: U.S. Bureau of the Census, 1996 Statistical Abstract of the United States, nos. 641, 649, 728.

to learn something from these courses. In addition, students learn informally. Whether you go to college in your hometown or across the country, college will introduce you to a greater diversity of people than you're likely to have experienced before. This diversity will challenge your mind and broaden your horizons. As a result of formal and informal learning, college graduates are more knowledgeable about the world around them, more tolerant and less prejudiced, more active in public and community affairs, less traditional in their religious and gender-role beliefs, and more open to new ideas than those who don't have a college degree (Weil 1985; Funk & Willits 1987).

Higher education is an excellent investment, one that would be justified on the basis of the monetary returns alone. When one also considers the nonmonetary rewards—the contributions to intellectual and social growth—there is "no doubt that American higher education is well worth what it costs" (Bowen 1977, 448).

A college education goes far beyond training for a specific career—or it should. The average American in the twenty-first century will experience several major career changes over his or her lifetime. Consequently, college should encourage creative and critical thinking and broaden one's view of the world. Some of this learning will come from elective courses such as sociology. Most college graduates do in fact believe that their college education makes them more open to new ideas and other cultures.

The Sociological Study of Religion

Unlike education, which we are forced by law to take part in, we have a choice about participation in religious organizations. Nevertheless, most people in the United States choose to participate, and religion is an important part of social life. It is intertwined with politics and culture, and it is intimately concerned with integration and conflict. At the micro level, sociologists examine the consequences of religious belief and involvement for the lives of ordinary individuals. At the macro level, sociologists examine how society affects religion and how religion affects society. Of particular concern is the contribution of religion to social order and social change.

What is Religion?

How can we define *religion* so that our definition includes the contemplative meditation of the Buddhist monk, the speaking in tongues of a modern Pentecostal, the worship of nature in Native American cultures, and the formal ceremonies of the Catholic church? Sociologists define **religion** as a system of beliefs and practices related to sacred things that unites believers into a moral community (Durkheim [1915] 1961, 62). This emphasis on the sacred allows us to include belief systems that invoke supernatural forces as explanations of earthly struggles (Stark & Bainbridge 1979, 121), as well as those that attribute personality and will to nature. It does not include, however, belief systems such as Marxism or science that do not emphasize the sacred.

Sociologists who study religion treat it as a set of values. As with monogamy or democracy, the concern is not whether the values are true or false. The scientific study of religion does not ask whether God exists, whether salvation is really possible, or which is the true religion. Rather it examines the ways in which culture, society, and class relationships affect religion and the ways in which religion affects individuals and social structure.

Religion is a system of beliefs and practices related to sacred things that unites believers into a moral community.

Ritual is an important part of all religions. Ritual occasions such as bar mitzvahs and first communions bring people together as a community of believers and reaffirm their shared values. Even people who are not particularly devout find the traditional ritual comforting. For many of us, as for these Jews, our religion is also part of our ethnic heritage; participating in religious ritual is an important means of tying us to our cultural heritage.

Why Religion? Some Theoretical Answers

Religion is a fundamental feature of all societies. Whether primitive or advanced, each society has forms of religious activity and expressions of religious behavior. Why? The answer appears to lie in the fact that every individual and every society must struggle to find explanations for events and experiences that go beyond personal experience. The poor man suffers in a land of plenty and wonders, "Why me?" The woman whose child dies in its sleep wonders, "Why mine?" The community struck by flood or tornado wonders, "Why us?" Beyond these personal dilemmas, people may wonder why the sun comes up every morning, why there is a rainbow in the sky, and what happens after death.

Religion helps us interpret and cope with events that are beyond our control and understanding; tornadoes, droughts, and plagues become meaningful when they are attributed to the workings of some greater force (Spilka, Shaver, & Kirkpatrick 1985). Beliefs and rituals develop as a way to control or appease this greater force, and eventually they become patterned responses to the unknown. Rain dances may not bring rain, and prayers may not lead to good harvests, but both provide a familiar and comforting context in which people can confront otherwise mysterious and inexplicable events. Regardless of whether they are right or wrong, religious beliefs and rituals help people cope with the extraordinary events they experience.

Within this general sociological approach to religion, there are two distinct theoretical perspectives. One school, associated with Durkheim, sees religion as a thinly disguised worship of society, serving to create and maintain social solidarity. The second school, associated with Weber, views religion as an intellectual force that may challenge society as well as support it.

Durkheim: Religion as the Worship of Society

Durkheim's approach to the study of religion is based on the structural-functional perspective. He assumed that if religion is universal, then it must meet basic needs of society; it must serve important functions. Durkheim began his analysis of religion by trying to identify what was common to all religions.

The Elementary Forms of Religion

Durkheim (1915/1961) concluded that all religions share three elements, which he called the elementary forms of religion: (1) a distinction between the sacred and the profane, (2) a set of beliefs, and (3) a set of rituals.

THE SACRED AND THE PROFANE.   All religions tend to divide human experience into the sacred and the profane. The **profane** represents all that is routine and taken for granted in the everyday world, things that are known and familiar, that we can control, understand, and manipulate. The **sacred,** by contrast, consists of the events and things that we hold in awe and reverence—what we can neither understand nor control.

In the cultures of premodern societies, a large proportion of the world is viewed as sacred. Many events are beyond control and manipulation. As advances in human knowledge increase a society's ability to explain and even control what was previously mysterious, fewer and fewer events require supernatural explanations; less is held sacred. This process of transferring things, ideas, or events from the sacred to the profane is called **secularization.** Science and technology have been major contributors to

The **profane** represents all that is routine and taken for granted in the everyday world, things that are known and familiar and that we can control, understand, and manipulate.

The **sacred** consists of events and things that we hold in awe and reverence—what we can neither understand nor control.

Secularization is the process of transferring things, ideas, or events from the sacred realm to the profane.

secularization. They have given us explanations for lightning, rainbows, and death that rely on physical rather than supernatural forces.

BELIEFS, MYTHS, AND CREEDS.   A second common dimension to all religions is a set of beliefs about the supernatural. Religious beliefs center around uncertainties associated with birth, death, creation, success, failure, and crisis. They become part of the worldview constructed by culture as a rationale for the human condition and the recurrent problems experienced. As beliefs become organized into an interrelated set of assumptions about the supernatural, they form the basis for official religious doctrines.

RITUALS.   Religion is a practice as well as a belief system. It brings people together to express through ritual the things they hold sacred. In contemporary Christianity, rituals are used to mark such events as births, deaths, weddings, Christ's birth, and the resurrection. In an earlier era, when most people lived off the land and life was more uncertain, planting and harvest were occasions for important rituals in the Christian church; they are still important ritual occasions in many religions.

Like all religions, voodoo beliefs help followers to cope with events that are beyond human understanding. Through a variety of rituals, adherents attempt to control such common experiences as death, drought, and the sometimes destructive consequences of human anger, willfulness, and sexuality. As Durkheim suggests, voodoo also provides followers with a sense of belonging to a moral community. Thus, magic is not only the stereotypic label that outsiders apply to voodoo, it is also the term that voodoo in-groups use to differentiate their own rituals (which they often call "African") from those of out-groups (Brown 1992).

The Functions of Religion

Durkheim argued that religion serves functions for individuals who believe in it and for society as a whole. For individuals, the beliefs and rituals of religion offer support, consolation, and reconciliation in times of need. On ordinary occasions, many people find satisfaction and a feeling of belongingness in religious participation. This feeling of belongingness is the moral community, or community of believers, that is part of the definition of religion.

On a societal level, Durkheim argued, the major function of religion is that it gives tradition a moral imperative. Most of the central values and norms of any culture are taught and reinforced through its religions. These values and norms cease to be merely the usual way of doing things and become the only moral way of doing them. They become sacred. When a tradition is sacred, it is continually affirmed through ritual and practice and is largely immune to change.

Within the functionalist perspective, the worship of God is seen as a barely disguised worship of society (Durkheim, in O'Dea 1966, 12). Religion is seen as a means of lending supernatural authority to traditional practices, a way of giving usual practice the unchallengeable standing of supernaturally established laws.

Weber: Religion as an Independent Force

Durkheim looked at the forms of religion and asked about their consequences for individuals and society. Weber shared this interest, but he was also concerned with the processes through which religious answers are developed and how their content affects society.

For most people, religion is a matter of following tradition; people worship as their parents did before them. To Weber, however, the essence of religion is the search for knowledge about the unknown. In this sense, religion is similar to science; it is a way of coming to understand the world around us, and as with science, the answers provided may be uncomfortable; they may challenge the status quo as well as support it.

Where do the answers to questions of ultimate meaning come from? Often they come from a charismatic religious leader. **Charisma** refers to extraordinary personal qualities that set the individual apart from ordinary mortals. Because these extraordinary characteristics are often thought to be supernatural in origin, charismatic leaders are often able to be the agents of dramatic change in individuals and society. Charismatic leaders include Christ, Muhammad, and, more recently, Joseph Smith (Mormonism), and David Koresh (Branch Davidians). Such individuals give answers that often disagree with traditional answers. Thus, Weber sees religious inquiry as a potential source of instability and change in society.

In viewing religion as a process, Weber gave it a much more active role than did Durkheim. This is most apparent in Weber's analysis of the Protestant Reformation.

The Protestant Ethic and the Spirit of Capitalism

In a classic analysis of the influence of religious ideas on other social institutions, Weber ([1904–1905] 1958) argued that the Protestant Reformation paved the way for bourgeois capitalism. The values of early Protestant religion, which Weber called the Protestant ethic, included the belief that work, rationalism, and plain living are moral virtues, whereas idleness and indulgence are sinful. What happens to a person who follows this ethic? Someone who works hard, who makes business decisions on rational rather than emotional criteria (for example, firing inefficient though needy employees), and who is frugal rather than self-indulgent will grow rich. According to Weber, it was not long before wealth became an end in itself. At this point, the moral values underlying early Protestantism had become the moral values underlying early capitalism.

Charisma refers to extraordinary personal qualities that set an individual apart from ordinary mortals.

In the more than 80 years since Weber's analysis, other scholars have explored the same issues, and many have come to somewhat different conclusions. Nevertheless, 80 years of scholarship have not changed Weber's major contribution to the sociology of religion: that religious ideas can be the source of change in social institutions.

Modern Conflict Theory: Beyond Marx

Like Durkheim, Marx saw religion as a supporter of tradition. This support ranges from injunctions that the poor and oppressed should endure rather than revolt (blessed be the poor, blessed be the meek, and so on) and that everyone should pay taxes (give unto Caesar) all the way to the endorsement of inequality implied by a belief in the divine right of kings.

Marx differed from Durkheim by interpreting the support for tradition in a negative light. Marx, an atheist, saw religion as a delusion deliberately fostered by the elite—a sort of shell game designed to keep the eyes of the downtrodden on the hereafter so that they would not notice their earthly oppression. This position is hardly value-free, and much more obviously than either Weber's or Durkheim's, it does make a statement about the truth or falsity of religious doctrine.

Modern conflict theory goes beyond Marx's interpretation of religion as the opiate of the masses. The major contribution of Marxism to the analysis of religion is the idea of the dialectic—that contradictions build up between existing institutions and that these contradictions lead to change. Sometimes, contradictions between other institutions erupt into religious expression. On other occasions, the basic tension is between society and religion. This tension is addressed in the next section of the chapter.

Tension Between Religion and Society

Each religion is confronted with two contradictory yet complementary tendencies: the tendency to reject the world and the tendency to compromise with the world (Troeltsch 1931). When a religion denounces adultery, homosexuality, and fornication, does the church categorically exclude adulterers, homosexuals, and fornicators, or does it adjust its expectations to take common human frailties into account? If "it is easier for a camel to go through the eye of a needle than for a rich man to enter the kingdom of God," must the church require that all its members forsake their worldly belongings?

How religions resolve these dilemmas is central to their eventual form and character. Scholars distinguish two general types of religious organizations: church and sect. The *church* represents the successful compromisers, and the *sect* represents the virtuous outsiders.

Churches are religious organizations that have become institutionalized. They have endured for generations, are supported by and support society's norms and values, and have become an active part of society. Their involvement in society does not necessarily mean that they have compromised essential values. They still retain the ability to protest injustice and immorality. From the abolition movement of the 1850s to the Civil Rights struggle of the 1960s to the sanctuary movement for Central American refugees or to modern antinuclear movements, churchmen and women have been in the forefront of social protest. Nevertheless, churches are generally committed to working with society. They may wish to improve it, but they have no wish to abandon it.

Sects are religious organizations that reject the social environment in which they exist (Johnson 1957). Religions that reject sexual relations (Shakerism), automobiles (the Amish), or monogamy (nineteenth-century Mormonism) are examples of sects

Churches are religious organizations that have become institutionalized. They have endured for generations, are supported by and support society's norms and values, and have become an active part of society.

Sects are religious organizations that reject the social environment in which they exist.

that differ so much from society's norms that their relationships with the larger society are often hostile. They reject major elements of the larger culture and are in turn rejected by it.

The categories of church and sect are what Weber referred to as *ideal* types. The distinguishing characteristics of each type are summarized in the Concept Summary (see page 300). Although no church or sect may have all of these characteristics, the ideal types serve as useful benchmarks against which to examine actual religious organizations.

Churchlike Religions

Within the general category of religions that exhibit a low degree of tension with society and a high degree of integration with society's institutions are two major types: the ecclesia and the denomination.

Ecclesia

An **ecclesia** is a churchlike religious organization that automatically includes every member of society.

The most institutionalized of all religious structures is an **ecclesia**—a religious organization that automatically includes every member of a society. Membership comes with citizenship (Becker, in Yinger 1957, 149). The Roman Catholic church in Europe was an ecclesia during the Middle Ages; Iran and Israel today have many of the characteristics of a modern ecclesia. Ecclesiae represent the highest degree of religious institutionalization. Little tension exists between the religion and society—the religion is fully assimilated into society. The fate of the church and the fate of the nation are wrapped up in each other, and the church is vitally involved in supporting the dominant institutions of society.

Denominations

Denominations are churchlike religious organizations that have accommodated to society and to other religions.

Religious organizations that have accommodated to society and to other religions are **denominations** (Robertson 1970). Most of the largest religious organizations in the United States fit this definition: Jewish, Catholic, Lutheran, Methodist, and Episcopalian. Their clergy meet together in ecumenical councils, they pray together at commencements, and they generally adopt a live-and-let-live policy toward one another. Denominations have adjusted to the existing social structure of society. They support and are supported by the other institutional structures. This endorsement of the broad and basic fabric of the social order assures members that the ways of both their religion and their society are moral and just.

Structure and Function of Churchlike Religions

Ecclesiae and denominations tend to be formal bureaucratic structures with hierarchical positions, specialization, and official creeds specifying their religious beliefs. Leadership is provided by a professional staff of ministers, rabbis, or priests, who have received formal training at specialized schools. These leaders are usually arranged in a hierarchy from the local to the district to the state and even the international level. Religious services almost always prescribe formal and detailed ritual, repeated in much the same way from generation to generation. Congregations often function more as audiences than as active participants. They are expected to stand up, sit down, and sing on cue, but the service is guided by ceremony rather than by the emotional interaction of participants.

Generally, people are born into churchlike religions rather than being converted to them. People who change churches, who become Methodists instead of Lutherans, Catholics instead of Presbyterians, usually do so because they marry somebody of the other faith, the other church is nearer, or their friends go to the other church; changing religions is seldom an emotional experience.

Denominations tend to be large in size and to have well-established facilities, financial security, and a predominantly middle-class membership. As part of their accommodation to the larger society, Jewish, Catholic, and Protestant denominations usually allow the Scriptures to be interpreted in ways that are relevant to modern culture. Because of these characteristics, denominations are frequently referred to as *mainline churches,* a term denoting their centrality in society.

Sectlike Religion

Within the category of religious organizations that have greater tension with society, there is a great deal of variability. We can distinguish three levels of tension. First are cults, with the greatest tension, then sects, and finally established sects. The latter begin to approach institutionalization.

Cults

A religious organization that is independent of the religious traditions of society is a **cult** (Stark & Bainbridge 1979). Examples of cults in the United States are Scientology, the Moonies, and the Hare Krishna. Each of these religions is rather foreign to the Judeo-Christian tradition of the United States: They have a different God or Gods or no God at all; they don't use the Old Testament as a text.

Cults tend to arise in times of societal stress and change, when established religions do not seem adequate to explaining the upheavals that individuals experience. Because they are so alien to society's institutions, however, cults are of little assistance in helping people cope with their everyday lives. Instead, they often urge their members to alter their lives radically and to withdraw from society altogether. Because of the radical changes they demand, cults generally remain small, and many fail to survive more than a few years.

Sects

Within the general category of sectlike religions, those called sects occupy a medium position. They reject the social world in which they live, but they embrace the religious heritage of society. The Amish are an excellent example: They base their lives on a strict reading of the Bible and remain aloof from the contemporary world.

Sects often view themselves as restoring true faith, which has been mislaid by religious institutions too eager to compromise with society. They see themselves as preservers of religious tradition rather than innovators. Like the Reformation churches of Calvin and Luther, they believe they are cleansing the church of its secular associations. However offbeat it may be in comparison to mainline churches, if a religious group in the United States uses the Bible as its source of inspiration and guidance, then it is probably a sect rather than a cult.

Elizabeth Claire Prophet is the charismatic spiritual leader of the New Age religious cult Church Universal and Triumphant (CUT). A blend of Christianity, Buddhism, Hinduism, and other religions, CUT doctrine is spelled out in weekly church publications and in the 50+ books authored by Prophet, who followers believe serves as "the channel" for such deceased spiritual leaders as Jesus, Buddha, and Mark Prophet, Elizabeth's second husband. The faithful also believe in astrology, UFOs, conspiracies of scientists, doctors, and communists, karma, reincarnation, lost continents, lost planets, and doomsday. Many followers have sold all of their possessions in order to buy space in the fallout shelters CUT is constructing just outside Yellowstone National Park in preparation for the coming end of the world.

A **cult** is a sectlike religious organization that is independent of the religious traditions of society.

Established Sects

A sect that has adapted to its institutional environment has become an **established sect.** An established sect often retains the belief that it is the one true church, but it is less antagonistic to other faiths than are sects.

Whereas sects often withdraw from the world in order to preserve their spiritual purity, established sects are active participants. Frequently, the motivation for this participation in the world is to spread their message, to make converts, and to change social institutions. To the extent that they are successful at these goals, they

An **established sect** is a sect that has adapted to its institutional environment.

reduce the tension between themselves and society. To effect social change, they must have lobby groups and participate in political, economic, and educational institutions. To spread their message and make converts, they must associate with many outsiders.

The Mormons are a classic example of an established sect. In the last 160 years, they have increased their accommodation to the larger society: They have abandoned plural marriage, left the seclusion of a virtual ecclesia in Utah, and spread throughout the world seeking converts. Mormons nevertheless retain many characteristics of sectlike religions, including lack of a paid clergy and an emphasis on conversion. These characteristic organizational features are covered in the next section.

Structure and Function of Sectlike Religions

The hundreds of cults and sects in the United States exhibit varying degrees of tension with society, but all are opposed to some basic societal institutions. Not surprisingly, these organizations tend to be particularly attractive to people who are left out of or estranged from society's basic institutions—the poor, the underprivileged, the handicapped, and the alienated. For this reason, sects have been called "the church of the disinherited" (Niebuhr [1929] 1957). Not all of the people converted to sectlike religions are poor or oppressed. Many are middle-class people who are spiritually rather than materially deprived. They are individuals who find established churches too bureaucratic; they seek a moral community that will offer them a feeling of belongingness and emotional commitment (Barker 1986).

Sect membership is often the result of conversion or emotional experience. Members do not merely follow their parents into the church; they are reborn or born again. Religious services are more informal than for churches. Leadership remains largely unspecialized, and there is little, if any, professional training for the calling. The religious doctrines emphasize otherworldly rewards, and the Scriptures are considered to be of divine origin and therefore subject to literal interpretation.

CONCEPT SUMMARY	*Distinctions Between Churches and Sects*

Church and sect are ideal types against which we can assess actual religious organizations. Many religious organizations combine some characteristics of both. Nevertheless, Catholicism and Lutheranism are obviously churches, whereas the Nation of Islam has many of the characteristics of a sect.

	CHURCHES	SECTS
Degree of tension with society	Low	High
Attitude toward other institutions and religions	Tolerant	Intolerant; rejecting
Type of authority	Traditional	Charismatic
Organization	Bureaucratic	Informal
Membership	Establishment	Alienated
Examples	Catholics, Lutherans	Jehovah's Witnesses, Amish, Nation of Islam

Sects and cults share many of the characteristics of primary groups: small size, informality, and loyalty. They are relatively closely knit groups that emphasize conformity and maintain significant control over their members.

Many, if not all, of the churches in the world today started out as sects. Over the centuries, they grew and became part of the institutional structures of society. Not all sects, of course, adjust and become assimilated in this way. Some remain established sects, antagonistic to many institutions in the general society; others suffer eventual extinction.

A Case Study of Islam: Churchlike or Sectlike Religion?

Islam was founded in the seventh century A.D. by an Arab prophet named Muhammad, near Mecca, in what is now Saudi Arabia. Estimates are that it is the fastest-growing religion in the world, encompassing nearly one-fifth of the world's population. No matter where they are found, believers in Islam share a set of common beliefs. Like Christians and Jews, Muslims believe in a single all-powerful God whose word is revealed to the faithful in the Koran, a book similar to the Christian Bible or the Jewish Torah. All Muslims must follow the Five Pillars of Islam. They must: (1) profess faith in one almighty God and Muhammad his prophet; (2) pray five times daily; (3) give alms to the Muslim community and the poor; (4) fast during daylight hours during the month of Ramadan, the time when the Koran was revealed to Muhammad; and (5) if possible, make at least one pilgrimage to Mecca. Prayer is usually in a mosque (an Islamic house of worship) and is lead by an Imam (a religious scholar). There is no formal central authority. Apart from this basic doctrine, however, there is considerable variation across countries in the relationship between Islamic clergy and the government and in the interaction between followers of the faith and members of the larger community. In some nations Islam more closely resembles a church, and in others it more closely resembles a sect.

Islam as a Churchlike Religion: Iran and Egypt

Iran, with its Islamic government, is a good example of a modern-day ecclesia. Church and state are intertwined, with every citizen being bound by religious law. For instance, all Iranian women, regardless of religion or nationality, must cover their hair and all skin, save the hands and face. Failure to comply with this rule is punishable by up to 80 lashes. Since Islamic clergy came to political power in Iran on the tide of a popular revolution, there is generally little tension revealed between religion and the larger society.

In contrast, Egypt has a secular government even though 90 percent of its population belongs to the Islamic church. Although there has been a recent upsurge in antigovernment violence by radical fundamentalist sects, Islam in Egypt is still more like a denomination than a sect. Christians and Muslims share the streets and businesses in relative peace. The government tacitly allows more moderate Islamic groups, like the Islamic Brotherhood, to exist in spite of official bans. Recently, the Egyptian government has also increased public discussion of moderate Islamic values and promised to improve services to the less fortunate, suggesting a willingness of church and state to work together in supportive roles (Murphy 1994).

Islam as a Sectlike Religion: The United States

Islam in the United States can be traced back to the importation of African Muslim slaves in the eighteenth and nineteenth centuries. In the late 1800s, new Muslim immigrants arrived and began to settle in the Midwest, especially North Dakota and

Muslims throughout the world believe that, if possible, they should make at least one pilgrimage to the Sacred Mosque in Mecca. Aside from this commitment and a shared belief in the four other pillars of Islamic faith, however, there is considerable variation both within and between countries in the way that Islam is practiced and in the relationship between the church and the state.

Iowa. Today, there are an estimated 4–6 million Muslims living in the United States, a substantial proportion of whom are African American (Buchsbaum 1993).

Of the many sects of Islam practiced in the United States, perhaps the most widely recognized is that of the Nation of Islam, African American Muslims. As an established sect, they are active participants in their social environment without being antagonistic to other religions. Membership is growing most rapidly among poor and disenfranchised inner city residents, for whom Islam may provide a sense of hope and community. The Muslim emphasis on community activism—anti-drug campaigns and economic development—and on discipline and modest dress provide the sense of order and belonging commonly provided by all sects.

Islamic Fundamentalism

Recent years have seen a worldwide increase in Islamic fundamentalism. As is true of fundamentalist churches in the United States, fundamentalist sects within Islam have been particularly appealing to individuals who lack economic and political power. Islamic fundamentalists have called for a rejection of the excesses and corruption of modern, secular culture and a return to "true" religious principles; only the most radical Islamic fundamentalist sects have advocated violence as a means of restoring religious values and law. Most Muslims, in fact, say the concept of the *jihad*—the holy war— ". . . refers not to battlefield wars, but to the inner spiritual struggle of Muslims for self-control in order to do good." (Sudo 1993: 5) Islamic fundamentalism does not uniformly endorse violence nor is it a single unified movement or sect. It varies from country to country and includes both religious and political elements; acts of violence do not derive from core Islamic fundamentalist beliefs but from the problematic behavior of a limited number of individuals, sects, and political regimes (Gordon 1993).

Religion in the United States

When asked what religion they belong to, only 5 percent of the people in the United States say they belong to no church. Most people are able to identify themselves not only as religious but also as affiliated with some particular religious organization. Most (58 percent) call themselves Protestants, but 27 percent are Catholics and 3 percent are Jews (see Table 12.3). Despite their differences, the three major religions in the United States embrace a common Judeo-Christian heritage. They accept the Old Testament, and they worship the same God. They rely on a similar moral tradition (the Ten Commandments, for example), which reinforces common values. This common religious heritage supplies an overarching sense of unity and character to U.S. society—providing a framework for the expression of our most crucial values concerning family, politics, economics, and education.

U.S. Civil Religion

Civil religion is the set of institutionalized rituals, beliefs, and symbols sacred to the U.S. nation.

Americans also share what has been called a civil religion (Bellah 1974, 29; 1985). **Civil religion** is a set of institutionalized rituals, beliefs, and symbols sacred to the U.S. nation. These include giving the pledge of allegiance and singing the national anthem,

TABLE 12.3
Religious Affiliation in the United States, 1997
Although nearly 90 percent of Americans call themselves Protestant, Catholic, or Jewish, there are more than 200 religious organizations in the United States and as many as 1,000 cults and sects.

		Percent Identifying As
Catholic		27%
Protestant		58
Baptist	19%	
Methodist	9	
Lutheran	6	
Presbyterian	5	
Church of Christ	3	
Episcopalian	2	
Other Protestant	14	
Jewish		3
Mormon		1
Other		6
None		5

SOURCE: http://www.gallup.com/poll/news/970329.html.

as well as folding and displaying the flag in ways that protect it from desecration. In many U.S. homes, the flag or a picture of the president is displayed along with a crucifix or a picture of the Last Supper.

Civil religion has the same functions as religion in general: It is a source of unity and integration, providing a sacred context for understanding the nation's history and current responsibilities (Wald 1987). We have made liberty, justice, and freedom sacred principles, and, as a result, the U.S. way of life—our economic and political system—has become not merely the usual way of doing things but also the only moral way of doing them, a way of life that is blessed by God.

A great deal of the sacredness of U.S. civil religion depends on the tie many people believe exists between God and the United States of America. The motto on our currency, our pledge of allegiance, and our national anthem all bear testimony to the belief that God has blessed our nation and that the United States operates "under God" with God's direct blessing. For many U.S. people, civil religion and regular religion are virtually inseparable.

Trends and Differentials in Religiosity

Many scholars have argued that religion is playing an increasingly less important role in social life in Western societies. The evidence for this point of view, however, is mixed. In 1776, it is estimated that only 10 to 12 percent of the U.S. population belonged to a church (Stark & Finke 1988). In 1997, approximately 67 percent of all U.S. people belonged to a church or synagogue. So in contrast to 1776, the United States is a very religious society. Since the 1950s, however, there has been some decline in individual religiosity (see Table 12.4). Church attendance appears to have changed little, although some studies suggest church attendance may actually be 50 percent lower than opinion polls indicate (Hadaway, Marler, & Chaves 1993). Furthermore,

TABLE 12.4
Changing Religious Commitment, 1947–1997
During the last 40 years, there has been little decline in outward religious observance. There has, however, been a substantial drop in the proportion who say that religion is very important to their lives, and there has been a sharp decrease in the proportion who think that the Bible is the actual word of God.

	1947–1952	1993–1997
Belong to a church or synagogue	76%	67%
Attended church last week	46	45
Have no religion	6	5
Religion is very important to their own lives	75	61
Believe Bible is actual word of God, to be taken literally word for word[a]	65	35

SOURCE: Gallup Report 1987, 1993, 1997.

[a] The first measure on this variable was taken in 1963.

THINKING

CRITICALLY

Drawing on the experiences of your own family or someone close to you, would you say Americans are becoming less religious? Can you foresee a time in your own life when at least some aspects of religion might be more important or less important than they now are? Why do you say so?

there has been a substantial drop in the proportion of persons who say that religion is very important in their lives and in the proportion who believe that the Bible is the actual word of God.

Although almost everybody believes in God, some people place more emphasis on religion than others (Table 12.5). The most striking differences in religiosity are related to age and sex. Older people and women report greater attachment to religion than do younger people and men; people who live in the Midwest and the South report greater religiosity than people who live in the Midwest, West, or East.

One interesting question is the relationship between socioeconomic status and religiosity. To many scholars, it has seemed logical that religion should appeal disproportionately to the poor, who may be in greater need of hope and help in dealing with the world. As the data in Table 12.5 indicate, however, people with a college education are as likely to attend church as people with a grade school education. Nevertheless, they are significantly less likely to say that religion is important to them.

Consequences of Religiosity

Because religion teaches and reinforces values, it has consequences for attitudes and behaviors. People who are more religious tend to hold more conservative attitudes on sexuality and personal honesty; they are also likely to hold more conservative attitudes about family life, being more likely, for instance, to support the use of corporal punishment in disciplining children (Ellison, Bartkowski, & Segal 1996). They also tend to be happier and more satisfied with their lives (Ellison 1991) and their marriages (Thomas & Cornwall 1990) and to be friendlier, more cooperative people (Ellison 1992).

Although data such as these generally support the view that religious training teaches and reinforces conventional behavior, religion and the church can be forces that promote social change. In the United States, African American churches and

In 1776, when the American colonies declared independence, the Founding Fathers were deeply afraid of investing power in any single man, but they needed a general if they wanted to fight a war. With little military experience or charisma, George Washington's major qualification for the job was that he didn't want it. Within weeks of his appointment as commander of the army, however, Washington became an object of near worship. Why did the cult of Washington develop? It emerged, in part, because Washington symbolized the fledgling nation's unity and, in part, because his disdain for power made him a hero. Perhaps as Durkheim suggested, in worshiping Washington, the colonists were worshiping themselves—their nation and the virtues they believed it embodied (Schwartz 1983). (ART WORK: "Memorial to George Washington" painting on glass.)

clergy played a significant role in the Civil Rights Movement of the 1950s and 1960s; religion also played a supportive role in the struggle of Appalachian coal miners to unionize in the 1920s and 1930s (Billings 1990). In Latin America, liberation theology aims at the creation of democratic Christian socialism that eliminates poverty, inequality, and political oppression.

The New Christian Right

One of the most striking changes in religion today is the vitality and growth of the fundamentalist churches compared with mainline denominations. Fundamentalists can be found in all religions; there are fundamentalist Catholics, Baptists, Presbyterians,

TABLE 12.5
Religious Participation and Attitudes, 1995
There are some pronounced patterns in U.S. religiosity. For example, men are less religious than women, and young people are less religious than their elders. The well educated go to church, but they are not otherwise as religious as the less well educated.

	PERCENTAGE OF ADULTS WHO	
	Attend Church or Synagogue Weekly	Say Religion Is Very Important
National	30%	60%
Region		
Midwest	29	55
South	38	70
East	27	56
West	22	54
Age		
Below 30	23	48
30–49	26	58
50 and older	39	70
Sex		
Male	26	50
Female	33	68
Education		
No college	32	69
College incomplete	26	56
College graduate	31	45
Postgraduate	29	39
Race		
White	29	56
Black	35	82

SOURCE: The Gallup Poll: Public Opinion 1995. 1996. Wilmington: Scholarly Resources.

and Lutherans. Their common aim is to restore what they perceive to be a healthy tension between church and society. One outgrowth of this is the development of the New Christian Right.

The New Christian Right is a loose coalition of fundamentalists who believe that the U.S. government and social institutions must be made to operate according to Christian principles. They believe the United States is God's chosen instrument to fight Communism and that it is a Christian obligation to be active politically in making the United States a Christian nation. Two quotations give a flavor of this mixture of Christianity and political activism:

The idea that religion and politics don't mix was invented by the Devil to keep Christians from running their own country. (Jerry Falwell, cited in Bollier 1982, 54)

Not voting is a sin against God. . . . Perverts, radicals, leftists, Communists, liberals, and humanists have taken over the country because

Are American Catholics secularized? Pope John Paul II is afraid so. That's why he visited New York's Giants Stadium in 1995, where he stood in the rain while the throng greeted him with the wave. Whether U.S. Catholics are secularized depends on how you define the term. The U.S. Roman Catholic Church has gained about 25 percent in membership over the past 40 years (Monroe 1995, 383), but these Catholics aren't as likely to attend Sunday Mass as Catholics used to be. And they don't follow their leader the way they once did. The Pope sees artificial birth control as sinful, but 82 percent of U.S. Catholics disagree; 64 percent disagree that abortion is always wrong. Just one-third agree that only males should be priests. Nevertheless, 86 percent say they approve of their Pope (Sheler 1995)—a sign that in the American Catholic church, members take what they like and leave the rest.

technology

focus on

THE ELECTRONIC CHURCH

Along with a growing world audience, people in the United States can take part in religion's *electronic church* with the touch of a dial. Programs from radio and television ministries are aired virtually 24 hours a day every day on network and cable stations and channels. People who tune in to these religious broadcasts are similar to those who watch the most TV generally (Bruce 1990) and also to those who attend churches: They are older, mostly female, Protestant, and disproportionately from rural areas and the southern and midwestern regions of the country. They also tend to have less education and lower socioeconomic status (Alexander 1994).

Because their audience is less well off and less well educated and because an important part of the electronic ministry is the solicitation of tithes and offerings—

according to one calculation up to 21 percent of airtime is used to raise funds (Abelman 1990; Hoover 1990)—critics of televangelism have argued that electronic preachers are using their persuasive powers to bilk naive audiences.

This criticism, however, has come almost entirely from nonviewers and at least some research suggests that the charge is largely unfounded. Studies show, for instance, that the average contribution to electronic ministries is $32 per month and that most of that money comes from middle-income viewers. In addition, some ministries such as Pat Robertson's 700 Club raise most of their funds from a small group of wealthy supporters (Hoover 1990).

Critics have also charged that televangelism does not fulfill one of the major functions of religion: bringing people together into a moral community. Because most people who listen to televangelists also attend church, this criticism has relatively little merit. However,

the electronic church does reach one sizeable group of nonchurchgoers: older people and people with physical disabilities, people who are simply unable to attend regular church services. The cost of airtime makes televangelism enormously expensive and the religious needs of these people could not, perhaps, be met if the electronic church did not devote considerable time to fund raising.

In the late 1980s, the televangelist industry was rocked by scandals involving sex and financial fraud: Jimmy Swaggart and Jim and Tammi Bakker fell from grace (Mertz 1987). Far from alienating televangelists' chief audiences, however, the scandals had strong negative effects only on those who already had negative opinions (Gallup Reports 1989d). Also, these scandals were only a temporary embarrassment to televangelism. The electronic church has actually grown since (Alexander 1994) and continues to be an important means by which to reach a religious audience.

Many observers, including Marx, have thought that religion ought to appeal disproportionately to the poor and disadvantaged. Although the poor and disadvantaged are more likely than others to be fervent in their beliefs, studies show that church-going is correlated more strongly with being conventional than with being disadvantaged. It is characteristic of people who are involved in their communities, belong to other voluntary associations as well, and hold traditional values.

Christians didn't want to dirty their hands in politics. (Pat Robertson, cited in Bollier 1982, 70)

The New Christian Right is best understood as a political rather than a religious movement. It uses normal political processes—lobbying, campaign contributions, and getting out the vote—as means of influencing public policy. In doing so, of course, it is building onto the existing foundations of civil religion.

Despite this link to established values and despite the growth in fundamentalist religions, the movement has had more publicity than power. Fundamentalists differ among themselves on matters of religious conviction; they are also divided on political issues such as abortion, gun control, military defense, and pornography (Moore & Whitt 1986). Interestingly, conservative Protestants do not differ from more liberal religious groups in their attitudes toward racial equality and in many ways hold more progressive positions on economic issues (Davis & Robinson 1996). Furthermore, with the exception of abortion, studies show that during the last 15 years, the social attitudes of evangelical Protestants have become more like those of Roman Catholics, Jews, and the more moderate to liberal Protestant denominations (DiMaggio, Evans, & Bryson 1996). The net result is that the actual political power of the New Christian Right has been relatively limited.

Conclusion

Durkheim and Weber were both right. On the one hand, the church is a conserver of tradition, an upholder of older values of community. It stresses commitment, unquestioning loyalty, and tradition over rationalism and individualism. On the other hand, the church is in the forefront of contemporary issues (Hammond 1988). Nowhere are the battles over abortion, women's equality, homosexuality, or Third World oppression fought more bitterly than in the councils of our major churches. Whatever the role of religion in your own life, religion plays a vital role in our society.

Summary

1. The structural-functional model of education suggests that education has been a beneficial response to changing needs. Among the positive outcomes of education are cultural reproduction, social control, the teaching of specific skills, the selection of students for future adult roles, and the promotion of change. Education is also recognized as having latent functions and dysfunctions.

2. Conflict theory suggests that education helps to maintain and reproduce the stratification structure through three mechanisms: training a docile labor force that accepts inequality (the hidden curriculum); using credentialism to save the best jobs for the children of the elite; and maintaining an educational system in which success is highly correlated with social class.

3. Education is the central link in the inheritance of social class. Key elements of this transmission are differences in cultural capital that children enter school with; fewer opportunities for cognitive development offered to disadvantaged students; and tracking programs. Unless carefully tailored, new demands for excellence in the schools will increase the problems of students from disadvantaged backgrounds.

4. About 40 percent of U.S. high school graduates between 18 and 24 are enrolled in college. There are few sex differences in college enrollments, and minority enrollments are increasing, with the notable exception of African American men.

5. Increased competition means that a college education is no longer a guarantee of a good job. Nevertheless, education

pays off handsomely in terms of increased income, better jobs, and lower unemployment.

6. The scientific study of religion concerns itself with the consequences of religious affiliation for individuals and with the interrelationships of religion and other social institutions. It is not concerned with evaluating the truth of particular religious beliefs.

7. There are two distinct viewpoints about the role of religion in society. One, associated with Durkheim, suggests that religion provides support for the traditional practices of a society and is a force for continuity and stability. The other, associated with Weber, suggests that religion generates new ideas and challenges the institutions of society.

8. All religions are confronted with a dilemma: the tendency to reject the secular world and the tendency to compromise with it. The way a religion resolves this question determines its form and character. Those that make adaptations to the world are called churches, whereas those that reject the world are called sects.

9. The primary distinction between a cult and a sect is that a cult is outside a society's religious heritage whereas a sect often sees itself as restoring the true faith of a society. Both tend to be primary groups characterized by small size, intense we-feeling, and informal leadership.

10. U.S. civil religion is an important source of unity for the U.S. people. It is composed of a set of beliefs (that God guides the country), symbols (the flag), and rituals (the pledge of allegiance) that many people of the United States of all faiths hold sacred.

11. Despite an apparently uniform endorsement of religion by U.S. people, the proportion who say religion is very important to them is declining. Women and older people are most likely to be religious; education reduces religious commitment but not church attendance.

12. A major development in the contemporary church is the growth of fundamentalism and its political arm, the New Christian Right.

Sociology on the Net

Education, like the family, is a hot topic. U.S. students often compare poorly to students from other nations on standardized tests. Teachers' unions are criticized, and dropout rates are far too high. Our inner-city schools are crumbling. How well is the United States educating its young?

`http://aft.org/index.htm`

The American Federation of Teachers is one of the largest and most influential teachers' unions in the country. Browse around the home page, and note the services for members and the issues of importance to the members. Click on the highlighted title *The Research Department;* now click on the *Publication, Reports, and Surveys* section. Scroll down to the *International* section, and at long last click on the highlighted title *How and How Much the United States Spends on K-12, etc.* Browse through this extensive document paying special attention to the charts, summary statements, and the conclusion section. Why is how we spend our money as important a question as how much we spend? How does the United States spend its money differently than other nations? What alternatives de we see when we look to other nations?

Religion can be a contentious subject that brings out strong feelings. Religious tolerance can on occasion be hard to come by. Let's see what The Ontario Centre For Religious Tolerance has to say on the topic.

`http://web.canlink.com/ocrt/ocrt_hp.htm`

You will find this a very useful home page if you have a personal interest in religion or if you are writing a term paper on religion. Click on the highlighted index of topics. Scroll through the list of religions until you come to the *Spiritual Topics Menu.* Click on the section entitled *What is Religious Tolerance?* Read through this section and go back to the next selection on the *Spiritual Topics Menu* entitled *Test Your Religious Tolerance.* Do you agree with this definition of *tolerance?* Are you a tolerant person by the standards of this test?

Return to the Main Menu and click on *Religiously Hot Topics.* Browse through some of the topics. Have any of these topics come up in discussions with your friends or family? How do your own religious views inform you about how to approach these topics? Before you leave this excellent web site, you might wish to browse around and satisfy your own curiosity about a particular religion or religious group. You may wish to visit the *Glossary of Confusing Religious Terms* that can be found under the *Spiritual Topics Menu.* Don't forget to take the time to browse through the list of 35 Religious and Ethical Systems.

FIND IT ON INFOTRAC COLLEGE EDITION

One area in which educational and religious institutions often disagree is in the inclusion of sex education in the public schools. Using InfoTrac College Edition, look up the following article:

"Education for Intimacy." Amitai Etzioni. *Educational Leadership,* May 1997, v54 n8 p20.

(Hint: Enter the search term *Amitai Etzioni* using the Key Words Guide.)

Etzioni makes the argument that sex education should be part of the public school curriculum. Do you agree? Why or why not? The education of children regarding sex has traditionally been assumed to be the responsibility of parents. Is the inclusion of sex education in the public school curriculum an indication that parental influence and responsibility are declining in our society? Defend your position.

Suggested Readings

Alexander, Bobby C. 1994. *Televangelism Reconsidered: Ritual in Search for Human Community.* Atlanta: Scholars Press. A thoughtful research study and analysis of televangelism and its current function in viewers' lives after the 1987 Bakker and Swaggart scandals.

Fraser, Steven (ed.). 1995. *The Bell Curve Wars: Race, Intelligence, and the Future of America.* New York: Basic Books. Essays on both sides of the controversial issue of whether IQ tests accurately measure intelligence and whether African Americans' average scores are lower than those of non-Hispanic whites because of innate abilities or sociocultural factors.

Jacoby, Russel, and Glaubeiman, Naomi (eds.). 1995. *The Bell Curve Debate: History, Documents, Opinions.* New York: Random House/Times Books. Similar to Fraser's book, described above, but with more explanation of the history of this issue along with excerpts from Richard Herrnstein and Charles Murray's book *The Bell Curve.*

Kephart, William M., and Zellner, William. 1994. *Extraordinary Groups: An Examination of Unconventional Lifestyles,* 5th ed. New York: St. Martin's Press. A very good overview of specific utopian groups and religious sects in the United States. The book includes coverage of modern communes and gypsies, as well as religious sects: the Amish, Jehovah's Witnesses, and the Father Divine movement.

Kozol, Jonathan, 1991. *Savage Inequalities: Children in America's Schools.* New York: Crown. A comparison with detailed examples of schools and school districts across the United States that illustrates serious inequalities, with minority children generally attending the U.S.'s most crowded and least well-equipped schools.

Oakes, Jeannie, and Quartz, Karen Hunter (eds.). 1995. *Creating New Educational Communities.* Chicago: University of Chicago Press. A collection of essays and research reports by respected sociologists of education that give concrete suggestions for improving the U.S. educational system.

Pope, Liston. 1942. *Millhands and Preachers.* New Haven, Conn.: Yale University Press. A classic study of the confrontation between economic and religious forces in a textile mill strike in the South in 1929. The narrative illuminates the very real tension between religion and society.

Politics and the Economy

Outline

Political and economic institutions are so closely related that earlier generations referred to them as a single institution—the political economy. Although we can treat them separately to some extent, we will see that many issues, such as the political power of organized labor, cut across both institutions. Throughout the chapter, a primary concern is to provide a sociological perspective that may help you to interpret your own political and economic experiences as well as the headlines in the national news.

Political Institutions and the State

Power and Political Institutions

Lisa wants to watch the Fresh Prince, while John wants to watch *Return of the Vampire Bats;* fundamentalists want prayer in the schools and the American Civil Liberties Union wants it out; state employees want higher salaries and the citizens want lower taxes. Who decides?

Whether the decision maker is mom or the Supreme Court, those who are able to make and enforce decisions have power. Formally, **power** is the ability to direct others' behavior, even against their wishes. As these examples illustrate, power occurs in all kinds of social groups, from families to societies.

Although both mothers and courts have power, there are obvious differences in the basis of their power, the breadth of their jurisdiction, and the means they have to compel obedience. The social structure most centrally involved with the exercise of power is the state, and that is the focus of this section. Before we begin, however, we give a broad overview of two kinds of power, coercion and authority.

Coercion

The exercise of power through force or the threat of force is **coercion.** The threat may be physical, financial, or social injury. The key is that we do as we have been told only because we are afraid not to. We may be afraid that we will be injured, but we may also be afraid of a fine or of rejection.

Authority

Threats are sometimes quite effective means of making people follow your orders. They tend to create conflict and animosity, however, and it would be much easier if people would just agree that they were supposed to do whatever it was you told them. This is not as rare as you might suppose. This kind of power is called **authority;** it refers to power that is supported by norms and values that legitimate its use. When you have authority, your subordinates agree that, in this matter at least, you have the right to make decisions and they have a duty to obey.

In a classic analysis of power, Weber distinguished three bases on which this agreement is likely to rest: tradition, extraordinary personal qualities (charisma), and legal rules.

TRADITIONAL AUTHORITY.   When the right to make decisions is based on the sanctity of time-honored routines, it is called **traditional authority** (Weber [1910] 1970c, 296). Monarchies and patriarchies are classic examples of this type of

---

Power is the ability to direct others' behavior even against their wishes.

Coercion is the exercise of power through force or the threat of force.

Authority is power supported by norms and values that legitimate its use.

Traditional authority is the right to make decisions for others that is based on the sanctity of time-honored routines.

authority. For example, only 30 or 40 years ago, the majority of women and men in our society believed that husbands ought to make all the major decisions in the family; husbands had authority. Today, most of that authority has disappeared. Traditional authority, according to Weber, is not based on reason; it is based on a reverence for the past.

CHARISMATIC AUTHORITY.   An individual who is given the right to make decisions because of perceived extraordinary personal characteristics is exercising **charismatic authority** (Weber [1910] 1970c, 295). These characteristics (often an assumed direct link to God) put the bearer of charisma on a different level from subordinates. Gandhi's authority was of this form. He held neither political office nor hereditary position; yet he was able to mold national policy in India.

Charismatic authority is the right to make decisions that is based on perceived extraordinary personal characteristics.

RATIONAL-LEGAL AUTHORITY.   When decision-making rights are allocated on the basis of rationally established rules, we speak of **rational-legal authority.** This ranges all the way from a decision to take turns to a decision to adopt a constitution. An essential element of rational-legal authority is that it is impersonal. You do not need to like or admire or even agree with the person in authority; you simply follow the rules.

Rational-legal authority is the right to make decisions that is based on rationally established rules.

Rational-legal authority is the kind on which our government is based. When we want to know whether the president or the Congress has a right to make certain decisions, we simply check our rule book: the Constitution. As long as they follow the rules, most of us agree that they have the right to make decisions and we have a duty to obey.

SUMMARY.   Analytically, we can make clear distinctions among these three types of authority. In practice, the successful exercise of authority usually combines two or more (Wrong 1979). An elected official who adds charisma to the rational-legal authority stipulated by the law will have more power; the successful charismatic leader will soon establish a bureaucratic system of rational-legal authority to help manage and direct followers.

All types of authority, however, rest on the agreement of subordinates that someone has the right to make a decision about them and that they have a duty to obey it. This does not mean that the decision will always be obeyed or even that each and every subordinate will agree that the distribution of power is legitimate. Rather, it means that society's norms and values legitimate the inequality in power. For example, if a parent tells her teenagers to be in at midnight, they may come in later. They may even argue that she has no right to run their lives. Nevertheless, most people, including children, would agree that the parent does have the right.

Because authority is supported by shared norms and values, it can usually be exercised without conflict. Ultimately, however, authority rests on the ability to back up commands with coercion. Parents may back up their authority over teenagers with threats to ground them or take the car away. Employers can fire or demote workers. Thus, authority rests on a legitimation of coercion (Wrong 1979).

Political Institutions

Power inequalities are built into almost all social institutions. In institutions as varied as the school and the family, roles associated with status pairs such as student/teacher and parent/child specify unequal power relationships as the normal and desirable standard.

The dominant political institution is the state. Although fathers have power and biology teachers have power, only the state can legitimately arm its authority figures with M-16s. Coercion and threats of coercion are important weapons used by the state in backing up its authority. Although most of us pay our taxes and obey the laws without any direct threat, we are all aware of the state's ability to fine or imprison should we stray from the rules.

CONCEPT SUMMARY	*Power*	
CONCEPT	**DEFINITION**	**EXAMPLE FROM FAMILY**
Power	Ability to get others to act as one wishes in spite of their resistance; includes coercion and authority	"I know you don't want to mow the lawn, but you have to do it anyway."
Coercion	Exercise of power through force or threat of force	"Do it or else."
Authority	Power supported by norms and values	"It is your duty to mow the lawn."
Traditional authority	Authority based on sanctity of time-honored routines	"I'm your father, and I told you to mow the lawn."
Charismatic authority	Authority based on extraordinary personal characteristics of leader	"I know you've been wondering how you might serve me,…"
Rational-legal authority	Authority based on submission to a set of rationally established rules	"It is your turn to mow the lawn; I did it last week."

Political institutions are institutions concerned with the social structure of power; the most prominent political institution is the state.

In a very general sense, **political institutions** are all those institutions concerned with the social structure of power. This general definition includes many of the institutions of society. The family, the workplace, the school, and even the church or synagogue have structured social inequality in decision making. The most prominent political institution, however, is the state.

The State

The **state** is the social structure that successfully claims a monopoly on the legitimate use of coercion and physical force within a territory.

The **state** is the social structure that successfully claims a monopoly on the legitimate use of coercion and physical force within a territory. It is usually distinguished from other political institutions by two characteristics: (1) its jurisdiction for legitimate decision making is broader than that of other institutions, and (2) it controls the use of coercion in society.

Jurisdiction

Whereas the other political institutions of society have rather narrow jurisdictions (over church members or over family members, for example), the state exercises power over the society as a whole.

Generally, states have been considered to be responsible for gathering resources (taxes, draftees, and so on) to meet collective goals, arbitrating relationships among the parts of society, and maintaining relationships with other societies (Williams 1970). As societies have become larger and more complex, the state's responsibilities have grown. A recent poll (see Figure 13.1) indicates that the majority of people think the U.S. government is also responsible for such things as providing day care and reducing income differences between rich and poor.

Coercion

The state claims a monopoly on the legitimate use of coercion. To the extent that other institutions use coercion (for example, the family or the school), they do so with the approval of the state, and as the state giveth, the state also taketh away. Thus, the state

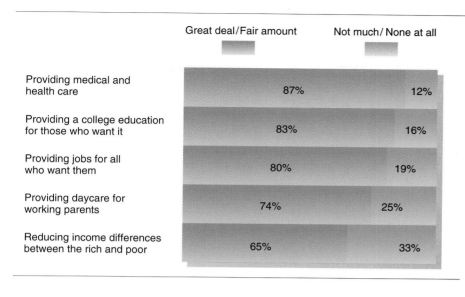

FIGURE 13.1
Question: How Much Responsibility Should the Government Take for:
SOURCE: Gallup Poll 1993.

Great deal/Fair amount   Not much/ None at all

	Great deal/Fair amount	Not much/None at all
Providing medical and health care	87%	12%
Providing a college education for those who want it	83%	16%
Providing jobs for all who want them	80%	19%
Providing daycare for working parents	74%	25%
Reducing income differences between the rich and poor	65%	33%

has withdrawn approval of physical coercion between husband and wife and has sharply restricted the amount of physical punishment that parents may legitimately administer to children. Similarly, the state has generally declared physical punishment illegitimate within the school system.

The state uses three primary types of coercion. First, the state uses its police power to claim a monopoly on the legitimate use of physical force. It is empowered to imprison people and even impose the death penalty. This claim to a monopoly on legitimate physical coercion has been strengthened in recent years by the declining legitimacy of coercion in other institutions, such as the home and the school. Second, the state uses taxation, a form of legitimated confiscation. Finally, the state is the only unit in society that can legally maintain an armed force and that is empowered to deal with foreign powers.

A variety of social structures can be devised to fulfill these functions of the state. Here we review two basic political forms: authoritarian systems and democracy.

THINKING

CRITICALLY

The family and the classroom are more often authoritarian than democratic. Try to explain this in sociological terms.

Democratic Versus Authoritarian Systems

Authoritarian Systems

Most people in most times have lived under **authoritarian systems.** Authoritarian governments go by a lot of other names: dictatorships, military juntas, despotisms, monarchies, theocracies, and so on. What they have in common is that the leadership was not selected by the people and legally cannot be changed by them.

Authoritarian structures vary in the extent to which they attempt to control people's lives, the extent to which they use terror and coercion to maintain power, and the purposes for which they exercise control. Some authoritarian governments, such as monarchies and theocracies, govern through traditional authority; others have no legitimate authority and rest their power almost exclusively on coercion.

Authoritarian systems are political systems in which the leadership is not selected by the people and legally cannot be changed by them.

Democracy

There are several forms of **democracy,** many of them rather different from that of the United States. All democracies, however, share two characteristics: there are regular, constitutional procedures for changing leaders, and these leadership changes reflect the will of the majority.

In a democracy, two basic groups exist: the group in power and one or more legal opposition groups that are trying to get into power. The rules of the game call for

Democracy is a political system that provides regular, constitutional opportunities for a change in leadership according to the will of the majority.

sportsmanship on both sides. The losers have to accept their loss and wait until the next constitutional opportunity to try again, and the winners have to refrain from eliminating or punishing the losers. Finally, there has to be public participation in choosing among the competing groups.

CONDITIONS FOR DEMOCRACY.   Why are some societies governed by democracies and others by authoritarian systems? The answer appears to have less to do with virtue than with economics. Democracy is found almost exclusively in the wealthier nations of the world. The key factor, however, is not the overall wealth of the nation but the way the wealth is distributed; democracy generally is found in nations that do not have extremes of income inequality (Simpson 1990). A large and relatively affluent middle class is especially important. Members of the middle class usually have sufficient social and economic resources to organize effectively; their economic power and organization enable them to hold the government accountable.

Democracy may also flourish only in societies with many competing groups, each of which comprises less than a majority (Williams 1970, 271). In such a situation, no single group can win a majority of voters without negotiating with other groups; because each group is a minority, safeguarding minority political groups protects everybody.

Although democratic stability depends on competing interest groups, two additional conditions must be met. First, if minority political groups are so divided or ineffective that there is little chance they can win an election, the public may become disillusioned with the democratic process (Weil 1989). Second, if competing interest groups do not share the same basic values, they are not likely to be able to abide by the rules of the game. Fundamental differences in the values of Israelis and each of the different factions of Palestinians make it difficult for democracy to exist in the West Bank and Gaza Strip, for instance.

Nelson Mandela was elected president of South Africa in the first election open to all citizens—black and white. Contrary to many scholars' predictions, the transition has been relatively peaceful and smooth. The question, of course, is whether or not South Africa will be able to sustain a democratic government. South Africa meets some of the prerequisites of democracy: It is one of the wealthier nations of the world and has a democratic tradition, even if until recently that tradition was limited to white citizens. However, racially based income inequality will pose a challenge to the continuation of the democratic process.

Who Governs? Models of U.S. Democracy

Everyone agrees that the United States is a democracy. Political parties that have at least moderately different economic and social agendas vie for public support, and every two, four, or six years there are opportunities to "turn the rascals out" and replace the leadership. There is substantial debate, however, about whether the decisions made by our leaders really reflect the will of the majority. This section outlines four models of how decisions are made: pluralist, power-elite, conflict/dialectic, and state autonomy.

Pluralist Model

The pluralist model of U.S. government focuses on the processes of coalition and competition that take place in state and federal governments. A vital part of this model is the hypothesis of shifting allegiances. According to pluralist theorists, different coalitions of interest groups arise for each decision. For example, labor unions will ally themselves with automakers in favoring tariffs against Japanese imports; when it comes to domestic issues such as wages, however, these two groups will oppose each other. This pattern of shifting allegiances keeps any interest groups from consistently being on the winning side and keeps political alliances fluid and temporary rather than allowing them to harden into permanent and unified cliques (Dahl 1961, 1971). As a result of these processes, pluralists see the decision-making process as relatively inefficient but also relatively free of conflict, a process in which competition among interest groups keeps any single group from gaining significant advantage.

Critics believe that the pluralist model is naïve at best. They argue that, although all of these interest groups may be skirmishing in Congress, the real decisions are being made in higher circles. At worst, critics argue, the pluralist model "obscures and shelters the citadels of domination" by refusing to recognize the controlling hand of the ruling class (Bowles & Gintis 1986).

The Power-Elite Model

Theorists associated with the power-elite model wave aside the competition among organized interest groups as the middle levels of power. Power-elite theorists contend that there is a higher-level of decision making where an elite makes all the major decisions—in its own interests. In his classic work, *The Power Elite,* C. Wright Mills (1956) defined the **power elite** as the people who occupy the top positions in three bureaucracies: the military, industry, and the executive branch of government. From these "command posts of power" and through a complex set of overlapping cliques, these people share decisions having at least national consequences (Mills 1956, 18).

The power-elite theory is a positional theory of power. It argues that individuals have power by virtue of the positions they hold in key institutions. If the interests of these individuals and institutions were in competition with one another, this model would not be significantly different from the pluralist model. The key factor in elite theory is the unity of purpose and outlook that top position holders have as a result of common membership in the upper class. In Mill's view, shared *social* class is as important as shared economic interests (class). The elite have gone to the same prep schools, joined the same clubs, and summered at the same resorts. As a result of this shared background, they share a common worldview and common interests (Orum 1987). This means that members of the elite will cooperate rather than compete with each other.

The **power elite** is comprised of the people who occupy the top positions in three bureaucracies—the military, industry, and the executive branch of government—and who are thought to act together to run the United States in their own interests.

That an upper class exists in the United States and that this upper class is highly overrepresented in the power elite seems unarguable (Domhoff 1983). The critical question is whether there is any evidence that this collection of top position holders acts together to promote the interests of the upper class.

The Conflict/Dialectic Model

A Marxist version of class conflict is at the root of the conflict/dialectic model. Like the power-elite model, the conflict/dialectic model features an elite that runs the show. The Marxist elite differs from the power elite in two ways. First, the Marxist elite is made up of a much smaller group of people: the people who actually own the means of production. Managers, bureaucrats, and generals are not considered to be members of the Marxist elite; they are rather tools of the elite. Second, Marxists do not require a unified elite tied together by social custom and tradition. Instead, they recognize that there will be factions within the elite with competing economic interests. For example, financiers like high interest rates; automobile manufacturers do not (Roy 1984). The Marxist elites include more of what political scientist Dye (1986) calls cowboys: They are aggressive about making and keeping their money, and they don't care much for fancy manners and old school ties. As a result, the Marxist elite includes built-in competition that may ultimately weaken it and lead to change.

Another difference between the conflict/dialectic and power-elite models is the tension they see between the elite and nonelites. Marxists argue that the working class has its own resources for power: class consciousness and the possibility of class action. The power of the subordinate class can be likened to that of a sleeping rattlesnake; the snake is not hurting you now, but you want to be certain not to awaken it. Thus, the conflict/dialectic model sees underlying tensions between the elite (dominant class) and the nonelite that are largely missing from the power-elite model.

A further element of this model is its emphasis on the dialectic as the process of social change. As noted in earlier chapters, the dialectic suggests that social change will emerge as a result of contradictions and conflicts within and between social institutions. Marxists believe that

> social institutions, economic systems, and political institutions contain inherent contradictions. These produce conflicts and strains that eventually lead to the transformation of those institutions and systems. *Contradictions* are thus engines of social change and their analysis is central also to understanding the dynamics of political power. (Whitt 1979, 84)

In terms of the U.S. political structure, the dialectic suggests that the elite has to be constantly on its toes to ward off the potential consciousness and power of the subordinate class in a climate of shifting economic and political conditions. Change rather than stability is the key to the conflict model. Whereas the power-elite model sees the elite striving to maintain privilege, the conflict model envisions a more rough-and-tumble battle in which both sides strive to structure change for their benefit. This conflict occurs within as well as between classes.

The State Autonomy Model

A growing number of scholars argue that the government bureaucracy is a powerful, independent actor in political decisions. The federal government employs 3 million people directly. In addition, its policies indirectly determine the employment of tens of millions of people who work for national defense contractors, state and local governments, schools, and social welfare agencies. In 1992, the federal government collected more than $1 trillion in revenue and spent more than $1.4 trillion! It seems only common sense that the state (meaning the federal bureaucracy) is in a powerful position to get what it wants. Its agenda is linked not to class (as in the power-elite model) but to the maintenance and extension of bureaucratic power.

A good example of this approach is Hooks's (1990) analysis of the profound effect that Pentagon and Defense Department policies have had on the development of the microelectronics and aeronautics industries. Hooks's findings show that the competitiveness of U.S. high-technology firms has been seriously jeopardized by defense policy. Because the Pentagon's goals have been strategic rather than commercial, the state has pursued its own interests at the expense of the capitalist class.

Summary and Evaluation: Who Does Govern?

Research in the United States supports two conclusions: First, there is little evidence for the pluralist expectation of shifting allegiances. Instead, research shows that business elites are relatively unified. Studies of political action committee (PAC) contributions show that business interests are unified by a shared conservatism: They act together as a class to support probusiness candidates (Burris & Salt 1990; Clawson & Su 1990). Second, there are only a few issues on which other interest groups in the United States have the unity or the resources to challenge the power of business effectively (Korpi 1989). In the United States, redistributive programs—such as civil rights or Social Security—have passed only under two conditions: A sense of crisis caused the elite to favor the change, or the members of the elite disagreed among themselves (Jenkins & Brent 1989).

A final point is important. Although there are some important differences among the four major models of decision making in the United States (see the Concept Summary), a notable feature of all four is that organized entities—businesses, unions, PACs, government agencies—rather than individuals are the key actors. Individuals play a very small role unless they represent or are represented by one of these organizations (Laumann, Knoke, & Kim 1985).

CONCEPT SUMMARY	*Comparison of Four Models of American Political Decision Making*			
	PLURALIST	**POWER-ELITE**	**CONFLICT/ DIALECTIC**	**STATE AUTONOMY**
Basic units of analysis	Interest groups	Institutional elites	Classes	Government bureaucracy
Source of power	Situational; depends on issue	Positional; top positions in bureaucracies	Class based; ownership of means of production	Control of personnel and budget of government
Distribution of power	Dispersed among competing diverse groups	Concentrated in relatively homogeneous elite	Held by dominant class, potentially available to lower class	Held by bureaucrats
Limits of power	Limited by shifting and cross-cutting loyalties	No identifiable limits to elite domination	Limited by class conflict and contradiction among social institutions	Limited if elite is unified and nonelites are unorganized
Role of the state	Arena where interest groups compete	One of several sources of power	Captured by the ruling class	A major source of power

Individual Participation in U.S. Government

Democracy is a political system that explicitly includes a large proportion of adults as political actors. Yet it is easy to overlook the role of individual citizens while concentrating on leaders and organized interests. This section describes the U.S. political structure and process from the viewpoint of the individual citizen.

Who Participates?

The average citizen is not politically oriented. A significant proportion of the voting-age population (approximately 38 percent in 1994) does not even register to vote; of those people who do register, many do not vote. In recent presidential elections, 40 to 50 percent of the voting-age population has not bothered to go to the polls. Electoral participation declines markedly as one gets closer to the local level, and often only 20 to 25 percent vote in local elections.

Voting is in many ways the easiest and most superficial means of participating in politics. If we include letter writing, returning congressional questionnaires, and making campaign contributions as elements of political activity, we will have to conclude that fewer than 20 percent of U.S. citizens take an active part in politics, and, of course, only a very small proportion take part to the extent of running for or occupying elective office.

The studies demonstrating low levels of political participation and involvement pose a crucial question about the structure of power in U.S. democracy. Who participates? If they are not a random sample of citizens, then some groups probably have more influence than others. Studies show that voters differ from non-voters by social class and age.

Social Class

One of the firmest findings in social science is that political participation (indeed, participation of any sort) is strongly related to social class. Whether we define participation as voting or letter writing, people with more education, more income, and more prestigious jobs are more likely to be politically active. They know more about the issues, have stronger opinions on a wider variety of issues, and are much more likely to try to influence the nature of political decisions. This conclusion is supported by data on voting patterns from the 1994 election (see Table 13.1). The higher the level of education, the greater the likelihood of voting; those who have graduated from college are twice as likely to vote as those who have not completed high school.

Age

Another significant determinant of political participation is age. Political interest, knowledge, opinion, and participation steadily increase with age. Nearly one-half of all voters in the 1994 election were 45 or older. Even in the turbulent years of the Vietnam War, when young antiwar demonstrators were so visible, young adults were significantly less likely to vote than were the middle aged. In that period, young adults engaged in other forms of political participation that did, in fact, influence political decisions. In most time periods, however, the low participation of younger people at the polls is a fair measure of their overall participation.

In the United States, all citizens over the age of 18 have the right to vote. A surprisingly large proportion of the population choose not to exercise this right. The pattern of political decisions reflects the fact that the people who are most apt to vote are establishment types of people: middle aged, better off, and well educated. If less-advantaged segments of the population would increase their political participation, the nature of U.S. political decisions might change.

TABLE 13.1
Participation in the 1994 Election
Political participation is greater among people who are older, better educated, and non-Hispanic.

	Percentage Registered	Percentage Reporting Actually Voted
Total	62.0	44.6
Education		
8 years or less	40.1	23.2
9–11 years	44.7	27.0
12 years	58.9	40.5
13–15 years (college)	68.4	49.1
16 or more years (college graduate)	76.3	63.1
Race/Ethnicity		
White	64.2	46.9
Black	58.3	37.0
Hispanic	30.0	19.1
Age		
18–20	37.2	16.5
21–24	45.5	22.3
25–34	51.5	32.2
35–44	63.3	46.0
45–64	71.0	56.0
65+	75.6	60.7

SOURCE: U.S. Bureau of the Census, 1996 Statistical Abstract of the United States, no. 456.

Race and Ethnicity

Racial differences in political participation have virtually disappeared. In fact, after social class has been taken into consideration, it seems likely that being African American increases political participation. Blacks are more apt than whites to want changes made in the system, and they turn to political participation as a means to effect these changes (Guterbock & London 1983, 440). Low Hispanic participation is partly traceable to low socioeconomic status, but much of the apparent low participation of Hispanics is an artifact of the measurement procedure: One-third of the voting-age population of Hispanics are aliens (legal as well as illegal) and are not eligible to register or to vote.

Differentials in Office Holding

By law, almost all those native-born to the United States who are older than 35 are eligible to hold any office. In practice, elected officials tend to be white men from the professional classes. Thus, for the most part, the political activities of other groups (women, minorities, nonelites) have been directed at choosing the white elite males to represent them. This practice has been changing, however; African Americans, Hispanics, and women are increasingly holding elected office, especially at local levels. Still, only about 6 percent of all state legislators are African American, and only 21 percent are female. The proportions in Congress are much smaller (U.S. Bureau of the Census 1995a).

TABLE 13.2
Party Identification, 1994
Although American parties are not closely tied to social class, the better-off tend to be Republicans, and the poor and nonwhite tend to be Democrats. The growing proportion who identify themselves as Independent tend to be young, well educated, and Anglo.

	Democrat	Republican	Independent
Race/Ethnicity			
White	43	46	10
Black	81	9	8
Education			
Grade School	59	24	13
High School	51	34	13
College	43	50	7
Sex			
Male	42	46	11
Female	52	37	10
Age			
17–24	51	37	10
25–34	44	43	12
35–44	45	43	12
45–54	46	45	7
55–64	48	43	8
65–74	51	42	8
75–99	54	35	9
Total	47	43	10

SOURCE: U.S. Bureau of the Census, 1996 Statistical Abstract of the United States, no. 455.

Party Affiliation

Although both major political parties in the United States are basically centrist, there are philosophical distinctions between them. Traditionally, the Democratic Party has been more likely to support the interests of the poor and the working class, and the Republican Party has been more likely to favor policies favoring economic growth. Because of these characteristics, the Republicans tend to attract people with higher incomes, and the Democrats tend to attract minority and lower-income voters as well as some highly educated liberals (see Table 13.2).

A growing proportion of voters align themselves with neither party but declare an intention to vote on the basis of issues rather than party loyalty. When the 33 percent of voters who call themselves independent go to the polls, however, they usually have to choose between a Democratic and a Republican candidate.

Why Doesn't the United States Have a Workers' Party?

The United States is almost alone among Western democracies in not having a party that explicitly supports the interests of the working class. As a direct result, it is almost alone among Western democracies in not having national health insurance, family allowances, and a comprehensive system of unemployment benefits (Quadagno 1990). If the working class outnumbers the upper class (and it does), why don't its members vote to change the system?

Three answers are usually given. First, U.S. labor unions have organized around specific occupations and have had little success in creating superunions that would represent the interests of labor as a whole (Quadagno 1990). One reason for this is that U.S. labor law has given unions the right to negotiate only over wage and wage-related issues (McCammon 1993). As a result, unionized workers earn four times more than those working at the minimum wage. Coupled with cross-cutting loyalties based on race and ethnicity, religion, and geographic region, the union focus on wages and the inequality in working-class income that it produces means that U.S. workers seldom recognize or vote for a common economic and political agenda (Form 1985; Western 1993).

Second, despite significant income disparities, the U.S. standard of living is luxurious by almost any comparison. Describing the failure of U.S. workers to sustain a socialist movement, one nineteenth-century socialist noted: "On the reefs of roast beef and apple pie, socialist utopias of every sort are sent to their doom" (Sombart [1906] 1974, 87).

Finally, of course, there is the American Dream (see Chapter 6). The U.S. worker has not been interested in absolute equality but rather in equality of opportunity. Because workers believe they can make it within the current system, they have not wanted to reduce the privileges associated with success.

Very early in this century, labor unions often aligned themselves with radical platforms favoring redistribution of wealth; at mid-century, they were less radical but solidly democratic. In the early 1990s, only about half of union members are members of the Democratic Party. This collapse of working-class liberalism has led some observers to speculate that any future pressure for redistribution will come from the minimum-wage service sector and from the unemployed, groups that so far have shown little political muscle (Form 1985).

Modern Economic Systems

Our description of the U.S. political process has crossed over into discussions of economic processes again and again. From the role of the working class to the role of the power elite, we find that understanding government requires understanding the economic relationships that underlie it. At this point we turn to an explicit assessment of economic relationships and how they affect the individual worker and the political economy.

Economic institutions are social structures concerned with the production and distribution of goods and services. Such issues as scarcity or abundance, guns or butter, craftwork or assembly lines, are all part of the production side of economic institutions. Issues of distribution include what proportion goes to the worker versus the manager, who is responsible for supporting nonworkers, and how much of society's production is distributed on the basis of need rather than effort or ability. The distribution aspect of economic institutions intimately touches the family, stratification systems, education, and government.

In the modern world, there are basically two types of economic systems: capitalism and socialism. Because economic systems must adapt to different political and natural environments, however, we find few instances of pure capitalism or pure socialism. Most modern economic systems represent some variation on the two and often combine elements of both.

Capitalism

The economic system in which most wealth (land, capital, and labor) is private property, to be used by its owners to maximize their own gain, is **capitalism.** This economic system

Economic institutions are social structures concerned with the production and distribution of goods and services.

Capitalism is the economic system in which most wealth (land, capital, and labor) is private property, to be used by its owners to maximize their own gain; this economic system is based on competition.

is based on competition. Each of us seeks to maximize our own profits by working harder or devising more efficient ways to produce our goods. Such a system encourages hard work, technical innovation, and a sharp eye for trends in consumer demand. Because self-interest is a powerful spur, such economies can be very productive.

Even when it is very productive, a capitalist economy has two drawbacks. First, it neglects the aspect of distribution. The capitalist system at its most ideal represents a competitive bargain between labor and capital, both of whom control a necessary resource. What happens, however, to those who have neither? With nothing to exchange, they are outside the market. Although the family may continue to care for members who cannot sell their labor (children, the elderly, the disabled), what happens when whole families, indeed whole communities, have no one who is willing to buy their labor? In theory, it is assumed that labor, like capital, will move to a new area of demand. An unemployed steelworker in Youngstown, Ohio, however, cannot easily transform himself into a computer repairperson or a frogman for an off-shore oil rig (Thurow 1980). Second, pure capitalism does not provide for public goods: streets, watersheds, sewers, defense. These goods must be produced, even if they offer profit to no one. Thus capitalist systems must have some means of distribution other than the market.

Socialism

Socialism is an economic structure in which productive tools (land, labor, and capital) are owned and managed by the workers and used for the collective good.

If capitalism is an economic system that maximizes production at the expense of distribution, socialism is a system that stresses distribution at the expense of production. As an ideal, **socialism** is an economic structure in which productive tools are owned and managed by the workers and used for the collective good.

In theory, socialism has several major advantages over capitalism. First, societal resources can be used for the benefit of society as a whole rather than for individuals. This advantage is most apparent in regard to common goods such as the environment. A related advantage is that of central planning. Because resources are controlled by the group, they can be deployed to help reach group goals. This may mean diverting them from profitable industries (say, those making bicycles, televisions, or compact discs) to industries that are viewed as more likely to benefit society in the long run: education, agriculture, or steel. The major advantage claimed for socialism, however, is that it produces equitable (though not necessarily equal) distribution.

The creed of pure socialism is from each according to ability, to each according to need. An explicit goal of socialism is to eliminate unequal reward as the major incentive to labor. Cuban revolutionary Che Guevera argued that "one of the fundamental objectives of Marxism is to remove interest, the factor of individual interest and gain from men's psychological motivations" (cited in Hollander 1982). Workers are expected to be motivated by loyalty to their community and their comrades. Unfortunately, the childless woman is not likely to be motivated to do her best when the incompetent worker next to her takes home a larger paycheck simply because she has several children and thus a greater need. Nor is the farmer as likely to make the extra effort to save the harvest from rain or drought if his rewards are unrelated to either effort or productivity. Because of this factor, production is usually lower in socialist economies than in capitalist economies.

The Political Economy

Both capitalism and socialism can coexist with either authoritarian or democratic political systems. Many Western European nations combine socialism with democracy. Communist societies, however, graft socialism onto an authoritarian political system. It is a socialist economy guided by a political elite and enforced by a military elite. The goals of socialism (equality, efficiency) are still there, but the political form is authoritarian rather than democratic.

These contrasting faces of socialism remind us that it is important to keep in mind the political structure in which an economy exists. Both capitalism and socialism are less congenial institutions when they exist within authoritarian regimes.

Mixed Economies

Most Western societies in the late twentieth century represent a mixture of both capitalist and socialist economic structures within a more-or-less democratic framework. In many nations, services such as the mail and the railroads and key industries such as steel and energy have been socialized. These moves to socialism are rarely the result of pure idealism. Rather, public ownership is often seen as the only way to ensure continuation of vital services that are not profitable enough to attract private enterprise. Other services—for example, health care—have been partially socialized because societies have judged it unethical for these services to be available only to those who can afford to pay for them. Education is a socialized service, but it went public so long ago that few recognize the public schools as one of the first socialized industries.

In the case of many socialized services, general availability and progressive tax rates have gone far toward meeting the maxim from each according to ability, to each according to need. There are still inequalities in education and health care, but many fewer than there would be if these services were available on a strictly cash basis. The United States has done the least among major Western powers toward creating a mixed economy, and our future direction is unclear. Although the long-term shift in the United States has been toward more socialized services, some of this commitment was reduced during the Reagan-Bush years. The future mix of socialist and capitalist principles will reflect political rather than strictly economic conditions.

The U.S. Economic System

The Postindustrial Economy

In the late twentieth century, the United States has what is called a postindustrial economy. To understand this concept requires a brief historical review of economic forms.

In a preindustrial economy, the vast majority of the labor force is engaged in **primary production,** extracting raw materials from the environment. Prominent among primary production activities are farming, herding, fishing, foresting, hunting, and mining. Preindustrial economic structures were characteristic of Europe until 500 years ago and are still typical of many societies.

Primary production is extracting raw materials from the environment.

Industrialization meant a shift from primary to **secondary production,** the processing of raw materials. For example, ore, cotton, and wood are processed by the steel, textile, and lumber industries, respectively; other secondary industries turn these materials into automobiles, clothing, and furniture. The shift from primary to secondary production is characterized by enormous increases in the standard of living.

Secondary production is the processing of raw materials.

Postindustrial development rests on a third stage of productivity, **tertiary production.** This stage is the production of services. The tertiary sector includes a wide variety of occupations: physicians, schoolteachers, hotel maids, short-order cooks, and police officers. It includes everyone who works for hospitals, governments, airlines, banks, hotels, schools, or grocery stores. None of these organizations produces tangible goods; they all provide service to others. They count their production not in barrels or tons but in numbers of satisfied customers.

Tertiary production is the production of services.

a global perspective

DEMOCRATIC SOCIALISM IN SWEDEN

What would it be like to live and work in Sweden? You would have a guaranteed job and income (or pension, if you were retired or disabled), guaranteed access to comfortable housing, free education through college, and free medical care. After you or your partner gave birth to or adopted a baby, you would be entitled to a full year of paid parental leave (Haas 1990). Once you went back to work, you could use a state-funded or cooperative day-care center, and you would give more than half your paycheck to the government in taxes. Some Swedish politicians want to expand social benefits—and raise taxes to as much as 70 percent of everyone's income (Helco & Madsen 1987).

Sweden is a democratic socialist society with an economy that mixes corporate capitalism with significant welfare benefits for workers and non-workers alike. Because Sweden is a democracy, the majority of Swedes have voted to receive these benefits and to pay high taxes for them. But Sweden's economy wasn't always arranged this way.

Sweden owes its economic organization to the rise of a strong labor movement (Koblik 1975). As industrialization began in Sweden in the 1870s, labor union members worked to create the Social Democratic Party, a political party dedicated to equi-

table wages, job security, and welfare programs for the entire society. While Communists in Russia were fighting and winning the Russian Revolution in 1914–1917, members of Sweden's Socialist Democratic Party were politicking for seats in parliament. After holding power on and off during the 1920s, the Socialist Democratic Party won an important election in 1932 and then retained political power for over 40 years (Olsen 1992). Today, the government is controlled by more conservative politicians, but the welfare state that emerged during 40 years of socialist majority government remains in place.

One reason the Swedish labor movement has been so successful in getting workers' benefits is that Swedish culture has coupled the ideas of development and progress, and faith in the inherent goodness of all individuals. The majority of Swedes believe, then, that virtually all future problems can

be resolved and all wants supplied (Nilsson 1975).

Not everyone in Sweden is a member of the Socialist Democratic Party, of course. Conservative groups favor a freer market economy. Furthermore, Socialist Democrats today are worried about whether Sweden's welfare society can survive the international economic downturn of the 1970s (Olsen 1996). Controlling Sweden's transnational corporations so that they do not export jobs and so that they continue to pay high taxes at home is proving to be more and more difficult. Some economists are beginning to point out that Sweden's market socialism is based on an inherent irony: Strong and profitable capitalist businesses are necessary so that workers can be employed and taxes for welfare benefits can be collected. However, insisting on generous worker benefits and full employment eats into capitalist profits (Olsen 1992, 1996).

Swedish Lesser Coat of Arms.

Primary production involves direct contact with natural resources—fishing, hunting, farming, and forestry. Until a few hundred years ago, the vast majority of human beings were involved in extracting these raw materials. Because primary production is almost necessarily rural production, it usually entails small, kin-based communities and a close association between work and other aspects of life.

The tertiary sector has grown very rapidly in the last half-century and is projected to grow still more. As Figure 13.2 illustrates, only 19 percent of the labor force was involved in tertiary production in 1920; by 1956, the figure had grown to 49 percent, and by 2000 it is expected to include a full 76 percent of the labor force. Simultaneously, the portion of the labor force employed in primary production has been reduced to almost nil, and the proportion employed in secondary production has halved. These shifts do not mean that primary and secondary production are no longer important. A large service sector depends on primary and secondary sectors that are so efficiently productive that large numbers of people are freed from the necessity of direct production.

The expansion of the tertiary sector is largely a post-World War II phenomenon, and the consequences of this change for societies and individuals are not yet fully understood. Some of these changes are addressed later in this chapter when we discuss the changing nature of work in the United States.

The Dual Economy

The U.S. economic system can be viewed as a **dual economy.** Its two parts are the complex giants of the industrial core and the small, competitive organizations that form the periphery. They are distinguished from each other on two dimensions: the complexity of their organizational forms and the degree to which they dominate their economic environment (Baron & Bielby 1984).

A **dual economy** consists of the complex giants of the industrial core and the small, competitive organizations that form the periphery.

The Industrial Core: Corporate Capitalism

Although there are more than 250,000 businesses in the United States, most of the nation's capital and labor are tied up in a few giants that form the industrial core. The top 20 U.S. companies control billions of dollars of assets, many of them employing hundreds of thousands of individuals. These giants loom large on both the national and international scene.

FIGURE 13.2
Changing Labor Force in the United States
Since 1820, the labor force in the United States has changed drastically. The proportion of workers engaged in primary production has declined sharply while the proportion engaged in service work has expanded greatly.

SOURCE: Graph adapted from Graham T. T. Molitor. 1981. *The Futurist.* With permission from Public Policy Forecasting, 9208 Wooden Bridge Road, Potomac, Maryland 20854.

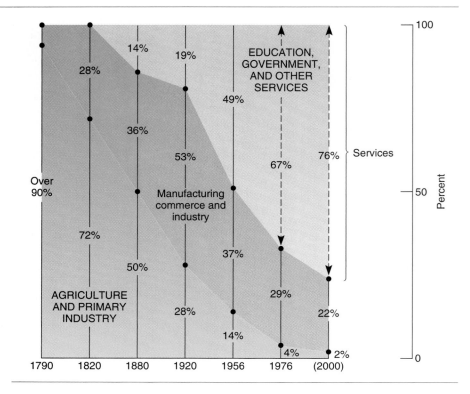

INTERDEPENDENCE AND TIES TO GOVERNMENT.   At the local level, we are familiar with the situation where a region's one major employer holds city and county government hostage and bargains for tax advantages and favorable zoning regulations in exchange for increasing or retaining jobs. Research strongly suggests that the growing size and interdependence of firms in the industrial core are causing this scene to be reenacted at the federal level.

Wealthy capitalists are linked to each other by shared ownership of large firms; large firms are linked to one another by having common members on their boards of directors and through their dealings with the same financial institutions. As a result of this interdependence, relations among large firms have become more cooperative than competitive. In strictly economic terms, decreased competition reduces productivity and efficiency. At the same time, decreased competition means that large firms and the capitalist class become a more potent political force (Mizruchi 1989). There is good evidence that large firms generally support the same political candidates and seldom engage in direct political opposition (Mizruchi 1990). Furthermore, one study shows that as the proportion of the nation's assets held by the top 100 firms increased, the effective rate of corporate taxation underwent a corresponding decrease (Jacobs 1988). The fact that individual income tax rates have increased over the last 40 years while corporate income tax rates have actually declined reflects both the increasing wealth and power of the capitalist elite and the erosion of organized labor as a significant political force (Campbell & Allen 1994).

Transnationals are large corporations that operate internationally and that have power that transcends the laws of any particular nation state.

Transnationals   Corporate size and the link between political and economic interests are of particular concern when the economic actors enter the international arena as **transnationals.** These large international companies (mostly of U.S. origin), such as International Telephone and Telegraph, IBM, and General Motors, are so large that

U.S.-owned transnational corporations have invested heavily in the economies and in the politics of many least-developed nations. Interested in access to new natural resources or new markets for their products, companies search for oil in the ecologically fragile Amazon River basin in Peru and distribute Coca Cola to consumers in New Guinea. In these instances and in all others, transnational corporations have a tremendous impact on the environment, the economy, and the lifestyles of the less- and least-developed nations.

they dwarf many national governments in size and wealth. Their ability to move capital, jobs, and prosperity from one nation to another gives them power that transcends the law of any particular country (Michalowski & Kramer 1987).

There is considerable debate about the possible effects, good and ill, of such international economic enterprises. A few observers hope that ties of international finance will create a more interdependent (and peaceful) world (Tannenbaum [1968] 1979). Others are concerned that transnationals are exercising a thinly veiled imperialism. They allege that the practice of having labor-intensive work done in the Third World exposes workers in those countries to dangers that are banned by law in most Western nations (Michalowski & Kramer 1987); this practice works to the disadvantage of U.S. workers at the same time that it helps to perpetuate economic dependence in less-developed countries (Barnet & Muller 1974). Yet another concern about transnationals is their influence on U.S. foreign policy. In the 1920s U.S. fruit, copper, and oil companies dominated the economic life of Latin America. These companies used bribes, guns, and cannons to affect national and international policy (Patterson, Clifford, & Hagan 1983). Occasionally, they even brought in the Marines; during the 1920s, the United States invaded 12 Latin American countries, including Nicaragua and El Salvador. Could economic interest lead us to invade those countries, or others, again?

The Competitive Sector: Small Business

The competitive sector of the U.S. economy is made up largely of small businesses that are family owned or operated by a small group of partners. They are usually characterized by few employees, economic uncertainty, relatively little bureaucratization of management and authority, and particularism in hiring, firing, and promotion. The chief examples of these kinds of businesses are farming, small banks and retail stores, and restaurant and repair services. Some small manufacturing companies continue to meet these criteria.

This segment of the competitive sector is what Marx called the petit (pronounced *petty*) bourgeois. The **petit bourgeois** are those who use their own modest capital to establish small enterprises in which they and their family provide the primary labor

The **petit bourgeois** are those persons who use their own modest capital to establish small enterprises in which they and their family provide the primary labor.

The competitive sector has provided important economic opportunities for minority Americans, who may encounter significant barriers to being hired or promoted in corporate United States. In cities such as Miami and Los Angeles, Cuban and Korean Americans earn modest profits by using blue-collar or low-skill white-collar talents plus the labor of all family members to run grocery stores, fast-food franchises, laundries, or other small businesses in neighborhoods that are considered too risky or too costly for larger business firms to serve.

(Bechhofer & Elliott 1985). This class plays an important role in Marxist theory. Its continued existence challenges Marx's prediction that the population would be polarized into only two classes.

Recent evaluations show that the class is smaller than it was 100 years ago. Nevertheless, it has remained a vigorous part of the economy. It continues to furnish jobs for a substantial portion of the population and to be a flexible and creative part of both the industrial and service sectors (Bechhofer & Elliott 1985). It has been an especially important avenue of opportunity for minorities in the United States. Koreans, Hispanics, and African Americans who doubt their ability to be hired and make a successful career with a major corporation may nevertheless achieve moderate prosperity by operating neighborhood grocery stores, laundries, and fast-food franchises.

INFORMAL ECONOMY. An important sector of the periphery is the underground or **informal economy.** This is the part of the economy that escapes the record keeping and regulation of the state. It includes illegal activities such as prostitution, selling drugs, and running numbers, but it also includes a large variety of legal but unofficial enterprises such as home repairs, house cleaning, and garment subcontracting. Often referred to disparagingly as "fly-by-night" businesses, enterprises in the informal sector are nevertheless an important source of employment. This is especially true for those segments of the population who would like to avoid federal record keeping: illegal aliens, those aliens whose visas do not permit working, senior citizens and welfare recipients who don't want their earnings to reduce their benefit levels, adolescents too young to meet work requirements, and many others (Portes & Sassen-Koob 1987).

The **informal economy** is the part of the economy that escapes the record keeping and regulation of the state; also known as the underground economy.

The Segmented Labor Market

The **segmented labor market** parallels the dual economy. Hiring, advancement, and benefits vary systematically between the industry core and the periphery.

Parallel to the dual economy is a dual labor market, generally referred to as a **segmented labor market,** in which hiring, advancement, and benefits vary systematically between the industrial core (the corporate sector) and the periphery (the competitive sector).

In the industrial core, firms generally rely on what are called internal labor markets. Almost all hiring is done at the entry level, and upper-level positions are filled from below. At all levels, credentials are critical for hiring and promotion. Within core firms, there are predictable career paths for both blue- and white-collar workers. Employment is generally secure, and benefits are relatively good. Wages and benefits are best in the very largest firms (Villemez & Bridges 1988).

In the competitive sector, on the other hand, credentials are less important, career paths are short and unpredictable, security is minimal, and benefits are low or nonexistent. However, bureaucratization and red tape are less in the competitive sector, and both workers and managers have more freedom in their work.

The competitive sector offers a haven of employment for those who do not meet the demands of the corporate sector: those who do not have the required credentials, who have spotty work records, or who want to work part-time. As a result, a disproportionate number of youths, minorities, and women work in the competitive sector. Because keeping a job and getting a promotion are governed almost exclusively by personal factors rather than by seniority or even ability, however, this sector is less likely to promote minorities or women; there is no affirmative-action officer at Joe's Café. As a result, the gender gap in wages is significantly larger in the competitive sector than it is in core industries (Coverdill 1988).

Work in the United States

From the individual's point of view, economic institutions mean jobs. For some, jobs are just jobs; for others, they are careers. But 40 years of involvement in the world of work is central to most people's lives.

Occupations

Aside from the simplest consequences of working (income and filling up 40 hours of your time), what you do at work is probably as important as whether you work. Here some of the important differences between the professions and white- and blue-collar work are described.

Professions

Occupations that demand specialized skills and creative freedom are **professions.** Their distinctive characteristics include (1) the production of an unstandardized product, (2) a high degree of personal involvement, (3) a wide knowledge of a specialized technique, (4) a sense of obligation to one's art, (5) a sense of group identity, and (6) a significant service to society (Gross 1958). The definition of professions was originally developed for the so-called learned professions (law, medicine, college teaching). It applies equally well, however, to actors, dancers, and potters.

> **Professions** are occupations that demand specialized skills and creative freedom.

The rewards that professionals achieve vary considerably. Some, such as physicians and lawyers, receive very high income: dancers and potters may earn very little. The major reward that all professionals share, however, is substantial freedom from supervision. Because their work is nonroutine and requires personal judgments, professionals have been able to demand—and get—the right to work their own hours, do things their own way, and arrange their own work lives.

Freedom from supervision remains the most outstanding reward of professional work, but it is a reward that is being eroded. Increasingly, people in the professions work for others within bureaucratic structures that constrain many of the most characteristic aspects of professionalism.

White-collar workers appear across nearly the entire spectrum of occupational prestige. They range from the top executives of major firms down to minimum-wage clerks in government offices. This worker is probably typical of the upper-end of the white-collar spectrum, with a job that demands a college education, the ability to do independent thinking, and excellent communication skills. Although it may be high-stress, it also probably pays well, providing intrinsic and extrinsic rewards.

What Color is Your Collar?

Fifty years ago, the color of your collar was a pretty good indication of the status of your job. People who worked with their hands wore blue (or brown or flannel) collars; managers and others who worked in clean offices wore white collars. Those days are past. The labor force is far more diversified, and some of the old guidelines no longer work. The bagger at Safeway wears a white shirt and tie; the librarian wears blue jeans and sandals. Yet the librarian is a white-collar worker and the bagger is not.

Traditional white-collar workers are managers, professionals, typists, sales-people—those who work in clean offices and are expected to be able to think independently. Blue-collar workers are people in primary and secondary industry who work with their hands; they farm, assemble telephones, build houses, and weld joints. Although some blue-collar workers earn more than some white-collar workers, their jobs are characterized by lower incomes, lower status, lower security, closer supervision, and more routine.

Fifty years ago, this simple, two-part division of the labor force included most workers. These days it leaves out a growing category of low-skilled, low-status workers. Some have called these people the pink-collar workers, but they as often appear in company-supplied brown polyester suits or turquoise jackets. They fry hamburgers, stock K-Mart shelves, and collect money at the "U-Serv" gas station. Characteristically, these workers hold jobs that have a short or nonexistent career ladder, and they earn the minimum wage or close to it.

Occupational Outlook

As the graph in Figure 13.2 indicated, the outlook for the future is for greater expansion of the tertiary sector and even more reductions in employment in secondary and primary production. What will this mean for the kinds of jobs that are available in the future? Some of the changes projected between 1994 and 2005 are illustrated in Table 13.3.

Some traditional occupational categories are expected to suffer major declines. The occupations with the largest projected decreases include both blue- and white-collar jobs: Typists and word processors will have a harder time finding jobs, as will sewing-machine operators and farmers. The declining opportunities in these occupations reflect a variety of factors: changing age structure, loss of U.S. jobs due to migration of industry overseas, and new technology.

The issue of most controversy is what kind of new jobs the economy will offer. Optimistic observers point to the fact that executive and professional jobs are growing faster than average and point to the high quality and good pay of these new jobs as indicators of what awaits today's college graduates. Others focus on the rapid increase in what one critic has called "McJobs." Although not all these jobs entail selling hamburgers, many are low-status jobs with low wages and no benefits: health aides, personal and home-care aides, and cashiers. The facts about the occupational outlook are presented in Table 13.3, which shows the five jobs that are expected to grow the most in absolute number and the five jobs that will grow the most in percentage terms (that is, in relation to their size in 1994).

Both the optimists and the critics are correct in their expectations for the future. Good jobs for college graduates and those with technical training—computer engineers and scientists, registered nurses, and system analysts—are growing rapidly. At the same time, however, bad jobs that pay very poorly—for home health aides, waiters and waitresses, and personal care assistants—are also growing rapidly. Thus, the fastest-growing occupations require either years of advanced education or almost

TABLE 13.3
The Shifting Job Market: Projected Changes Between 1994 and 2005[a]
The demand for labor is expected to grow between 1994 and 2005, but opinion differs over the kinds of jobs that will be available for future workers. Although some observers note with satisfaction the growth of high-skill positions, others point with concern to the fact that many of the fastest growing jobs are low-skill and low-wage jobs.

	CHANGES 1994–2005	
Jobs	Percentage Increase	Number of New Jobs
The five fastest-growing jobs relative to their size in 1994:		
Personal and home-care aides	119	212,000
Home health aides	102	428,000
Systems analysts	92	445,000
Computer engineers	90	117,000
Physical and corrective therapy assistants and aides	83	64,000
The five fastest-growing jobs in absolute number:		
Cashiers	19	562,000
Janitors and cleaners, including maids and housekeeping cleaners	18	559,000
Salespersons, retail	14	532,000
Waiters and waitresses	26	479,000
Registered nurses	25	473,000

SOURCE: Silvestri, George. *Monthly Labor Review,* November, 1995.

[a] These projections are from the moderate series, which assumes no major changes in general economic trends.

no skill at all, and the latter offer very little reward. The losers in the transformation of the labor market are likely to be the traditional working class: men and women who did skilled manual labor.

The Meaning of Work

For most people, work is essential as the means to earn a livelihood. As noted in Chapter 6, one's work is often the most important determinant of one's position in the stratification structure and, consequently, of one's health, happiness, and lifestyle.

Work is more than this, however. It is also the major means that most of us use to structure our lives. It determines what time we get up, what we do all day, who we do it with, and how much time we have left for leisure. Thus, the nature of our work and our attitude toward it can have a tremendous impact on whether we view our lives as fulfilling or painful. If we are good at it, if it gives us a chance to demonstrate competence, and if it is meaningful and socially valued, then it can be a major contributor to life satisfaction.

Work Satisfaction

U.S. surveys consistently find that the large majority (80 percent) of workers report satisfaction with their work. Although such a report may represent an acceptance of one's lot rather than real enthusiasm, it is remarkable that so few report dissatisfaction.

Studies of job satisfaction concentrate on two kinds of rewards that are available from work. **Intrinsic rewards** arise from the process of work; you experience them when you enjoy the people you work with and feel pride in your creativity and

Intrinsic rewards are rewards that arise from the process of work; they include enjoying the people you work with and pride in your creativity and accomplishments.

The fastest growing jobs in the United States today are in the service sector. Many of these are what have been called "McJobs." They are minimum-wage jobs offering few benefits and very limited career ladders. Our future job market appears to be splitting into two very different components: high-technology jobs that require advanced education, and low-skill service jobs. The decline of good-paying working-class jobs is a major concern.

Extrinsic rewards are tangible benefits such as income and security.

accomplishments. **Extrinsic rewards** are more tangible benefits, such as income and security; if you hate your job but love your paycheck, you are experiencing extrinsic rewards.

Generally, the most-satisfied workers are those in the learned professions, people such as lawyers, doctors, and professors. These people have considerable freedom to plan their own work, to express their talents and creativity, and to work with others; furthermore, their extrinsic rewards are substantial. The least-satisfied workers are those who work on automobile assembly lines. Although their extrinsic rewards are good, their work is almost completely without intrinsic reward; they have no control over the pace or content of the work and are generally unable to interact with co-workers. A survey of automotive assembly-line workers showed that only 8 percent would choose the same occupation again, whereas 93 percent of urban university professors would choose the same occupation (Kohn 1972). In between these extremes, professionals and skilled workers generally demonstrate the greatest satisfaction; semiskilled, unskilled, and clerical workers indicate lower levels of satisfaction.

Alienation

Alienation occurs when workers have no control over the work process or the product of their labor; they are estranged from their work.

Another dimension of the quality of work life is explicitly Marxist: **Alienation** occurs when workers have no control over their labor. Workers are alienated when they do work that they think is meaningless (push papers or brooms) or immoral (build bombs) or when their work takes their physical and emotional energies without giving any intrinsic rewards in return. Alienated workers feel *used*.

The factory system of the mid-nineteenth century was the ultimate in alienation. In 1863, a mother gave the following testimony to a committee investigating child labor:

> When he was seven years old I used to carry him [to work] on my back to and fro through the snow, and he used to work 16 hours a day. . . . I have often knelt down to feed him, as he stood by the machine, for he could not leave it or stop. (cited in Hochschild 1985, 3)

This child was truly an instrument of labor. He was being used, just as a hammer or a shovel is, to create a product that would belong to someone else.

Although few of us work on assembly lines any more, modern work can also be alienating. Service work, in fact, has its own forms of alienation. In occupations from nursing to teaching to working as flight attendants, not merely our bodies but also our emotions become instruments of labor. To turn out satisfied customers, we must smile and be cheerful in the face of ill humor, rudeness, or actual abuse. Studies of individuals in these occupations show that many have trouble with this emotional component of their work. After smiling for eight hours a day for pay, they feel that their smiles have no meaning at home. They lose touch with their emotions and feel alienated from themselves (Hochschild 1985). This is especially true when workers feel that they have no control over the conditions of their job (Bulan, Erickson, & Wharton 1997).

Alienation is not the same as job dissatisfaction (Erikson 1986.) Alienation occurs when workers lack control. It is perfectly possible that workers with no control, but with high wages and a pleasant work environment, will express high job satisfaction. Marxist scholars believe that job satisfaction in such circumstances is a sign of false consciousness (Halaby 1986).

Technology, Globalization, and the Future of Work

The productivity of workers and the quality of their work experience are often tied directly to the technologies they work with. Some work technologies, such as the assembly line, increase alienation while they increase productivity. Others, such as the photocopier, appear to be unmixed blessings. Although computerization and automation have increased productivity per worker, many people argue that the new technology is inescapably antilabor. These critics point out three negative effects of technology on labor: deskilling, displacing workers, and greater supervision.

1. *Deskilling.* Many observers believe that increased mechanization has reduced the skill level needed for many jobs to the point where it is difficult to take pride in craft or a job well done. The process of automating a job so that it takes much less skill than it used to is called *deskilling.* Deskilling occurs at all levels of labor, not just on the assembly line. For example, in the days before word processors, photocopiers, and self-correcting typewriters, a good typist could take pride in the work. With the new technologies, almost anyone can turn out decent-looking copy.

An important element of the deskilling process is that it reduces the scope for individual judgment. In hundreds of jobs across the occupational spectrum, computers make decisions for us. In the sawmill industry, for example, a computer now assesses the shape of a log and decides how it should be cut to get the most board feet of lumber from it. An important element of skill and judgment honed from years of experience is now made worthless. According to the chief proponent of this argument, the central process in deskilling is the separation of mind and hand (Braverman 1974).

2. *Displacement of the labor force.* One of the most critical complaints about automation is that it replaces people with machines. A few examples suffice. Computerization in grocery stores has resulted in sharp reductions in employment by eliminating inventory clerks and pricing/repricing personnel, as well as reducing the skill level in cashiering to the point that an average 15-year-old could do it. In the automobile industry, it is estimated that one of the new robots can replace 1.5 humans per shift—and the robot can work three shifts a day (U.S. Department of Labor 1985; 1986b).

In industry after industry, more sophisticated technology has made sharp inroads into the number of hours of labor necessary to produce goods and services. Many have concluded that fear of job loss is one of the reasons employees seldom complain about

THINKING CRITICALLY

How has technology affected your schoolwork in the last ten years? Has it given you new tools or robbed you of old skills or simply given your instructors an excuse to demand more of you?

technology

THE LUDDITES: DOWN WITH MACHINES

Since the dawn of the industrial era, there has been tension between labor and technology. In 1675, weavers rioted against the introduction of looms that could allegedly do the work of 20 people; in 1768, sawyers in London destroyed a mechanized sawmill. The most widely known revolt of labor against machinery, however, was the Luddite uprising in England between 1811 and 1816.

Wool was a major part of the English economy in the early nineteenth century. It was largely a home industry; and in Lancashire, nearly every home was engaged in wool production. The work was tedious and difficult. Particularly difficult was the last stage, in which a worker wielding 50-pound shears finished the fabric by cutting off all the nubs. Being able to handle these shears for 88 hours a week (the standard work week) required great strength and skill. It was an esteemed occupation.

In 1811, finishing machines were introduced to do this work. Each machine replaced six men. Not only were the men out of a job, but also the skills developed over a lifetime were made worthless. As use of the machines spread, large numbers of men were thrown out of work, and their families starved. On the horizon were more machines to take over other phases of wool production. Added to this, England was engaged in the Napoleonic Wars, and associated trade embargoes made the price of food high. The classic ingredients for insurrection were in place. The focus of the workers' anger was the machines, and their response was to destroy them.

In 1811, a young man named Ned Ludd, or Lud, or maybe Ludlam, is alleged to have broken up his father's hosiery loom because he resented a rebuke. The incident, which may have been imaginary, coincided with the eruption of machine-breaking demonstrations, and the labor movement came to be called the Luddites.

In the early nineteenth century, any organization of labor was illegal. Nevertheless, laborers met secretly at taverns and later in the woods to plan well-organized attacks. A body of men with blackened faces would break into a shop and destroy all of the machinery. As the movement progressed, it became less disciplined, and owners too were assaulted and their homes looted. It was, said one observer, collective bargaining by riot.

The government was uncertain how to respond. There were a few liberals who were sympathetic with labor. Lord Byron, for example, wrote that "however much we may rejoice in any improvement in ye arts which may be beneficial to mankind; we must not allow mankind to be sacrificed to improvements in Mechanism" (cited in Reid 1986). The hardliners won, however, and troops were sent in to restore order. Leaders and alleged leaders of the Luddites were hanged or deported to the far corners of the empire.

> **Now, every year many new technologies are developed that can reduce even further the dependence of industry on human labor.**

The Luddite movement caused hardly a pause in the increasing use of machines to replace workers. Now, every year many new technologies are developed that can reduce even further the dependence of industry on human labor. Although we have unemployment insurance, early retirement schemes, and welfare to cushion the blows, the process still causes human misery as valuable skills are debased and employment is lost. If we use the term *Luddite* to include all of those who "resist mechanization, automation, and the like, and who are the supposed enemies of 'progress' where the adoption of labor-saving devices and information technology is concerned (Thomis 1970, 12), there are probably plenty of people who remain Luddites in spirit.

the deskilling aspects of their jobs. If they still have a job, they are happy about it (Vallas & Yarrow 1987).

3. *Greater supervision.* Computerization and automation give management more control over the production process. More aspects of the production process are determined by management through its computerized instructions, and fewer aspects are determined by the employees. Computers also keep more complex and thorough records on employees. For example, the scanner machines used in grocery stores do more than keep inventory records and add up your grocery bill. They also keep tabs on the checker by producing statistics such as number of corrections made per hour, number of items run through per hour, and average length of time per customer. It is not

surprising, therefore, that studies show that computers have increased work alienation among the cashiers and typists who use them (Vallas 1987).

Whether new technologies are an enemy of labor may depend on which laborer we ask. From the standpoint of professionals in the knowledge industries (education, communications), new technology is undoubtedly a boon. Computers have expanded their job opportunities and enhanced their lives. Those whose work is being replaced rather than aided by computers, however, are less likely to see anything wonderful about it. Those most adversely affected by these developments are women and less-skilled workers (Gill 1985).

One scholar has argued that technology by itself is a neutral force: It can aid management or it can aid the workers (Davies 1986). Which technologies are implemented and the way they are implemented reflects a struggle between labor and management, and this struggle, not the technology itself, will determine the outcome.

Protecting U.S. Jobs

The consequences of "de-laborization"—loss of jobs—are substantial. Not only individual workers but entire communities are impoverished as new technologies facilitate corporate decisions to move factories to other parts of the world where labor is much cheaper. In fact, the national economy is undergoing a process of reverse development: Like a Third World country, we export raw materials such as logs and wheat and import manufactured products such as VCRs and automobiles. People in Mexico and Japan and Korea have jobs manufacturing products for the U.S. market while U.S. workers are making hamburgers.

What can public policy do to protect jobs in the United States? There are three general policy options: the conservative free-market option, new industrial policies, and the social welfare option (Hooks 1984).

THE CONSERVATIVE FREE MARKET APPROACH.   Generally, business leaders and conservatives argue that the way to keep jobs in the United States is to reduce wages and benefits. If labor is cheap, they argue, business will have less incentive to automate or to move assembly plants to Mexico or Indonesia.

By default, this policy has been implemented. In communities across the nation, managements have used threats of plant closings to force wage concessions and reduce benefits. Labor unions are reduced to negotiating benefit protection in the face of wage reductions. Because so many workers have been afraid of losing their jobs, organized labor's power has been sharply reduced. Thus, one result of "de-laborization" is the reduced economic circumstances of workers who still have jobs.

NEW INDUSTRIAL POLICIES.   Liberals argue that private profit should not be the only goal of economic activity and that the state should see to it that economic decisions protect communities' and workers' interests (Genovese 1989). Among the specific policies recommended are: (a) federal trade policies that make U.S.-made goods more competitive in international markets and that reduce the advantage that foreign-made products have in the United States, (b) vigorous state investment in industries that will provide the largest number of decent jobs, (c) government oversight of mergers and plant closings to make sure plants behave responsibly, and (d) state support for worker efforts to buy and manage their own industries.

SOCIAL WELFARE POLICIES.   New industrial policies are designed to keep people working; social welfare policies are aimed at protecting those who are thrown out of

THINKING

CRITICALLY

How will a postindustrial economy affect *your* working and economic future, do you think? In what ways is a postindustrialized economy a global economy? Which of the three general policy options outlined here do you think the United States will follow, and why? Which, in your opinion, would be the best one to follow, and why?

work. Among the policies recommended are (Blakely & Shapira 1984): (a) six-month notification of plant closings, (b) paid leave for soon-to-be displaced employees to look for jobs, (c) retraining programs for displaced workers, (d) relocation assistance for displaced workers, and (e) substantially more generous unemployment benefits. Of course, such suggestions are open to the same concerns that Sweden is now experiencing in regard to its welfare economy.

Conclusion

As you sit in the classroom in order to prepare yourself for a good position in the labor force, the economy itself is changing. Indeed, it is changing so fast that you may need to retool several times before your work life is complete. Increasingly, rapid developments in technology have dramatically changed the workplace from what it was 20 or even 10 years ago. The globalization of the economy has further changed the job situation for Americans. While the specter of unemployment still haunts ethnic minorities and blue-collar workers the most, the middle class is also experiencing the pangs of job insecurity and alienation. One political approach to the changing job situation involves social welfare policies that would provide greater security to workers. In the face of large budget deficits, it seems unlikely that the United States will expand its welfare policies. It is more likely that U.S. workers will increasingly need to learn new skills and be creative in finding and keeping jobs.

Summary

1. Power may be exercised through coercion or through authority. Authority may be traditional, charismatic, or rational-legal.

2. Any ongoing social structure with institutionalized power relationships can be referred to as a political institution. The most prominent political institution is the state. It is distinguished from other political institutions because it claims a monopoly on the legitimate use of coercion and it has power over a broader array of issues.

3. Democracy is not just a matter of having the right values; it also requires a supportive institutional environment. Such an environment is characterized by a large and relatively affluent middle class.

4. Although the Democrats tend to attract working-class and minority voters and Republicans tend to attract better-off voters, both U.S. political parties tend to have middle-of-the-road platforms with broad appeal. Unlike many other democracies, the United States has no working-class party.

5. Four major models are used in describing the U.S. political process: the pluralist model, the power-elite model, the conflict/dialectic model, and the state autonomy model. None of them suggests that the average voter has much power to influence events.

6. Political participation is rather low in the United States; fewer than half of the people of voting age vote in most national elections, and fewer yet take an active role in politics. Political participation is greater among those with high social status and among middle-aged and older people—establishment types who are more likely to support the status quo.

7. Capitalism is an economic system that maximizes productivity but tends to neglect aspects of distribution; socialism emphasizes distribution and neglects aspects of production. The actual operation of an economic system depends on the political structure in which it operates. Communism is a socialist economy in an authoritarian political system.

8. Changes from preindustrial to industrial to postindustrial economies have had profound effects on social organization. The tertiary sector of the economy is expected to occupy three-quarters of the labor force by the year 2000; it includes highly paid professional occupations as well as maids and waitresses.

9. The United States has a dual economy containing two distinct parts: the industrial core and the competitive sector or the periphery. These are paralleled by a segmented labor market.

10. Economic projections show substantial changes in occupations in just the next 10 years. The largest number of new jobs will be low-status, low-wage service positions. The major losers will be those who have occupied traditional blue-collar jobs.

11. Scholars look at the individual meaning of work from two perspectives: work satisfaction and alienation. Although most U.S. workers report satisfaction with their work, Marxists argue that they are nevertheless alienated because they are estranged from the products of their labor.

12. Critics argue that automation and computerization have had three ill effects on labor: deskilling jobs, reducing the number of jobs, and increasing control of workers. Nevertheless, some occupations have grown or been made easier through new technology.

Sociology on the Net

Over the past few decades politics have changed. Organized political parties have lost a great deal of power and political action committees (PACs) had emerged as a central player in the political world. The Federal Election Commission keeps track of all PAC contributions.

```
http://www.fec.gov/
```

Once you have arrived at the Federal Election Commission home page, open the section on *Financial Information About Candidates, Parties and PACs*. Scroll down to the section dealing with *PACs* and click on the other selection called *PAC Summary Financial Information*. Browse though the selections. What organizations seem to be the biggest contributors? Who gets more money from PACs, incumbents or challengers? Why would these organizations be interested in influencing politics?

Return to the section of *Financial Information About Candidates, Parties and PACs* and click on the highlighted phrase *U.S. House Candidates*. Go to your home state and find your local representative. How much money does that person have? For more detailed information on any candidate, you might wish to try *Campaign Central*.

```
http://www.clark.net/central/
```

While the United States has maintained a stable democratic government, Mexico has struggled with the democratic process. Today, the Chiapas rebels stand in opposition to the current policies of the Mexican government. Thanks to the Internet, we can view the struggle from their perspective.

```
http://www.peak.org/~justin/ezln/ezln.html
```

Begin your browsing by scrolling down to the background section and clicking on *EZLN FAQ*. When you have acquainted yourself with the EZLN, return to the background section and open the highlighted section entitled *The Southeast in Two Winds by Subcomandante Marcos*. What is the EZLN? What is the history of this struggle? How does conflict theory help you understand the rebel position? You may wish to find a few shorter statements under the heading of *Communicados*.

FIND IT ON INFOTRAC COLLEGE EDITION

A 1996 survey conducted by Brown University contends that most U.S. citizens doubt the effectiveness of U.S. democracy. A discussion of the Brown survey can be found in the following article in InfoTrac College Edition:

> "How Effective is Democracy?" *USA Today (Magazine)*, August 1996, v125 n2615 p3. (Hint: Enter the search term *democracy and effective* using the Key Words Guide.)

How effective is the United States's system of political democracy? Your text describes four models of democracy: the pluralist, power-elite, conflict/dialectic, and the state autonomy models. Which model best describes the current system in the United States? If the Brown survey is an accurate reflection of popular sentiment, which model do you believe would citizens see as more effective?

Suggested Readings

Domhoff, G. Williams. 1993. *Who Rules America?* Englewood Cliffs, N.J.: Prentice Hall. A classic by a scholar who spent 20 years chronicling the antics of the power elite.

Elshtain, Jean Bethke. 1995. *Democracy on Trial.* New York: Basic Books. A work in which the author, a professor of social and political ethics at the University of Chicago, argues that people in the United States have become too self-absorbed and politically alienated and that as a result U.S. democracy is in serious crisis.

Gamst, Frederocl C. (ed.). 1995. *Meanings of Work: Considerations for the Twenty-first Century.* Albany, N.Y.: State University of New York Press. An up-to-date analysis of what it means to work today, given all the changes in the workplace and workforce over the past several decades.

Hochschild, Arlie R. 1985. *The Managed Heart: The Commercialization of Human Feelings.* Berkeley: University of California Press. A study of alienation in service occupations, with detailed examination of how flight attendants handle emotional work. This book is becoming a classic.

Mills, C. Wright. 1956. *The Power Elite.* New York: Oxford University Press. One of the most important discussions of the relationship among three areas of power in the United States: government, business, and the military. Mills, a conflict theorist, argues that the members of a small elite make the decisions that control U.S. society.

Romero, Mary. 1992. *Maid in the U.S.A.* New York: Routledge. Research monograph, written mainly from a conflict, or critical, perspective, on the occupation of domestic workers in the United States today.

Rubin, Beth A. 1996. *Shifts in the Social Contract: Understanding Change in American Society.* Thousand Oaks, Calif.: Pine Forge Press. A thoughtful analysis of how changes in the U.S. and world economies have resulted in changes in politics, in the institution of family, and in the way we can expect to live our everyday lives.

Population and Urban Life

Outline

Birth and death. Nothing in our lives quite matches the importance of these two events. Naturally, each of us is most intimately concerned with our own birth and death, but to an important extent our lives are also influenced by the births and deaths of those around us. Do we live in large or small families, large or small communities? Is life predictably long or are families, relationships, and communities periodically and unpredictably shattered by death?

In this chapter we take a historical and cross-cultural perspective on the relationship between social structures and population. We look at fertility and mortality and how they relate to social structures, and also at the effect of community size on social relationships. We are interested in questions such as how birthrates or community size affect social roles and, conversely, how changing social institutions affect fertility and residence patterns.

The Demographic Transition

In 1997, the world population was 5.8 billion, give or take a couple hundred million. This is more than twice as many people as lived in 1950. In part because of this growth, millions are poor, underfed, and undereducated. In part because of this growth, the world economic system is in danger of bankruptcy. Perhaps no other issue is so vitally connected to so many of our era's crises. This section first describes the current world population and then examines the process by which it was reached.

Although population is concerned with such intimate human experiences as birth and death, the big picture of population growth and change can be understood only if we use statistical summaries of human experience. Three measures are especially important: the crude birthrate, the crude deathrate, and the natural growth rate:

The **crude birthrate** (CBR) is the number of births divided by the total midyear population and then multiplied by 1,000; it is read as births per thousand.

The **crude deathrate** (CDR) is the number of deaths divided by the total midyear population and then multiplied by 1,000; it is read as deaths per thousand.

The **natural growth rate** is the crude birthrate minus the crude deathrate and then divided by 10; it is read as the percentage growth rate.

$$\text{Crude birthrate (CBR)} = \frac{\text{Number of births in a year}}{\text{Total population}} \times 1{,}000$$

$$\text{Crude deathrate (CDR)} = \frac{\text{Number of deaths in a year}}{\text{Total population}} \times 1{,}000$$

$$\text{Natural growth rate} = \frac{\text{CBR} - \text{CDR}}{10}$$

Table 14.1 shows these rates in 1997. For the world as a whole, the crude birthrate in 1997 was 24 births per 1,000 population; the crude deathrate was a much lower 9 per 1,000. Because the number of births exceeded the number of deaths by 15 per 1,000, the crude natural growth rate of the world's population was 1.5 percent. If your savings were growing at the rate of 1.5 percent per year, you would undoubtedly think that the growth rate was very low. A growth rate of 1.5 percent in population, however, translates into a doubling time of 47 years. This means that *if* this growth rate continues, the population will double to 11.6 billion in just 47 years.

The frightening prospect of welcoming another 5.8 billion people in our lifetime is complicated by the fact that the growth is uneven. As Table 14.1 shows, growth rates

TABLE 14.1

The World Population Picture, 1997

In 1997, the world population was 5.8 billion and growing at a rate of 1.5 per year. Growth was uneven, however; the less developed areas of the world were growing much more rapidly than the more developed areas. As a result, most of the additions to the world's population were in poor nations.

Area	Crude Birth-Rate	Crude Death-Rate	Crude Natural Growth Rate	Total Population (in millions)	Doubling Time (in years)
World	24	9	1.5%	5,840	47
Africa	40	14	2.6	743	26
Asia	24	8	1.6	3,552	44
Latin America	25	7	1.8	490	38
North America	14	9	0.6	298	117
Europe	10	12	−.01	729	—

SOURCE: Population Reference Bureau, 1997 World Population Data Sheet.

are startlingly different across the areas of the world. Africa is the world's fastest-growing continent. At a growth rate of 2.6 percent per year, it will double its population size in only 26 years. In Europe, by contrast, deaths actually exceed births.

These differentials in growth are of tremendous importance. Almost all the additions to world population in the next several decades will take place in the less-developed nations. As a result, the world is likely to be proportionately poorer in 2030 than it is now. How did we get into this fix?

Population in Former Times

For most of human history, **fertility** (childbearing) was barely able to keep up with **mortality** (death), and the population grew little or not at all. Historical demographers estimate that in the long period before population growth exploded, both the birth- and deathrates hovered around 40–50 per 1,000. (Birthrates are still 40 in Africa). Translated into personal terms, this means that the average woman spent most of the years between 20 and 45 pregnant or nursing. If both she and her husband survived until they were 45, she would produce an average of 6–10 children. The average life expectancy was perhaps 30 or 35. Such a low life expectancy was largely due to very high infant mortality: perhaps one-quarter to one-third of all babies died before they reached their first birthday. Both birth and death were frequent occurrences in most preindustrial households.

Fertility is the incidence of childbearing.

Mortality is the incidence of death.

The Transition in the West

The industrial revolution set in motion a whole series of events that revolutionized population in the West. First, mortality dropped; then, after a period of rapid population growth, fertility declines followed. Because studies of population are called **demography,** this process is called the **demographic transition.**

Decline in Mortality

General malnutrition was an important factor supporting high levels of mortality. Though few died of outright starvation, poor nutrition increased the susceptibility of the population to disease. Improvements in nutrition were the first major cause of

Demography is the study of population—its size, growth, and composition.

The **demographic transition** is the process of moving from the traditional balance of high birth- and deathrates to a new balance of low birth- and deathrates.

the decline in mortality. New crop varieties from the United States (corn and potatoes especially), new agricultural methods and equipment, and increased trade all helped improve nutrition. The second major cause of the decline in mortality was a general increase in the standard of living: better shelter and clothing . . . and soap. Changes in hygiene were vital in reducing communicable disease, especially those affecting young children, such as typhoid fever and diarrhea (Razzel 1974). Because of these factors, the deathrate gradually declined between 1600 and 1850. Despite this decline, the life expectancy for women in the United States was only 40 years at the time of the Civil War.

In the late nineteenth century, public-health engineering led to further reductions in communicable disease by providing clean drinking water and adequate sewage. Medical science did not have an appreciable effect on life expectancy until the twentieth century, but its contributions have sparked a remarkable and continuing increase in life expectancy. In the first 85 years of this century, the life expectancy of U.S. women increased from 49 to 78 years. Thus, although mortality began a steady decline in about 1600, the fastest decreases have occurred in the twentieth century. This decline reflects the almost total elimination of deaths from infectious disease and the steady progress in eliminating deaths caused by poor nutrition and an inadequate standard of living (McKeown & Record 1962; McKeown, Record, & Turner 1975).

Decline in Fertility

The industrial revolution also affected fertility, though less directly. The reduction in fertility was not a response to the drop in mortality or even a direct response to industrialization itself. Rather, it appears to have been a response to changed values and aspirations triggered by the whole transformation of life (Coale 1973).

Industrialization meant increasing urbanization, greater education, the real possibility of getting ahead in an expanding economy, and, most important, a break with tradition—an awareness of the possibility of doing things differently than they had been done by previous generations. The idea of controlling family size to satisfy individual goals spread even to areas that had not experienced industrialization, and by the end of the nineteenth century, the idea of family limitation had gained widespread currency (van de Walle & Knodel 1980). In England and Wales, the average number of children per family fell from 6.2 to 2.8 between 1860 and 1920, the space of just two generations (Wrigley 1969).

As Figure 14.1 shows, the demographic transition in the West will come nearly full circle by the end of the twentieth century. Throughout Europe and North

FIGURE 14.1
The Demographic Transition in the West
Preindustrial populations were characterized by fluctuating deathrates and relatively stable birthrates. Mortality rates gradually stabilized and fell below the fertility rate. Because the decline in mortality was slow and because many of the excess people moved to North America or Australia, this growth did not cause dramatic problems for Europe.

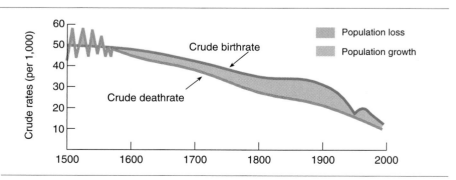

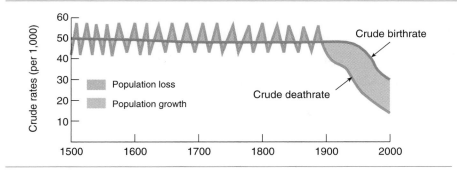

FIGURE 14.2
The Demographic Transition in the Third World
In the Third World, mortality rates fell suddenly while fertility continued at very high levels. The result was dramatic growth. Fertility rates have begun to decline but are still far higher than mortality rates.

America, birth- and deathrates are again about even, and there is little population growth. In the non-West, Africa especially, the transition is taking a very different course; in many of its nations, the transition from high birth- and deathrates to low ones has just begun.

The Transition in the Non-West

In the non-West, birth- and deathrates remained at roughly preindustrial levels until World War II. After that, modern medicine and public-health engineering caused deathrates to tumble. Unlike the mortality decline in the West, this change was not caused by changes in social structure or standard of living, nor was it gradual. It was sudden change brought in from outside. The changes in social structure and culture that are necessary to cause fertility decline are occurring more slowly. With AIDS reaching epidemic proportions in many countries, further fertility declines may come slowly. Despite the near doubling of the deathrate in countries such as Uganda, Zambia, Kenya, and Zimbabwe where the AIDS toll is especially high, the population of these countries is still expected to double in the next 25 years (gopher://gopher.undp.org70/00/ungophers/popin/wdtrends/AIDS). As Figure 14.2 demonstrates, the result has been massive and sudden population growth.

The following sections explore contemporary relationships between social structure and population and then return to the question of what this tells us about solutions to the world population problem.

Fertility, Mortality, and Social Structure

The Effects of Social Structure on Fertility and Mortality

Social Structure and Fertility

In Kenya, the average woman has five or six children; in Italy, the average woman has only one or two. These differences are the product of values, roles, and statuses in very different societies. The average woman in Kenya *wants* five or six children, and the average woman in Italy *wants* only one or two.

The level of fertility in a society is strongly related to the roles of women. Generally, fertility is higher where women marry at younger ages, where they have less access to education, and where their roles outside the household are limited. Fertility also reflects the development of society's institutions. When the family is the source of security, income, social interaction, and even salvation, fertility is high.

Social Structure and Mortality

As noted in Chapter 10, the single most important social factor affecting mortality is the standard of living—access to good nutrition, safe drinking water, protective housing, and decent medical care. Differences in living standards almost entirely account for the fact that the average American can expect to be healthier and to live 43 years longer than the average person living in Sierra Leone; differences in living standards and access to health care also explain the findings from Britain (Marmot, Kogevinas, & Elston 1987) and the United States that show death rates are higher and health is poorer among members of the lower social classes. However, by focusing on prevention and not-for-profit medicine, countries like China have achieved high life expectancy with low per capita income.

More subtly, social structure affects mortality through its structuring of social roles and lifestyle. Race, socioeconomic status, and gender all affect exposure to unhealthy or dangerous lifestyles. People with less education, for example, get less physical exercise and are more likely to smoke than those with more education, and young men are more likely to die in automobile accidents than young women.

The Effects of Fertility and Mortality on Social Structure

Fertility Effects

Fertility has powerful effects on the roles of women. The greater the number of children a woman has, the less likely she is to have any involvement in social structures outside the family. When the average woman bears only two children, fertility places much less restriction on her social involvement.

AGE STRUCTURE.   In addition to affecting women's roles, fertility has a major impact on the age structure of the population: The higher the fertility, the younger the population. This is graphically shown in the population pyramid in Figure 14.3. When fertility is low, the number of young people is about the same as the number of adults; when fertility is high, there are many more children than adults, and the age structure takes on a pyramidal shape.

This pyramidal age structure has both short- and long-term consequences. In the short term, it means that a large proportion of the population is too young to work, and this reduces society's productivity. In the long run, it translates into potential for explosive population growth. In sub-Saharan Africa, the number of girls aged 0–4 exceeds the number of women aged 20–24 by a ratio of two to one. This means that in the next generation there will be twice as many mothers as there are today.

Mortality Effects

Like fertility, mortality has particularly strong effects on the family. A popular myth about the preindustrial family is that it was a multiple-generation household, what we call an extended family. A little reflection will demonstrate how unlikely it is that many children lived with their grandparents when life expectancy was only 30 to 35 years and when fertility was seven to eight children per woman. Reconstructions tell us that only a small percentage of all families could have been three-generational. Quite often, if the children lived with their grandparents, it was because their parents were dead. The household of a high-mortality society was probably as fractured, as full of stepmothers, half-sisters, and stepbrothers, as is the current household of the high-divorce society.

THINKING

CRITICALLY

Because African Americans are more likely to be poor than white Americans, their lower life expectancy is not surprising. Women are more likely to be poor than men; yet their life expectancy is higher. Is the reason biology or social structure?

FIGURE 14.3

A Comparison of Age Structures in Low- and High-Fertility Societies

When fertility is high, the number of children tends to be much larger than the number of parents. When this pattern is repeated for generations, the result is a pyramidal age structure. When fertility is low, however, each generation has a similar size, and a boxier age structure results.

SOURCE: Ashford 1995, 13.

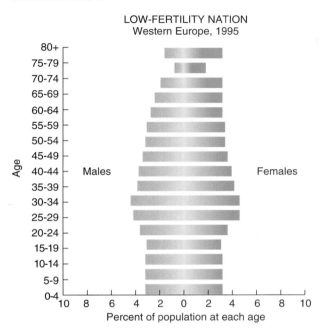

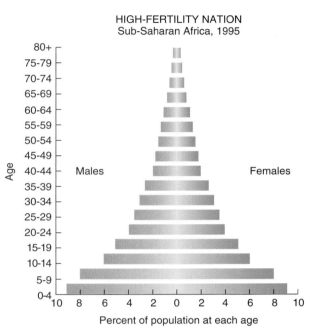

Although children impose burdens on parents, they also bring joy and entertainment. Moreover, children are one of the few sources of joy and reward equally available to rich and poor. As this photograph indicates, women and men with few economic rewards may still find pleasure in their children. Thus, poverty and children are not incompatible and poor people find it rational to have children.

Population and Social Structure: Two Examples

Kenya: A Case Study

Kenya is an example of a society where traditional social structures encourage high fertility. It is also an example of a society where high fertility may ensure continuing traditionalism—and poverty.

The Effects of Social Roles on Fertility

Kenya still has a crude birthrate of 38 per 1,000 population. Mortality, however, is down to 12 per 1,000. This means that the population of Kenya is growing at a rate of 2.6 percent per year. If that rate continues, the population will double in just 27 years. An aggressive family-planning program is unlikely to reduce this growth: The average woman in Kenya wants five or six children.

The high value placed on fertility in Kenyan society is a reflection of several pressures. Among them are tribal loyalties, women's roles, and the need for economic security (Mott & Mott 1980).

TRIBAL LOYALTIES.   Kenya is a diverse nation in which there is jealous competition among tribal groups. Because the size of each tribe's population is an important factor in political power, large families are regarded as politically advantageous. This is a common reason for high fertility in any diverse society. Among French Canadians, it is called the revenge of the cradle.

WOMEN'S NEED FOR CHILDREN.   Regardless of the needs of their tribe or nation, most women give first consideration to how another child will affect them and their family. For the 80 percent of Kenyan women who are responsible for family farms, children are an asset. Three-quarters of Kenyan women list "help with work" as a reason for having children. In addition to helping with the work, children are an important, perhaps the only, source of esteem and power open to women. This is especially true of the 30 percent of Kenyan women who live in polygamous unions. The number of children, especially the number of sons, is an important determinant of a woman's position relative to that of other wives.

ECONOMIC SECURITY.   Children add to their parents' economic security in a number of ways. They are the only form of old-age insurance available. When they grow up and marry, they may also add to the family's economic and political security by their marriages. The greater the number of children, the greater the number of in-laws. A family that can bind itself to many other families has greater political power and more security.

In short, a family's income, status, and long-term security are all enhanced by having many children. There are comparatively few rewards for having a small family. Children are virtually cost-free—no expensive medical treatment is available, what schooling there is has no direct cost to the parents, and children's desires do not run to designer jeans or $130 tennis shoes. With a cost/benefit ratio of this sort, it is not surprising the Kenyans desire many children.

The Effect of Population Growth on Society

Although high fertility may appear to be in the best interests of individual women, its consequences for society are less beneficial. At current rates of fertility and mortality, Kenya's population is doubling every 27 years. As a result, development goals are

In Kenya, the family continues to be the center of economic and social relationships. Women and men find that having many children enhances their prestige, helps with their work, and provides them with economic security for their later years. Because there are few costs associated with having children, the average woman in Kenya desires six and has seven children.

shooting at a moving target. To double the proportion of children getting an elementary school education (from 45 to 92 percent), the government has to raise the dollars spent on education fourfold because the total number of children needing schooling will double. Simply to maintain that level over the next 27 years, the Kenyan government would have to double the dollars spent on education again. Unfortunately, there are other demands on the budget—defense, highways, development, agriculture. All these areas face the same problem of escalating demand.

Thus, a decision that is rational on the individual level turns out to be less wise on the societal level. Occasionally, people in the West make remarks of the sort: "Are they stupid? Can't they figure out they would be better off if they had fewer children?" Unfortunately for the argument, nations don't have children; women do. High fertility continues to be a rational choice for individual Kenyans.

Policy Response

Kenya was the first sub-Saharan African nation to establish an official family-planning policy. The program has had some notable successes, and Kenya recently relinquished the title of having the highest fertility in the world to a Middle Eastern area, the Gaza strip, and to an African nation, Niger. The program has more than doubled the percentage of people using contraception, but the figure still stands at a mere 33 percent of married women (United Nations 1994). Like most family-planning programs, Kenya's is voluntary. It tries to make services available, conveniently and inexpensively to women who want them. When women want six or seven children, though, contraceptives aren't very relevant. As long as social institutions continue to make large families rewarding, government policies are unlikely to reduce fertility substantially.

Europe: Is Fertility Too Low?

In a world reeling from the impact of doubling populations in the less-developed world, it is ironic that many developed countries are worried that fertility is too low.

With modern levels of mortality, fertility must average 2.1 children per woman if the population is to replace itself: two children so that the woman and her partner are replaced and a little extra to cover unavoidable childhood mortality. This is called

Replacement level fertility requires that each woman bear approximately two children, one to replace herself and one to replace her partner. When this occurs, the next generation will be the same size as the current generation of parents.

replacement level fertility. If fertility is less than this, the next generation will be smaller than the current one.

Currently in Europe, the average woman is having 1.4 children. This means that Europe as a whole will begin to lose population over the next few generations. The problem is more severe in some nations than others (Table 14.2). In Spain, Italy, Latvía, and Bulgaria, the average woman is having only 1.2 children. This means that the next generation will be only two-thirds the size of the current one. If this situation continues for several generations, the populations of these countries will be sharply reduced.

Given the serious worldwide dilemmas posed by population growth and the very high density of many European nations, why should this be a problem? There are three broad areas of concern:

1. *Population suicide.* In 1984, French Prime Minister Chirac said, "[the prospect] is terrifying. In demographic terms, Europe is vanishing. Twenty years or so from now, our countries will be empty, and no matter what our technological strength, we shall be incapable of putting it to use" (Teitelbaum 1987).

2. *Too many old people.* A very low-fertility society takes on an age structure that looks like an inverted pyramid. Because each generation is smaller than the preceding one, the older generation is larger than the younger generations on whom it relies for support. This age structure will cause a major dilemma for the old-age portions of many nations' social welfare programs—including that of the United States.

3. *Labor-force shortages.* The decline in fertility has already caused labor-force shortages in many European nations. These shortages have been felt in industry and in the armed forces. The industrial shortages have, until recently, been made up by importing workers from the Middle East or the Mediterranean. No nation, however, wants to staff its army with outsiders. In any case, concern for cultural dilution has caused many European nations to reduce the importation of workers and to urge their guest workers to return home.

THINKING

CRITICALLY

One alternative to the problems posed by population aging is for the current 20-something generation to have more children. Another is to postpone retirement until 70 or so. Would you yourself rather have more kids or work longer? Which problems of population aging will longer lasting careers not solve?

Policy Response

In response to these concerns, many European nations have established incentives to encourage fertility. Among them are paid maternity leave, cash bonuses for extra chil-

TABLE 14.2
Fertility and Population Growth in Europe, 1997
Overall births continue to exceed deaths in Europe, but the margin is increasingly slight and some nations are already experiencing population decline.

Country	CBR	CDR	Crude Natural Growth Rate	Average Number of Children per Woman	Population Change, 1995 to 2025
Europe, total	10	12	−.01%	1.4	+2
Austria	11	10	.1	1.4	+1
Denmark	13	12	.1	1.7	+4
Germany	10	11	−.1	1.3	−3
Hungary	11	14	−.4	1.5	−8
Italy	9	9	.0	1.2	−5
Romania	10	13	−.2	1.3	−6
Spain	9	9	.1	1.2	−12
United Kingdom	13	11	.2	1.7	+6

SOURCE: Population Reference Bureau, 1997 World Population Data Sheet.

Il paraît que je suis un phénomène socio-culturel.

CAMPAGNE RÉALISÉE PAR AVENIR.DAUPHIN GIRAUDY

LA FRANCE A BESOIN D'ENFANTS.

Many countries in Europe have reduced fertility to levels below replacement and are now concerned about what this will mean for the future. Countries such as France are attempting to deal with this problem by actively encouraging fertility through billboard campaigns such as this one. In this picture, the infant portrays itself as a sociocultural phenomenon!

dren, longer vacations for mothers, and graduated family allowances. In some countries, new mothers are eligible for six months of paid maternity leave, and many benefits (family allowances, housing subsidies, and even a lower age at retirement) are graded according to the number of children. Low-interest loans for buying and furnishing homes are available to newly married couples; and with each additional child, an increasing proportion of the loan is written off.

Studies suggest that these incentive plans have had modest effects. In some countries, birthrates jumped after the incentives were introduced, but subsequent analysis suggests that this was because some couples had their children earlier than they intended; the incentives did not prompt very many people to have third or fourth children (David 1982). The difficulty is that children are an expensive, intensive 20-year project. The incentives being offered—for example, Quebec offers a one-time payment of $3,000 for a third child ("Quebec" 1988)—are simply not enough to tempt a sensible person to plan another child. Quebec's payment works out to 50 cents a day for the child's first 18 years. Because current estimates suggest that it may cost as much as $150,000 to raise a middle-class child, any serious attempt to defray child-rearing costs would be prohibitively expensive (Keyfitz 1987). As long as women (and men) have attractive alternatives outside the home, children are seen as being somewhat less central to life (Jones and Brayfield 1997). For this reason, it is unlikely that governments can afford to bribe them into voluntarily taking on more than one or two children.

Population Growth, Environment, and Poverty

Mention the word *population* and many people immediately think of *population problems.* Certainly, the fact that the population is likely to grow to 10 to 11 billion in their lifetimes seems like a problem to most people. There are, however, many population problems: high mortality in Africa, low fertility in Europe; environmental devastation; the redistribution of populations through urbanization and migration; and Third World poverty. In this section, we address two of these problems—the environment and poverty—and analyze the role of population growth and redistribution in creating and resolving these problems.

Environmental Devastation: A Population Problem?

All around the world, there are signs of enormous environmental destruction. In the developed world, we have acid rain and oil spills; in South America, the Amazon forest is being destroyed; in Africa, desert environments are spreading rapidly due to deforestation and overgrazing. Although all of these pose serious threats to the natural order, only the last one is truly a population problem.

It is estimated that the United States, which contains only 5 percent of the world's population, consumes one quarter of the world's resources and produces nearly three-quarters of the world's hazardous waste (Ashford 1995). Our affluent, throwaway lifestyle requires large amounts of petroleum and other natural resources. This unceasing demand for more lies behind oil exploration of fragile lands and subsequent events such as the *Exxon Valdez* oil spill in Prince William Sound, Alaska, in 1989. Our unwillingness to pay the price for emission controls lies behind the acid rain and smog that are killing our eastern forests and polluting lakes and rivers. Although these problems would be less severe if there were half as many of us (and hence half as many cars, factories, and Styrofoam cups), they are not really population problems. They stem from our way of life rather than our numbers.

The destruction of the Amazon forest is also not a population problem. It is a poverty problem (Durning 1989). Brazil needs export dollars to pay its foreign debt and to establish an industrial economy. To get these dollars, it sells what it has most of: trees. Reducing the number of Brazilians would not make a serious dent in problems of internal or international poverty.

In sub-Saharan Africa and on the Indian subcontinent, however, population pressure is a major culprit in environmental destruction. In rural areas, the typical scenario runs like this: Population pressure forces farmers to try to plow marginal land and to plant high-yielding crops in quick succession without soil-enhancing rotations or fallow periods. The marginal lands and the overworked soils produce less and less food, forcing farmers to push the land even harder. They cut down forests and windbreaks to free more land for production. Soon, water and wind erosion becomes so pervasive that the topsoil is borne off entirely, and the tillable land is replaced by desert or barren rock.

This cycle of environmental destruction—which destroys forests, topsoil, and the plant and animal species that depend upon them—is characteristic of high population growth in combination with poverty. When one's children are starving, it is hard to make long-term decisions that will protect the environment for future generations.

In the Philippines, Cameroon, Malaysia, and the Amazon basin of Brazil, tropical rainforests are being felled at an alarming rate. Tropical hardwoods are sold to generate hard currency to pay huge foreign debts. Land is cleared to plant quick-growing crops or to raise cattle, and the remaining trees are felled to make charcoal, often a country's primary source of fuel. As a result of this deforestation, soil is rapidly eroding and global temperatures are rising. In none of these instances is environmental destruction truly a population problem. Instead, Third World poverty and international debt are the root causes. Although the people of Southeast Asia, Africa, and South America will pay the highest price for deforestation, environmental scientists have made it abundantly clear that the nations of the industrialized West will also experience the fallout from the ecological havoc it is creating.

Summary

Reducing population growth would reduce future pressure on natural resources, but it would not solve the current problem. The solution rests in an international moral and financial commitment to reducing rural poverty, improving farming practices, reducing Third World debt, and curbing our own wasteful and destructive practices.

Third World Poverty

There are nearly 6 billion people in the world. Three-quarters of them live in less-developed nations, where the gross national product per capita is one-twentieth that of the developed world. Perhaps 500 million are seriously undernourished, and each year outbreaks of famine and starvation occur in Africa and Asia; a billion more are poorly nourished, poorly educated, and poorly sheltered. These people live in the same nations that have high population growth.

Some observers blame Third World poverty on high fertility, thus neatly laying the entire fault at the victim's door. It is clear, however, that high fertility is not the only or even the primary cause of Third World poverty. Nigeria, for example, has become poorer because of the collapse of international oil prices in the 1980s, not because of its growing population. Rwandans are dying of starvation because of war rather than because of high fertility.

Nevertheless, it is clearly true that rapid population growth makes the reduction of poverty more difficult. Despite relatively rapid economic gain, many less-developed economies are barely able to keep up with their growing populations. This means that little is left over for investment in a more productive future.

Programs to Reduce Fertility

Almost all world leaders recognize that lower fertility is an important step toward increasing the standard of living in the poorer nations of the world. The most successful programs to reduce fertility have combined an aggressive family-planning program with a push toward economic development (Poston & Gu 1987).

1. *Family-planning programs.* These programs are designed to make modern contraceptives and sterilization available inexpensively and conveniently to individuals who desire to limit the number of their children. Between 1975 and 1991, an aggressive family-planning program increased contraceptive use in Bangladesh by 500% and decreased the average number of children per woman from 7.0 to 4.9 in just 16 years (Kalish 1994a). However, there is still a great need for family-planning programs; a 1989 survey indicated that 41% of the Haitian women who were not using contraceptives desired to do so (Barberis 1994).

2. *Economic development.* Experience all over the world shows that fertility declines as education increases and the country undergoes economic development. For example, South Korea's fertility has plummeted from 6.0 children per woman in 1960 to only 1.7 in the wake of its dramatic economic development (Haub 1991).

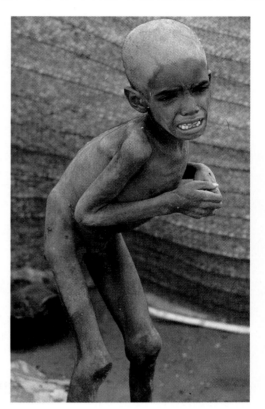

Perhaps half a billion people in the world are seriously malnourished, and some, such as this starving child, face permanent physical and intellectual damage or even death. Few such cases are directly related to overpopulation. Instead, they are due to war, drought, and poverty—and a capitalist world economic system in which rich countries have storehouses of grain and poor nations starve. When the world population reaches over 10 billion in 40 to 50 years, however, there could actually be too little food to go around, even with equal distribution.

Summary

Analysis of world population growth reveals a good-news/bad-news situation. The good news is that fertility is declining in every part of the world (see Table 14.3). The bad news is that the population of the world will double within 40 to 50 years anyway. The reason for this gloomy prediction lies in the current population age structure. The next generation of mothers is already born—and there are a lot of them. Thus, we must continue to plan for a world that will soon hold 11 to 12 billion people.

TABLE 14.3
Changes in Average Family Size
In the last 20 years, the average number of children being born per woman has decreased 36 percent worldwide. This decrease has been noticeable in all parts of the world. Because mortality has fallen too, however, and because of the momentum of the age structure, population is still growing rapidly.

	AVERAGE NUMBER OF CHILDREN PER WOMAN		
Area	1968–1972	1997	Percentage Change
World	4.7	3.0	−36%
Africa	6.4	5.6	−11
Asia	5.4	2.9	−46
Latin America	5.5	3.0	−44
Europe and North America	2.8	1.8	−36

SOURCE: World Population Data Sheet, 1997. Population Reference Bureau.

Population in the United States

The U.S. population picture is very much the same as that of Western Europe: low mortality and below-replacement fertility. In this section, we briefly describe fertility, mortality, growth, and migration issues in the United States.

Fertility

For nearly 20 years now, the number of children per woman in the United States has stood at 1.8 to 2.0—slightly less than the 2.1 necessary to replace the population. This low fertility has been accompanied by sharp reductions in social-class, racial, and religious differences in fertility. Some women will have their children as teenagers and some when they are 30, but increasingly they will stop at two.

Mortality

Death is almost a stranger to U.S. families. The average age at death is now in the late 70s, and many people who survive to 65 expect to live another 20 years. Parents can feel relatively secure that their infants will survive; if they don't divorce, young newlyweds can safely plan on a golden wedding anniversary.

In the last 20 years, we have added 4 years to life expectancy. These increases are due to better diagnosis and treatment of the degenerative diseases (such as heart disease and cancer) that strike the elderly. In addition, increases in life expectancy have been made possible by reducing racial and social-class differentials in mortality. At the time of World War II, black women lived a full 12 years less than white women; by 1992, the gap was down to slightly less than 6 years.

The AIDS epidemic has given death a new face. First recognized in 1981 when 185 cases were reported in the United States, AIDS is now the leading cause of death among American men aged 25 to 44 ("The AIDS Front" 1995). Greater awareness of the risks of AIDS over the last several years has substantially reduced the number of AIDS cases among homosexual men. The fastest-growing categories of victims are heterosexuals and children, who contracted the disease from their infected mothers during gestation. Often spread through intravenous drug use, AIDS is increasingly becoming a disease of the poor and disadvantaged.

Although a majority of U.S. women continue to want children, there is growing variability in fertility behavior. Perhaps as many as 15 to 20 percent will decide to get a dog instead of having children. Among those who do have children, a growing proportion will stop at one; nearly one quarter will become mothers before they become wives.

Migration

Demographers are interested in the study of migration patterns for two reasons: First, migration is one of the major determinants of population size. Although it can safely be ignored as a factor in world population growth, migration often has dramatic effects on the growth of individual nations. The United States is one of the nations where **immigration,** the permanent movement of people into another country, has had an important impact. Second, patterns of internal migration have enormous consequences for most of our social institutions. In the United States, the processes of urbanization, suburbanization, and migration from Rustbelt to Sunbelt have created a unique set of problems and have dramatically changed our political landscape.

Immigration is the permanent movement of people into another country.

Immigration

An estimated 1 million people enter the United States each year. Almost all of the recent immigrants come from Latin America or Asia. Perhaps as many as half are illegal immigrants, most of whom are from Mexico or other countries in Central America.

At the present time, immigration accounts for approximately one-third of U.S. population growth. As a consequence, the United States does not need to fear population decline. The racial and ethnic composition of the nation will change substantially, however. In fact, by 2050, it is estimated that the combination of migration and low fertility will reduce the proportion of our population that is Anglo from 74 to 53 percent (see Figure 14.4).

The causes of immigration are primarily economic; immigrants, legal and illegal, are pushed from their native lands by poor local economies and are pulled by an unmet demand in the United States for low-skill, low-paid labor. The consequences of current immigration trends are likely to be both economic and cultural. From the standpoint of economics, three generalizations seem to be supported: (1) Immigrants are not taking jobs away from U.S. citizens; but (2) the availability of low-wage illegals does depress wages in some economic sectors; and (3) Hispanics and other minorities are the ones hardest hit by this. From the standpoint of culture, it is likely that the United States will become a more pluralistic society, perhaps one that is multilingual and has no majority population.

FIGURE 14.4
Changing Composition of U.S. Population

If annual migration remains at 1 million and if fertility remains low, the racial and ethnic composition of the U.S. population will change substantially in the decades ahead. The most noticeable effect will be a sharp rise in the proportion who are Hispanic and Asian and a corresponding decrease in the proportion who are Anglo.

SOURCE: U.S. Bureau of the Census Population Projections of the United States by age, sex, race, and Hispanic origin, 1996.

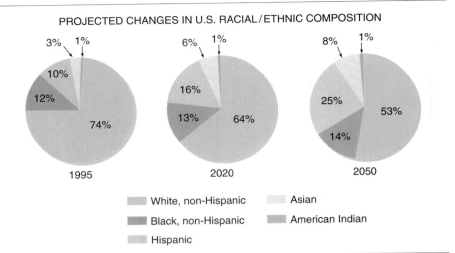

PROJECTED CHANGES IN U.S. RACIAL/ETHNIC COMPOSITION

1995: 3%, 1%, 10%, 12%, 74%
2020: 6%, 1%, 16%, 13%, 64%
2050: 8%, 1%, 25%, 14%, 53%

White, non-Hispanic
Black, non-Hispanic
Hispanic
Asian
American Indian

Urbanization is the process of population concentration in metropolitan areas.

Internal Migration

Until 1970, the story of the U.S. population was one of progressive **urbanization,** the process of population concentration. For most of our history, urban areas grew faster than rural areas, with the largest urban areas growing the most. During the last 25 years, however, there have been three major variations on this pattern: shrinking central cities, Sunbelt growth, and nonmetropolitan resurgence.

Because central cities have grown much less rapidly than their suburban rings, the proportion of the U.S. population that lives in central cities has decreased since 1970. Not only has the proportion of central-city residents dwindled, in the Midwest and the North, several large cities have actually decreased in size. In fact, in the last two decades, almost all metropolitan growth has occurred in the South and the West. At the same time that central cities have been shrinking, nonmetropolitan areas have experienced some modest growth (Johnson 1989). However, there has not been a big boom in rural Iowa or Nebraska (see Map 14.1); most nonmetropolitan growth has also occurred in Sunbelt states, often in areas within a few hours' drive of a big city and in the last several years, even that growth has slowed.

Patterns of internal migration have significantly changed the geographic concentration of the U.S. population. As a result of this relocation, the United States must confront three additional problems: the urbanization of poverty, a declining central-city tax base, and environmental hazards.

1. *Urbanization of poverty.* There has been a sharp increase in urban poverty over the last 20 years (Wilson & Aponte 1985), primarily as a result of the exodus of jobs to the suburbs.

2. *Declining tax base.* As retail businesses, jobs, and more affluent property owners have moved outside the city limits, the loss of tax dollars has left many cities teetering on the edge of bankruptcy. In an attempt to remain solvent, cities have cut some services and eliminated others.

3. *Environmental hazards.* Much of the geographic relocation of the U.S. population over the last 20 years has been to those regions of the country that are least able to withstand the ecological impact of a large population. In many areas of Florida, California, and the Southwest, the demand for water already outstrips the supply. As states argue over water rights, political tensions are likely to increase; within states, competition for access to water may increase conflict between agricultural and urban interests.

MAP 14.1
Rural Rebound Redux

Nonmetropolitan counties are gaining population in much of the nation, except for the Plains states.

SOURCE: U.S. Bureau of the Census, 1996 Statistical Abstract of the United States, no. 28. Adapted from *American Demographics* magazine with permission. ©1995 American Demographics, Inc. Ithaca, NY.

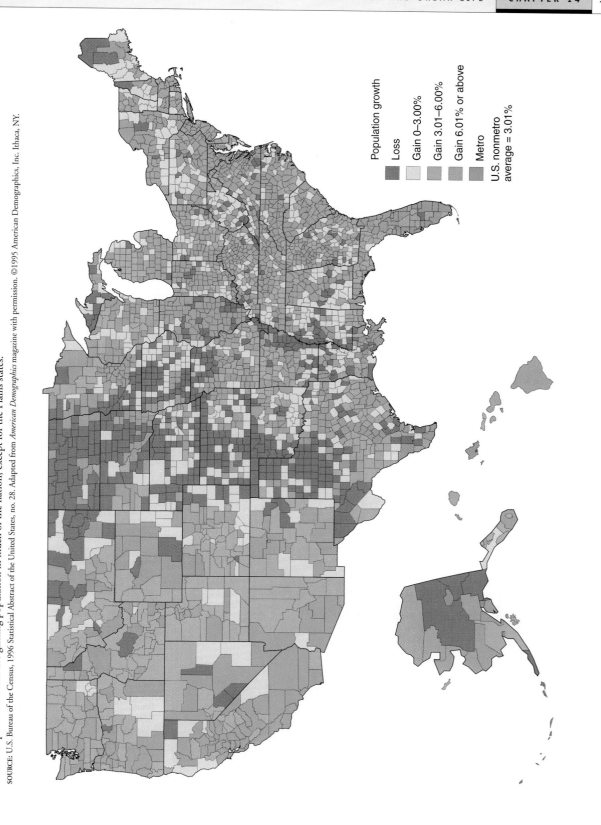

Population growth

- Loss
- Gain 0–3.00%
- Gain 3.01–6.00%
- Gain 6.01% or above
- Metro

U.S. nonmetro average = 3.01%

a global perspective

INTERNATIONAL MIGRATION

Some are pushed out of their homeland by violence and drought, while others are pulled by new opportunities. Perhaps as many as 40 percent have left their homes involuntarily. They are refugees or "forced migrants"—individuals for whom ethnic conflict or changes in national boundaries have produced refugee-like situations. Although the United States has long grappled with the question of how many immigrants and refugees our own educational and economic institutions can successfully absorb, what do we know about international migration? Map 14.2 shows the countries with the highest migration rates.

Estimates are that in the mid-1980s, 100 million people were "living outside of their countries of birth or citizenship." Of these, a high proportion were refugees, approximately two-thirds of whom had moved to other developing countries. Demographers believe that the political turmoil and violence in Eastern Europe, Southeast Asia, and sub-Saharan Africa have substantially increased the numbers of involuntary migrants living in other Third World nations. The result, of course, is that enormous strain is placed on the already limited capacities of host countries to sustain their own growing populations.

Although push factors such as war and famine account for much of the movement between developing countries, some migrants are also pulled by the economic growth and employment opportunities in newly industrializing nations, such as South Korea, Singapore, and Malaysia. Pull factors also account for much of the immigration from less to more developed countries. Strong European economies provide increasing numbers of jobs to a growing non-Western labor force. Between 1978 and 1983, for instance, migration provided employment for one-third of the growth in Pakistan's labor force. Migrants traditionally have been young men, but women and girls now make up 40 to 60 percent of the international migrant stream; many of these are mothers who, in growing numbers, seek employment opportunities in more affluent neighboring countries in order to send money to the family, friends, or neighbors who are raising their children.

The money sent back home by migrants—both men and women—is a large and growing source of revenue for many nations. In 1990, payments amounted to $71 billion, a sum second in value globally only to trade in crude oil. In the late 1980s, remittances to developing countries alone ($37 billion) amounted to almost two-thirds the value of official development assistance. It is not yet clear, however, who profits most from the international migrant stream. Although countries such as Germany, France, and Italy do face new challenges stemming from an ethnically diverse population, workers from developing nations are, in fact, helping to sustain the continued expansion of the European economy. Low birth rates have led to smaller labor forces and aging populations in much of Europe. Thus, migrants from countries such as Turkey and Pakistan fill the demand for more workers, particularly those at the low end of the labor hierarchy. Whether the money that migrants send home will significantly improve the quality of life in Third World nations remains an open question.

SOURCE: Kalish 1994b.

While the labor that immigrants provide is often a major source of economic growth, during times of economic retrenchment nations find it more difficult to provide the jobs and services that immigrants may initially require. As California's recent passage of Proposition 187 suggests, immigration policy often becomes more restrictive. Nations like the United States can, in fact, afford to provide for relatively large immigrant populations. Two-thirds of the world's displaced persons, however, are found in countries of the Third World, where problems are only slightly less severe than those of the refugees' homelands. Even with Western assistance, these Rwandan refugees will divert resources sorely needed by the people of Tanzania for their own subsistence.

MAP 14.2
Countries With the Highest Rates of Net International Migration (1985–1995)

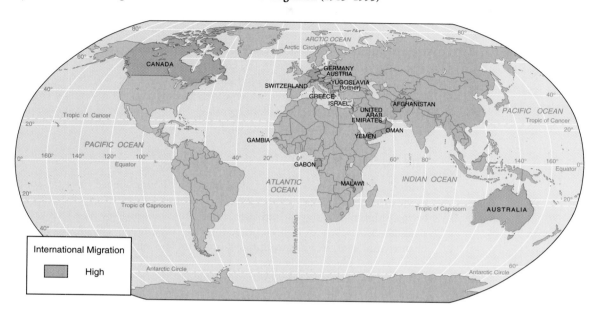

International Migration
High

Average Annual Net Migration
(per 1,000 population)

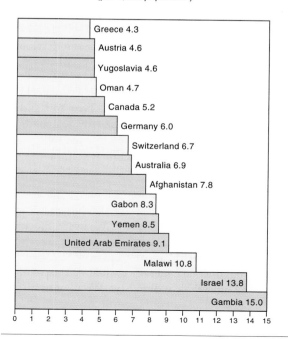

Greece 4.3
Austria 4.6
Yugoslavia 4.6
Oman 4.7
Canada 5.2
Germany 6.0
Switzerland 6.7
Australia 6.9
Afghanistan 7.8
Gabon 8.3
Yemen 8.5
United Arab Emirates 9.1
Malawi 10.8
Israel 13.8
Gambia 15.0

0 1 2 3 4 5 6 7 8 9 10 11 12 13 14 15

Population Growth Due to International Migration
(percentage)

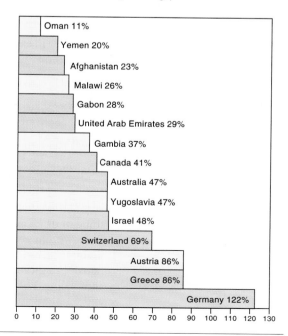

Oman 11%
Yemen 20%
Afghanistan 23%
Malawi 26%
Gabon 28%
United Arab Emirates 29%
Gambia 37%
Canada 41%
Australia 47%
Yugoslavia 47%
Israel 48%
Switzerland 69%
Austria 86%
Greece 86%
Germany 122%

0 10 20 30 40 50 60 70 80 90 100 110 120 130

Fertility, mortality, and migration patterns in the United States provide clear examples of the interrelationships between population and social institutions. Social class, women's roles, and race and ethnic relationships are all intimately connected to changes in population. One additional element of population that is especially important for social relationships is community size, an issue to which we now turn.

Urbanization

Most of our social institutions evolved in agrarian societies, where the vast bulk of the population lived and worked in the countryside. As late as 1850, it is estimated that only 2 percent of the world's population lived in cities of 100,000 or more (Davis 1973). Today, nearly a quarter of the world's population and more than two-thirds of the U.S. population live in cities larger than 100,000. What are the consequences of these trends for individuals and social institutions?

The Industrial City

With the advent of the industrial revolution, production moved from the countryside to the urban factory, and industrial cities were born. These cities were mill towns, steel towns, shipbuilding towns, and later, automobile-building towns; they were home to slaughterers, packagers, millers, processors, and fabricators. They were the product of new technologies, new forms of transportation, and vastly increased agricultural productivity that freed most workers from the land.

Fired by a tremendous growth in technology, the new industrial cities grew rapidly during the nineteenth century. In the United States, the urban population grew from 2 to 22 million in the half-century between 1840 and 1890. In 1860, New York was the first U.S. city to reach 1 million. The industrial base that provided the impetus for city growth also gave the industrial city its character: tremendous density and a central business district.

Density

A critical factor in explaining the character of the industrial city as it developed in the nineteenth century is that most people walked to work—and everywhere else, for that matter. The result was dense crowding of working-class housing around manufacturing plants. Even in 1910, the average New Yorker commuted only two blocks to work. Entire families shared a single room; and in major cities such as New York and London, dozens of people crowded into a single cellar or attic. The crowded conditions, accompanied by a lack of sewage treatment and clean water, fostered tuberculosis, epidemic diseases, and generally high mortality. A glimpse of these conditions is provided by a letter that appeared in the London *Times* in 1849:

> Sur,—May we beg and besearch your proteckshion and power. We are Sur, as it may be, livin in a Wilderness, so far as the rest of London knows anything of us, or as the rich and great people care about. We live in muck and filth. We aint got no priviz, no dust bins, no drains, no water-splies, and no drain or suer in the hole place. The Suer Company, in Greek St., Soho Square, all great, rich powerfool men take no notice whasomdever of our complaints. The Stenche of a Gulley-hole is disguistin. We all of us suffer, and numbers are ill, and if the Cholera comes Lord help us.
>
> Some gentlemans comed yesterday. . . . They was much surprized to see the sellar in No. 12, Carrier St., in our lane, where a child was dying from fever, and would not believe that Sixty persons sleep in it every night.

During the nineteenth and early twentieth centuries, many cities grew up around manufacturing plants. These industrial cities were characterized by high density. Because the working class walked to work (and everywhere else), working-class housing was crowded into the blocks immediately surrounding the plants. The middle and upper classes lived in the suburbs. Although the plants are largely gone, this residential social-class pattern persists.

This here seller you couldent swing a cat in, and the rent is five shillings a week; but theare are great many sich deare sellers. . . .

Praeye Sir com and see us, for we are living like piggs, and it aint faire we should be so ill treted. (Cited in Thomlinson 1976)

Central Business District (CBD)

The lack of transportation and communication facilities also contributed to another characteristic of the industrial city, the central business district (CBD). The CBD is a dense concentration of retail trade, banking and finance, and government offices, all clustered close together so messengers could run between offices and businessmen could walk to meet one another. By 1880, most major cities had electric streetcars or railway systems to take traffic into and out of the city. Because most transit routes offered service only into and out of the CBD rather than providing crosstown routes, the earliest improvements over walking enhanced rather than decreased the importance of the CBD as the hub of the city.

The Postindustrial City

The industrial city was a product of a manufacturing economy plus a relatively immobile labor force. Beginning about 1950, these conditions changed and a new type of city began to grow. Among the factors prominent in shaping the character of the postindustrial city are the change from secondary to tertiary production and greater ease of communication and transportation. These changes have led to a much diminished role for the central business districts, a dispersion of retail, manufacturing, and residential areas, and a much lower urban density.

Change from Secondary to Tertiary Production

As we noted in Chapter 13, the last decades have seen a tremendous expansion of jobs in tertiary production and the subsequent decline of jobs in secondary production. The manufacturing plants that shaped the industrial city are disappearing. Many of those that remain have moved to the suburbs where land is cheaper, taking working-class jobs, housing, and trade with them.

Instead of manufacturing, the contemporary central city is dominated by medical and educational complexes, information-processing industries, convention and entertainment centers, and administrative offices. These are the growth industries. They are also white-collar industries. These same industries, plus retail trade, also dominate the suburban economy.

Easier Communication and Transportation

Development of telecommunications and good highways has greatly reduced the importance of physical location. The central business district of the industrial city was held together by the need for physical proximity. Once this need was eliminated, high land values and commuting costs led more and more businesses to locate on the periphery, where land was cheaper and housing more desirable. Many corporate headquarters moved all the way to Arizona or Texas.

A key factor in increasing individual mobility was the automobile. Without the automobile, workers and businesses could not have moved to the city periphery, and space-gobbling single-family homes would not have been built. In this sense, the automobile has been the chief architect of U.S. cities since 1950. It has given them a freedom of form that older cities could not have.

Urban Sprawl

The new cities are much larger in geographical area than the industrial cities were. The average city in 1940 was probably less than 15 miles across; now many metropolitan areas are 50 to 75 miles across. No longer are the majority of people bound by subway and railway lines that only go back and forth to downtown. Retail trade is dominated by huge, climate-controlled, pedestrian-safe suburban malls. A great proportion of the retail and service labor force has also moved out to these suburban centers, and many of the people who live in the suburbs also work in them.

Urbanization in the United States

What is considered urban in one century or nation is often rural in another. To impose some consistency in usage, the U.S. Bureau of the Census has replaced the common words *urban* and *rural* with two technical terms: *metropolitan* and *non-metropolitan*.

A **metropolitan area** is a county that has a city of 50,000 or more in it plus any neighboring counties that are significantly linked, economically or socially, with the core county. Some metropolitan areas have only one county; others, such as New York, San Francisco, or Detroit, include half a dozen neighboring counties. In each case, the metropolitan area goes beyond the city limits and includes what is frequently referred to as, for example, the Greater New York area. A **nonmetropolitan area** is a county that has no major city in it and is not closely tied to a county that does have such a city.

Figure 14.5 shows the current distribution of the U.S. population by type of residence. A total of 77 percent of the population lives in metropolitan areas. This metropolitan population is divided between those who live in the central city (within the actual city limits) and those who live in the balance of the county or counties, the suburban ring. More than half of the metropolitan population live in the suburbs rather than in the central city itself. Although they are judged to have access to a metropolitan way of life, they may live as far as 30 or 50 miles from the city center.

The nonmetropolitan population of the United States has shrunk to 23 percent. Although there are nonmetropolitan counties in every state of the Union except New Jersey, the majority of the nonmetropolitan population lives in either the Midwest or the South. Few of these people are farmers; many live in small towns and cities of 10,000 or 30,000.

A **metropolitan area** is a county that has a city of 50,000 or more in it plus any neighboring counties that are significantly linked, economically or socially, with the core county.

A **nonmetropolitan area** is a county that has no major city in it and is not closely tied to a county that does have such a city.

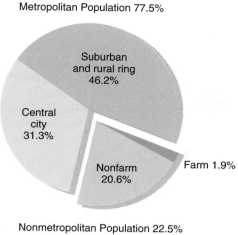

Metropolitan Population 77.5%

Nonmetropolitan Population 22.5%

FIGURE 14.5
The Urbanization of the United States Population 1990
More than three-fourths of the U.S. population live in metropolitan areas, and nearly half live in areas with more than 1 million people. Nevertheless, less than one-third actually live in central cities. The remainder live in suburbs and small towns.
SOURCE: U.S. Bureau of the Census 1990.

Third World Cities

The growth of large cities and an urban way of life has occurred everywhere very recently; in the less and least developed cities it is happening almost overnight (see Figure 14.6). Mexico City, São Paulo, Bogotá, Seoul, Kinshasa, Karachi, Calcutta—these and other Third World cities are growing at a rate of 5 to 8 percent per year. This means that their populations will double in approximately a decade. The roads, the schools, and the sewers that used to be sufficient no longer are; neighborhoods triple their populations and change their character from year to year. These problems are similar to the problems that plagued Western societies at the onset of the industrial revolution, but on a much larger scale.

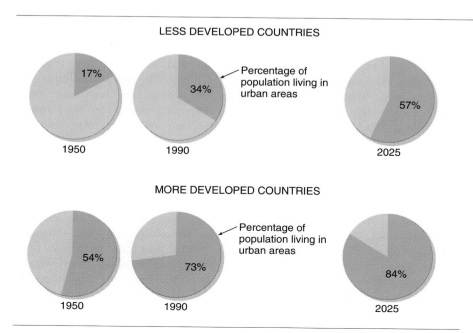

FIGURE 14.6
Urbanization Trends in the Developed and Less-Developed World, 1950–2025
Although the world is still more rural than urban, this is changing within our lifetime. Urbanization is growing particularly quickly in the less-developed world. In 1960, only three of the world's ten largest cities were in developing countries: Shanghai, Buenos Aires, and Calcutta. In 1990, all except three (Tokyo, New York, and Los Angeles) were in the developing world. By the year 2025, three quarters of the world's urban population will live in less-developed countries.
SOURCE: Haub 1993.

Third World urbanization differs from that of the developed world, not only in pace, but also in causal factors. First, more than half of the growth in Third World cities arises from a very high excess of births over deaths. Where overall population growth is 3 or even 4 percent annually, cities grow rapidly even without migration from the countryside. Second, many of the large and growing Third World cities have never been industrial cities. They are government, trade, and administrative centers. More than one-third of the regular full-time jobs in Mexico City are government jobs. These cities offer few working-class jobs, and the growing populations of unskilled men and women become part of the informal economy—artisans, peddlers, bicycle renters, laundrywomen, and beggars.

Place of Residence and Social Relationships

Urban Living

Although it is important to understand the factors that encourage urban growth, most sociologists are interested in **urbanism**—a distinctly urban mode of life that has developed in the city though it is not confined there (Wirth 1938). They are concerned with the extent to which social relationships and the norms that govern them differ between rural and urban places.

Urbanism is a distinctly urban mode of life that is developed in cities but not confined there.

Theoretical Views

The Western world as a whole has an antiurban bias. Big cities are seen as haunts of iniquity and vice, corruptors of youth and health, and destroyers of family and community ties. City dwellers are characterized as sophisticated but artificial; rural people are characterized as possessing homegrown goodness and warmth.

EARLY WRITERS.   This general antiurban bias (which has been around at least since the time of ancient Rome), coupled with the very real problems of the industrial city, had considerable influence on early sociologists. Ferdinand Tönnies (1855–1930) offered one of the earliest sociological descriptions of the differences between urban and rural society. He argued that rural society was characterized by **gemeinschaft,** personal and permanent ties associated with primary groups. Urban society was characterized by **gesellschaft,** the impersonal and instrumental ties associated with secondary groups. Durkheim saw the essence of urbanization as a shift from social cohesion built on similarity (mechanical solidarity) to a cohesion built on a complex division of labor and high interdependence (organic solidarity). Weber spoke of a shift from tradition to rationalism as a guide to social activities.

Gemeinschaft refers to society characterized by the personal and permanent ties associated with primary groups.

Gesellschaft refers to society characterized by the impersonal and instrumental ties associated with secondary groups.

These early writers were not blind to the drawbacks of rural society. They recognized that rural society was static and confining, that tradition bound individuals to a station in life and to ways of thinking that left little room for innovation or individualism. Their preference for rural life was based on the security it provided—the security of knowing exactly what was expected of you, what your place in the social order was and what your neighbor's place was. Many of the early social observers believed that this certainty was vital to individual happiness and social cohesion. In addition, the long-lasting personal relationships characteristic of rural society were thought to be essential to informal social control. Many were concerned that when people did not have

Big cities such as New York offer glamour, sophistication, and wonderful museums, music, and entertainment. Nevertheless, surveys show that the vast majority of Americans would rather visit big cities than live in them. Anxiety about crime is an important factor in this preference for suburban and small-town living.

to worry about what the neighbors would think, deviance would become commonplace and the social order would be threatened.

WIRTH: URBAN DETERMINISM.   The classic statement of the negative consequences of urban life for the individual and for social order was made by Louis Wirth in 1938. In his influential work "Urbanism as a Way of Life," Wirth suggested that the greater size, heterogeneity, and density of urban living necessarily led to a breakdown of the normative and moral fabric of everyday life.

Greater size means that many members of the community will be strangers to us. Greater density means that we will be forced into close and frequent contact with these strangers. Wirth postulated that individuals would try to protect themselves from this crowd by developing a cool personal style that would allow them to ignore some people (including people who were physically close, such as in a crowded elevator) and to interact with others, such as salesclerks, in an impersonal style so that their personalities would not be engaged. Wirth did not suggest that urbanites had no friends or primary ties, but he did think that the city bred a personal style that was cold and calculating (Fischer 1976).

Wirth also believed that faced with a welter of differing norms, the city dweller was apt to conclude that anything goes. Such an attitude, coupled with the lack of informal social control brought on by size, would lead to greater crime and deviance and a greater emphasis on formal controls.

In sum, Wirth argued that city living brought negative consequences for individuals and society; that is, he believed that if a well-integrated, warm, and conforming person from the farm moved to the city, that person would change and become calculating, indifferent, and nonconforming.

THE COMPOSITIONAL MODEL.   Later theorists have had a more benign view of the city. Compositional theorists suggest that individuals experience the city as a mosaic of small worlds that are manageable and knowable. Thus, the person who lives in New York City does not have to cope with 9 million people and 500 square miles of city; rather the individual's private world is made up of family, a small neighborhood, and an immediate work group. Compositional theorists argue that the primary group lives on in cities and that the quality of interpersonal ties is not affected even though the number of impersonal contacts is much greater than in rural areas.

The compositional model does recognize that deviance, loneliness, and other problems are greater in cities than in rural areas. It suggests, however, that deviants, singles, people without children, the lonely, and the alienated are attracted to the cities rather than created by them. Those with families and those willing to conform are attracted to the suburbs.

THE SUBCULTURAL VIEW.   In Wirth's view, the city has essentially negative effects; in the compositional view, urban environment has few direct consequences. The subcultural view straddles the two positions and presents a more moderate picture of the city. The essential idea of the subcultural view is that of critical mass. Special subcultures—intellectuals, radicals, gays—cannot develop until there are a relatively large number of people sharing some relatively uncommon set of norms or values. For example, one homosexual in a small community will be under constant pressure to conform to general standards; only when there are many others will it be possible to sustain a gay community with its own set of norms and values. Similarly, a symphony orchestra, a football team, and a synagogue all await the development of a critical mass of people who share the same interest. Once they identify one another, they will have group support for their identities and standards. In this way, the greater diversity and size of the city lead to the development of subcultures with different, perhaps even deviant, norms and values. Wirth might interpret these subcultures as evidence of a lack of moral integration of the community, but they can also be seen as private worlds within which individuals find cohesion and primary group support.

THINKING

CRITICALLY

Campuses often bring large numbers of strangers together in crowded circumstances. Which theoretical view best describes the outcomes of this interaction? Can you find some examples to support all three perspectives?

Empirical Consequences of Urban Living

One theory suggests that urban living has negative consequences, another that it has few consequences, and still another that it leads to the development of subcultures. This section reviews the evidence about the effects of urban living on social networks, neighborhood integration, and quality of life.

SOCIAL NETWORKS.   The effects of urban living on personal integration are rather slight. Surveys asking about social networks show that urban people have as many intimate ties as rural people. There is a slight tendency for urban people to name fewer kin and more friends that rural people, but the kin omitted from the urban lists are not parents, children, and siblings but more distant relatives (Fischer 1981; Amato 1993). There is no evidence that urban people are disproportionately lonely, alienated, or estranged from family and friends.

THE NEIGHBORHOOD.   Empirical research generally reveals the neighborhood to be a very weak group. Most city dwellers, whether central city or suburban, find that city living has freed them from the necessity of liking the people they live next to and has given them the opportunity to select intimates on a basis other than physical proximity; this freedom is something that people in rural areas do not have. There is growing consensus among urban researchers that physical proximity is no longer a primary basis of intimacy (Flanagan 1993). Rather, people form intimate networks on the basis of kin, friendship, and work groups; and they keep in touch by telephone rather than relying solely on face-to-face communication. In short, urban people do have intimates, but they are unlikely to live in the same neighborhood with them. When in trouble, they call on their good friends, parents, or adult children for help. In fact, one study of neighborhood interaction in Albany-Schenectady-Troy, New York, found that a substantial share—15 to 25 percent—of all interaction with neighbors was with *family* neighbors—parents or adult children who happen to live in the same neighborhood (Logan & Spitze 1994).

Neighbors are seldom strangers, however, and there are instances in which being nearby is more important than being emotionally close. When we are locked out of the house, we need a teaspoon of vanilla, or we want someone to accept a United Parcel Service package, we still rely on our neighbors (Wellman and Wortley 1990). Although we generally do not ask large favors of our neighbors and don't want them to rely heavily on us, most of us expect our neighbors to be good people who are willing to help in a pinch. This has much to do with the fact that neighborhoods are often segregated by social class and stage in the family life cycle. We know that our neighbors will be people pretty much like us.

QUALITY OF LIFE.   Big cities are exciting places to live. People can choose from a wide variety of activities, 24 hours a day, seven days a week. The bigger the city, the more it offers in the way of entertainment, libraries, museums, zoos, parks, concerts, and galleries. The quality of medical services and police and fire protection also increase with city size. These advantages offer important incentives for big-city living.

On the other hand, there are also disadvantages: more noise, more crowds, more expensive housing, and more crime. The latter is a particularly important problem for many people. More than 50 percent of those living in cities over 1 million report that they would be afraid to walk alone at night in their neighborhood; only 29 percent of the rural population would be afraid. Data on crime rates suggest this fear might be justified: Crime rates are strongly correlated with city size. This is especially true of the kind of crime that people fear most—violence against the person.

Because of these disadvantages, many people would rather live close to a big city than actually in it. For Americans, the ideal is a three-bedroom house on a spacious lot in the suburbs, but close enough to a big city that they can spend an evening or afternoon there. Some groups, however, prefer big-city living, in particular, childless people who work downtown. Many of these people are decidedly pro-urban and relish the entertainment and diversity that the city offers. Because of their affluence and childlessness, they can afford to dismiss many of the disadvantages of city living.

Sociological attention has been captured by cities such as Manhattan and San Francisco with their bright lights, ethnic diversity, and crowding. Nevertheless, only a quarter of our population actually lives in these big-city centers. The rest live in suburbs and small towns. How does their experience differ?

Suburban Living

The classic picture of a suburb is a development of very similar single-family detached homes on individual lots. This low-density housing pattern is the lifestyle to which a majority of people in the United States aspire; it provides room for dogs, children, and barbecues. This is the classic picture of suburbia. How has it changed?

The Growth of the Suburbs

The suburbs are no longer bedroom communities that daily send all their adults elsewhere to work. They are increasingly major manufacturing and retail trade centers. Most people who live in the suburbs work in the suburbs. Thus, many close-in suburban areas have become densely populated and substantially interlaced with retail trade centers, highways, and manufacturing plants.

The suburbs are the fastest growing part of America, now encompassing almost half of the entire U.S. population. Suburbs like this one in Orange County are quite affluent; others, such as the first ones built in the 1950s, are beginning to show signs of age. In both types, however, each family has its own house, its own yard, and two cars. House structure, income, lifestyle, and values tend to be very similar within each subdivision. Because of this, suburbs emphasize conformity.

These changes have altered the character of the suburbs. Suburban lots have become smaller, and neighborhoods of townhouses, duplexes, and apartment buildings have begun to appear. Childless couples, singles, and retired couples are seen in greater numbers. Suburbia has become more crowded and less dominated by the station-wagon set.

With expansion, suburbia has become more diverse. Although each suburban neighborhood tends to have its own style, stemming in large part from the fact that each development includes houses of similar size and price, there are a wide variety of styles. In addition to the neighborhoods of classic suburbia are spacious miniestate suburbs where people have horses and riding lawn mowers, as well as dense suburbs of duplexes, townhouses, and apartment buildings. Some of the first suburbs are now 45 years old. Because people tend to age in place, these suburbs are more often characterized by retirees than young families (Fitzpatrick & Logan 1985). Many of the older suburbs are becoming rundown, and renting is more common than home owning.

Suburban Problems

Many of the people who moved to suburbia did so to escape urban problems: They were looking for lower crime rates, less traffic, less crowding, and lower tax rates. The growth of the suburbs, however, has brought its own problems, including the following (Adler 1995b).

1. *Housing costs.* The increased demand for suburban housing has driven housing costs up to a level that cripples the finances of many families and is simply beyond the reach of people who could have afforded a home 10 years ago.
2. *Weak governments.* The county governments and municipal governments of the small cities in the suburban ring are fragmented and relatively powerless. One result of this is the very haphazard growth associated with weak and inadequate zoning authority.
3. *Higher density.* The increased density of the suburbs recreates the urban problems of crowding and traffic congestion.
4. *Transportation.* Living in suburbia depends on access to automobiles. People who don't have them are basically excluded from the suburban lifestyle and suburban

jobs. If you are a central-city resident who doesn't own a car (about one-third of all minority residents of the central city), suburban jobs and opportunities are simply out of reach.

Small-Town and Rural Living

Approximately 25 percent of the nation's population lives in small towns (less than 2,500) or rural areas. Some of these rural and small-town people are included in the metropolitan population count because they live within the orbit of a major metro area, but most live in nonmetropolitan areas—in South Dakota and Alabama, but also in Vermont and Pennsylvania.

The nonmetropolitan population of the United States continues to grow. Although young people often leave to go to school or get jobs elsewhere, enough come back to keep populations growing. In addition, small-town growth is maintained by a small, but steady stream of people seeking refuge from the problems of urban and suburban living.

People find small-town living attractive for a number of reasons: It offers lots of open space, low property taxes, affordable housing, and relative freedom from worry about crime. In addition, research supports the popular perception that small towns provide greater opportunities for neighboring (Freudenberg 1986) and community involvement. While knowing your neighbors may promote social integration, it is not an unalloyed blessing. The fact that everybody knows everybody else does help keep crime down, but for some people the lack of privacy and enforced conventionality are oppressive (Johansen & Fuguitt 1984). Indeed, several studies have found small-town and rural residents to be in somewhat poorer mental health than residents of large metropolitan areas (Beeson & Johnson 1987).

Rubes and Hicks?

According to stereotype, rural people, especially farmers, are hicks, rednecks, and rubes. They use bad grammar and think that if it was good enough for grandpa, it is good enough for me, that a woman's place is in the home, and that children should be seen

If you live in Boston, New York, Miami, or Houston, it is probably pretty hard to imagine spending a Saturday morning at a country auction. As the picture suggests, lifestyles do differ depending on region and size of place. Rural residents may be somewhat more conservative (family oriented and self-reliant) and less sophisticated (less likely to go to the symphony and more likely to attend a community gathering) than their urban counterparts. However, rural/urban differences are smaller than they used to be. Farm kids probably pick up MTV and their parents, David Letterman, with their satellite dish.

THINKING

CRITICALLY

We all hold stereotypes about what people are like if they're from California or North Dakota or Texas. Although surveys of actual populations find little difference on many social issues, does that mean these stereotypes are without foundation? On what kinds of matters do you expect these groups to really differ?

and not heard. In contrast, the stereotype of the sophisticated city dweller suggests someone who is aware of current events, innovative, and upbeat.

These portraits are very much exaggerated. Although rural and small-town residents are somewhat less tolerant than city dwellers (Wilson 1991), on most social issues, from church going to support for welfare, there are almost no size-of-place differentials (Camasso & Moore 1985). One recent study found no size-of-place or metropolitan–nonmetropolitan differences on the importance that people attached to any of the following values: working hard, achievement, personal freedom, helping others, salvation, or leisure (Christenson 1984). All but the remotest cabin dwellers have access to television, radio, movies, and news magazines. The automobile and a good highway system have also increased the access of rural people to urban culture.

Although they watch the same television shows and shop from the same catalogs as their urban counterparts, rural and urban lifestyles do differ. The city continues to be the major source of innovation and change. New dress styles, music, educational philosophies, and technologies originate in the city and spread to the countryside. Thus, the rural-urban difference is constantly created anew and seems unlikely ever to be totally eliminated (Fischer 1979). Because the speed of cultural diffusion is now much more rapid than before, however, rural-urban differences are far less profound that they were in the past (Flanagan 1993).

Summary

1. For most of human history, fertility was about equal to mortality and the population grew slowly or not at all. Childbearing was a lifelong task for most women, and death was a frequent visitor to most households, claiming one-quarter to one-third of all infants in the first year of life.

2. The demographic transition in the West took centuries. Mortality declined because of better nutrition, an improved standard of living and hygiene, improved public sanitation, and more recently, modern medicine. Somewhat later, changes in social structure associated with industrialization caused fertility to decline.

3. Social structure, fertility, and mortality are interdependent; changes in one affect the others. Among the most important consequences of high fertility are a restricted role for women and a very young age structure.

4. The level of fertility in a society has much to do with the balance of costs and rewards associated with childbearing. In traditional societies, such as that of Kenya, most social structures (the economy, religion, and the family, for example) support high fertility. In many modern societies, such as those of Europe and the United States, social structure imposes many costs on parents.

5. Population growth is not the only or even the primary cause of environmental devastation or Third World poverty, nor is reducing it the primary solution to these problems. Continued population growth does contribute to these problems, however, and reducing growth will make it easier to seek solutions.

6. In the United States, life expectancy is high and continues to increase. Childlessness is increasing and fertility is below replacement level. Because of high immigration rates, however, the U.S. population is unlikely to decline. Because many of the new Americans are Asian and Latino, the racial and ethnic composition of the U.S. population is likely to change substantially.

7. Three-quarters of U.S. residents live in metropolitan areas, but most of these live in the suburban ring rather than the central city. Big cities in the Northeast and Midwest have lost population, and almost all recent growth has occurred in the Sunbelt. Changing patterns of internal migration have led to three problems: the urbanization of poverty, a declining central-city tax base, and environmental hazards.

8. The industrial city has high density and a central-city business district; the postindustrial city is characterized by lower density and urban sprawl.

9. Urbanization is exploding in the less-developed world; many of its large cities will double in size in a decade. This urban growth is less the result of industrialization than of high urban fertility.

10. There are three major theories about the consequences of urban living. Wirth's urban determinism theory suggests that urban living will lead to nonconformity and indifference to others; compositional theory suggests few consequences; and subcultural theory suggests that the size of the city allows for the development of unconventional subcultures.

11. Urban living is associated with less reliance on neighbors and kin and more reliance on friends, with greater fear of crime.

12. Suburban living has become more diverse. Retail trade and manufacturing have moved to the suburbs, and the suburbs are now more densely populated, more congested, and less dominated by the station-wagon set.

13. Small-town and rural living is characterized by more emphasis on family and neighborliness, less crowding, more informal social control, and somewhat poorer mental health. There are fewer cultural and lifestyle differences between rural and urban areas than there used to be.

Sociology on the Net

The world's population is rapidly changing in ways that will influence the lives of each and every one of us. There are many organizations throughout the world that are carefully monitoring these changes and their impact on our lives. One of these organizations is the Population Reference Bureau.

```
http://www.igc.apc.org/prb/
```

Let's begin our tour by playing population jeopardy. Click on *Play population jeopardy!* How well did you do? Now go back to the home page and browse around for a bit. What is the Population Reference Bureau and what do they do? Scroll down to *Read Population Today* and open this section. Scroll down and open the April 1996 issue. Find the *Earth's Future.* What resources do the top fifth of the world's population consume? What is the environmental impact caused by a birth in the United States versus a birth in India or sub-Saharan Africa? What does the future hold in terms of natural resources and population?

FIND IT ON INFOTRAC COLLEGE EDITION

According to many estimates, the population of the world will reach 8 billion around 2025. One aspect of global population growth, according to *Society,* will be a dramatic increase in the urban population. Using InfoTrac College Edition, look up the following article:

"The Urban Future." *Society,* January, February, 1997, v34 n2 p2. (Hint: Enter the search terms *urban future* using the Key Words Guide.)

What problems will the world face as the urban and overall populations increase during the next 20–30 years? What specific problems might you expect the United States to face? Do you believe the United States should take an active stance in slowing world population growth? Defend your position.

Suggested Readings

Ashford, Lori S. 1995. "New Perspectives on Population: Lessons from Cairo." *Population Bulletin* 50 (1) (March). Washington, D.C.: Population Reference Bureau. A pamphlet that discusses current population growth and its implications and explains events and controversies at the 1994 International Conference on Population and Development (ICPD) in Cairo.

Conger, Rand D., and Elder, Glen H. Jr. (1994). *Families in Troubled Times: Adapting to Change in Rural America.* New York: Aldine de Gruyter. A close look at the emotional and behavioral consequences of the 1980s farm crisis for several hundred Iowa families who experienced the severe economic hardship of this period.

Falkenmark, Malin, and Widstrand, Carl. 1992. "Population and Water Resources: A Delicate Balance." *Population Bulletin* 47 (3) (November). Washington, D.C.: Population Reference Bureau. A pamphlet that explores one aspect of the impact of population growth on the environment by examining in detail the world's water supply and how population growth affects this absolutely necessary resource.

Groth, Paul. 1994. *Living Downtown: The History of Residential Hotels in the United States.* Berkeley: University of California Press. A fascinating social history, augmented with wonderful photographs, of what it meant and means to live in the residential hotels and boardinghouses of downtown U.S. cities.

Kelly, Barbara M. (ed.). 1989. *Suburbia Re-examined.* New York: Greenwood Press. A collection of theory-based readings, empirical studies, and policy essays that document the development of and challenges to suburban quality of life created by increasing population pressure.

Knox, Paul L. (ed.). 1994. *The Restless Urban Landscape.* Englewood Cliffs, N.J.: Prentice Hall. Readings in urban sociology from the conflict and symbolic interactionist perspectives, as well as discussions on topics such as the globalization of the city, Third World cities, urban aesthetics, and housing policies.

Riis, Jacob A. (1971). *How the Other Half Lives.* New York: Dover Publications. (Original work published in 1901.) A liberally illustrated essay on conditions in U.S. urban slums at the turn of the century. Riis's early photographs provide ample documentation of the poverty and filth of the industrial city.

U.S. Congress, Office of Technology Assessment. 1995. *The Technological Reshaping of Metropolitan America.* Washington, D.C.: U.S. Government Printing Office. A good book that describes in detail the impact of technology, from the automobile to the Internet, on metropolitan spatial distribution in the United States.

Social Movements, Technology, and Social Change

Outline

Social institutions do not stand still. Often, things change without our knowing how or why. For example, in the years between 1982 and 1988, the United States went from being the largest creditor to the largest debtor nation in the world. In less than a decade, relations between the United States and Russia changed dramatically from that of antagonists in the Afghanistan conflict to that of allies in the Desert Storm operation. Eastern Europe, South Africa, and Nicaragua lurch toward new economic and political forms, while the fortunes of all nations are increasingly dependent upon an international political and economic system.

What is going on? Many Americans shake their heads in confusion. Social change in the last decade has brought such wonderful developments as the eradication of smallpox, the destruction of the Berlin Wall, and the development of sophisticated new computer technologies. Balanced against these positive changes, however, are the scourge of AIDS, civil war and starvation in much of the Third World, the destruction of the Amazon forest, and depletion of the ozone layer. The rapid pace of social change and the complexity of late twentieth-century problems lead many individuals to feel a sense of both urgency and helplessness.

We take two approaches to understanding social change. In this chapter, we begin at the level of social movements—deliberate attempts by people and organizations to create change, and conclude with a discussion of the role of technology in promoting change.

Social Movements

- In June 1990, a triumphant crowd of Detroit Pistons fans poured into the streets after the Pistons won the National Basketball Association championship. Jubilant (and drunk) fans got so carried away that seven people died in the ensuing melee.
- In June 1989, a month-long demonstration by 100,000 pro-democracy students in Tiânanmen Square was violently squashed by the Chinese government. Several hundred students and soldiers died in the confrontation. The government followed its victory with a rigorous campaign against activists, and the short-run effect was to reduce the likelihood of reform.

Sociology divides these kinds of activities into two related but distinct topics: collective behavior and social movements. **Collective behavior** is nonroutine action by an emotionally aroused gathering of people who face an ambiguous situation (Lofland 1985, 29). It includes situations such as the impromptu celebration in Detroit. These are unplanned, relatively spontaneous actions, where individuals and groups improvise some joint response to an unusual or problematic situation (Zygmunt 1986).

In contrast, a **social movement** is an ongoing, goal-directed effort to change social institutions from the outside. Examples include the antichoice, gay rights, and Civil Rights movements. A social movement is extraordinarily complex. It may include sit-ins, demonstrations, and even riots, but it also includes meetings, fundraisers, legislative lobbying, and letter-writing campaigns (Marwell & Oliver 1984).

Both collective behavior and social movements challenge the status quo. The primary distinction between the two is that social movements are organized, relatively

Collective behavior is nonroutine action by an emotionally aroused gathering of people who face an ambiguous situation.

A **social movement** is an ongoing, goal-directed effort to change social institutions from the outside.

broad based, and long term; collective behavior, on the other hand, is un-planned and spontaneous. Although the two are conceptually distinct, in prac-tice they are related in at least two ways. First, social movements need and encourage some instances of collective behavior simply to keep issues in the public eye (Delgado 1986). There is nothing like a riot or police breaking up an illegal demonstration to get people's attention. Second, even though col-lective behavior is usually limited to a particular place and time, it can be part of a repeated mass response to arousing conditions. When this happens, col-lective behavior at a grass-roots level may be a driving force in mobilizing so-cial movements (Ash 1972; Rule & Tilly 1975; Zygmunt 1986).

In the following section, we focus on the processes by which disorganized protests become organized, politicized social movements. We adopt the per-spective that social movements are best understood as political processes (Tarrow 1988). Some social movements will be closely allied with traditional political groups, such as parties, and others will seek to overthrow or radically change the state. Although all movements have the goal of changing the con-ventional ways of doing things by affecting public policy decisions, the most successful social movements are those that have been able, through one tac-tic or another, to mobilize the government, the courts, and the law on their behalf (Burstein 1991). Thus, social movements are a part of the political process, and social movement members tend to be the same kinds of people who vote and write letters to their congressional representatives.

As we documented in Chapter 13, most people in the United States have relatively little interest in politics. Why, then, do some people shake off this lethargy and try to change the system? Under what circumstances do so-cial movements succeed or fail?

Theories of Social Movements

Two major theories explain the circumstances in which social movements arise: relative-deprivation theory and resource mobilization theory. Both theories suggest that social movements arise out of inequalities and cleavages in society, but they offer somewhat different scenarios of how and why protest develops.

Relative-Deprivation Theory

Poverty and injustice are universal phenomena. Why is it that they so seldom lead to social movements? According to **relative-deprivation theory,** social movements arise when we experience an intolerable gap between our rewards and what we believe we have a right to expect. What we believe we have a right to expect, in turn, is usually de-termined by comparing ourselves to other groups or to other times. Because the the-ory refers to deprivation relative to other groups or other times rather than to absolute deprivation, it is called *relative*-deprivation theory.

Figure 15.1 diagrams three conditions for which relative-deprivation theory would predict the development of a social movement. In Condition A, disaster or tax-ation suddenly reduces the absolute level of living. If there is no parallel drop in what people rightfully expect, they will feel that their deprivation is illegitimate. In Condition B, both expectations and the real standard of living are improving, but ex-pectations continue to rise even after the standard of living has leveled off. Consequently, people feel deprived relative to what they had anticipated. Finally, in Condition C, expectations rise faster than the standard of living, again creating a gap between reality and expectations. Relative-deprivation theory has the merit of pro-viding a plausible explanation for the fact that many social movements occur in times

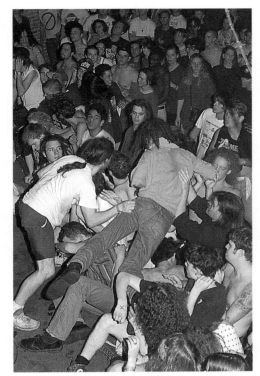

Collective behavior differs from a social move-ment in being more spontaneous and relatively unplanned. Although stage-diving has taken place at more than one rock concert, in more than one place, and at more than one point of time, it is likely to be a relatively short-lived fad, at least in part because participants lack a clearly defined social agenda and the resources and organization to affect public policy. It is these latter characteristics that are generally necessary to transform isolated instances of col-lective behavior into an organized and enduring social movement.

Relative-deprivation theory argues that social movements arise when people ex-perience an intolerable gap between their rewards and what they believe they have a right to expect; also known as break-down theory.

FIGURE 15.1
The Gap Between Expectations and Rewards
Relative-deprivation theory suggests that whenever there is a gap between expectations (E) and rewards (R), relative deprivation is created. It may occur when conditions are stable or improving as well as when the real standard of living is declining.

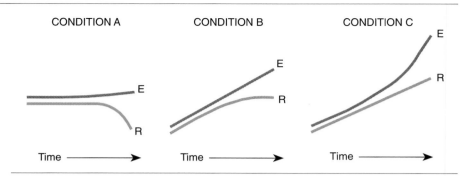

CONDITION A   CONDITION B   CONDITION C

Time ⟶   Time ⟶   Time ⟶

when objective conditions either are improving (Condition C) or at least showing a major improvement over the past (Condition B). Because relative-deprivation theory relies ultimately on the disorganizing effects of social change, it is often referred to as *breakdown theory.*

CRITICISMS OF RELATIVE-DEPRIVATION THEORY. There are two major criticisms of relative-deprivation theory. First, empirical evidence does not bear out the prediction that those who are most deprived, absolutely or relatively, will be the ones most likely to participate in social movement. Often, social movement participants are the best off in their groups rather than the worst off. In many other situations, individuals participate in and lead social movements on behalf of groups or even species to which they do not belong. An example is the animal-rights movement. Second, the theory fails to specify the conditions under which relative deprivation will lead to social movements. Why do some relatively deprived groups form social movements and others not? In general, empirical studies suggest that relative deprivation by itself is not a good predictor of the development of social movements (Gurney & Tierney 1982).

Resource Mobilization Theory

Resource mobilization theory suggests that social movements develop when organized groups are competing for scarce resources; also known as solidarity theory.

According to **resource mobilization theory,** social movements develop when organized groups are competing for scarce resources. This theory differs from relative-deprivation theory in two important ways. First, it argues that deprivation and competition are universal and thus relatively unimportant as predictors of social movements (Oberschall 1973). Second, it assumes that the spark for turning deprivation into a movement is not anger and resentment but rather organization.

Research shows that the most effective social movements emerge from preestablished groups that share two characteristics: relative homogeneity and many overlapping ties (Tilly 1978, 63). This implies that a black civil-rights group that admits whites will be less cohesive than one consisting only of blacks. Groups will be stronger if, in addition to homogeneity, their members share a strong network of ties—if they belong to the same clubs, if they work together, or if they live in the same neighborhood. Research on the 1871 Paris Commune revolt shows, for instance, that neighborhood-based insurgent groups were much more cohesive, effective, and long-lived than units that drew volunteers from all parts of the city (Gould 1991).

Mobilization theory is often referred to as *solidarity theory* because it suggests that the building blocks of social movements are organized groups, not alienated, discontented individuals.

CRITICISMS OF RESOURCE MOBILIZATION THEORY. One major criticism of resource mobilization theory is that it underestimates the importance of grievance and spontaneity as triggers for social movements (Klandermas 1984; Zygmunt 1986). A second criticism is that it overlooks the importance of ideology in translating vague individual dissatisfactions into organized political agendas (Jasper & Poulsen 1995; Williams 1995). Participants in the women's movement often say that "The personal is political." Derived from feminist ideology, this proposition nicely illustrates a point that critics say resource mobilization theory generally seems to miss (Buechler 1993).

Integration

Recent research suggests that both relative-deprivation and resource mobilization theories have merit. Some social movements do develop out of a strongly felt sense of grievance; shared sentiment leads previously unacquainted people to join together to address their concerns. Several of the protest groups that developed after the near disaster at the nuclear energy plant at Three Mile Island in 1980 were of this form (Cable, Walsh, & Warland 1988). On the other hand, some social movements have more to do with the strength of previously existing social networks than with the strength of their grievance. The League of Women Voters, for example, is one of the many civic organizations that adopts a new cause each year.

Resource mobilization theory is clearly the dominant theoretical perspective in contemporary accounts of social movements. If it is broadened to take into account the important role that spontaneous outbursts, triggering events, and a strong sense of grievance play, it provides a useful model of social movement development. Figure 15.2 compares the two models.

How Do Social Movements Operate?

Social Movement Organization

A social movement is the product of the activities of dozens and even hundreds of groups and organizations, all pursuing, in their own way, the same general goals. For example, there are probably 50 different social movement organizations (SMOs) within the environmental movement, ranging from the relatively conventional Audubon Society to the radical Greenpeace movement. The organizations within a movement may be highly divergent and may compete with each other for participants and supporters. Because this assortment of organizations provides avenues of participation for people with a variety of goals and styles, however, the existence of diverse SMOs is functional for the social movement.

SMOs can be organized in one of two basic ways: as professional or as volunteer organizations. On the one hand, we have organizations such as the Audubon Society or the Sierra Club that have offices in Washington, D.C., and a relatively large paid staff, many of whom are professional fund raisers or lobbyists who develop an interest

RELATIVE-DEPRIVATION/BREAKDOWN MODEL

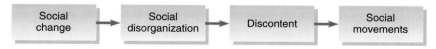

RESOURCE MOBILIZATION/SOLIDARITY MODEL

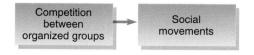

FIGURE 15.2
A Comparison of Relative-Deprivation/Breakdown Models with Resource Mobilization/Solidarity Models
SOURCE: Adapted from Useem 1980.

Greenpeace has been particularly successful at keeping the environmental movement in the public eye by combining elements of both professional and indigenous social-movement organizations. Local chapters have engaged in a variety of direct-action tactics that have publicized such causes as creating and enforcing an international ban on whaling or the slaughter of baby seals for the fur industry. In recent years, however, the parent organization has adopted a somewhat less confrontational stance, at least in part so that it can maintain the support of the organizations and individuals who form its larger conscience constituency.

in the environment after being hired. At the other extreme is the SMO staffed on a volunteer basis by people who are personally involved—for example, neighbors who organize in the church basement in order to prevent a nuclear power plant from being built in their neighborhood. These two types of SMO are referred to, respectively, as the *professional* SMO and the *indigenous* SMO.

Evidence suggests that the existence of both types of organizations facilitates a social movement. The professional SMO is usually more effective at soliciting resources from foundations, corporations, and government agencies. It appeals to what is called a *conscience constituency,* those who are ideologically or morally committed to the group's cause. On the other hand, because they themselves are not underprivileged and because they work daily with the establishment, professional SMOs sometimes lose the sense of grievance that is necessary to motivate continued, imaginative efforts at change. A social movement also requires sustained indigenous organizations (Jenkins & Eckert 1986). Indigenous organizations perform two vital functions. By keeping the aggrieved group actively supportive of the social movement, they help to maintain the sense of urgency necessary for sustained effort; and their anger and grievance propel them to more direct-action tactics (sit-ins, demonstrations, and the like) that publicize the cause and keep it on the national agenda.

The women's movement is an excellent example of a social movement that combines both professional and indigenous SMOs. Informal networks kept and continue to keep the discussion of equal rights and equal opportunities alive, even in periods when SMOs are nonexistent or marginalized. The most successful periods of feminist activism have been when SMOs, such as the National Organization for Women (NOW), have worked in close cooperation with informal networks and conscience constituencies (Buechler 1993). However, in the absence of direct actions—women's suffrage marches, bra burnings, or ERA rallies—pressure from both professional and indigenous SMOs produce modest results, at best.

Mobilization

Mobilization is the "process by which a unit gains significantly in the control of assets it previously did not control" (Etzioni 1968, 388). These assets may be weapons,

Mobilization is the process by which a unit gains significantly in the control of assets it did not previously control.

Mass demonstrations, such as this early women's suffrage march, serve to gain publicity for social movements. The real work of most social movements, however, goes on in committee meetings, at fundraisers, and in other methodical tasks, designed to mobilize resources and influence public opinion. When a movement is successful in reaching some of its major goals—passage of the Nineteenth Amendment, for instance—it often becomes indistinguishable from any other humanitarian or political professional organization. As new issues arise, however, the organization is likely to rely again on protests and demonstrations to gain new supporters or to reinvigorate existing rank-and-file members.

technologies, goods, money, or members. The resources available to a social movement depend on two factors: the amount of resources controlled by group members and the proportion of their resources that the members are willing to contribute to the movement. Thus, mobilization can proceed by increasing the size of the membership, increasing the proportion of assets that members are willing to give to the group, or recruiting richer members.

Mobilization proceeds through two tactics: the recruitment of individual adherents (micromobilization) and the recruitment of supportive organizations (bloc mobilization).

MICROMOBILIZATION   A strategy through which SMOs attract individual new members is **frame alignment.** It is a process of convincing individuals that their interests, values, and beliefs are complementary to those of the social movement

Frame alignment is a strategy of micromobilization in which SMOs attract individual new recruits; it seeks to convince individuals that their interests, values, and beliefs are complementary to those of the social movement organization.

organization. According to Snow and associates (1986), SMOs use four tactics of frame alignment:

1. *Frame bridging* targets people who have similar, though perhaps not identical, organized interests and attempts to convince them that the similarities are great enough that they should support your organization too. Part of this strategy involves the use of computerized mailing lists of related organizations. Thus, the Sierra Club may appeal to members of the National Wildlife Federation or the Audubon Society as people who have the same values and goals.

2. *Frame amplification* is the equivalent to consciousness raising. This strategy gives structure to unfocused dissatisfaction by offering the SMO's frame as an explanation. It tries to convince people that their problems are caused by patriarchy or racism or whatever definition of the situation is used by the SMO.

3. *Frame extension* extends the frame of the social movement so that more and more problems and concerns are included within its definition of the situation. For example, some peace movements have tried to attract more recruits by suggesting that the struggle for peace is also a struggle for social justice against racism, sexism, and poverty.

4. *Frame transformation* is equivalent to a religious conversion. It requires convincing individuals that the way they have seen things is entirely wrong. This strategy is used by the cults, the Black Muslims, and radical feminists—groups that provide a radically different definition of a person's entire past and future and that frequently demand not just one evening a week but total, full-time dedication.

Who is most likely to be affected by these mobilization strategies? Studies of social movement activists show that, although ideology and grievances are important, the key factor is the strength of their personal ties with other movement activists and the absence of ties with significant others who oppose their activism. No matter how deeply committed to the movement they are ideologically, if they are not part of a network of others with similar convictions or if they are embedded in networks that give conflicting messages, people are unlikely to be more than token members (McAdam 1986; McAdam & Paulsen 1993).

Bloc mobilization is a strategy whereby social movement organizations recruit other organizations to support their cause rather than trying to recruit single individuals.

BLOC MOBILIZATION. In addition to converting or mobilizing adherents one at a time through frame alignment processes, SMOs also use a strategy called **bloc mobilization** to recruit other organizations to support their cause. For example, the antipornography movement has asked for and received support from PTAs, fundamentalist churches, and feminist organizations.

Bloc mobilization is a very effective way to expand a movement's resources: It means access to other organizations' newsletters, members, and funds. It is, however, only suitable for social movements that make very low demands on their members. It would not be an effective means of recruiting members to a radical religious sect or to an extremist political group.

Factors Associated with Movement Success

Empirical analysis of social movements in the United States and around the world suggests that a number of factors are important to the success of a movement. First, in the face of resistance from the larger society, movements must be able to develop an ideology capable of sustaining the enthusiasm and continued participation of current members, while at the same time providing the motivation for new members to join.

Research indicates that to succeed in this task, social movement organizations must create ideological frameworks that can convince potential and current participants that a particular grievance is serious, that there is an urgent need to address the problem, that taking action is proper, and most importantly that such action will be effective (Benford 1993).

The relationship between the movement and third parties and the nature of movement demands also influence whether a movement will succeed or fail. Marx (1971) identifies the following variables as important:

1. The demands of the movement are seen to be consistent with the broader values of society. For example, the movement seeks to increase freedom or reduce injustice.
2. The movement has the support of influential third parties or can demonstrate that its demands will benefit other groups as well. For example, the abolitionist movement gained the support of the early feminist movement because women believed that extending suffrage to blacks would help women gain suffrage.
3. The movement's demands are concrete and focused. A protest against a specific urban renewal project is more likely to succeed than is a general protest against poor housing.
4. The movement is able to exert pressure directly on the responsible party without harming uninvolved third parties. For example, a fruit boycott that hurts truckers as well as fruit growers will generate more opposition and less support.
5. The movement adopts techniques with which the authorities have had little experience. The nonviolent sit-in had tremendous impact when it was first employed during the early Civil Rights movement; in 1980, however, hundreds of protestors sitting in at the Seabrook Nuclear Plant were hauled away with little publicity and little effect. The police now know how to deal with the tactic, and the media no longer find it newsworthy.
6. Neutral third parties who have an interest in restoring harmony are present.
7. The movement's demands are negotiable rather than absolute.
8. The movement's demands involve a request for acceptance of social diversity, equal treatment, or inclusion, rather than a fundamental redistribution of income and power.
9. The movement seeks to veto proposed policies rather than implement new ones.
10. The movement is large enough to organize itself for conflict but not so large as to be perceived as a serious threat to the dominant group.

THINKING CRITICALLY
Given what you know about the animal-rights movement and about the factors associated with movement success, how likely do you think it is that this social movement will achieve its goals?

Countermovements

A major and growing category of social movements is the **countermovement,** which seeks to reverse or resist change advocated by a social movement (Lo 1982; Mottl 1980). Countermovements are almost always right-wing in orientation. They seek to maintain traditional structures of status, power, and values.

Resource mobilization theory is particularly appropriate for understanding countermovements. Because they defend the status quo, they are often closely tied to already established groups (Lo 1982; Mottl, 1980). Bloc mobilization is a chief means of recruiting members and resources.

The anti-ERA movement is an excellent case study of a countermovement. The movement made heavy use of bloc mobilization strategies, tapping into the organizational resources of the antichoice movement, fundamentalist Protestant churches, the

A **countermovement** seeks to reverse or resist change advocated by a social movement.

Whenever a social movement starts to get a lot of publicity and it appears that it may affect public policy, a countermovement tends to develop to support the status quo. These pro-life/pro-choice demonstrators are an example. Because countermovements generally support existing or traditional social arrangements, their members tend to be more conservative than the members of social movements. These are the same people who benefit from the current arrangements and thus want to maintain them.

John Birch Society, and the American Farm Bureau. One of the most interesting features of the movement was the strategy used to portray its goals. As noted in the preceding section, successful movements claim to be supporting social values, not denying them. Thus, the anti-ERA movement could hardly campaign on the basis of being against equality. Instead, its rhetoric claimed to be supporting traditional values of family, motherhood, and womanhood, which were under attack by lesbians and other dissatisfied women (Marshall 1985).

A Case Study: The Gay Rights Movement

In recent Western thought, homosexuality has been considered a sin or a sickness. As a result, it has often been furtive and concealed. An active social movement now exists to change this situation. Like many social movements, it is diverse and fragmented, replete with competing SMOs and even a countermovement.

History of the Gay Rights Movement

Homosexual acts are illegal in most states. Homosexuals have been barred from service in the military and dishonorably discharged if discovered; they are often barred from teaching in public schools; and they may be denied custody of their children and, in some cases, even visiting rights. Until a few years ago, they were barred from employment in the federal civil service and from immigration to the United States. In addition, they are often shunned by their family and co-workers, forced out of their jobs, and subjected to taunts and jeers.

As a result, most homosexuals have concealed their sexual preference. As long as they did so, there could be no social movement. For a movement to exist, there must be a group of people who acknowledge that they are members of the same group and share a common interest.

The beginning of the gay rights movement came when sufficient numbers of prominent individuals were willing to step forward and define themselves as homosexuals. This development began in Germany at the end of the nineteenth century. It was abruptly halted by Hitler, who included homosexuals among the undesirables of the world and sent known homosexuals to concentration camps. In the United States, the gay rights movement began after World War II when the two founding SMOs of the gay rights movement—the Mattachine Society for male homosexuals and the Daughters of Bilitis for female homosexuals—were established.

The Current Movement

The gay rights movement is not unified. It is divided by sex, class, race, and political ideology. Broadly, however, the movement seeks to do four things (Altman 1983, 122):

1. *To define a gay community and a gay identity.* The movement seeks to help gay individuals realize they are not alone. As C. Wright Mills might have put it, they want homosexuals to recognize that their problems are not merely personal troubles but are shaped by social structure.
2. *To establish the legitimacy of a gay identity.* The movement seeks to reduce the shame, to overcome the internalized self-hatred and doubts, of people who have been socialized to believe they were wicked and sick.
3. *To achieve civil rights for homosexuals.* The movement seeks to decriminalize homosexual acts and to establish antidiscrimination laws to protect homosexuals.
4. *To challenge the general ascription of gender roles in society.* The movement seeks to give people the right to choose roles rather than being forced to act out a role thrust on them by reason of their sex.
5. *To secure family rights.* The most recent goal of the movement is to obtain for gay couples the same legal rights as other couples enjoy, such as health insurance or survivor's benefits for a partner.

Conflict Within the Movement

There are several major schisms within the gay rights movement. The most important is that between men and women. Homosexual men (gays) and women (lesbians) have some goals in common—in particular, civil rights goals. However, lesbians face a situation of double jeopardy. They may be discriminated against on the basis of sex as well

Public demonstrations by the gay rights movement are designed to bring homosexuality out of the closet, to make it seem a less deviant and dangerous practice. By publicly acknowledging their sexual preference, men and women such as the ones in this parade force society to acknowledge that there are relatively large numbers of people—many of them apparently normal and decent people—who are homosexuals. Thus, the public demonstration is a particularly important weapon in the social movement for homosexual rights.

as sexual preference. Lesbian women often believe that they will make more progress working with straight women than with gay men; they believe that improvements in the status of women (especially in economic terms) will be more beneficial than will general improvements for homosexuals. As a result, there is relatively little cooperation between male and female homosexual groups.

A second schism is of class and politics. On the one side are the middle-class professionals who wear gender-appropriate suits and insist that homosexuals are respectable, decent people—good parents, good credit risks, good neighbors. On the other side are people who insist that gay rights means the freedom to wear lavender and leather and who wish to dismantle the entire system of gender roles and status politics. These are, respectively, the people who want to tinker with the system—to extend the basic rights package just a little further—and the people who think that the whole system is a sham and want to overthrow it.

Successes, Failures, and Prospects

The gay rights movement has seen some notable successes. The American Psychological Association voted in 1974 to declare that homosexuality is not a sickness; Wisconsin and Massachusetts have passed laws making discrimination on the basis of sexual orientation illegal; and acknowledged homosexuals have been elected to public office, including that of U.S. senator. Public opinion polls show substantial increases in support for homosexual rights. In offices and families around the country, it is becoming possible to be open about one's homosexuality without losing the respect of others.

In many homes, offices, and neighborhoods, however, the position of an acknowledged lesbian or gay man would still be awkward at best. The military continues to discharge individuals who engage in homosexual activity, and in 23 states homosexual acts are still a crime. In addition to these assaults on their dignity, homosexuals also have a high risk of incurring physical assaults (Jenness 1995). "Gay bashing" is a relatively frequent recreational activity of young toughs, and homosexual activists claim that gays are seven times more likely than straights to be assaulted (Singer and Deschamps 1994). Furthermore, gay and lesbian activists are more likely than those who are simply "out" to report discrimination in hiring and promotion processes (Taylor & Raeburn 1995).

For more than a decade, now, the male homosexual community has been galvanized by the specter of AIDS (acquired immunodeficiency syndrome). Many gays believe that proposals for mandatory AIDS testing are another basis for stigmatizing gays. Coupled with realistic health concerns, this fear has provided an impetus to organization. AIDS hotlines and information meetings have triggered gay men's networks that cut across class, race, and political cleavages. By increasing the solidarity of the group, these new networks may increase the effectiveness of the gay rights movement.

The Role of the Mass Media

A social movement is a deliberate attempt to create change. To do so, it must reach the public and create the appearance that public opinion is on its side. The relationship between the media and social movements is one of mutual need. The movements need publicity, and the media need material. Sometimes both needs can be met satisfactorily. In this mutual exchange, however, most of the power belongs to the media. The media can affect a social movement's success by giving or withholding publicity and by slanting the story positively or negatively. What the media choose to cover "not only affects the success of the movement, but also shapes its leadership and its meaning to the general public and to its own adherents—in short, what the movement actually is" (Molotch 1979, 81; Mulcahy 1995).

THINKING CRITICALLY

Based on what you know about the gay rights movement, what social structural conditions in the larger society helped to spark this social movement? What countermovements do you know of regarding gay rights? Of the five goals listed, which do you think are the most likely to be achieved, and why? Which least likely, and why?

A particularly interesting example of the importance of the media can be seen in the Tiânanmen Square uprising. Two major events in the spring of 1989 brought an unusually large contingent of international journalists to Beijing—the historic visit of Mikhail Gorbachev and the Asian Development Bank's first meeting in the People's Republic of China. Aware that the whole world would be watching and that state authorities would feel somewhat constrained to act repressively, student activists used these journalists to create a global stage. Thus, foreign media played a crucial role in the movement, not only by reporting events to their audiences back home, but also by keeping Chinese citizens informed regarding movement developments (Zuo & Benford 1995).

In both the West and East, then, media coverage appears to be a vital mechanism through which resource-poor organizations can generate public debate over their grievances. What does an organization have to do to get news coverage? Empirical studies show that four factors are critical (Kielbowicz & Scherer 1986):

1. *Dramatic, visible events.* Sit-ins and demonstrations are more newsworthy than news conferences, and both are more newsworthy than a pamphlet.

2. *Authoritative sources.* Journalists want to save time and gain credibility by going straight to the horse's mouth. This means they tend to rely on established figures who have public recognition.

3. *Timing.* News is published or aired according to regular deadlines. If you want to be on the evening news, your action should be scheduled (so that the news cameras can be there) before three in the afternoon (so that it can be edited before the news hour) on a day when not much else is going on.

4. *News nets.* Most reporters have a beat, a particular area of the news they cover. If your movement falls in the cracks, you are less likely to get coverage. A prerequisite may be a little frame alignment on the part of the journalists.

Because they rely on the media, social movements find themselves changing to maximize news coverage. The link between action and newsworthiness encourages direct action rather than more quiet forms of activity such as lobbying or letter writing, and because the public tires of watching demonstrators hauled off, the degree of extremism necessary to get news coverage may escalate. This may lead to inflamed rhetoric, greater conflict within the movement, and a disproportionate attention to publicity rather than to other movement goals. Of course, if the alternative is no publicity at all, these changes may be a price the movement has to pay. Without free publicity, the cost to a social movement of spreading its message is greatly increased.

Technology

The pervasive influence of technology in our daily lives today is obvious. Perhaps you woke up to an alarm this morning, to find coffee already brewed in your preset electric coffeemaker, used a modem to read your E-mail, and listened to your favorite CD, all before you made it to your first class. Not only are our tools far more pervasive than they once were, but they are also potentially far more powerful and dangerous. It is vitally important, then, that we understand the relationship between technology and social change.

Technology can be defined as the human application of knowledge to the making of tools and the use of natural resources. It is important to note that the term *technology* refers not only to the tools themselves but also to our beliefs, values, and attitudes toward them. While we may be inclined to think of technology in terms of today's

THINKING CRITICALLY
Suppose you were interested in mobilizing public opinion against the death penalty. What kind of activity or event would you try to use to get the media's attention?

Technology involves the human application of knowledge to the making of tools and to the use of natural resources.

focus on

An Application
TREES, OWLS, AND THE ENVIRONMENTAL MOVEMENT

Being in favor of protecting the environment sounds like an innocuous position to take. After all, who is in favor of dirty air, dirty water, and disappearing species? By default, nearly all of us.

Ruining our environment is part of the status quo; it is part of our accepted way of life, of manufacturing and packaging merchandise, and of dealing with garbage. The average American produces 35 pounds of garbage each week, only a tiny fraction of which is recycled. Environmental protection will entail costs: higher-priced goods, more bother over recycling, more regulation, and fewer consumer goods. It is also likely to result in the loss of some jobs.

The Battle for Trees and Owls

The battle over environmental policy is being fought on many fronts—over nuclear power, hazardous wastes, forests, and habitat. One of the most contentious in recent years has been the battle over protecting old-growth forests. These forests provide a unique habitat, and some species—notably the spotted owl—do not live anywhere else. Recently, lumber companies won the right to clear-cut 130,000 acres of old-growth forests in the Pacific Northwest. To environmentalists this is a

tragic loss, but for loggers it creates much-needed jobs. A bumper sticker distributed during this highly acrimonious controversy says, "Save a logger. Kill an owl."

Radical environmentalists refer to lumber companies and developers as "eco-thugs." Environmentalists are not much more complimentary about the U.S. Forest Service, which frequently finds itself caught in the middle. The service was set up to administer the logging and lumbering of U.S. forestland. Its major purpose was to grow and protect forests and to create and maintain logging roads *so that loggers could cut down the trees.* The service and the loggers were colleagues, not opponents.

In recent years, the environmental movement has challenged the U.S. Forest Service to protect the forests *from* the loggers, not *for* the loggers. The professional SMOs of the environmental movement—the Sierra Club, the National Audubon Society, and the Wilderness Society—write letters to congressional representatives to urge support for clean-air laws or to lobby against dam projects; they throw a battery of lawyers into the effort to get court injunctions against development projects. The militants, who are unwilling to compromise, use sit-ins

and demonstrations (mostly legal) and "monkey wrenching" or sabotage (mostly very illegal). They have cut power lines, spiked trees to prevent them from being logged, and damaged bulldozers. The forest service refers to these people as the "violent fringe," but supporters claim that their only crime is "to protect 4 billion years of evolution [and] I'm proud to be associated with that sort of criminal element" ("Trying" 1990, 25).

Although militants have done much to publicize and galvanize the environmental movement, they cannot succeed by themselves. Throwing one's body in front of the bulldozers may buy a few days, but permanent victory involves court orders and legal battles. Thus, both professional and indigenous, both conservative and radical SMOs are helpful in pushing the movement forward.

Increasingly, the movement has infiltrated government organizations. The forest service has expanded its mission to include providing recreation and protecting wildlife. This new mission puts the forest service in the middle of two angry groups, neither of which is willing to compromise. As a result, one forest-service ranger jokes about being a "combat biologist" ("Oregon's" 1989).

"high-tech" advances, it also includes such relatively simple tools as pottery and woven baskets. Thus, technology has been a component of culture from the beginning of human society.

Social change is any significant modification or transformation of social structures or institutions over time; technology is one major cause of social change. Chapter 2 describes how technology helped to transform hunting-and-fishing-and-gathering societies to horticultural, then agricultural, and then industrial societies.

Because technology defines the limits of what a society can do, technological innovation is a major impetus to social change. Meanwhile, new technologies are created to meet the new needs created by a changing culture and society. The result is a never-ending cycle in which social change both causes and results from new technology. In this section, we briefly review two theories of technologically

Social change is any significant modification or transformation of social structures and sociocultural processes over time.

The Environmental Movement Assessed

One reason federal agencies have changed their policies on trees is that concern for the environment has increased markedly over the last decade; a large majority of Americans now say they are willing to pay more taxes to clean up the environment. Not all of today's activists are leftover hippies. Many of those involved in campaigns against hazardous waste and nuclear power, for example, are "just moms and dads who are willing to protect their children at any cost" ("Trying" 1990, 24).

The environmental movement has had some notable successes. Afraid of being buried in their own waste, many communities in the United States are experimenting with mandatory recycling. Nevertheless, there are many controversial issues left to be negotiated. It is estimated that 2 percent of the world's species will disappear every year. If every one of them produces battles the size of that now being fought over the spotted owl in the Pacific Northwest, we are in for a long and bloody war. As environmental protectionism starts to threaten the lifestyles and livelihoods of people other than isolated loggers, we are likely to see more controversy rather than less (Cable & Cable 1995).

Earth First! is one of the most radical SMOs within the environmental movement. It performs a useful function for the movement by making all other environmental organizations seem rather reasonable by comparison. In recent years, even Earth First! leader, Dave Foreman, has worried that there are "too many hippies and leftists in Earth First!" and he has begun developing an alternative SMO.

induced social change and present a case study of how information technology may change society.

Two Theories of Technologically Induced Social Change

Social change is a central topic in sociology. As discussed in Chapter 1, the early sociologists were bent on understanding the consequences of the Industrial Revolution, an event that triggered dramatic social change. Auguste Comte, the founder of sociology, argued that any understanding of society required not only an understanding of the sources of order (statics) but also of the process of change (dynamics). Throughout this text, we examine many important aspects of social change such as global inequality (Chapter 7) and population growth (Chapter 14) to name but a few. This section explores how structural functionalism and conflict theory explain social change.

Structural-Functional Theory: Social Change as Evolutionary

While structural-functional theory primarily asks how social organization is maintained in an orderly way, the theory does not ignore the fact that societies and cultures change. As pointed out in Chapter 1, according to the structural-functional perspective, change occurs through evolution: Social structures adapt to new needs and demands in an orderly way while outdated patterns, ideas, and values gradually disappear. Often, the new needs and demands that prompt this evolution are technological advances.

As noted in the Focus on Technology Box in Chapter 10, the sociologist William T. Ogburn (1922) added the concept of cultural lag to the idea of evolutionary change. Because the components of a society are interrelated, Ogburn reasoned, changes in one aspect of the culture invariably affect other aspects. The society will adapt, but only after some time has passed. As an illustration, Ogburn noted that by 1870, large numbers of U.S. industrial workers were being injured in factory accidents, but workers' compensation laws were not passed until the 1920s—a cultural lag of about 50 years. Ogburn pointed out that a society can hardly adapt to a new technology before it is introduced. Hence, cultural lag is a temporary period of maladjustment during which the social structure adapts to new technologies.

Conflict Theory: Power and Social Change

While structural functionalism sees social change as orderly and generally consensual, conflict theorists contend that change results from conflict between competing interests. Furthermore, conflict theorists assert that those with greater power actually direct social change to their own advantage. In a process characterized by conflict and disruption, social structure changes (or does not change) as powerful groups act either to alter or to maintain the status quo.

According to Thorstein Veblen (1919), those for whom the status quo is profitable are said to have a vested interest in maintaining it. **Vested interests** represent stakes in either maintaining or transforming the status quo; people or groups who would suffer from social change have a vested interest in maintaining the status quo, while those who would profit from social change have a vested interest in transforming it. We can think of many examples of people or groups with a vested interest in maintaining the status quo. Communities with a military post have a vested interest in retaining it because the inflow of government money and jobs is good for local business and many university students have a vested interest in the U.S. government's retaining federally guaranteed loans.

We can also think of instances in which a vested interest in maintaining the status quo involves efforts to halt technological innovations. For example, automobile and oil companies have blocked widespread production and marketing of the electrically powered car, even though this technology has been available for a long time. As another example, movie theater operators did their best to impede cable television.

Meanwhile, other groups in a society have a vested interest in changing things. Those who would benefit from an innovation have a vested interest in working to see that it is introduced and typically appear as its promoters. In the early decades of the twentieth century, the American Medical Association worked hard to officially and legally replace midwives with licensed physicians (Starr 1982). In a current example, Bill Gates, the Microsoft mogul, has a vested interest in promoting future software and Internet innovations, along with the ongoing diffusion of computers throughout the world.

Just as the benefits of a particular social innovation are unevenly distributed, so also are the costs. Benefits go to the more powerful, while costs tend to go to the less

Vested interests are stakes in either maintaining or transforming the status quo.

THINKING CRITICALLY
How might you analyze the current debate over affirmative-action policies and programs in terms of various groups' vested interests?

powerful. As one observer has pointed out, corporate managers and others who put new technologies into place usually do not take into account what economists like to call "external costs"—costs that will not be paid by the corporation itself. This situation can be seriously problematic—and unfair—because society at large then pays the external costs (Mesthene 1993, 79). The costs to the environment caused by the gasoline-powered engine, for instance, are not directly paid by motor or petroleum companies. Instead, these costs are paid by us all in the form of air pollution.

Like evolutionary theories, the conflict perspective on social change makes intuitive sense to many, and there is empirical evidence to support it. However, a general assumption of the conflict perspective—which dates back to its founder, Karl Marx—is that those with a disproportionate share of society's wealth, status, and power have a vested interest in preserving the status quo. In today's rapidly changing society, this may no longer be the case, as powerful factions may be just as likely to find reason to push for more and more technological innovations. Furthermore, some scholars have argued that technology is virtually "autonomous." That is, once the necessary supporting knowledge is developed, a particular invention will be created by someone, even if this invention is terribly costly to nearly everyone in the society, including many of the powerful. In other words, technological changes may be caused by social forces beyond our effective control. We return to this issue near the end of this chapter. At this point, we look at some ways in which today's information technology affects our society.

A Case Study: Information Technology and Social Change

Consider the student in 1967 who is assigned the task of writing a term paper on the consequences of parental divorce. She goes to the library and walks through the periodicals section until she stumbles on the *Journal of Marriage and the Family,* in which she eventually finds five articles—the number her professor requires—on her topic. She takes notes on three-by-five-inch cards (there are no photocopying machines) and goes home to draft her paper on her new electric typewriter. She cuts and tapes her draft copy until it looks good, checks words of dubious spelling in her dictionary, and then retypes a final copy. She uses a carbon paper so that she will have a copy for herself. When she makes a mistake, she erases it carefully and tries to type the correction in the original space.

Now consider the student in 1997. This student starts her paper by logging onto SOCIOFILE©, an electronic bibliography of more than 100,000 sociology articles on a single CD-ROM disk. When she enters the keywords *divorce* and *parental,* the program prints out full citations and a summary for 41 articles. After identifying and photocopying the 5 articles she wants, she drafts a report on her word processor, edits it to her satisfaction, runs it through her spelling checker, and adjusts the vocabulary a bit by using the built-in thesaurus. She also runs the report through her new grammar checker, which will catch errors in punctuation, capitalization, and so forth. Finally, she sends the whole thing to her mother (who lives 2,000 miles away) by electronic mail and asks her to read it for logic and organization. She receives the edited version from her mother in an hour, prints two copies, and hands in the report. If she is taking an off-campus course, she may send the paper to her instructor via electronic mail, or she may fax it.

Information technology—computers and telecommunication tools for storing, using, and sending information—has changed many aspects of our daily lives. Over the past few decades, the United States has become an "information society." More and more workers are employed in information acquisition, processing, and communication (Beninger 1993). More important in its social implications has been the convergence of

Information technology comprises computers and telecommunication tools for storing, using, and sending information.

focus on *technology*

BIOTECHNOLOGY'S POWER, PROMISE, AND PERIL

Plants glow like fireflies, and goats give milk containing human medicine. This may sound like science fiction, but it's not. These innovations and others like them exist today because of *biotechnology*—scientists' purposeful and direct manipulation of genetic material in animals and plants. Sometimes a gene is simply altered through biotechnology. To invent the "Flavr Savr" tomato, first marketed in 1994, scientists identified the tomato gene that promotes softening and changed it. The engineered, or "designer", tomatoes can stay on the vine longer, ripening and gaining flavor without getting too mushy to ship (Shapiro 1994). Other times, a gene from one organism is introduced into another; this process creates a different, or "transgenic," plant or animal that never existed before. Scientists invented plants that glow by inserting firefly genes into the plants' genetic materials.

Originating with pioneering work in the 1960s, biotechnology is a powerful new tool that promises dramatic improvements in human beings' lives. Through biotechnology, we can create longer-lasting foods; plants that yield more food; new plants that can be grown specifically for fuel; animals that secrete medicines, such as insulin for diabetes and TPA for blood clots; and laboratory animals that allow scientists to better study and perhaps find cures for diseases like AIDS, cancer, and sickle-cell anemia.

While acknowledging the promised benefits of biotechnology, environmentalists point out serious risks. Most scientists dismiss the fear sometimes voiced in the popular culture that they will inadvertently create monsters. There are concerns more real, immediate, and pressing, however. Some risks remain unknown because it is still too early even to imagine all the possible consequences of biotechnology. There are knowable risks to our health and ecosystem as well. For instance, scientists can now design crops with pesticides in their genes. These plants would kill damaging insects "automatically." However, eating the food from such crops would require ingesting the pesticide, a situation possibly hazardous to health. Beside pesticides, the expanded use of herbicides causes concern.

Biotechnology can now create plants that are genetically resistant to weed killers. Farmers growing such plants as crops can more effectively spray herbicides over their fields without damaging the crop itself. Meanwhile, environmentalists point out that herbicide-resistant crops run directly counter to the goal of reduced dependence on agricultural chemicals that pollute land and groundwater.

More direct risks to our ecosystem involve potential catastrophes that could result if transgenic organisms are introduced into the environment. Nonnative animals introduced into a new environment (for example, rabbits taken into Australia from Britain) have often caused havoc in their new environment. Similar to nonnative species inasmuch as they have traits novel to the ecosystem, engineered species pose similar serious risks. As an example, scientists are now creating engineered fish, such as a carp with an "antifreeze" gene that can live in very cold water. If introduced into an ecosystem, the carp might expand its feeding range and thereby displace or destroy some or all of the native fish species.

In addition to these issues, animal rights activists have raised concerns that biotechnology sometimes causes cruel treatment of animals (Varner 1994). About 10 years ago, for instance, scientists genetically implanted pigs with a growth hormone to increase meat production and reduce fat. The pigs experienced serious and very likely painful health complications that precluded their immediate commercial use. In a second example, laboratory mice are genetically engineered to be born with humanlike diseases such as cancer and cystic fibrosis. Producing such mice allows scientists to research new, effective ways to treat such diseases. But the mice suffer in this role, a morally objectionable situation in the view of animal rights activists.

> **How society will weave its way among the two faces of biotechnology remains to be seen.**

How society will weave its way among the two faces of biotechnology remains to be seen. There is currently evidence of cultural lag. In the opinion of many scientists, neither federal law nor the Food and Drug Administration (FDA), the federal agency responsible for regulating engineered organisms, is yet equipped to handle all the possible issues or situations that could emerge. Is biotechnology potentially more dangerous than helpful? No, most scientists say, but the new technology must be understood, debated, monitored, and effectively regulated (Donnelley, McCarthy, Singleton, Jr. 1994; Mellon 1993).

various information technologies—mass media, telecommunications, and computing—to form a unified system. Aside from enabling us to write term papers more easily, how will information technology change our lives? Will it reduce or increase social class inequality? Make life safer and better? Or make life more stressful and isolated?

The answer is likely to be some of each. As shown in Map 15.1, information technology links us to the rest of the world. We are linked to distant family and friends, to doctors and medical information, to libraries and data banks, and to world events. In 1995, for instance, the news media carried a story describing how a young refugee from Bosnia had found asylum in the United States by locating a California sponsor via the Internet. Information technology may allow us to participate more fully in the political process by making it possible to communicate more effectively and directly with our elected official. By linking us to distant work sites, computers and electronic mail may allow more employees to work from their homes in small cities, simultaneously reducing the amount of time they need to spend commuting to a job and increasing the flexibility of working hours in order to have more time to spend with their families.

On the downside, advances in information technology have introduced new forms of crime (hacking and electronic theft), new forms of social control (information about when you buy or sell a car, pay taxes, receive income or a speeding ticket, register a birth, or marry will increasingly be available for

People in the academic community—students and teachers—have been prime beneficiaries of developments in information technology. Our ability to access and process information has been expanded enormously during the past 20 years. Because today's hardware and software are both much better and much cheaper than those of even 10 years ago, a very large portion of all college students have access to personal computers and sophisticated word processing programs.

MAP 15.1
Access to the Internet
SOURCE: Matrix Information Directory Services, Inc. (MIDS), Austin, Tex. (http://www.mids.org/) Data used with permission.

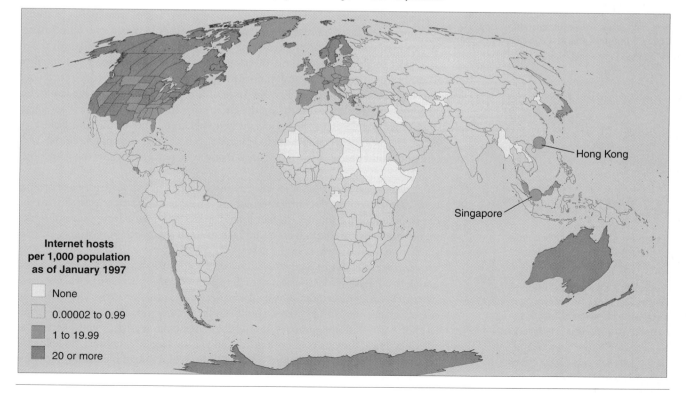

Internet hosts per 1,000 population as of January 1997

None
0.00002 to 0.99
1 to 19.99
20 or more

Hong Kong

Singapore

Joel Schatz travels all over the world in the course of his job. To keep in touch with his office, he simply hooks his portable computer up to any telephone and dials an electronic mailbox to check for messages.

Summary

The effect of information technology on society will depend as much on social institutions as it does on the technological capacities of computers and telecommunications. Information technology offers us more freedom of residence, and more input into local and federal legislative bodies; but we simultaneously lose some privacy and autonomy. Whether the blessings or costs will be predominant will depend on how these technologies are implemented in schools, workplaces, and government bureaucracies. To the extent that they affect relationships among work, class, neighborhood, and family, the new technologies are of vital interest to those concerned with social institutions.

Making the best use of advancing technology and helping assure that advances prompt desirable social changes, requires social planning—the conscious and deliberate process of investigating, discussing, and coming to some agreement regarding desirable action based on common values.

"bureaucratic surveillance"), new defense worries (breaches of defense data systems and faulty software programs that may inadvertently launch World War III), and new inefficiencies ("I'm sorry; the system is down").

The Sociologist's Contribution

Whether it originates with a social movement or a change in technology, any social change will have opponents. Every winner potentially produces a loser. This means that change creates a situation of competition and conflict.

In Chapter 1, we discussed the appropriate role of sociologists in studying social issues. Should they be value-free, or should they take a stand? Issues of inequality, war, and peace bring this question into sharp focus. Although most sociologists go about their business as if sociology were unrelated to global conflict, an increasingly vocal minority argue that it is shortsighted to restrict our forces to divorce, educational equity, the state of U.S. race relations or the gay rights movement when people are destroying the environment and one another. These people believe that sociologists should be actively involved in issues of war and peace.

What can sociologists contribute to enhancing justice and protecting both the environment and world peace? A few of the areas that can be pursued include the following:

1. *The study of conflict resolution.* A growing number of universities have special courses or programs on conflict resolution. These courses are concerned with the development of techniques for handling disputes and negotiating peaceful settlements. Sociological research on topics such as small group decision making and organizational culture are relevant here.

2. *Developing social justice perspectives.* The cold war didn't fizzle because the United States and the Soviet Union scared each other to death or because Gorbachev was nicer than his predecessors. The cold war fizzled because the Soviet Union, and to a lesser extent the United States, couldn't afford it any more (Bundy 1990); similarly, if East/West antagonisms should flare anew, money again will likely be at the root. Thus, the causes of both peace and war are largely economic. Sociological research can help us document the extent, causes, and role of inequality in the international arena.

THINKING

CRITICALLY

If you were to run for office, how might you use E-mail in your campaign? Which of your constituents would you be more likely to hear from via the Internet? How would you know whether they were actually U.S. citizens with the legal right to vote—and would it matter? How might you make sure that other voices, those without modems, were heard as well?

3. *Modeling practical development strategies.* Sociological research may lead to the development of programs that are more effective in improving the well-being of Third World people. Studies of the consequences of transnational investment, foreign aid, and investment loans are examples of the kinds of research likely to be most useful.

The involvement of sociologists in issues of justice and conflict resolution is not likely to be the crucial factor that brings about peace. We can be sure, however, that scholarly neglect of these issues is both shortsighted and immoral. To the extent that knowledge of the principles of human behavior bears on issues of international conflict, we have an obligation—as scholars and citizens—to apply our knowledge to what is clearly one of the most critical policy issues of our century.

Summary

1. Collective behavior and social movements, although related are distinct activities. Collective behavior is spontaneous and unplanned; a social movement is organized, relatively broad based, and long term.

2. Relative-deprivation and resource mobilization theory offer two different scenarios for why inequalities and cleavages result in social movements. Relative-deprivation theory stresses the eruption of anger over a sense of deprivation; resource mobilization theory stresses the importance of organizational solidarity as a precondition for a social movement.

3. Each social movement contains a variety of social movement organizations (SMOs). Some are professional SMOs and some are indigenous SMOs. A variety of SMOs tend to enhance a movement's ability to mobilize resources.

4. Mobilization occurs at two levels: the recruitment of individual adherents (micromobilization) and the recruitment of sympathetic organizations (bloc mobilization). Bloc mobilization is faster but is less likely to attract committed activists.

5. Movements are more successful if their demands are consistent with a society's larger values, if they have an innovative, attention-getting strategy, and if they can gain the support of, or at least not harm, third-party organizations.

6. When a social movement threatens to be at least partially successful, a countermovement often develops to protect the status quo. Countermovements depend highly on bloc mobilization.

7. The mass media play an important and growing role in collective behavior and social movements. Because resource-poor social movements depend on free publicity, they are often forced to become more extremist and less democratic to continue attracting media coverage.

8. In its effort to affect public policy, the environmental movement uses a variety of tactics, ranging from courtroom battles to ecoterrorism. Among the reasons for the movement's growing successes are the wide variety of SMOs within the movement and the size of its public constituency.

9. Technology is the human application of knowledge to the making of tools and hence to humans' use of natural resources. The term refers not only to the tools themselves (aspects of material culture) but also to people's beliefs, values, and attitudes regarding those tools (aspects of nonmaterial culture).

10. Social change is any significant modification or transformation of social structures or institutions over time. Technology and social change are closely related concepts; technology is one important cause of social change.

11. Structural-functional theory primarily asks how social organization is maintained in an orderly way and sees social change as occurring through gradual evolution. Cultural lag is the time interval between the arrival of a change in society and the completion of the adaptations that this change prompts.

12. While structural functionalism sees social change as orderly and generally consensual, conflict theorists contend that change results from conflict between competing interests. Furthermore, conflict theorists assert that those with greater power actually direct social change to their own advantage. People or groups who would either suffer or profit from social change have vested interests—stakes in either maintaining or transforming the status quo.

13. Information technology—computers and telecommunication tools for storing, using, and sending information—has changed many aspects of our daily lives. Four social implications of information technology involve social integration, social control, work, and politics.

Sociology on the Net

One of the more widespread and visible social movements is the women's rights movement. One of the mainstream organizations is the Feminist Majority Foundation.

    http://www.feminist.org/home.html

Check out the *Feminist Majority* home page. Click on the bar entitled *About the Feminist Majority Foundation.* Browse through the first four items: *The Feminist Majority, Why Are We Named the Feminist Majority? What We're Doing For You* and *How to Contact Us.* What is this organization and who do they claim to represent? Now return to the home page and click on the *Take Action* bar. Is this an indigenous SMO or is it a professional SMO? Who is their conscience constituency? Is this organization best understood using relative deprivation theory or resource mobilization theory? Hint: Return to the Internet exercise in Chapter 3 and refresh your memory from the "Feminist Chronicles."

The environmental movement has taken many forms and can be found almost anywhere in the world. Earth First is a vocal part of this movement.

    http://www.telalink.net/~zoomst/earth-
    first/index.html

This is the *Earth First Home Page.* Is this a social movement or an organization according to your text? Is this a professional or an indigenous SMO? Go to the bottom of the page and click on the highlighted phrase *. . . the need for action!* Browse through the information and when you reach the bottom of the page, click on the highlighted *Earth First Journal.* Open a few of the selections and see what the organization is all about. You might even go to the Gopher Archives. How is this organization engaging in frame align-

ment? Review the factors for social movement success from the text. How well does Earth First measure up?

Countermovements seek to reverse or resist change. One such organization is the Family Research Council.

    http://www.heritage.org/frc/

Once you are at the home page, click on the highlighted phrase *Who is the Family Research Council?* Return to the home page and browse around a bit. Now open *Washington Watch* and scroll down to *Washington Watch, The Newsletter.* Open a newsletter in this section and skim through the articles. What is this countermovement against? What values do they believe should be restored? Return to the home page and click on the *Hot List of Links.* Scroll through the offerings. Can you find any other organizations that are part of this countermovement?

FIND IT ON INFOTRAC COLLEGE EDITION

With the increase of technology also comes an increase in the venues available to criminals. One area in which crimes are occurring more frequently is in cyberspace. Using InfoTrac College Edition, look up the following article:

"Policing Cyberspace." Vic Sussman. *U.S. News & World Report,* January 23, 1995, v118 n3 p54.

(Hint: Enter the search term *Policing cyberspace,* using the Subject Guide.)

Describe the crimes that are becoming more common in cyberspace. How do constitutional freedoms of speech enter into the debate surrounding the policing of cyberspace? What strategies can you offer in the policing of cyberspace?

Suggested Readings

Cable, Sherry, and Cable, Charles. 1995. *Environmental Problems, Grassroots Solutions: The Politics of Grassroots Environmental Conflict.* New York: St. Martin's Press. A description of the environmental movement with details on how various regional, grass-roots organizations address environmental issues and concerns.

Jasper, James M., and Nelkin, Dorothy, 1992. *The Animal Rights Crusade: The Growth of a Moral Protest.* New York: Free Press. A rich description of the origins and development of the animal-rights movement, including an analysis of the link between the animal-rights movement, environmentalism, vegetarianism, and other New Age movements.

Marx, Gary T., and McAdam, Douglas. 1994. *Collective Behavior and Social Movements: Structure and Process.* Englewood Cliffs, N.J.: Prentice-Hall. A concise, well-written review of the major ideas and issues in the study of collective behavior and social movements. Written primarily from a resource mobilization perspective, the text pays close attention to the relationship between spontaneous and more sustained forms of collective action.

Morris, Aldon. 1984. *The Origins of the Civil Rights Movement.* New York: Free Press. A powerful account of the modern Civil Rights movement. Utilizing personal interviews and original documents, Morris draws particular attention to the importance of local African American community groups, in addition to the efforts of national leaders and organizations.

Teich, Albert H. 1993. *Technology and the Future,* 6th ed. New York: St. Martin's Press. A collection of essays by recognized authors from many fields addressing the impact of technology on social change. Topics include biomedical technology, environmental and energy issues, and technology and women.

Union of Concerned Scientists. 1993. *Nucleus. Nucleus* is a quarterly publication of the Union of Concerned Scientists devoted to environmental and other social is-sues resulting from scientific and technological research and innovation. (The address is 26 Church Street, Cambridge, MA 02238.)

Wriston, Walter. 1993. *The Twilight of Sovereignty: How the Information Revolution Is Transforming Our World.* New York: Scribner's. An argument that the microchip and satellite technologies have created a world community, with political and economic ramifications far beyond our imagination. The author believes that the new technology fosters democracy and can be used to improve economic conditions throughout the world.

Glossary

A

Absolute poverty is the inability to provide the minimum requirements of life.

Accommodation occurs when two groups co-exist as separate cultures in the same society.

Accounts are explanations of unexpected or untoward behavior. There are two kinds: excuses and justifications.

Acculturation occurs when the minority group adopts the culture of the majority group.

An **achieved status** is an optional one that a person can obtain in a lifetime.

Achievement motivation is the continual drive to match oneself against standards of excellence.

An **aggregate** is people who are temporarily clustered together in the same location.

Alienation occurs when workers have no control over the work process or the product of their labor; they are estranged from their work.

Anglo-conformity is the process of acculturation in which new immigrant groups adopt English language and English customs.

Anomie is a situation where the norms of society are unclear or no longer applicable to current conditions.

Anticipatory socialization is role-learning that prepares us for roles we are likely to assume in the future.

An **ascribed status** is fixed by birth and inheritance and is unalterable in a person's lifetime.

Assimilation is the full integration of the minority group into the institutions of society and the end of its identity as a subordinate group.

Authoritarian systems are political systems in which the leadership is not selected by the people and legally cannot be changed by them.

Authoritarianism is the tendency to be submissive to those in authority coupled with an aggressive and negative attitude toward those lower in status.

Authority is power supported by norms and values that legitimate its use.

B

Bloc mobilization is a strategy whereby social movement organizations recruit other organizations to support their cause rather than trying to recruit single individuals.

The **bourgeoisie** is the class that owns the tools and materials for their work—the means of production.

Bureaucracy is a special type of complex organization characterized by explicit rules and a hierarchical authority structure, all designed to maximize efficiency.

C

Capitalism is the economic system in which most wealth (land, capital, and labor) is private property, to be used by its owners to maximize their own gain; this economic system is based on competition.

Caste systems rely largely on ascribed statuses as the basis for distributing scarce resources.

A **category** is a collection of people who share a common characteristic.

Charisma refers to extraordinary personal qualities that set the individual apart from ordinary mortals.

Charismatic authority is the right to make decisions that is based on perceived extraordinary personal characteristics.

Churches are religious organizations that have become institutionalized. They have endured for generations, are supported by and support society's norms and values, and have become an active part of society.

Civil religion is the set of institutionalized rituals, beliefs, and symbols sacred to the U.S. nation.

Class refers to a person's relationship to the means of production.

Class consciousness occurs when people are aware of their relationship to the means of production and recognize their true class identity.

Class systems rely largely on achieved statuses as the basis for distributing scarce resources.

Coercion is the exercise of power through force or the threat of force.

Cohesion refers to the degree of attraction members feel to the group.

Collective behavior is nonroutine action by an emotionally aroused gathering of people who face an ambiguous situation.

Complex organizations are large, formal organizations with complex status networks.

Conflict is a struggle over scarce resources that is not regulated by shared rules; it may include attempts to destroy or neutralize one's rivals.

Conflict theory addresses the points of stress and conflict in society and the ways they contribute to social change.

A **control group** is the group in an experiment that does not receive the independent variable.

Cooperation is interaction that occurs when people work together to achieve shared goals.

Core societies are rich, powerful nations that are economically diversified and relatively free of outside control.

Correlation occurs when there is an empirical relationship between two variables.

Countercultures are groups having values, interests, beliefs, and lifestyles that are opposed to those of the larger culture.

A **countermovement** seeks to reverse or resist change advocated by a social movement.

Credentialism is the use of educational credentials to measure social origins and social status.

Crimes are acts that are subject to legal or civil penalties.

A **cross-sectional design** uses a sample (or cross-section) of the population at a single point in time.

The **crude birthrate** (CBR) is the number of births divided by the total population and then multiplied by 1,000; it is read as births per thousand.

The **crude deathrate** (CDR) is the number of deaths divided by the total population and then multiplied by 1,000; it is read as deaths per thousand.

A **cult** is a sectlike religious organization that is independent of the religious traditions of society.

Cultural capital refers to social assets such as familiarity and identification with elite culture.

Cultural lag is the time interval between the arrival of a technological change in society and the completion of the structural and cultural adaptations to it.

Cultural relativity requires that each cultural trait must be evaluated in the context of its own culture.

Culture is the total way of life shared by members of a society. It includes language, values, and symbolic meanings, but also technology and material objects.

The **culture of poverty** is a set of values that emphasizes living for the moment rather than thrift, investment in the future, and hard work.

D

Deduction is the process of moving from theory to data by testing hypotheses drawn from theory.

Democracy is a political system that provides regular, constitutional opportunities for a change in leadership according to the will of the majority.

The **demographic transition** is the process of moving from the traditional balance of high birth- and deathrates to a new balance of low birth- and deathrates.

Demography is the study of population—its size, growth, and composition.

Denominations are churchlike religious organizations that have accommodated to society and to other religions.

The **dependent variable** is the effect in cause-and-effect relationships. It is dependent on the actions of the independent variable.

Deterrence theories suggest that deviance results when social sanctions, formal and informal, provide insufficient rewards for conformity.

Development refers to the process of increasing the productivity and standard of living of a society—longer life expectancies, more adequate diets, better education, better housing, and more consumer goods.

Deviance refers to norm violations that exceed the tolerance level of the community and result in negative sanctions.

Dialectic philosophy views change as a product of contradictions and conflicts between the parts of society.

Differential association theory argues that people learn to be deviant when their associates favor deviance more than conformity.

Differentials are differences in the incidence of a phenomenon across subcategories of the population.

Disclaimers are verbal devices employed to ward off and defeat in advance doubts and negative reactions that might result from one's conduct.

Discrimination is the unequal treatment of individuals on the basis of their membership in categories.

The **divorce rate** is calculated as the number of divorces each year per 1,000 married women.

Double or triple jeopardy means having low status on two or three different dimensions of stratification.

Dramaturgy is a version of symbolic interaction that views social situations as scenes manipulated by the actors to convey the desired impression to the audience.

A **dual economy** consists of the complex giants of the industrial core and the small, competitive organizations that form the periphery.

Dysfunctions are consequences of social structures that have negative effects on the stability of society.

E

An **ecclesia** is a churchlike religious organization that automatically includes every member of society.

Economic determinism means that economic relationships provide the foundation on which all other social and political arrangements are built.

Economic institutions are social structures concerned with the production and distribution of goods and services.

The **educational institution** is the social structure concerned with the formal transmission of knowledge.

An **established sect** is a sect that has adapted to its institutional environment.

An **ethnic group** is a category whose members are thought to share a common origin and to share important elements of a common culture.

Ethnocentrism is the tendency to view the norms and values of our own culture as standards against which to judge the practices of other cultures.

Exchange is voluntary interaction from which all parties expect some reward.

Excuses are accounts in which one admits that the act in question is bad, wrong, or inappropriate, but claims that one couldn't help it.

The **experiment** is a method in which the researcher manipulates independent variables to test theories of cause and effect.

An **experimental group** is the group in an experiment that experiences the independent variable. Results for this group are compared with those for the control group.

Expressive describes activities or roles that provide integration and emotional support.

Extrinsic rewards are tangible benefits such as income and security.

F

False consciousness is a lack of awareness of one's real position in the class structure.

The **family** is a relatively permanent group of persons linked together in social roles by ties of blood, marriage, or adoption, who live together and cooperate economically and in the rearing of children.

Fertility is the incidence of childbearing.

Folkways are norms that are customary, normal, habitual ways a group does things.

Formal social controls are administrative sanctions such as fines, expulsion, and imprisonment.

A **frame** is an answer to the question, what is going on here? It is roughly identical to a definition of the situation.

Frame alignment is a strategy of micromobilization in which SMOs attract individual new recruits; it seeks to convince individuals that their interests, values, and beliefs are complementary to those of the social movement organization.

Functions are consequences of social structures that have positive effects on the stability of society.

G

Gemeinschaft refers to society characterized by the personal and permanent ties associated with primary groups.

Gender refers to the expected dispositions and behaviors that cultures assign to each sex.

Gender roles refer to the rights and obligations that are normative for men and women in a particular culture.

The **generalized other** is the composite expectations of all the other role players with whom we interact; it is Mead's term for our awareness of social norms.

Gesellschaft refers to society characterized by the impersonal and instrumental ties associated with secondary groups.

A **group** is two or more people who interact on the basis of interdependent statuses, roles, and norms and recognize mutual dependency.

The **guinea-pig effect** occurs when subjects' knowledge that they are participating in an experiment affects their response to the independent variable.

H

The **hidden curriculum** socializes young people into obedience and conformity.

Homogamy is the tendency to choose a mate similar to oneself.

A **hypothesis** is a statement about relationships that we expect to find if our theory is correct.

I

The **I** is the spontaneous, creative part of the self.

Identity salience hierarchy is a ranking of an individual's various role identities in order of their importance to him or her.

An **ideology** is a set of norms and values that rationalizes the existing social structure.

Immigration is the permanent movement of people into another country.

Incidence is the frequency with which an attitude or behavior occurs.

The **independent variable** is the variable that does the causing in cause-and-effect relationships.

The **indirect inheritance model** argues that children have occupations of a status similar to that of their parents because the family status and income determine children's aspirations and opportunities.

Induction is the process of moving from data to theory by devising theories that account for empirically observed patterns.

The **informal economy** is that part of the economy that escapes the record keeping and regulation of the state; also known as the underground economy.

Informal social control is self-restraint exercised because of fear of what others will think.

Information technology comprises computers and telecommunication tools for storing, using, and sending information.

Institutionalized racism occurs when the normal operation of apparently neutral processes systematically produces unequal results for majority and minority groups.

Institutions are enduring social structures that provide ready-made arrangements to meet basic human problems.

Instrumental describes activities or roles that are task oriented.

The **interaction school of symbolic interaction** focuses on the active role of the individual in creating the self and self-concept.

Intergenerational mobility is the change in social class from one generation to the next.

Internalization occurs when individuals accept the norms and values of their group and make conformity to these norms part of their self-concept.

Intragenerational mobility is the change in social class within an individual's own career.

Intrinsic rewards are rewards that arise from the process of work; they include enjoying the people you work with and pride in your creativity and accomplishments.

J

Justifications are accounts that explain the good reasons the violator had for choosing to break the rule; often these are appeals to some alternate rule.

L

Labeling theory is concerned with the processes by which the label *deviant* comes to be attached to specific people and specific behaviors.

Latent functions are consequences of social structures that are neither intended nor recognized.

Laws are rules that are enforced and sanctioned by the authority of government.

Less-developed countries include the former Soviet Union and the nations of the former Communist bloc plus several of the developing nations of Southeast Asia and Central and South America.

Least-developed countries include those nations that share a peripheral or marginal status in the world capitalist system.

Lifetime divorce probability is the estimated probability that a marriage will ever end in divorce.

Linguistic relativity hypothesis argues that the grammar, structure, and categories embodied in each language affect how its speakers see reality.

The **looking-glass self** is the process of learning to view ourselves as we think others view us.

M

Macrosociology focuses on social structures and organizations and the relationships between them.

A **majority group** is a group that is culturally, economically, and politically dominant.

Manifest functions are consequences of social structures that are intended and recognized.

Marriage is an institutionalized social structure that provides an enduring framework for regulating sexual behavior and childbearing.

The **me** represents the self as social object.

A **metropolitan area** is a county that has a city of 50,000 or more in it plus any neighboring counties that are significantly linked, economically or socially, with the core county.

Microsociology focuses on interactions among individuals.

A **minority group** is a group that is culturally, economically, and politically subordinate.

Mobilization is the process by which a unit gains significantly in the control of assets it did not previously control.

Modernization theory sees development as the natural unfolding of an evolutionary process in which societies go from simple to complex institutional structures and from primary to secondary and tertiary production.

Moral entrepreneurs are people who are in a position to create and enforce new definitions of morality.

Mores are norms associated with fairly strong ideas of right or wrong; they carry a moral connotation.

Mortality is the incidence of death.

Most-developed countries include those rich nations that have relatively high degrees of economic and political autonomy: the United States, Western Europe, Canada, Japan, Australia, and New Zealand.

N

The **natural growth rate** is the crude birthrate minus the crude deathrate and then divided by 10; it is read as the percentage growth rate.

A **nonmetropolitan area** is a county that has no major city in it and is not closely tied to a county that does have such a city.

Norms are shared rules of conduct that specify how people ought to think and act.

O

Operational definitions describe the exact procedures by which a variable is measured.

P

The **panel design** follows a sample over a period of time.

Participant observation includes a variety of research strategies—participating, interviewing, observing—that examine the context and meanings of human behavior.

Peripheral societies are poor and weak, with highly specialized economies over which they have relatively little control.

The **petit bourgeois** are those who use their own modest capital to establish small enterprises in which they and their family provide the primary labor.

Political institutions are institutions concerned with the social structure of power; the most prominent political institution is the state.

Power is the ability to direct others' behavior even against their wishes.

The **power elite** is the people who occupy the top positions in three bureaucracies—the military, industry, and the executive branch of government—and who are thought to act together to run the United States in their own interests.

Prejudice is irrationally based negative attitudes toward categories of people.

Primary groups are groups characterized by intimate, face-to-face interaction.

Primary production is extracting raw materials from the environment.

Primary socialization is personality development and role learning that occurs during early childhood.

The **profane** represents all that is routine and taken for granted in the everyday world, things that are known and familiar and that we can control, understand, and manipulate.

Professions are occupations that demand specialized skills and creative freedom.

The **proletariat** is the class that does not own the means of production. They must support themselves by selling their labor to those who own the means of production.

Propinquity is spatial nearness.

R

A **race** is a category of people whom we treat as distinct on account of physical characteristics to which we have assigned *social* importance.

Racism is a belief that inherited physical characteristics determine the presence or absence of socially relevant abilities and characteristics and that such differences provide a legitimate basis for unequal treatment.

Rational-legal authority is the right to make decisions that is based on rationally established rules.

Relative deprivation theory argues that social movements arise when people experience an intolerable gap between their rewards and what they believe they have a right to expect; also known as breakdown theory.

Relative poverty is the inability to maintain what your society regards as a decent standard of living.

Religion is a system of beliefs and practices related to sacred things that unites believers into a moral community.

Replacement level fertility requires that each woman bear approximately two children, one to replace herself and one to re-

place her partner. When this occurs, the next generation will be the same size as the current generation of parents.

Replication is repeating empirical studies with another investigator or a different sample to see if the same results occur.

Resocialization occurs when we abandon our self-concept and way of life for a radically different one.

Resource mobilization theory suggests that social movements develop when organized groups are competing for scarce resources; also known as solidarity theory.

Rites of passage are formal rituals that mark the end of one status and the beginning of another.

A **role** is a set of norms specifying the rights and obligations associated with a status.

Role identity is the image we have of ourself in a specific social role.

Role taking involves imagining ourselves in the role of the other in order to determine the criteria others will use to judge our behavior.

S

The **sacred** consists of events and things that we hold in awe and reverence—what we can neither understand nor control.

A **sample** is a systematic selection of representative cases from the larger population.

Sanctions are rewards for conformity and punishments for nonconformity.

Scapegoating occurs when people or groups who are blocked in their own goal attainment blame others for their failures.

Science is a way of knowing based on empirical evidence.

Secondary groups are groups that are formal, large, and impersonal.

Secondary production is the processing of raw materials.

Sects are religious organizations that reject the social environment in which they exist.

Secularization is the process of transferring things, ideas or events from the sacred realm to the profane.

The **segmented labor market** parallels the dual economy. Hiring, advancement, and benefits vary systematically between the industrial core and the periphery.

Segregation refers to the physical separation of minority and majority group members.

The **self** is a complex whole that includes unique attributes and normative responses. In sociology, these two parts are called the I and the me.

The **self-concept** is the self we are aware of. It is our thoughts about our personality and social roles.

Self-esteem is the evaluative component of the self concept; it is our judgment about our worth compared with others.

The **self-fulfilling prophecy** occurs when acting on the belief that a situation exists causes it to become real.

Sex is a biological characteristic, male or female.

Sexism is a belief that men and women have biologically different capacities and that these form a legitimate basis for unequal treatment.

Significant others are the role players with whom we have close personal relationships.

Social change is any significant modification or transformation of social structures and sociocultural processes over time.

Social class is a category of people who share roughly the same class, status, and power and who have a sense of identification with each other.

Social control is the forces and processes that encourage conformity, including self-control, informal control, and formal control.

Social-desirability bias is the tendency of people to color the truth so that they sound nicer, richer, and more desirable than they really are.

Social distance is the degree of intimacy in relationships between two groups.

Social mobility is the process of changing one's social class.

A **social movement** is an on-going, goal-directed effort to change social institutions from the outside.

A **social network** is an individual's total set of relationships.

A **social structure** is a network of statuses, or positions, whose interactions are regulated by social norms.

Socialism is an economic structure in which productive tools (land, labor, and capital) are owned and managed by the workers and used for their collective good.

Socialization is the process of learning the roles, statuses, and values necessary for participation in social institutions.

A **society** is the population that shares the same territory and is bound together by economic and political ties.

Sociobiology is the study of the biological basis of all forms of human behavior.

Socioeconomic status (SES) is a measure of social class that ranks individuals on income, education, occupation, or some combination.

The **sociological imagination** is the ability to see the intimate realities of our own lives in the context of common social structures; it is the ability to see personal troubles as public issues.

Sociology is the systematic study of human social interaction.

The **sociology of everyday life** focuses on the social processes that structure our experience in ordinary face-to-face situations.

Sustainable development is development that meets the needs of the present without compromising the ability of future generations to meet their own needs.

The **state** is the social structure that successfully claims a monopoly on the legitimate use of coercion and physical force within a territory.

A **status** is a specialized position within the social structure.

Status is a social honor, expressed in lifestyle.

Strain theory suggests that deviance occurs when culturally approved goals cannot be reached by culturally approved means.

Stratification is an institutionalized pattern of inequality in which social categories are ranked on the basis of their access to scarce resources.

Street-level justice is the decisions the police make in the initial stages of an investigation.

Strong ties are relationships characterized by intimacy, emotional intensity, and sharing.

Structural-functional theory addresses the question of social organization and how it is maintained.

The **structural school of symbolic interaction** focuses on the self as a product of social roles.

Subcultures are groups that share in the overall culture of society, but also maintain a distinctive set of values, norms, lifestyles, and even language.

Survey research is a method that involves asking a relatively large number of people the same set of standardized questions.

Symbolic interaction theory addresses the subjective meanings of human acts and the processes through which people come to develop and communicate shared meanings.

T

Technology involves the human application of knowledge to the making of tools and to the use of natural resources.

Tertiary production is the production of services.

A **theory** is an interrelated set of assumptions that explains observed patterns.

Total institutions are facilities in which all aspects of life are strictly controlled for the purpose of radical resocialization.

Tracking occurs when evaluations relatively early in a child's career determine the educational programs the child will be encouraged to follow.

Traditional authority is a right to make decisions for others that is based on the sanctity of time-honored routines.

Transnationals are large corporations that operate internationally and which have power that transcends the laws of any particular nation state.

Trends are changes in a variable over time.

U

The **underclass** is the group that is unemployed and unemployable, not an integrated part of the nation but a miserable substratum.

Urbanism is a distinctly urban mode of life that is developed in the cities but not confined there.

Urbanization is the process of population concentration.

V

Value-free sociology concerns itself with establishing what is, not what ought to be.

Values are shared ideas about desirable goals.

Variables are measured characteristics that vary from one individual or group to the next.

Vested interests are stakes in either maintaining or transforming the status quo.

Victimless crimes such as drug use, prostitution, gambling, and pornography are voluntary exchanges between persons who desire goods or services from each other.

Voluntary associations are nonprofit organizations designed to allow individuals an opportunity to pursue their shared interests collectively.

W

Weak ties are relationships that are characterized by low intensity and intimacy.

White-collar crime is crime committed by respectable people of high status in the course of their occupation.

World system theory is a conflict perspective of the economic relationships between developed and developing countries, the core and peripheral societies.

References

Abelman, Robert. 1990. "The Selling of Salvation in the Electronic Church." Pp. 173–183 in Robert Abelman and Stewart M. Hoover (eds.), Religious Television: Controversies and Conclusions. Norwood, N.J.: Ablex Publishing Company.

Achenbaum, W. Andrew. 1985. "Societal Perceptions of Aging and the Aged." In Robert Binstock and Ethel Shanas (eds.), Handbook of Aging and the Social Sciences. (2nd ed.) New York: Van Nostrand Reinhold.

Acock, Alan, and Kiecolt, Jill. 1989. "Is it Family Structure or Socioeconomic Status? Family Structure During Adolescence and Adult Adjustment." Social Forces 68:553–571.

Adams, Bert N. 1971. The American Family. A Sociological Interpretation. Chicago: Markham.

Adams, Bert N. 1979. "Mate Selection in the United States: A Theoretical Summarization." In W. R. Burr, Reuben Hill, F. Ivan Nye, and Ira L. Reiss (eds.), Contemporary Theories About the Family. Vol. 1. New York: Free Press.

Adams, Bert N. 1985. "The Family: Problems and Solutions." Journal of Marriage and the Family 47 (August): 525–529.

Adler, Jerrby B. 1995. "Bye-Bye, Suburban Dream." Newsweek (May 15): 40–53.

Affleck, Marilyn, Morgan, Carolyn, and Hayes, Maggie. 1989. "The Influence of Gender-Role Attitudes on Life Expectations of College Students." Youth and Society 20:307–319.

Agnew, Robert, and Petersen, David. 1989. "Leisure and Delinquency." Social Problems 36:322–350.

Ahlberg, Dennis A., and De Vita, Carol J. 1992. "New Realities of the American Family." Population Bulletin 47 (August): 2–43.

AIDS Horizons. February, 1983.

"The AIDS Front: Good News and Grim News." 1995. U.S. News and World Report (February 13): 7.

Akers, Ronald. 1968. "Problems in the Sociology of Deviance: Social Definitions and Behavior." Social Forces 46:455–465.

Albas, Daniel, and Albas, Cheryl. 1988. "Aces and Bombers: The Postexam Impression Management Strategics of Students." Symbolic Interaction 11:289–302.

Alesci, Nina. 1994. "Can't We Talk? Bertrice Berry Begins National TV Talk Show." Footnotes 21 (December): 5.

Alexander, Bobby C. 1994. Televangelism Reconsidered: Ritual in Search of Human Community. Atlanta, Ga.: Scholars Press.

Alexander, Karl, Entwisle, Doris, Cadigan, Doris, and Pallas, Aaron. 1987. "Getting Ready for First Grade: Standards of Deportment in Home and School." Social Forces 66:57–84.

Alexander, Karl L., Entwisle, Doris, and Thompson, Maxine. 1987. "School Performance, Status Relations, and the Structure of Sentiment: Bringing the Teacher Back In." American Sociological Review 52:665–682.

Allan, Emilie, and Steffensmeier, Darrell. 1989. "Youth Unemployment and Property Crime." American Sociological Review 54:107–123.

Altman, Dennis. 1983. The Homosexualization of America. Boston: Beacon Press. (Originally published 1982.)

Alwin, Duane G. 1986. "Religion and Parental Child-rearing Orientations: Evidence of a Catholic-Protestant Convergence." American Journal of Sociology 92:412–440.

Amato, Paul, and Keith, Bruce. 1991a. "Parental Divorce and Adult Well-Being: A Meta-Analysis." Journal of Marriage and the Family 53 (February): 43–58.

Amato, Paul R., and Keith, Bruce. 1991b. "Consequences of Parental Divorce for Children's Well-being: A Meta-Analysis." Psychological Bulletin 110:26–46.

Amato, Paul R. 1993. "Urban-Rural Differences in Helping Friends and Family Members." Social Psychology Quarterly 56:249–262.

American Council on Education. 1990. Personal Communication, April.

Andersson, Bengt-Erik. 1989. "Effects of Public Day Care: A Longitudinal Study." Child Development 60:857–866.

"Another Winter for the Homeless." 1989. Population Today 17 (February): 3–4

Aquilino, William S., and Supple, Khalil R. 1991. "Parent-Child Relations and Parents' Satisfaction with Living Arrangements When Adult Children Live at Home." Journal of Marriage and the Family 53:13–28.

Archer, Dane. 1985. "Social Deviance." In Gardner Lindzey and Elliot Aronson (eds.), The Handbook of Social Psychology Vol. 2. (3rd ed.) New York: Random House.

Aries, Philippe. 1974. Western Attitudes Toward Death. Baltimore: Johns Hopkins Press.

Asch, Solomon. 1955. "Opinions and Social Pressure." Scientific American 193:31–35.

Ash, Roberta. 1972. Social Movements in America. Chicago: Markham.

Ashford, Lori S. 1995. "New Perspectives on Population: Lessons from Cairo. Population Bulletin 50 (1) (March). Washington D.C.: Population Reference Bureau.

Atchley, Robert C. 1982. "Retirement as a Social Institution." American Review of Sociology 8:263–287.

Austin, Regina, and Schill, Michael. 1994. "Black, Brown, Red, and Poisoned." The Humanist 54 (4) (July–August): 9–17.

Austrom, D., and Hanel, N. 1985. "Psychological Issues of Single Life in Canada: An Exploratory Study." International Journal of Women's Studies 8:12–23.

Babbie, Earl R. 1986. The Practice of Social Research. (4th ed.) Belmont, Calif.: Wadsworth.

Babbie, Earl. 1995. The Practice of Social Research. (7th ed.) Belmont, Calif.: Wadsworth.

Baldus, D. C., Pulaski, C., and Woodworth, G. 1986. "Arbitrariness and Discrimination in the Administration of the Death Penalty: A Challenge to State Supreme Courts." Stetson Law Review 15:133–261.

Ball, Michael, R. 1990. Professional Wrestling as Ritual Drama in American Popular Culture. Lewiston, N.Y.: Mellon.

Barberis, Mary. 1994. "Spotlight on Haiti." Population Today 22:7.

Barker, Eileen. 1986. "Religious Movements: Cult and Anticult Since Jonestown." Annual Review of Sociology 12:329–46.

Barnet, Richard J., and Muller, Ronald E. 1974. Global Research: The Power of Multinational Corporations. New York: Simon and Schuster.

Baron, James N., and Bielby, William T. 1984. "The Organization of a Segmented Economy." American Sociological Review 49 (August): 454–473.

Baron, Stephen. 1989. "Resistance and Its Consequences. The Street Culture of Punks." Youth and Society 21:207–237.

Bechhofer, F., and Elliott, B. 1985. "The Petite Bourgeoisie in Late Capitalism." Annual Review of Sociology 11:181–207.

Becker, Howard S. 1963. Outsiders: Studies in the Sociology of Deviance. New York: Free Press.

Beckwith, Carol. 1983. "Niger's Wodaabe: 'People of the Taboo.'" National Geographic 164 (October): 482–509.

Beegley, Leonard. 1989. The Structure of Stratification in the United States. Newton, Mass.: Allyn & Bacon.

Beeson, Peter G., and Johnson, David R. 1987. "A Panel Study of Change (1981–1986) in Rural Mental Health Status: Effects of the Rural Crisis." Paper presented at the National Conference on Mental Health Statistics, Denver, Colorado, May 19.

Bell, Derrick. 1992. Race, Racism, and American Law. Boston: Little, Brown and Company.

Bell, Wendell, and Robinson, Robert V. 1978. "An Index of Evaluated Equality. Measuring Conceptions of Social Justice in England and the United States." In Richard F. Tomasson (ed.), Comparative Studies in Sociology. Vol. 1. Greenwich, Conn.: JAI Press

Bell-Fialkoff, Andrew. 1993. "A Brief History of Ethnic Cleansing." Foreign Affairs 72:110–121.

Bellah, Robert N. 1974. "Civil Religion in America." In Russell B. Richey and Donald G. Jones (eds.), American Civil Religion. New York: Harper & Row.

Bellah, Robert N., and Associates. 1985. Habits of the Heart: Individualism and Commitment in American Life. Berkeley: University of California Press.

Beller, Andrea H., and Graham, John W. 1986. "Child Support Awards. Differentials and Trends by Race and Marital Status." Demography 23 (May): 231–245.

Belsky, Jay. 1990. "Parental and Nonparental Child Care and Children's Socioemotional Development: A Decade in Review." Journal of Marriage and the Family 52:885–903.

Benford, Robert D. 1993. "'You Could Be the Hundredth Monkey': Collective Action Frames and Vocabularies of Motive within the Nuclear Disarmament Movement." The Sociological Quarterly 34:195–216.

Benjamin, Lois. 1991. The Black Elite: Facing the Color Line in the Twilight of the Twentieth Century. Chicago: Nelson-Hall.

Bensman, Joseph, and Lilienfeld, Robert. 1979. Between Public and Private: The Lost Boundaries of Self. New York: Free Press.

Berger, Peter L. 1963. Invitation to Sociology. A Humanistic Perspective. New York: Doubleday.

Berk, Laura. 1989. Child Development. Newton, Mass.: Allyn & Bacon.

Bertoli, Fernando, and Associates 1984. "Infant Mortality by Socioeconomic Status for Blacks, Indians, and Whites: A Longitudinal Analysis of North Carolina, 1868–1977." Sociology and Social Research 68:364–377.

Bian, Yanjie. 1997. "Bringing Strong Ties Back In: Indirect Ties, Network Bridges, and Job Searches in China." American Sociological Review 62:366–385.

Bian, Yanjie, and Soon, Ang. 1997. "Guanxi Networks and Job Mobility in China and Singapore." Social Forces 75:981–1005.

Biddle, B. J. 1986. "Recent Developments in Role Theory." Annual Review of Sociology 12:67–92.

Bielby, William T., and Baron, James N. 1986. "Men and Women at Work: Sex Segregation and Statistical Discrimination." American Journal of Sociology 91 (January): 759–798.

Bielby, Denise D., and Bielby, William T. 1988. "She Works Hard for the Money. Household Responsibilities and the Allocation of Work Effort." American Journal of Sociology 93:1031–1059.

Billings, Dwight B. 1990. "Religion as Opposition." American Journal of Sociology 96:1–31.

Blau, F. D. 1977. Equal Pay in the Office. Lexington, Mass.: Lexington Books.

Blau, Peter M. 1987. "Contrasting Theoretical Perspectives." In J. Alexander, B. Giesen, R. Munch, and N. Smelser (eds.), The Micro-Macro Link. Berkeley: University of California Press.

Blau, Peter M., and Meyer, Marshall W. 1971. Bureaucracy in Modern Society (2nd ed.) New York: Random House.

Blau, Peter M., and Schwartz, Joseph E. 1984. Cross-Cutting Social Circles. Orlando, Fla: Academic Press.

Blee, Kathleen M., and Billings, Dwight B. 1986. "Reconstructing Daily Life in the Past: An Hermeneutical Approach to Ethnographic Data." Sociological Quarterly 27 (Winter):443–462.

Bloom, Samuel. 1988. "Structure and Ideology in Medical Education. An Analysis of Resistance to Change." Journal of Health and Social Behavior 29:294–306.

Blumberg, Rae Lesser. 1978. Stratification: Socioeconomic and Sexual Inequality. Dubuque, Iowa: Brown.

Blumer, H. 1969. Symbolic Interactionism: Perspective and Method. Englewood Cliffs, N.J.: Prentice-Hall.

Blumstein, Phillip, and Schwartz, Pepper. 1983. American Couples. New York: William Morrow.

Bobo, Lawrence, and Hutchings, Vincent. 1996. "Perceptions of Racial Group Competition: Extending Blumer's Theory of Group Position to a Multiracial Social Context." American Sociological Review 61:951–972.

Bobo, Lawrence, and Kluegel, James R. 1993. "Opposition to Race-Targeting, Self-Interest, Stratification Ideology, or Racial Attitudes?" American Sociological Review 58:443–464.

Bohland, James R. 1982. "Indian Residential Segregation in the Urban Southwest, 1970 and 1980." Social Science Quarterly 63 (December): 749–761.

Bollier, David. 1982. Liberty and Justice for Some. New York: Frederick Ungar.

Booth, Alan, Johnson, David, White, Lynn, and Edwards, John. 1984. Marital Instability and the Life Course: Methodology Report. Lincoln: Bureau of Sociological Research.

Boritch, Helen, and Hagan, John. 1990. "A Century of Crime in Toronto: Gender, Class, and Patterns of Social Control, 1859 to 1955." Criminology 28:567–599.

Bose, Christine E., and Rossi, Peter H. 1983. "Gender and Jobs: Prestige Standings of Occupations as Affected by Gender." American Sociological Review 48 (June): 316–330.

Bourdieu, P. 1973. "Cultural Reproduction and Social Reproduction." In R. Brown (ed.), Knowledge, Education, and Cultural Change. London: Tavistock.

Bowen, Howard R. 1977. Investment in Learning. San Francisco, Calif.: Jossey-Bass.

Bowles, Samuel, and Gintis, Herbert. 1976. Schooling in Capitalist America: Educational Reform and the Contradictions of Economic Life. New York: Basic Books.

Bowles, Samuel, and Gintis, Herbert. 1986. Democracy and Capitalism: Property, Community, and the Contradictions of Modern Social Thought. New York: Basic Books.

Bowles, Samuel. 1972. "Unequal Education and the Reproduction of the Social Division of Labor." In M. Conroy (ed.), Schooling in a Corporate Society. New York: David McKay.

Boyle, John P., and O'Connor, Liz Clapp. 1996. "Leveraging Technology and Partnerships in Criminal Investigations." Police Chief 63:19–24.

Bozett, Frederick (ed.). 1987. Gay and Lesbian Parents. New York: Praeger.

Bradshaw, York. 1988. "Reassessing Economic Dependency and Uneven Development: The Kenyan Experience." American Sociological Review 53:693–708.

Braithwaite, John. 1981. "The Myth of Social Class and Criminality, Reconsidered." American Sociological Review 46 (February): 36–58.

Braithwaite, John. 1985. "White Collar Crime." Annual Review of Sociology 11:1–25.

Braverman, Harry. 1974. Labor and Monopoly Capital. New York: Monthly Review Press.

Breslow, Lester. 1987. "Setting Objectives for Public Health." Annual Review of Public Health 8:289–307.

Breuninger, Paul. 1995. "Crime Scene Reconstruction Using 3D Computer-Aided Drafting." Police Chief 62:61–62.

Bright, Chris. 1990. "Shipping Unto Others." E: The Environmental Magazine 1:30–35.

Brinkerhoff, Merlin B., and Kunz, Phillip R. (eds.). 1972. Complex Organizations and Their Environments. Dubuque, IA: Wm. C. Brown.

Brossi, Kathleen B. 1979. A Cross-City Comparison of Felony Case Processing. Washington, D.C.: U.S. Government Printing Office.

Brown, Karen McCarthy. 1987. "Voodoo." Pp. 296–301 in Mircea Eliade (ed.), The Encyclopedia of Religion. New York: Macmillan.

Brown, R. S., Moon, M., and Zoloth, B. S. 1980. "Incorporating Occupational Attainment in Studies of Male-Female Earnings Differentials." Journal of Human Resources 15:3–28.

Bruce, Steve. 1990. Pray TV: Televangelism in America. New York: Routledge.

Bryant, Howard. 1994. "Finding a Home for Orphaned Computers." The Alameda Newspaper Group, December 4:D-1, D-4.

Buchsbaum, Herbert. 1993. "Islam in America." Scholastic Update, October 22, pp. 15–18.

Buechler, Steven M. 1993. "Beyond Resource Mobilization? Emerging Trends in Social Movement Theory." The Sociological Quarterly 34:217–235.

Bulan, Heather Ferguson, Erickson, Rebecca J., and Wharton, Amy S. 1997. "Doing for Others on the Job: The Affective Requirements of Service Work, Gender, and Emotional Well-Being." Social Problems 44:235–255.

Bullard, Robert D. 1990. Dumping in Dixie: Race, Class, and Environmental Quality. Boulder, Colo.: Westview.

Bullard, Robert D. 1993. Confronting Environmental Racism: Voices from the Grassroots. Boston: South End Press.

Bumpass, Larry, Raley, R. K., and Sweet, J. 1994. "The Changing Character of Stepfamilies: Implications of Cohabitation and Nonmarital Childbearing." Paper presented at RAND conference, January 1994, Los Angeles.

Bumpass, Larry L., Raley, R. Karen, and Sweet, James. 1995. "The Changing Character of Stepfamilies: Implications of Cohabitation and Nonmarital Childbearing." Demography 32:425–436.

Bumpass, Larry L., and Sweet, James A. 1991. "The Role of Cohabitation in Declining Rates of Marriage." Journal of Marriage and Family 53:913–927.

Bumpass, Larry L., and Sweet, James A. 1995. "Cohabitation, Marriage, Nonmarital Childbearing, and Union Stability: Preliminary Findings from NSFH2." Population Association of America Annual Meeting, San Francisco, April.

Bundy, McGeorge. 1990. "From Cold War to Lasting Peace." Foreign Affairs 69 (1): 197–212.

Burke, Peter J. 1980. "The Self: Measurement Requirements from the Interactionist Perspective." Social Psychological Quarterly 43 (1): 18–29.

Burris, Val, and Salt, James. 1990. "The Politics of Capitalist Class Segments: A Test of Corporate Liberalism Theory." Social Problems 37:341–359.

Burstein, Paul. 1991. "Legal Mobilization as a Social Movement Tactic. The Struggle for Equal Opportunity Employment." American Journal of Sociology 96:1201–1225.

Bush, Corlann. Gee. 1993. "Women and the Assessment of Technology." Pp. 192–264 in Albert H. Teich (ed.), Technology and the Future. (6th ed.) New York: St. Martin's Press.

Cable, Sherry, and Cable, Charles. 1995. Environmental Problems, Grassroots Solutions: The Politics of Grassroots Environmental Conflict. New York: St. Martin's Press.

Cable, Sherry, Walsh, Edward, and Warland, Rex. 1988. "Differential Paths to Political Activism: Comparisons of Four Mobilization Processes After the Three Mile Island Accident." Social Forces 66:951–969.

Cahill, Spencer E. 1983. "Reexamining the Acquisition of Sex Roles: A Social Interactionist Perspective." Sex Roles 9:1–15.

Calavita, Kitty, and Pontell, Henry N. 1993. "Savings and Loan Fraud as Organized Crime: Toward a Conceptual Typology of Corporate Illegality." Criminology 31:519–548.

Call, Vaughn, Sprecher, Susan, and Schwartz, Pepper. 1995. "The Incidence and Frequency of Marital Sex in a National Sample." Journal of Marriage and the Family 57:639–652.

Callero, Peter L. 1985. "Role Identity Salience." Social Psychology Quarterly 48 (3): 203–215.

Camasso, Michael J., and Moore, Dan E. 1985. "Rurality and the Residualist Social Welfare Response." Rural Sociology 50 (Fall): 397–408.

Campbell, Ernest Q. 1969. "Adolescent Socialization." In David A. Goslin (ed.), Handbook of Socialization Theory and Research. New York: Russell Sage Foundation.

Campbell, John L., and Allen, Michael Patrick. 1994. "The Political Economy of Revenue Extraction in the Modern State: A Time-Series Analysis of U.S. Income Taxes, 1916–1986." Social Forces 72:643–669.

Campbell, Karen E. 1990. "Networks Past: A 1939 Bloomington Neighborhood." Social Forces 69:139–155.

Caplow, Theodore, and Chadwick, Bruce. 1979. "Inequality and Life-Style in Middletown, 1920–1978." Social Science Quarterly 60 (December): 367–386.

Chambliss, William. 1978. "Toward a Political Economy of Crime." In Charles Reasons and Robert Rich (eds.), The Sociology of Law: A Conflict Perspective. Toronto: Butterworths.

Canci, A. Silvia, Evans, T. Davic, and Maume, Jr., David J. 1996. "Reconsidering the Declining Significance of Race: Racial Differences in Early Career Wages." American Sociological Review 61:541–556.

Chafetz, Janet S. 1984. Sex and Advantage. Totowa, N.J.: Rowman and Allanheld.

Chapman, Jane R., and Gates, Margaret (eds). 1978. The Victimization of Women. Newbury Park, Calif.: Sage.

Charny, M. C., Lewis, P. A., and Farrow, S. C. 1989. "Choosing Who Shall Not Be Treated." Social Science and Medicine 28:1331–38.

Cherlin, Andrew. 1981. Marriage, Divorce, Remarriage. Cambridge, Mass.: Harvard University Press.

Cherlin, Andrew J. 1992. Marriage, Divorce, and Remarriage, Rev. ed. Cambridge, Mass.: Harvard University Press.

"China's Demographic Disaster of 1958–1962." 1985. Population Today 13 (March): 7.

Chirot, Daniel. 1977. Social Change in the Twentieth Century. San Francisco, Calif.: Harcourt Brace Jovanovich.

Chirot, Daniel. 1986. Social Change in the Modern Era. San Diego, Calif.: Harcourt Brace Jovanovich.

Chodak, Symon. 1973. Societal Development. Five Approaches with Conclusions from Comparative Analysis. New York: Oxford University Press.

Christenson, James A. 1984. "Gemeinschaft and Gesellschaft: Testing the Spatial and Communal Hypothesis." Social Forces 63 (September): 160–168.

Cicourel, Aaron V. 1985. "Text and Discourse." Annual Review of Anthropology 14:159–185.

Clawson, Dan, and Su, Tie-ting. 1990. "Was 1980 Special? A Comparison of 1980 and 1986 Corporate PAC Contributions." Sociological Quarterly 31:371–387.

Coale, Ansley. 1973. Cited in M. Tettelbaum. 1975. "Relevance of Demographic Transition Theory to Developing Countries." Science 188 (May 2): 420–425.

Cobarrubias, Juan. 1983. "Ethical Issues in Status Planning." In Juan Cobarrubias and Joshua Fishman (eds.), Progress in Language Planning: International Perspectives. Berlin: Mouton.

Cobb, Charles, E., Jr. 1987. "Haiti Against All Odds." National Geographic 172:645–671.

Cockerham, William C. 1989. Medical Sociology. (4th ed.) Englewood Cliffs, N.J.: Prentice-Hall.

Cockerham, William C. 1992. Medical Sociology. (5th ed.) Englewood Cliffs, N.J.: Prentice-Hall.

Cockerham, William C. 1997. "The Social Determinants of the Decline of Life Expectancy in Russian and Eastern Europe: A Lifestyle Explanation." Journal of Health and Social Behavior 38:117–130.

Cockerham, William, Kunz, Gerhard, and Lueschen, Guenther. 1988. "Social Stratification and Health Life-Styles in Two Systems of Health Care Delivery: A Comparison of the United States and West Germany." Journal of Health and Social Behavior 29:113–126.

Coleman, James. 1988. "Competition and the Structure of Industrial Society: Reply to Braithwaite." American Journal of Sociology 94: 632–636.

Collins, Patricia Hill. 1991. Black Feminist Thought. New York: Routledge, Chapman, and Hall.

Collins, Randall. 1979. The Credential Society. Orlando, Fla.: Academic Press.

Collins, Sharon M. 1993. "Blacks on the Bubble: The Vulnerability of Black Executives in White Corporations." Sociological Quarterly 34:429–447.

Collins, Sharon M. 1997. "Black Mobility in White Corporations: Up the Corporate Ladder but Out on a Limb." Social Problems 44:55–67.

Conger, Rand D., and Elder, Glen H., Jr. 1994. Families in Troubled Times: Adapting to Change in Rural America. New York: Aldine de Gruyter.

Conrad, John P. 1983. "Deterrence, the Death Penalty and the Data." In Ernest van den Haag and John P. Conrad (eds.), The Death Penalty: A Debate. New York: Plenum.

Cönrad, Peter, and Schneider, Joseph W. 1980. Deviance and Medicalization: From Badness to Sickness. Saint Louis: C. V. Mosby.

Cool, Linda, and McCabe, Justine. 1983. "The 'Scheming Hag' and the 'Dear Old Thing': The Anthropology of Aging Women." In Jay Sokolovsky (ed.), Growing Old in Different Cultures. Belmont, Calif.: Wadsworth.

Cooley, Charles Horton. 1902. Human Nature and the Social Order. New York: Scribner's.

Cooley, Charles Horton. 1967. "Primary Groups." In A. Paul Hare, Edgar F. Borgotta, and Robert F. Bales (eds.), Small Groups: Studies in Social Interaction. (Rev. ed.) New York: Knopf. (Originally published 1909.)

Cooney, Mark. 1997. "The Decline of Elite Homicide." Criminology 35:381–407.

Cornell, Claire, and Gelles, Richard. 1982. "Adolescent to Parent Violence." Urban Social Change Review 15 (Winter): 8–14.

Corrections Today. 1990. "Boot Camp Prison Offers Second Chance to Young Felons." July, p. 144.

Coverdill, James E. 1988. "The Dual Economy and Sex Differences in Earnings" Social Forces 66:97–993.

Cowan, Ruth Schwartz, 1993. "Less Work for Mother?" Pp. 329–339 in Albert H. Teich (ed.), Technology and the Future. (6th ed.) New York: St. Martin's Press.

Cowgill, Donald O. 1986. Aging Around the World. Belmont, Calif.: Wadsworth Publishing Company.

Crèvecoeur, J. Hector. 1974. "What Is an American?" In Richard J. Meister (ed.), Race and Ethnicity in Modern America. Lexington, Mass.: Heath. (Originally published 1782.)

Crimmins, Ellen M., Hayward, Mark D., and Saito, Yashuhiko. 1994. "Changing Mortality and Morbidity Rates and the Health Status and Life Expectancy of the Older Population." Demography 31:168–169.

Crozier, Michael, and Friedberg, Erhard. 1980. Actors and Systems: The Politics of Collective Action. Chicago: University of Chicago Press.

Crutchfield, Robert D. 1989. "Labor Stratification and Violent Crime." Social Forces 68:489–512.

Culver, John H. 1992. "Capital Punishment, 1977–1990. Characteristics of the 143 Executed." Sociology and Social Research 76:59–61.

Currie, Elliott. 1989. "Confronting Crime Looking Toward the 21st Century." Justice Quarterly 6:5–25.

Curry, Theodore. R. 1996. "Conservative Protestantism and the Perceived Wrongfulness of Crimes: A Research Note." Criminology 34:453–464.

Dahl, Robert. 1961. Who Governs? New Haven, Conn.: Yale University Press.

Dahl, Robert. 1971. Polarchy. New Haven, Conn.: Yale University Press.

Dale, Roger, 1977. "Implications of the Rediscovery of the Hidden Curriculum of the Sociology of Teaching." In Denis Gleeson (ed.), Identity and Structure: Issues in the Sociology of Education. Driffield, England: Nafferton Books.

Daly, Martin, and Wilson, Margo. 1983. Sex, Evolution, and Behavior. (2nd ed.) Boston: Willard Grant.

David, Henry. 1982. "Eastern Europe: Pronatalist Policies and Private Behavior." Population Bulletin 36 (6): 1–50.

Davies, Annette. 1986. Industrial Relations and New Technology. London: Croom Helm.

Davis, James Allan, and Smith, Tom W. 1986. General Social Surveys, 1972–1986. Chicago: National Opinion Research Center.

Davis, Kingsley. 1961. "Prostitution." In Robert K. Merton and Robert A. Nisbet (eds.), Contemporary Social Problems. San Francisco: Harcourt Brace Jovanovich.

Davis, Kingsley. 1973. "Introduction." In Kingsley Davis (ed.), Cities. New York: W. H. Freeman.

Davis, Kingsley, and Moore, Wilbert E. 1945. "Some Principles of Stratification." American Sociological Review 10 (April): 242–249.

Davis, Nancy J., and Robinson, Robert V. 1996. "Are Rumors of War Exaggerated? Religious Orthodoxy and Moral Progressivism in America." American Journal of Sociology 102:756–787.

Davis, James, and Stasson, Mark. 1988. "Small-Group Performance: Past and Future Research Trends." Advances in Group Processes 5:245–277.

Deegan, Mary Jo. 1987. Jane Addams and the Men of the Chicago School, 1892–1918. New Brunswick, N.J.: Transaction.

Delgado, Gary. 1986. Organizing the Movement: The Roots and Growth of ACORN. Philadelphia: Temple University Press.

Denzin, Norman K. 1984. "Toward a Phenomenology of Domestic, Family Violence." American Journal of Sociology 90 (November): 483–513.

Devine, Joel, Sheley, Joseph, and Smith, M. Dwayne. 1988. "Macroeconomic and Social Control Policy Influences in Crime Rate Changes, 1948–85." American Sociological Review 53:407–420.

DeWitt, J. L. 1943. Japanese in the United States. Final Report: Japanese Evacuation from the West Coast. P. 34 cited in Paul E. Horton and Gerald R. Leslie, Social Problems 1955. East Norwalk, Conn.: Appleton-Century-Crofts.

DiMaggio, Paul, Evans, John, and Bryson, Bethany. 1996. "Have Americans' Social Attitudes Become More Polarized?" American Journal of Sociology, 102:690–755.

DiMaggio, Paul, and Mohr, John. 1985. "Cultural Capital, Educational Attainment, and Marital Selection." American Journal of Sociology 90 (May): 1231–1261.

DiTomaso, Nancy. 1987. "Symbolic Media and Social Solidarity: The Foundations of Corporate Culture." Sociology of Organizations 5:105–134.

Dixon, William J., and Boswell, Terry. 1996. "Dependency, Disarticulation, and Denominator Effects: Another Look at Foreign Capital Penetration." American Journal of Sociology 102:543–562.

Doane, Ashley W., Jr. 1997. "Dominant Group Ethnic Identity in the United States: The Role of 'Hidden' Ethnicity in Intergroup Relations." The Sociological Quarterly 38:375–395.

Dolnick, Edward. 1993. "Deafness as Culture." The Atlantic Monthly (September):37–53.

Domholl, G. William. 1983. Who Rules America Now? A View for the 80s. Englewood Cliffs, N.J.: Prentice-Hall.

Donnelly, Strachan, McCarthy, Charles R., and Singleton, Rivers, Jr. 1994. "The Brave New World of Animal Biotechnology." Special Supplement. Hastings Center Report 24 (January/February).

Dono, John E., et al. 1979. "Primary Groups in Old Age: Structure and Function." Research on Aging 1 (December): 403–433.

Dore, Ronald P. 1973. British Factory, Japanese Factory. Berkeley, Calif.: University of California Press.

Dornbusch, Sanford. 1989. "The Sociology of Adolescence." Annual Review of Sociology 15:233–259.

Douglas, Jack D., and Waksler, Frances C. 1982. The Sociology of Deviance: An Introduction. Boston: Little, Brown.

Duncan, Greg. 1984. Years of Poverty, Years of Plenty. Ann Arbor: Institute for Social Research.

Duncan, Otis Dudley, Featherman, David L., and Duncan, Beverly. 1972. Socioeconomic Background and Achievement. New York: Seminar Books.

Dunlap, Riley E., Gallup, George H., Jr., and Gallup, Alec M. 1993. Health of the Planet: A George H. Gallup Memorial Survey. Princeton: Gallup International Institute.

Dunlap, Riley E., and Van Liere, Kent D. 1984. "Commitment to the Dominant Social Paradigm and Concern for Environmental Quality." Social Science Quarterly 65:1013–1028.

Durkheim, Emile. 1938. The Rules of Sociological Method. New York: Free Press. (Originally published 1895.)

Durkheim, Emile. 1951. Suicide: A Study in Sociology. New York: Free Press. (Originally published 1897.)

Durkheim, Emile. 1961. The Elementary Forms of the Religious Life. London: Allen & Unwin. (Originally published 1915.)

Durning, Alan. 1989. "Poverty and the Environment. Reversing the Downward Spiral." Worldwatch Paper 92. Washington, D.C.: Worldwatch Institute.

Dworkin, Andrea. 1981. Pornography: Men Possessing Women. New York: Putnam.

Dye, Thomas R. 1983. Who's Running America? The Reagan Years. (3rd ed.) Englewood Cliffs, N.J.: Prentice-Hall.

Dye, Thomas R. 1986. Who's Running America: The Conservative Years. (4th ed.) Englewood Cliffs, N.J.: Prentice-Hall.

Eckstein, Susan (ed.). 1988. Power and Popular Protest Latin American Social Movements. Berkeley, Calif.: University of California Press.

Editors, The Harvard Law Review. 1989. "Family Law Issues Involving Children." Reprinted in Dolores Maggiore (ed.), Lesbians and Child Custody, pp. 157–194. New York: Garland.

Edwards, Bob. 1995. "No Laws Protect Women in Russia from Domestic Violence." Morning Edition, Segment #6, September 27. Washington D.C.: National Public Radio.

Eglit, Howard. 1985. "Age and the Law." In Robert Binstock and Ethel Shanas (eds.), Handbook of Aging and the Social Sciences. (2nd ed.) New York: Van Nostrand Reinhold.

Eisenstadt, S. N. 1985. "Macrosocietal Analysis—Background, Development and Indications." In S. N. Eisenstadt and H. J. Helle (eds.), Macrosociological Theory: Perspectives on Sociological Theory. Newbury Park, Calif.: Sage.

Elder, G. H., Jr. 1969. "Appearance and Education in Marriage Mobility." American Sociological Review 34 (August): 519–533.

Elliott, Delbert S., and Agelon, Suzanne S. 1980. "Reconciling Race and Class Differences in Self-Reported Official Estimates of Delinquency." American Sociological Review 45 (February): 95–110.

Ellison, Christopher. 1991. "An Eye for an Eye? A Note on the Southern Subculture of Violence Thesis." Social Forces 69:1223–1239.

Ellison, Christopher. 1991. "Religious Involvement and Subjective Well-Being." Journal of Health and Social Behavior 32:80–99.

Ellison, Christopher. 1992. "Are Religious People Nice People? Evidence from the National Survey of Black Americans." Social Forces 71:411–430.

Elmer-Dewitt, Philip. 1994. "A Royal Pain in the Wrist." Time, October 24: 60–61.

Elmer-Dewitt, Philip. 1995. "Snuff Porn on the Net." Time (February 20): 69.

Ellison, Christopher G., Bartkowski, John P., and Segal, Michelle L. 1996. "Conservative Protestantism and the Parental Use of Corporal Punishment." Social Forces 74:1003–1028.

Ellison, Christopher G., and Sherkat, Darren E. 1993. "Conservative Protestantism and Support of Corporal Punishment." American Sociological Review 58:131–144.

Ember, Lois. 1994. "Minorities Still More Likely to Live Near Toxic Sites." Chemical & Engineering News 72 (36), September 5:19.

Engels, Friedrich. 1965. "Socialism: Utopian and Scientific." In Arthur P. Mendel (ed.), The Essential Works of Marxism. New York: Bantam Books. (Originally published 1880.)

Entwisle, Doris E., and Alexander, Karl L. 1992. "Summer Setback: Race, Poverty, School Composition and Mathematics Achievement in the First Two Years of School." American Sociological Review 57:72–84.

Entwisle, Doris R., Alexander, Karl, and Olson, Linda Steffel. 1994. "The Gender Gap in Math: Its Possible Origins in Neighborhood Effects." American Sociological Review 59:822–838.

Erickson, Kai. 1986. "On Work and Alienation." American Sociological Review 51 (February): 1–8.

Etzioni, Amitai. 1968. The Active Society. New York: Free Press.

The Europa Yearbook. 1997. A World Survey. London: Europa Publications.

Ezorsky, Gertrude. 1991. Racism and Justice. Ithaca, N.Y.: Cornell University Press.

Falkenmark, Malin, and Widstrand, Carl. 1992. "Population and Water Resources. A Delicate Balance." Population Bulletin 47:2–35.

Farkas, George, Grobe, Robert P., Sheehan, Daniel, and Shuan, Yuan. 1990. "Cultural Resources and School Success." American Sociological Review 55:127–142.

Feeley, Malcolm M., and Simon, Jonathan. 1992. "The New Penology: Notes on the Emerging Strategy of Corrections and Its Implications." Criminology 30:449–474.

Felson, Richard B., and Trudeau, Lisa. 1991. "Gender Differences on Mathematics Performance." Social Psychology Quarterly 54:113–126.

Figueira-McDonough, Josefina. 1985. "Gender Differences in Informal Processing: A Look at Charge Bargaining and Sentence Reduction in Washington, D.C." Journal of Research in Crime and Delinquency 22 (May): 101–133.

Fine, Gary Alan. 1984. "Negotiated Orders and Organizational Cultures." Annual Review of Sociology 10:239–262.

Fineman, Howard, and Thomas, Rich. 1994. "A New Generation." Newsweek, February 21, pp. 20–22.

Finkelhor, David. 1986. A Sourcebook on Child Sexual Abuse. Beverly Hills: Sage.

Firebaugh, Glenn. 1996. "Does Foreign Capital Harm Poor Nations? New Estimates Based on Dixon and Boswell's Measures of Capital Penetration." American Journal of Sociology 102:563–575.

Fischer, Claude S. 1976. The Urban Experience. San Diego, Calif.: Harcourt Brace Jovanovich.

Fischer, Claude S. 1979. "Urban-to-Rural Diffusion of Opinion in Contemporary America." American Journal of Sociology 84 (July): 151–159.

Fischer, Claude S. 1981. "The Public and Private Worlds of City Life." American Sociological Review 46 (June): 306–317.

Fischer, Claude S. 1982. To Dwell Among Friends: Personal Networks in Town and City. Chicago: University of Chicago Press.

Fisher, A. D. 1987. "Alcoholism and Race: The Misapplication of Both Concepts to North American Indians. Canadian Sociological and Anthropological Review 24:80–95.

Fishman, Joshua. 1985a. "Macrosociolinguistics and the Sociology of Language in the Early Eighties." Annual Review of Sociology 11:113–127.

Fishman, Joshua. 1985b. The Rise and Fall of the Ethnic Revival: Perspectives on Language and Ethnicity. Berlin: Mouton.

Fishman, Pamela M. 1978. "Interaction: The Work Women Do." Social Problems 25:397–406.

Fitzpatrick, Kevin M., and Logan, John. 1985. "The Aging of the Suburbs, 1960–1980." American Sociological Review 50 (February): 106–117.

Flanagan, William G. 1993. Contemporary Urban Sociology. New York: Cambridge University Press.

Flanagan, William G. 1993. Contemporary Urban Sociology. London and New York: Cambridge University Press.

"The Forbes Four Hundred." 1994. Forbes 154 (October 17): 103–111.

Form, William. 1985. Divided We Stand: Working Class Stratification in America. Urbana: University of Illinois Press.

Foster, John Bellamy. 1993. "Let Them Eat Pollution: Capitalism and the World Environment." Monthly Review 44:10–20.

Freeman, Sue J. M. 1990. Managing Lives: Corporate Women and Social Change. Amherst, Mass.: University of Massachusetts Press.

French, Hilary F. 1991. "Green Revolutions: Environmental Reconstruction in Eastern Europe and the Soviet Union." Columbia Journal of World Business 26:28–51.

Freudenburg, William R. 1986. "The Density of Acquaintanceship: An Overlooked Variable in Community Research." American Journal of Sociology 92 (July): 27–63.

Freyka, Tomas. 1973. The Future of Population Growth: Alternative Paths to Equilibrium. New York: Wiley.

Funk, Richard, and Willits, Fern. 1987. "College Attendance and Attitudinal Change: A Panel Study, 1970–81." Sociology of Education 60:224–231.

"The Future of Gay America." 1990. Newsweek. March 12, pp. 20ff.

Gaes, Gerald G., and McGuire, William J. 1985. "Prison Violence: The Contribution of Crowding Versus Other Determinants of Prison Assault Rates." Journal of Research in Crime and Delinquency 22 (February): 41–65.

Gagnon, J. H. 1977. Human Sexualities. Glenview, Ill.: Scott, Foresman.

Gagnon, J. H., Roberts, E., and Greenblat, C. 1978. "Stability and Change in Rates of Marital Intercourse." Paper presented at the annual meeting of the International Academy of Sex Research, Toronto, Canada, August.

Galen, Michelle. 1994. "White, Male, and Worried." Business Week 31 (January):50–55.

Gallup Poll Monthly. 1990. No. 297.

Gallup Poll, Public Opinion. 1993, 1994. Wilmington, Del. Scholarly Resources Inc.

Gallup Poll. 1996.

Gallup Report. 1984. Nos. 227/228.

Gallup Report. 1987.

Gallup Report. 1987. No. 259.

Gallup Report. 1989c. No. 286.

Gallup Report. 1989d. No. 288.

Gallup Report. 1989e. No. 289.

Gallup Report. 1992a. No. 318.

Gallup Report. 1992b.

Gamoran, Adam. 1992. The Variable Effects of High School Tracking." American Sociological Review 57:812–828.

Ganz, Alexander. 1985. "Where Has the Urban Crisis Gone? How Boston and Other large Cities Have Stemmed Economic Decline." Urban Affairs Quarterly 20 (June): 449–468.

Gardner, Howard. 1983. Frames of Mind. The Theory of Multiple Intelligences. New York: Basic Books.

Gardner, L. I. 1972. "Deprivation Dwarfism." Scientific American 227 (July): 76–82.

Gardner, LeGrande, and Shoemaker, Donald. 1989. "Social Bonds and Delinquency: A Comparative Analysis." Sociological Quarterly 30:481–500.

Garfinkel, H. 1967. Studies in Ethnomethodology. Englewood Cliffs, N.J.: Prentice-Hall.

Gecas, Viktor. 1981. "Contents of Socialization." In Morris Rosenberg and Ralph H. Turner (eds.), Social Psychology: Sociological Perspectives. New York. Basic Books.

Gecas, Viktor. 1989. "The Social Psychology of Self-Efficiency" Annual Review of Sociology 15:291–316.

Gecas, Viktor, and Schwalbe, Michael. 1983. "Beyond the Looking-Glass Self: Structure and Efficacy-Based Self-Esteem." Social Psychological Quarterly 46(2):77–88.

Geertz, Clifford. 1973. "Thick Description: Toward an Interpretive Theory of Culture." In Clifford Geertz, The Interpretation of Cultures: Selected Essays. New York. Basic Books.

Gelles, Richard J., and Straus, Murray. 1988. Intimate Violence. New York: Simon & Schuster.

Genovese, Frank. 1988. "An Examination of Proposals for a U.S. Industrial Policy." American Journal of Economics and Sociology 47:441–453.

Gerzon, Mark. 1982. A Choice of Heroes: The Changing Faces of American Manhood. Boston: Houghton Mifflin.

Giddens, Anthony. 1984. The Constitution of Society. Cambridge, England: Polity Press.

Gill, Colen. 1985. Work, Unemployment, and the New Technology. Cambridge, England: Polity Press.

Ginzberg, Eli. 1994. The Road to Reform: The Future of Health Care in America. New York: The Free Press.

Glick, Paul C. 1984. "American Household Structure in Transition." Family Planning Perspectives 16 (5): 205–11.

Goffman, Erving. 1959. The Presentation of Self in Everyday Life. New York: Doubleday.

Goffman, Erving. 1961a. Asylums: Essays on the Social Situation of Mental Patients and Other Inmates. New York: Doubleday.

Goffman, Erving. 1961b. Encounters: Two Studies in the Sociology of Interaction. Indianapolis, Ind.: Bobbs-Merrill.

Goffman, Erving. 1963a. Behavior in Public Places: Notes on the Social Organization of Gatherings. New York: Free Press.

Goffman, Erving. 1963b. Stigmas: Notes on the Management of Spoiled Identity. Englewood Cliffs, N.J.: Prentice-Hall.

Goffman, Erving. 1974a. Gender Advertisements. New York: Harper and Row.

Goffman, Erving. 1974b. Frame Analysis: An Essay on the Organization of Experience. New York: Harper and Row.

Goldscheider, Frances, and Goldscheider, Calvin. 1989. "Family Structure and Conflict: Nest-Leaving Expectations of Young Adults and Their Parents." Journal of Marriage and the Family 51:87–97.

Goldscheider, Frances, and Goldscheider, Calvin. 1994. "Leaving and Returning Home in 20th Century America." Population Bulletin 48:1–50.

Gordon, Myles. 1993. "Is There an Islamic Threat?" Scholastic Update, October 22, p. 11.

Gorman, Christine. 1992. "Sizing Up the Sexes." Time, January 20, pp. 42–51.

Goudy, Willis J., Powers, Edward A., Keith, Patricia, and Reger, Richard A. 1980. "Changes in Attitude Toward Retirement: Evidence from a Panel Study of Older Males." Journal of Gerontology, 35:942–948.

Gould, Roger V. 1991. "Multiple Networks and Mobilization in the Paris Commune, 1871." American Sociological Review 56:716–729.

Gove, Walter R., Ortega, Suzanne T., and Style, Carolyn. 1989. "The Maturational and Role Perspectives on Aging and Self Through the Adult Years: An Empirical Evaluation." American Journal of Sociology 94:1117–1145.

Granovetter, Mark. 1973. "The Strength of Weak Ties." American Journal of Sociology 78 (May): 1360–1380.

Granovetter, Mark. 1974. Getting a Job: A Study of Contacts and Careers. Cambridge, Mass.: Harvard University Press.

Greenberg, David F. 1985. "Age, Crime, and Social Explanation." American Journal of Sociology, 91 (July): 1–21.

Greenblat, Cathy Stein. 1983. "The Salience of Sexuality in the Early Years of Marriage." Journal of Marriage and the Family 45 (May):289–300.

Grimes, Michael D. 1989. "Class and Attitudes Toward Structural Inequalities: An Empirical Comparison of Key Variables in Neo- and Post-Marxist Scholarship." Sociological Quarterly 30:441–463.

Gross, Edward. 1958. Work and Society. New York: Crowell.

Gurney, Joan N., and Tierney, Kathleen, J. 1982. "Relative Deprivation and Social Movements: A Critical Look at Twenty Years of Theory and Research." Sociological Quarterly 23 (Winter): 33–47.

Guterbock, Thomas M., and London, Bruce. 1983. "Race, Political Orientation, and Participation: An Empirical Test of Four Competing Theories. American Sociological Review 48 (August): 439–453.

Haas, Linda. 1990. "Gender Equality and Social Policy: Implications of a Study of Parental Leave in Sweden." Journal of Family Issues 11:401–423.

Hadaway, C. Kirk, Marler, Penny Long, and Chaves, Mark. 1993. "What Polls Don't Show. A Closer Look at U.S. Church Attendance." American Sociological Review 58:741–752.

Hagan, John, Gillis, A. R., and Simpson, John. 1985. "The Class Structure of Gender and Delinquency: Toward a Power-Control Theory of Common Delinquent Behavior." American Journal of Sociology 90 (May): 1151–1178.

Halaby, Charles N. 1986. "Worker Attachment and Workplace Authority." American Sociological Review 51 (October): 634–649.

Halle, David. 1984. America's Working Man: Work, Home, and Politics Among Blue-Collar Property Owners. Chicago: University of Chicago Press.

Hallinan, M. 1988. "Equality of Educational Opportunity." Annual Review of Sociology 14:249–268.

Hallinan, Maureen T. 1994. "School Differences in Tracking Effects on Achievement." Social Forces 72:799–820.

Hallinan, Maureen T., and Sorenson, Aage B. 1986. "Student Characteristics and Assignment to Ability Groups: Two Conceptual Formulations." Sociological Quarterly 27 (1):1–13.

Hamilton, V. Lee, Broman, Clifford L., Hoffman, William S., and Renner, Deborah S. 1990. "Hard Times and Vulnerable People: Initial Effects of Plant Closing on Autoworkers' Mental Health." Journal of Health and Social Behavior 31:123–140.

Hammond, Phillip E. 1988. "Religion and the Persistence of Identity." Journal for the Scientific Study of Religion 27:1–11.

Hanks, Michael. 1981. "Youth, Voluntary Associations and Political Socialization." Social Forces 60 (September): 211–223.

Harlow, H. F., and Harlow, M. K. 1966. "Learning to Live." Scientific American 1:244–272.

Harris Survey. 1989a. April 9.

Harris Survey. 1989b. October 29.

Harris, Louis, and Associates. 1975. The Myth and Reality of Aging in America. Washington, D.C.: National Council on Aging.

Haub, Carl. 1991. "South Korea's Low Fertility Raises European-Style Issues." Population Today 19 (October): 3.

Haub, Carl. 1993. "Tokyo Now Recognized as World's Largest City." Population Today 21 (March): 1–2.

Haub, Carl. 1994. "Population Change in the Former Soviet Republics." Population Bulletin, Vol. 49. Washington, D.C.: Population Reference Bureau.

Hechter, Michael. 1987. Principles of Group Solidarity. Berkeley, Calif.: University of California Press.

Heclo, Hugh, and Madson, Henrik. 1987. Policy and Politics in Sweden: Principled Pragmatism. Philadelphia: Temple University Press.

Heidensohn, Frances. 1985. Women and Crime: The Life of the Female Offender. New York: New York University Press.

Hertzman, Clyde, Frank, J., and Evans, Robert G. 1994. "Heterogeneities in Health Status and the Determinants of Population Health." Pp. 67–92 in R. Evans, M. Barer, and T. Marmor (eds.), Why Are Some People Healthy and Others Not? The Determinants of Health of Populations. New York: Aldine de Gruyter.

Hess, Beth. 1985. "Aging Policies and Old Women: The Hidden Agenda." In Alice Rossi (ed.), Gender and the Life Course. Hawthorne, New York: Aldine.

Hetherington, E. M., and Jodl, K. M. 1993. "Stepfamilies as Settings for Child Development." Paper read at the National Symposium on Stepfamilies, October 14–15, 1993, Pennsylvania State University.

Hewitt, John, and Stokes, Randall. 1975. "Disclaimers." American Sociological Review 40:1–11.

Heyl, Barbara. 1979. The Madam as Entrepreneur. Career Management in House Prostitution. New Brunswick, N.J.: Transaction.

Hickok, Kathleen. 1981. "The Spinster in Victorian England: Changing Attitudes in Popular Poetry." Journal of Popular Culture 15 (3): 118–131.

Hirschi, Travis. 1969. Causes of Delinquency. Berkeley and Los Angeles: University of California Press.

Hirschi, Travis, and Goufredson, Michael. 1983. "Age and the Explanation of Crime." American Journal of Sociology 89 (November): 552–584.

Hochschild, Arlie R. 1985. The Managed Heart: The Commercialization of Human Feeling. Berkeley: University of California Press.

Hochschild, Arlie R. 1997. The Time Bind: When Work Becomes Home and Home Becomes Work. New York: Holt.

Hochschild, Jennifer. 1981. What's Fair? American Beliefs About Distributive Justice. Cambridge, Mass.: Harvard University Press.

Hodge, Robert W., Treiman, Donald J., and Rossi, Peter. 1966. "A Comparative Study of Occupational Prestige." In Reinhard Bendix and Seymour Martin Lipset (eds.), Class, Status, and Power. (2nd ed.) New York: Free Press.

Hogan, D. P. 1981. Transitions and Social Change: The Early Lives of American Men. Orlando, Fla.: Academic Press.

Hogan, Dennis P., and Astone, Nan Marie. 1986. "The Transition to Adulthood." Annual Review of Sociology 12:109–130.

Hogan, Dennis P., Eggebeen, David J., and Clogg, Clifford C. 1993. "The Structure of Intergenerational Exchanges in American Families." American Journal of Sociology 98:1428–1458.

Hoge, Dean R. and DeZulueta, Ernesto. 1985. "Salience as a Condition for Various Social Consequences of Religious Commitment." Journal for the Scientific Study of Religion 24:21–38.

Holden, Karen, Burkhauser, Richard, and Feaster, Daniel. 1988. "The Timing of Falls into Poverty After Retirement and Widowhood." Demography 25:405–414.

Holland, Walter W. 1994. "Commentary: Recent Reforms in the British National Health Service—Lessons for the United States." American Journal of Public Health 84:186–189.

Hollander, Paul. 1982. "Research on Marxist Societies: The Relationship Between Theory and Practice." Annual Review of Sociology 8:319–351.

Holloway, Marguerite. 1994. "Trends in Women's Health: A Global View." Scientific American (August): 76–83.

Homans, George. 1950. The Human Group. San Diego, Calif.: Harcourt Brace Jovanovich.

Hooks, Gregory. 1984. "The Policy Response to Factory Closings: A Comparison of the United States, Sweden, and France." Annals of the American Academy of Political and Social Science 475:110–124.

Hooks, Gregory. 1990. "The Rise of the Pentagon and U.S. State Building: The Defense Program as Industry Policy." American Journal of Sociology 96:358–404.

Hoover, Stewart M. 1990. "Ten Myths about Religious Broadcasting." Pp. 23–39 in Robert Abelman and Stewart M. Hoover (eds.), Religious Television: Controversies and Conclusions. Norwood, N.J.: Ablex Publishing Corporation.

"Hope for our Cities." 1994. People, January 17, pp. 81–83.

Hostetler, John. 1963. Amish Society. Baltimore, Md.: Johns Hopkins University Press.

Hout, Michael. 1986. "Opportunity and the Minority Middle Class: A Comparison of Blacks in the United States and Catholics in Northern Ireland." American Sociological Review 51:214–223.

Howery, Carla. 1983. "Sociologists Shaping Public Policy: Two Profiles." Footnotes 11 (August): 12.

Huber, Joan, and Form, William H. 1973. Income and Ideology: An Analysis of the American Political Formula. New York: Free Press.

Huber, Joan, and Spitze, Glenna. 1983. Sex Stratification: Children, Housework, and Jobs. Orlando, Fla.: Academic Press.

Hudson, Robert B., and Strate, John. 1985. "Aging and Political Systems." In Robert H. Binstock and Ethel Shanas (eds.), Handbook of Aging and the Social Sciences. (2nd ed.) New York: Van Nostrand Reinhold.

Hummer, Robert A. 1993. "Racial Differentials in Infant Mortality in the U.S.: An Examination of Social and Health Determinants." Social Forces 72:529–554.

Hurlbert, Jeanne S. 1989. "The Southern Region: A Test of the Hypothesis of Cultural Distinctiveness." Sociological Quarterly 30:245–266.

Hyde, Janet S., Fennema, Elizabeth H., and Lamon, Susan J. 1990. "Gender Differences in Mathematics Performance: A Meta-Analysis." Psychological Bulletin 107:139–155.

Jackson, Pamela Brody, Thoits, Peggy A., and Taylor, Howard F. 1995. "Composition of the Workplace and Psychological Well-Being: The Effects of Tokenism on America's Black Elite." Social Forces 74:543–557.

Jacobs, David. 1988. "Corporate Economic Power and the State: A Longitudinal Assessment of Two Explanations." American Journal of Sociology 93:852–881.

Jacobs, David, and Helms, Ronald E. 1996. "A Political Model of Incarceration: A Time-Series Examination of Multiple Explanations for Prison Admission Rates." American Journal of Sociology 102:323–357.

Jacobs, David, and Helms, Ronald E. 1997. "Testing Coercive Explanations for Order: The Determinants of Law Enforcement Strength over Time." Social Forces 75:1361–1392.

Jacobs, Jerry. 1989. "Long-Term Trends in Occupational Segregation." American Journal of Sociology 95:160–173.

Jacobs, Jerry. 1989. Revolving Doors: Sex Segregation and Women's Careers. Stanford, Calif.: Stanford University Press.

Janis, Irving. 1982. Groupthink: Psychological Studies of Policy Decisions and Fiascoes. Boston: Houghton Mifflin.

Jasper, James M., and Nelkin, Dorothy. 1992. The Animal Rights Crusade: The Growth of a Moral Protest. New York: Free Press.

Jasper, James M., and Poulsen, Jane D. 1995. "Recruiting Strangers and Friends: Moral Shocks and Social Networks in Animal Rights and Anti-Nuclear Protests." Social Problems 42:493–512.

Jenkins, J. Craig, and Brent, Barbara. 1989. "Social Protest, Hegemonic Competition, and Social Reform." American Sociological Review 54:891–909.

Jenkins, J. Craig, and Eckert, Craig, M. 1986. "Channeling Black Insurgency. Elite Patronage and Professional Social Movement Organizations in the Development of the Black Movement." American Sociological Review 51 (December): 812–829.

Jenness, Valerie. 1990. "From Sex to Sin to Sex as Work: COYOTE and the Reorganization of Prostitution as a Social Problem." Social Problems 37:403–420.

Jenness, Valerie. 1995. "Social Movement Growth, Domain Expansion, and Framing Processes: The Gay/Lesbian Movement and Violence Against Gays and Lesbians as a Social Problem." Social Problems 42:145–170.

Jensen, Holger. 1990. "The Cost of Neglect." Maclean's, May 7:54–55.

Jessor, R., and Jessor, S. 1977. Problem Behavior and Psychosocial Development: A Longitudinal Study of Youth. Orlando, Fla.: Academic Press.

Johansen, Harley, and Fuguitt, Glenn. 1984. The Changing Rural Village in America: Demographic and Economic Trends Since 1950. Cambridge, Mass.: Ballinger.

Johnson, Benton. 1957. "A Critical Appraisal of the Church Sect Typology." American Sociological Review 22 (1):88–92.

Johnson, Dirk. 1994. "Economies Come to Life on Indian Reservations." The New York Times (July 3):1Y,10Y–11Y.

Johnson, Kenneth. 1989. "Recent Population Redistribution Trends in Nonmetropolitan America." Rural Sociology 54:301–326.

Johnson, Richard E. 1980. "Social Class and Delinquent Behavior: A New Test." Criminology 18 (1):86–93.

Johnson, Robert C. 1993. "Science, Technology, and Black Community Development." Pp. 265–282 in Albert H. Teich (ed.), Technology and the Future. (6th ed.) New York: St. Martin's Press.

Joint Center for Political Studies. 1991. Black Elected Officials.

Jones, Rachel K., and Brayfield, April. 1997. "Life's Greatest Joy: European Attitudes Toward the Centrality of Children." Social Forces 75:1239–1270.

Juster, Susan, and Vinovskis, Maris. 1987. "Changing Perspectives on the American Family in the Past." Annual Review of Sociology 13:193–216.

Kalab, Kathleen. 1987. "Student Vocabularies of Motive Accounts for Absence." Symbolic Interaction 10:71–83.

Kalish, Susan. 1994a. "Culturally Sensitive Family Planning: Bangladesh Story Suggests It Can Reduce Family Size." Population Today 22 (February): 5.

Kalish, Susan, 1994b. "International Migration. New Findings on Magnitude, Importance." Population Today 22 (March): 1–2.

Kalish, Susan. 1995. "Multiracial Births Increase as U.S. Ponders Racial Definition." Population Today 23 (4):1–2.

Kaplan, Howard B., Martin, Steven S., and Johnson, Robert J. 1986. "Self-Rejection and the Explanation of Deviance: Specification of the Structure Among Latent Constructs." American Journal of Sociology 92 (September): 384–411.

Katel, Peter, Liu, Melinda, and Cohn, Bob. 1994. "The Bust in Boot Camps." Newsweek 123 (February 21): 26.

Katz, Neil, 1989. "Conflict Resolution and Peace Studies." The Annals 504 (July): 14–21.

"Kenya." 1989. Population Today 17 (10):5.

Kephart, William M. 1983 and 1987. Extraordinary Groups: The Sociology of Unconventional Life-Styles. (2nd and 3rd eds.) New York: St. Martin's Press.

Kephart, William M., and Zellner, William W. 1994. Extraordinary Groups: An Examination of Unconventional Life-Styles. (5th ed.) New York: St. Martin's Press.

Kerbo, Harold R. 1991. Social Stratification and Inequality: Class Conflict in Historical and Comparative Perspective. (2nd ed.) New York: McGraw-Hill.

Kerckhoff, Alan C., and Davis, Keith E. 1962. "Value Consensus and Need Complementarity in Mate Selection." American Sociological Review 27 (June): 295–303.

Keyfitz, Nathan. 1987. "The Family That Does Not Reproduce Itself." In Kingsley Davis, Mikhail Bernstam, and Rita Ricardo-Campbell (eds.), Below Replacement Fertility in Industrial Societies: Causes, Consequences, Policies. Cambridge, England: Cambridge University Press.

Kielbowicz, Richard B., and Scherer, Clifford. 1986. "The Role of the Press in the Dynamics of Social Movements." Research in Social Movements, Conflicts, and Change 9:71–96.

Kilner, John. 1988. "Selecting Patients: When Resources Are Limited: A Study of U.S. Medical Directors of Kidney Dialysis and Transplantation Facilities." American Journal of Public Health 78 (2):144–147.

King Moshoeshoe II. 1993. "Return to Self-Reliance: Balancing the African Condition and the Environment." Pp. 158–170 in Pablo Piacetini (ed.), Story Earth: Native Voices on the Environment. San Francisco, Calif.: Mercury House.

Kinsey, A. C. 1948. Sexual Behavior in the Human Male. Philadelphia: Saunders.

Kitson, Gay, and Sussman, Marvin. 1982. "Marital Complaints, Demographic Characteristics, and Symptoms of Mental Distress in Divorce." Journal of Marriage and the Family 44:87–101.

Klandermas, Bert. 1984. "Mobilization and Participation: Social-Psychological Expansion of Resource Mobilization Theory." American Sociological Review 49 (October): 583–600.

Kleck, Gary. 1982. "On the Use of Self-Report Data to Determine the Class Distribution of Criminal and Delinquent Behavior." American Sociological Review 47 (June): 427–433.

Klepper, Steven, and Nagin, Daniel, 1989. "The Deterrent Effect of Perceived Certainty and Severity of Punishment Revisited." Criminology 27:721–746.

Kluegel, James R., and Smith, Eliot R. 1983. "Affirmative Action Attitudes. Effects of Self-Interest, Racial Affect, and Stratification Beliefs on Whites' Views." Social Forces 61 (March): 170–181.

Knoke, David. 1981. "Commitment and Detachment in Voluntary Associations." American Sociological Review 46 (2):141–158.

Koblik, Steven. 1975. Sweden's Development from Poverty to Affluence 1750–1970. Minneapolis, Minn.: University of Minnesota Press.

Kohn, Melvin, and Schooler, Carmi, and Associates. 1983. Work and Personality: An Inquiry into the Impact of Social Stratification. Norwood, N.J.: Ablex.

Kohn, Robert L. 1972. "The Meaning of Work: Interpretation and Proposals for Measurement." In A. Campbell and P. Converse (eds.), The Human Meaning of Social Change. New York: Basic Books.

Kollock, Peter, Blumstein, Phillip, and Schwartz, Pepper. 1985. "Sex and Power in Conversation: Conversational Privileges and Duties." American Sociological Review 50:34–46.

Konig, René. 1968. "Auguste Comte." In David J. Sills (ed.), International Encyclopedia of the Social Sciences. Vol. 3. New York: Macmillan and Free Press.

Korpi, Walter. 1989. "Power, Politics, and State Autonomy in the Development of Social Citizenship." American Sociological Review 54:309–328.

Korte, C. 1980. "Urban-Nonurban Differences in Social Behavior: Social Psychological Models of Urban Impact." Journal of Social Issues 36 (1):29–51.

Kozol, Jonathan. 1991. Savage Inequalities. Children in America's Schools. New York: Crown Publisher.

Kraska, Peter B., and Kappeler, Victor E. 1997. "Militarizing American Police: The Rise and Normalization of Paramilitary Unites." Social Problems 44:1–18.

Krohn, Marvin D., Akers, Ronald L., Radosevich, Marcia J., and Lanza-Kaduce, Lonn. 1980. "Social Status and Deviance." Criminology 18:303–318.

Laumann, Edward O., Gagnon, John H., Michael, Robert T., and Michaels, Stuart. 1994. The Social Organization of Sexuality: Sexual Practices in the United States. Chicago: The University of Chicago Press.

Laumann, Edward O., Knoke, David, and Kim, Yong-hat. 1985. "An Organizational Approach to State Policy Formation: A Comparative Study of Energy and Health Domains." American Sociological Review 50:1–19.

Lavee, Yoar, McCubbin, Hamilton, I., and Patterson, Joan M. 1985. "The Double ABCX Model of Family Stress and Adaptation: An Empirical Test by Analysis of Structural Equations with Latent Variables." Journal of Marriage and the Family 47 (November): 811–825.

Lebergott, Stanley. 1975. Wealth and Want. Princeton, N.J.: Princeton University Press.

Leach, Penelope. 1994. Children First. New York: Alfred Knopf.

Lemert, Edwin, 1981. "Issues in the Study of Deviance." Sociological Quarterly 22 (Spring): 285–305.

Lenski, Gerhard. 1966. Power and Privilege: A Theory of Social Stratification. New York: McGraw-Hill.

Levin, William. 1988. "Age Stereotyping: College Student Evaluations." Research on Aging 10:134–148.

Levine, Daniel U., and Havighurst, Robert J. 1992. Society and Education. (8th ed.) Boston: Allyn and Bacon.

Lewis, Oscar. 1969. "The Culture of Poverty." In Daniel P. Moynihan (ed.), On Understanding Poverty. New York: Basic Books.

Lichter, Daniel T. 1988. "Racial Differences in Underemployment in American Cities." American Journal of Sociology 93:771–792.

Lichter, Daniel T., and Eggebeen, David J. 1993. "Rich Kids, Poor Kids: Changing Income Inequality among American Children." Social Forces 71:761–780.

Lichter, Daniel T., LeClere, Felicia B., and McLaughlin, Diane K. 1991. "Local Marriage Markets and the Marital Behavior of

Black and White Women." American Journal of Sociology 96:843–867.

Lichter, Daniel T., McLaughlin, Diane K., Kephart, George, and Landry, David J. 1992. "Race and the Retreat from Marriage: A Shortage of Marriageable Men?" American Sociological Review 57:781–799.

Lieberson, Stanley, and Waters, Mary. 1988. From Many Strands: Ethnic and Racial Groups in Contemporary America. New York: Russell Sage Foundation.

Lieberson, Stanley, and Waters, Mary C. 1993. "The Ethnic Responses of Whites: What Causes Their Instability, Simplification, and Inconsistency?" Social Forces 72:421–450.

Light, Donald W. 1988. "Toward a New Sociology of Medical Education." Journal of Health and Social Behavior 29:307–322.

Lin, Chien, and Liu, William T. 1993. "Intergenerational Relationships Among Chinese Immigrant Families from Taiwan." Pp. 271–286 in Harriett Pipes McAdoo (ed.), Family Ethnicity: Strength in Diversity. Newbury Park, Calif.: Sage.

Lin, Nan, Ensel, Walter M, and Vaughn, John C. 1981. "Social Resources and Strength of Ties: Structural Factors in Occupational Status Attainment." American Sociological Review 46:393–405.

Lincoln, James R., and McBride, Kerry. 1987. "Japanese Industrial Organization in Comparative Perspective." Annual Review of Sociology 13:289–312.

Lincoln, Yvonna, and Guba, Egan. 1985. Naturalistic Inquiry. Newbury Park, Calif.: Sage.

Link, Bruce. 1987. "Understanding Labeling Effects in the Area of Mental Disorders. An Assessment of the Effects of Expectations of Rejection." American Sociological Review 52 (February): 96–112.

Liska, Allen E., Chamlin, Mitchell B., and Reed, Mark. 1985. "Testing the Economic Production and Conflict Models of Crime Control." Social Forces 64 (September): 119–138.

Litwak, Eugene. 1960. "Geographic Mobility and Extended Family Cohesion." American Sociological Review 25:385–394.

Litwak, Eugene. 1961. "Voluntary Association and Neighborhood Cohesion." American Sociological Review 26 (April): 266–271.

Litwak, Eugene, and Kulis, Stephen. 1988. "Technology, Proximity, and Measures of Kin Support." Journal of Marriage and the Family 49:649–661.

Lo, Clarence, Y. H. 1982. "Countermovements and Conservative Movements in the Contemporary U.S." Annual Review of Sociology 8:10–34.

Lofland, John. 1985. Protest: Studies of Collective Behavior and Social Movements. New Brunswick, N.J.: Transaction.

Logan, John R., and Spitze, Glenna D. 1994. "Family Neighbors." American Journal of Sociology 100:453–476.

London, Bruce, and Robinson, Thomas. 1989. "The Effect of International Dependence on Income Inequality and Political Violence." American Sociological Review 54:305–308.

Love, Douglas, and Torrence, William. 1989. "The Impact of Worker Age on Unemployment and Earnings After Plant Closings." Journal of Gerontology 44:S190–S195.

Lurigio, Arthur. 1990. "Introduction." Crime and Delinquency 36:3–5.

Lydall, Harold. 1989. Yugoslavia in Crisis. Oxford: Clarendon Press.

Lynch, J. J. 1979. The Broken Heart: The Medical Consequences of Loneliness. New York: Basic Books.

MacKenzie, Doris Layton, Brame, Robert, McDowall, David, and Souryal, Claire. 1995. "Boot Camps and Recidivism in Eight States." Criminology 33:327–357.

MacKenzie, Doris Layton, Shaw, James W., and Gowdy, Voncile B. 1993. "An Evaluation of Shock Incarceration in Louisiana." National Institute of Justice Research in Brief (June). Washington D.C.: National Institute of Justice.

MacLeod, Jay. 1987. Ain't No Making It: Leveled Aspirations in a Low-Income Neighborhood. Boulder, Colo.: Westview.

Maher, Lisa, and Daly, Kathleen. 1996. "Women in the Street-Level Drug Economy: Continuity or Change?" Criminology 34:465–491.

Mannheim, Karl. 1929. Ideology and Utopia: An Introduction to the Sociology of Knowledge. San Diego, Calif.: Harcourt Brace Jovanovich.

Marger, Martin. 1994. Race and Ethnic Relations. Belmont, Calif.: Wadsworth.

Marini, Margaret. 1989. "Sex Differences in Earnings in the United States." Annual Review of Sociology 15:343–380.

Marmot, M. G., Kogevinas, M., and Elston, M. 1987. "Social/Economic Status and Disease." Annual Review of Public Health 8:111–135.

Marsden, Peter V. 1987. "Core Discussion Networks of Americans." American Sociological Review 52 (February): 122–131.

Marsh, Robert M., and Mannari, Niroshi. 1976. Modernization and the Japanese Factory. Princeton, N.J.: Princeton University Press.

Marshall, Patrick G. 1991. "The Greening of Eastern Europe." CO Researcher 1:851–871.

Marshall, Susan E. 1985. "Ladies Against Women: Mobilization Dilemmas of Antifeminist Movements." Social Problems 32 (April): 348–362.

Martin, Teresa, and Bumpass, Larry. 1989. "Recent Trends in Marital Disruption." Demography 26:37–51.

Marwell, Gerald, and Oliver, Pamela. 1984. "Collective Action Theory and Social Movement Research." Research in Social Movements, Conflicts, and Change 7:1–27.

Marx, Gary T. (ed.). 1971. Racial Conflict. Boston: Little, Brown.

Marx, Gary T., and McAdam, Douglas. 1994. Collective Behavior and Social Movements: Process and Structure. Englewood Cliffs, N.J.: Prentice-Hall.

Marx, Karl, and Engels, Friedrich. 1965. "The Communist Manifesto." In Arthur Mendel (ed.), Essential Works of Marxism. New York: Bantam Books. (Originally published 1848.)

Masatsugu, Mitsuyuki. 1982. The Modern Samurai Society: Duty and Dependence in Contemporary Japan. New York: American Management Association.

Massey, Douglas S. 1990. "American Apartheid: Segregation and the Making of the Underclass." American Journal of Sociology 96:329–357.

Massey, Douglas S., and Butterman, Brooks. 1985. "Explaining the Paradox of Puerto Rican Segregation." Social Forces 64 (December): 306–331.

Massey, Douglas S., and Mullan, Brendan P. 1984. "Processes of Hispanic and Black Spatial Assimilation." American Journal of Sociology 89 (January): 836–873.

Massey, Douglas, and Denton, Nancy. 1988. "Suburbanization and Segregation in U.S. Metropolitan Areas." American Journal of Sociology 94:592–626.

McAdam, Doug. 1986. "Recruitment to High-Risk Activism." American Journal of Sociology 92 (July): 64–90.

McAdam, Doug, and Paulsen, Ronnelle. 1993. "Specifying the Relationship between Social Ties and Activism." American Journal of Sociology 99:640–667.

McCammon, Holly. 1993. "From Repressive Intervention to Integrative Prevention: The U.S. State's Legal Management of Labor Militancy, 1881–1978." Social Forces 71:569–601.

McConnell, Harvey. 1977. "The Indian War on Alcohol." Social Resources Series, Alcohol I 76:72–81.

McDill, Edward L., Natriello, Gary, and Pallas, Aaron. 1986. "A Population at Risk: Potential Consequences of Tougher School Standards for School Dropouts." American Journal of Education 94 (February): 135–181.

McFarlane, S. Neil. 1985. Superpower Rivalry and Third World Radicalism: The Idea of National Liberation. London: Croom Helm.

McGhee, Jerrie L. 1985. "The Effect of Siblings on the Life Satisfaction of the Rural Elderly." Journal of Marriage and the Family 47 (February): 85–90.

McGinn, Robert E. 1991. Science, Technology, and Society. Englewood Cliffs, N.J.: Prentice Hall.

McKeown, T., and Record, R. G. 1962. "Reasons for the Decline of Mortality in England and Wales During the Nineteenth Century." Population Studies 16 (March): 94–122.

McKeown, T., Record, R. G., and Turner, R. D. 1975. "An Interpretation of the Decline of Mortality in England and Wales During the Twentieth Century." Population Studies 29 (November): 390–421.

McLanahan, Sara. 1985. "Family Structure and the Reproduction of Poverty." American Journal of Sociology 90 (January): 873–901.

McLanahan, Sara, and Adams, Julia. 1987. "Parenthood and Psychological Well-being." Annual Review of Sociology 13:237–257.

McLanahan, Sara, and Booth, Karen. 1989. "Mother-Only Families: Problems, Prospects, and Policies." Journal of Marriage and the Family 51:557–580.

McLaughlin, Steven, and Associates. 1988. The Changing Lives of American Women. Chapel Hill: University of North Carolina Press.

McPherson, J. Miller, and Smith-Lovin, Lynn. 1982. "Women and Weak Ties: Differences by Sex in the Size of Voluntary Organizations." American Journal of Sociology 87:883–904.

McPherson, J. Miller, Popielarz, Pamela A., and Drobnic, Sonja. 1992. "Social Networks and Organizational Dynamics." American Sociological Review 57:153–170.

McPherson, Miller. 1983. "The Size of Voluntary Organizations." Social Forces 61:1044–1064.

Mead, George Herbert. 1934. Mind, Self, and Society: From the Standpoint of a Social Behaviorist (Charles W. Morris, ed.). Chicago: University of Chicago Press.

Mellon, Margaret. 1993. "Altered Traits." Nucleus (Fall): 4–6, 12.

Menken, Jane L. 1985. "Age and Fertility: How Late Can You Wait?" Demography 22 (November): 469–484.

Merton, Robert. 1949. "Discrimination and the American Creed." In Robert MacIver (ed.), Discrimination and National Welfare. New York: Harper and Row.

Merton, Robert. 1957. Social Theory and Social Structure. (2nd ed.) New York: Free Press.

Messenger, John C. 1969. Inis Beag: Isle of Ireland. New York: Holt, Rinehart & Winston.

Messner, Steven F. 1989. "Economic Discrimination and Societal Homicide Rates: Further Evidence on the Cost of Inequality." American Sociological Review 54:597–611.

Messner, Steven F., and Krohn, Marvin D. 1990. "Class, Compliance Structures, and Delinquency: Assessing Integrated Structural-Marxist Theory." American Journal of Sociology 96:300–328.

Michalowski, Raymond J., and Kramer, Ronald C. 1987. "The Space Between Laws: The Problem of Corporate Crime in a Transnational Context." Social Problems 34:34–53.

Miller, Karen A., Kohn, Melvin L., and Schooler, Carmi. 1985. "Educational Self-Direction and the Cognitive Functioning of Students." Social Forces 63 (June): 923–944.

Miller, Karen, Kohn, Melvin, and Schooler, Carmi. 1986. "Educational Self-direction and Personality." American Sociological Review 51:372–390.

Mills, C. Wright. 1940. "Situated Actions and Vocabularies of Motives." American Sociological Review 5:904–913.

Mills, C. Wright. 1956. The Power Elite. New York: Oxford University Press.

Mills, C. Wright. 1959. The Sociological Imagination. Oxford, England: Oxford University Press.

Mizrahi, Terry. 1986. Getting Rid of Patients: Contradictions in the Socialization of Physicians. New Brunswick, N.J.: Rutgers University Press.

Mizruchi, Mark. 1989. "Similarity of Political Behavior Among Large American Corporations." American Journal of Sociology 95:401–424.

Mizruchi, Mark S. 1990. "Determinants of Political Opposition Among Large American Corporations." Social Forces 68:1065–1088.

Moen, Phyllis, Dempster-McClain, Donna, and Williams, Robin. 1989. "Social Integration and Longevity: An Event-History Analysis of Women's Roles and Resilience." American Sociological Review 54:635–647.

Molotch, Harvy. 1979. "Media and Movements." In M. Zald and J. McCarthy (eds.), The Dynamics of Social Movements. Cambridge, Mass.: Winthrop.

Moore, Gwen S. 1990. "Structural Determinants of Men's and Women's Personal Networks." American Sociological Review 55:726–735.

Moore, Helen A., and Whitt, Hugh P. 1986. "Multiple Dimensions of the Moral Majority Platform: Shifting Interest Group Coalitions." The Sociological Quarterly 27 (3): 423–439.

Morgenthau, Tom. 1995. "What Color is Black?" Newsweek (February 13): 63–70.

Mortimer, Jeylan T. 1979. Changing Attitudes Toward Work. Scarsdale, N.Y.: Work in America Institute.

Mortimer, Jeylan, T., and Simmons, R. G. 1978. "Adult Socialization." Annual Review of Sociology 4:421–454.

Mott, Frank, and Mott, Susan. 1980. "Kenya's Record Population Growth: A Dilemma of Development." Population Bulletin 35 (3):1–45.

Mottl, Tahi L. 1980. "The Analysis of Countermovements." Social Problems 27 (June): 620–635.

Mulcahy, Aogan. 1995. "Claims-Making and the Construction of Legitimacy: Press Coverage of the 1981 Northern Irish Hunger Strike." Social Problems 42:449–467.

Munch, Allison, McPherson, J. Miller, and Smith-Lovin, Lynn. 1997. "Gender, Children, and Social Contact: The Effects of Childrearing for Men and Women." American Sociological Review 62:509–520.

Murdock, George Peter. 1949. Social Structure. New York: Free Press.

Murphy, Caryle. 1994. "Egypt: An Uneasy Portent of Change." Current History 93:78–82.

Mushane, Michael, Palumbo, Dennis, Maynard-Moody, Steven, and Levine, James. 1989. "Community Correctional Innovation: What Works and Why?" Journal of Research on Crime and Delinquency 26:136–167.

Mutran, Elizabeth, and Reitzes, Donald C. 1984. "Intergenerational Support Activities and Well-Being Among the Elderly: A Convergence of Exchange and Symbolic Interaction Perspectives." American Sociological Review 49 (February): 117–130.

Myrdal, Gunnar. 1962. Challenge to Affluence. New York: Random House.

Nagel, Joane. 1994. "Constructing Ethnicity: Creating and Recreating Ethnic Identity and Culture." Social Problems 41:152–176.

Nardi, Peter. 1992. Men's Friendships: Research on Men and Masculinities. Newbury Park, Calif.: Sage.

Nathanson, C. A. 1984. "Sex Differences in Mortality." Annual Review of Sociology 10:191–213.

Navarro, Vicente. 1989. "Why Some Countries Have National Health Insurance, Others Have National Health Service, and the U.S. Has Neither." Social Science and Medicine 28:887–898.

Nelan, Bruce W. 1993. "Is Haiti Worth It?" Time 142 (November 1): 26–29.

Nemeth, Charlan J. 1985. "Dissent, Group Process, and Creativity: The Contribution of Minority Influence." Advances in Group Processes 2:57–75.

Neugarten, Bernice L., and Neugarten, Dail A. 1986. "Changing Meanings of Age in the Aging Society." In Alan Pifer and Lydia Bronte (eds.), Our Aging Society: Paradox and Promise. New York: Norton.

Newcomb, Chad C. 1994. "Fearful and Punitive Responses to Concern about Crime." Unpublished manuscript. University of Nebraska–Lincoln.

Niebuhr, H. Richard. 1957. The Social Sources of Denominationalism. New York: Holt, Rinehart & Winston. (Originally published 1929.)

Nilsson, Goran B. 1975. "Swedish Liberalism at Mid-Nineteenth Century." Pp. 141–166 in Steven Koblik (ed.), Sweden's Development from Poverty to Affluence 1750–1970. Minneapolis, Minn.: University of Minnesota Press.

O'Bryant, Shirley. 1988. "Sibling Support and Older Widow's Well-Being." Journal of Marriage and Family 50:173–183.

O'Dea, Thomas F. 1966. The Sociology of Religion. Englewood Cliffs, N.J.: Prentice-Hall.

Oakes, Jeannie. 1985. Keeping Track: How Schools Structure Inequality. New Haven: Yale University Press.

Oberschall, A. 1973. Social Conflict and Social Movements. Englewood Cliffs, N.J.: Prentice-Hall.

Oberschall, Anthony, and Leifer, Eric J. 1986. "Efficiency and Social Institutions: Uses and Misuses of Economic Reasoning in Sociology." Annual Review of Sociology 12:233–253.

Oliver, Melvin L., and Shapiro, Thomas M. 1997. Black Wealth/White Wealth. New York: Routledge.

Olsen, Gregg M. 1992. The Struggle for Economic Democracy in Sweden. Aldershot, Great Britain: Avebury.

Olsen, Gregg M. 1996. "Re-modeling Sweden: The Rise and Demise of the Compromise in a Global Economy." Social Problems 43:1–20.

"Oregon's Not-So-Sweet Home." 1989. Newsweek (December 12): 55.

Orenstein, Peggy. 1994. School Girls: Young Women, Self-Esteem, and the Confidence Gap. New York: Doubleday.

Orum, Anthony. 1987. "In Defense of Domhoff: A Comment on Manning's Review of Who Rules America Now." American Journal of Sociology 92 (January): 975–977.

Osgood, D. Wayne, and Wilson, Janet. 1989. "Role Transitions and Mundane Activities in Late Adolescence and Early Adulthood." Paper read at the 1989 meetings of the Midwest Sociological Society, St. Louis.

Ouichi, William G., and Wilkins, Alan L. 1985. "Organizational Culture." Annual Review of Sociology 11:457–483.

Painton, Frederick. 1990. "Where the Sky Stays Dark." Time, May 28:40–42.

Parsons, Talcott. 1951. The Social System. New York: Free Press.

Parsons, Talcott. 1964. "The School Class as a Social System: Some of Its Functions in American Society." In Talcott Parsons (ed.), Social Structure and Personality. New York: Free Press.

Paternoster, Raymond. 1989. "Absolute and Restrictive Deterrence in a Panel of Youth: Explaining the Onset, Persistence/Desistance, and Frequency of Delinquent Offending." Social Problems 36:289–309.

Patterson, Thomas, Clifford, J. G., and Hagan, Kenneth. 1983. American Foreign Policy. Lexington, Mass.: D.C. Heath.

Paz, Juan J. 1993. "Support of Hispanic Elderly." Pp. 177–183 in Harriett Pipes McAdoo (ed.), Family Ethnicity: Strength in Diversity. Newbury Park, Calif.: Sage.

Pebley, Anne R., and Westoff, Charles. 1982. "Women's Sex Preferences in the United States: 1970–1975." Demography 19:177–190.

Perrow, Charles. 1986. Complex Organizations: A Critical Essay. (3rd ed.) New York: Random House.

Peterson, Ruth. 1988. "Youthful Offender Designations and Sentencing in the New York Criminal Courts." Social Problems 35:111–130.

Pettigrew, Thomas F. 1982. "Prejudice." In Thomas F. Pettigrew, George M. Fredrickson, Dale T. Knobel, Nathan Glazer, and Reed Ueda (eds.), Prejudice: Dimensions of Ethnicity. Cambridge, Mass.: Harvard University Press.

Pifer, Alan. 1986. "The Public Policy Response." In A. Pifer and L. Bronte (eds.), Our Aging Society: Paradox and Promise (pp. 391–413). New York: Norton.

Piliavin, Irving, Gartner, Rosemary, Thornton, Craig, and Matsueda, Ross. 1986. "Crime, Deterrence, and Rational Choice." American Sociological Review 51:101–119.

Pitts, Jesse R. 1964. "The Structural-Functional Approach." In Harold T. Christensen (ed.), Handbook of Marriage and the Family. Skokie, Ill.: Rand McNally.

Plummer, Gayle, 1985. "Haitian Migrants and Backyard Imperialism." Race and Class 26:35–43.

Podolsky, Doug. 1986–1987. "NIAAA Minority Research Activities." Alcohol Health and Research World 11 (2):4–7.

Pollock, Philip H., III. 1982. "Organizations and Alienation. The Mediation Hypothesis Revisited." The Sociological Quarterly 23 (Spring):143–155.

Population Reference Bureau. 1992. The United States Population Data Sheet. Washington, D.C.: Population Reference Bureau.

Population Reference Bureau. 1997. World Population Data Sheet. Washington D.C.: Population Reference Bureau.

Portes, Alejandro, and Sassen-Koob, Saskia. 1987. "Making It Underground: Comparative Material on the Informal Sector in Western Market Economies." American Journal of Sociology 93:30–61.

Portes, Alejandro, and Truelove, Cynthia. 1987. "Making Sense of Diversity: Recent Research on Hispanic Minorities in the United States." Annual Review of Sociology 13:359–385.

Post, Tom. 1993. "Sailing into Big Trouble." Newsweek 122 (November 1):34–35.

Postel, Sandra. 1992. "Denial in the Decisive Decade." Pp. 3–8 in Lester Brown (ed.), State of the World 1992. New York: W. W. Norton.

Poston, Dudley, and Gu, Baochang. 1987. "Socioeconomic Development, Family Planning, and Fertility in China." Demography 24:531–551.

Preimsberger, Duane. 1996. "Cops and Space Scientists: New Crime-Fighting Partners." Police Chief 63:108–114.

Presser, Harriet. 1989. "Can We Make Time for Children?" Demography 26:523–543.

Preston, Samuel H. 1976. Mortality Patterns in National Populations with Special Reference to Recorded Causes of Death. Orlando, Fla.: Academic Press.

Preston, Samuel H. 1984. "Children and the Elderly: Divergent Paths for America's Dependents." Demography 21 (November): 435–458.

Provence, Sally, and Lipton, Rose. 1962. Infants in Institutions: A Comparison of Their Development with Family-Reared Infants During the First Year of Life. New York: International Universities Press.

Public Opinion. 1986a. Vol. 9, no. 1.

Public Opinion. 1986b. Vol. 9, no. 6.

Public Opinion. 1988.

Public Opinion. 1989. May–June, p. 33.

Pugliesi, Karen. 1995. "Work and Well-Being: Gender Differences in the Psychological Consequences of Employment." Journal of Health and Social Behavior 36:57–71.

Quadagno, Jill. 1990. "Race, Class, and Gender in the U.S. Welfare State: Nixon's Failed Family Assistance Plan." American Sociological Review 55:11–28.

"Quebec Encouraging Births with New Baby Bonuses." 1988. Population Today 16 (July–August): 8.

Quillian, Lincoln. 1996. "Group Threat and Regional Change in Attitudes toward African-Americans." American Journal of Sociology 102:816–860.

Quinn, Naomi. 1977. "Anthropological Studies of Women's Status." Annual Review of Anthropology 6:181–225.

Quinney, Richard. 1980. Class, State, and Crime. (2nd ed.) New York: Longman.

Radelet, Michael L. 1981. "Racial Characteristics and the Imposition of the Death Penalty." American Sociological Review 46:918–927.

Radelet, Michael L. 1989. "Executions of Whites for Crimes Against Blacks" Exceptions to the Rule?" The Sociological Quarterly 30:529–544.

Raley, J. Kelly. 1996. "A Shortage of Marriageable Men? A Note on the Role of Cohabitation in Black-White Differences in Marriage Rates." American Sociological Review 61:973–983.

Rawlings, Stephen. 1978. "Perspectives on American Husbands and Wives." Current Population Reports, Special Studies Series P-23 No. 77. U.S. Department of Commerce, Bureau of the Census. Washington, D.C.: U.S. Government Printing Office.

Razzell, P. 1974. "An Interpretation of the Modern Rise of Population in Europe: A Critique." Population Studies 28 (March): 5–15.

Reichman, Nancy. 1989. "Breaking Confidences: Organizational Influences on Insider Trading." Sociological Quarterly 30:185–204.

Reiss, Ira L. 1980. Family Systems in America. (3rd ed.) New York: Holt, Rinehart & Winston.

Renner, Craig, and Navarro, Vicente. 1989. "Why Is Our Population of Uninsured and Underinsured Persons Growing? The Consequences of the Deindustrialization of America." Annual Review of Public Health 10:85–94.

Reskin, Barbara. 1989. "Women Taking 'Male' Jobs Because Men Leave Them." IlliniWeek (July 20): 7.

Rich, Robert. 1977. The Sociology of Law. Washington, D.C.: University Press of America.

Ridgeway, Cecilia L., Berger, Joseph, and Smith, LeRoy. 1985. "Nonverbal Cues and Status: An Expectation States Approach." American Journal of Sociology 90 (March): 955–978.

Riedmann, Agnes. 1987. "Ex-Wife at the Funeral: Keyed Antistructure." Free Inquiry in Sociology 16:123–129.

Riedmann, Agnes. 1993. Science That Colonizes. Philadelphia: Temple University Press.

Rindfuss, Ronald R., Swicegood, C. Gray, and Rosenfeld, Rachel A. 1987. "Disorder in the Life Course: How Common and Does it Matter?" American Sociological Review 52:785–801.

Robertson, Roland. 1970. The Sociological Interpretation of Religion. Oxford, England: Blackwell.

Robinson, J. Gregg, and McIlwee, Judith. 1989. "Women in Engineering: A Promise Unfulfilled." Social Problems 36:455–472.

Robinson, Robert V. 1984. "Reproducing Class Relations in Industrial Capitalism." American Sociological Review 49:182–196.

Rogers, Everett M. 1960. Social Change in Rural Society: A Textbook in Rural Sociology. East Norwalk, Conn.: Appleton-Century-Crofts.

Ronan, Laura, and Reichman, Walter. 1987. "Back to Work." Alcohol Health and Research World 11:34.

Rose, Peter. 1981. They and We: Racial and Ethnic Relations in the United States. (3rd ed.) New York: Random House.

Rosecrance, John. 1985. "Compulsive Gambling and the Medicalization of Deviance." Social Problems 32 (February): 273–284.

Rosenbaum, Emily. 1996. "Racial/Ethnic Differences in Home Ownership and Housing Quality, 1991." Social Problems 43:403–426.

Rosenberg, Debra. 1994. "Men, Women, Computers." Newsweek (May 16): 48–55.

Rosenberg, Morris, 1979. Conceiving the Self. New York: Basic Books.

Rosenberg, Morris, Schooler, Carmi, and Schoenbach, Carrie, 1989. "Self-Esteem and Adolescent Problems Modeling Reciprocal Effects." American Sociological Review 54:1004–1018.

Rosenthal, Carolyn J. 1985. "Kinkeeping in the Familial Division of Labor." Journal of Marriage and the Family 47 (Nov.): 965–974.

Rossi, Alice. 1984. "Gender and Parenthood." American Sociological Review 49 (February): 1–19.

Rothschild, Joyce. 1986. "Alternatives to Bureaucracy. Democratic Participation in the Economy." Annual Review of Sociology 12:307–328.

Roy, William G. 1984. "Class Conflict and Social Change in Historical Perspective." Annual Review of Sociology 10:483–506.

Rubel, Maxmilien. 1968. "Karl Marx." In David Sills (ed.), International Encyclopedia of the Social Sciences. Vol. 10. New York: Macmillan and Free Press.

Rubin, Lillian Breslow. 1976. Worlds of Pain. N.Y.: Basic Books.

Rule, James, and Tilly, Charles. 1975. "Political Process in Revolutionary France 1830–1932." Pp. 41–85 in J. Merriman (ed.), 1830 in France. New York: New Viewpoints.

Ryan, William. 1981. Equality. New York: Pantheon Books.

Saigo, Roy. 1989. "The Barriers of Racism." Change (November–December): 8, 10, 69.

Saks, Michael J., and Krupat, Edward. 1988. Social Psychology and its Applications. New York: Harper and Sons.

Saltman, Juliet. 1991. "Maintaining Racially Diverse Neighborhoods." Urban Affairs Quarterly 26:416–441.

Sampson, Robert. 1987. "Urban Black Violence: The Effect of Male Joblessness and Family Disruption" American Journal of Sociology 93:348–382.

Sampson, Robert. 1988. "Local Friendship Ties and Community Attachment in Mass Society: A Multilevel Systemic Model." American Sociological Review 53: 766–779

San Miguel, Guadalupe, Jr. 1987. Let Them All Take Heed. Austin: University of Texas Press.

Sandefur, Gary D., and Sakamoto, Arthur. 1988. "American Indian Household Structure and Income." Demography 25:71–80.

Sarkitov, Nikolay. 1987. "From 'Hard Rock' to 'Heavy Metal': The Stupefaction Effect." Sotsiologicheskie-Issledovaniya 14:93–94.

Scanzoni, John. 1989. "Alternative Images for Public Policy: Family Structure Versus Families Struggling." Policy Studies Review 8:610–621.

Schaefer, Richard R. 1990. Racial and Ethnic Groups. (4th ed.) New York: HarperCollins.

Schlesinger, Arthur, Jr. 1965. A Thousand Days. Boston: Houghton Mifflin.

Schmidt, Hans. 1971. The U.S. Occupation of Haiti, 1915–1934. New Brunswick, N.J.: Rutgers University Press.

Schneider, David J. 1981. "Tactical Self-Presentations: Toward a Broader Conceptualization." In J. T. Tedeschi (ed.), Impression Management Theory and Social Psychological Research. Orlando, Fla.: Academic Press.

Schoen, Robert, Urton, William, Woodrow, Karen, and Baj, John. 1985. "Marriage and Divorce in Twentieth Century Cohorts." Demography 22:1–114

Schultz, T. Paul. 1993. "Investments in the Schooling and Health of Men and Women." The Journal of Human Resources 28:694–734.

Schur, Edwin M. 1979. Interpreting Deviance: A Sociological Introduction. New York: Harper and Row.

Schwartz, Barry. 1983. "George Washington and the Whig Conception of Heroic Leadership." American Sociological Review 48 (February): 18–33

Scott, Marvin B., and Lyman, Stafford M. 1968. "Accounts." American Sociological Review 33 (December): 46–62.

Sekulic, Dusko, Massey, Garth, and Hodson, Randy. 1994. "Who Were the Yugoslavs? Failed Sources of Common Identity in the Former Yugoslavia." American Sociological Review 59:83–97.

Sen, Gita, and Grown, Caren. 1987. Development, Crises, and Alternative Visions. New York: Monthly Review Press.

Serpe, Richard. 1987. "Stability and Change in Self: A Structural Symbolic Interactionist Explanation." Social Psychology Quarterly 50(1): 44–55.

Shafer, John. 1989. "Theories of Alcohol Abuse: What Do Native Americans Think?" Unpublished manuscript, Department of Sociology, University of Nebraska–Lincoln.

Shalin, Dmitri, 1986. "Pragmatism and Social Interaction." American Sociological Review 51 (February): 9–29.

Shapiro, Laura. 1994. "A Tomato with a Body that Just Won't Quit." Newsweek 6 (June): 80–82.

Shavit, Yossi. 1984. "Tracking and Ethnicity in Israeli Secondary Education." American Sociological Review 49 (April): 210–220.

Sherif, Muzafer. 1936. The Psychology of Social Norms. New York: Harper and Row.

Shkilnyk, Anastasia M. 1985. A Poison Stronger Than Love: The Destruction of an Ojibwa Community. New Haven: Yale University Press.

Shupe, Anson, and Stacy, William A. 1982. Born-Again Politics and the Moral Majority: What Social Surveys Really Show. New York: Mellen Press.

Siegel, Larry J. 1995. Criminology. (5th ed.) Minneapolis, Minn.: West Publishing.

Silvestri, George T. 1993. "Occupational Employment: Wide Variations in Growth." Monthly Labor Review 116:58–86.

Simons, Ronald, and Gray, Phyllis. 1989. "Perceived Blocked Opportunity as an Explanation of Delinquency Among Lower-Class Black Males." Journal of Research on Crime and Delinquency 26:90–101.

Simpson, Miles. 1990. "Political Rights and Income Inequality: A Cross-National Test." American Sociological Review 55:682–693.

Simpson, Richard L. 1985. "Social Control of Occupations and Work." Annual Review of Sociology 11:415–436.

Singh, Karan. 1993. "Let No Enemy Ever Wish Us Ill: The Hindu Vision of the Environment." Pp. 146–156 in Pablo Piacentini (ed.), Story Earth: Native Voices on the Environment. San Francisco: Mercury House.

Sjoberg, Gideon. 1960. The Preindustrial City. New York: Free Press.

Sloan, Irving. 1981. Youth and the Law. Dobbs Ferry, N.Y.: Oceana.

Smith, Douglas. 1987. "Police Response to Interpersonal Violence: Defining the Parameters of Legal Control." Social Forces 65 (March): 767–782.

Smith, Douglas A., and Visher, Christy A. 1981. "Street-Level Justice: Situational Determinants of Police Arrest Decisions." Social Problems 29 (2):167–177.

Smith, Kevin, and Stone, Lorence. 1989. "Rags, Riches, and Bootstraps." Sociological Quarterly 30:93–107.

Smith, Ryan A. 1997. "Race, Income, and Authority at Work: A Cross-Temporal Analysis of Black and White Men (1972–1994)." Social Problems 44:19–37.

Snow, David A., and Anderson, Leon. 1987. "Identity Work Among the Homeless: The Verbal Construction and Avowal of Personal Identities." American Journal of Sociology 92 (May): 1336–1371.

Snow, David A., Rochford, E. Burke, Jr., Worden, Steven K., and Benford, Robert D. 1986. "Frame Alignment Processes, Micromobilization, and Movement Participation." American Sociological Review 51 (August): 464–481.

Sohoni, Neera Kuckreja. 1994. "Where are the Girls?" Ms. (July/August): 96.

Sombart, Werner 1974 "Why Is There No Socialism in the U.S.?" Excerpted in John Laslett and S. M. Lipset (eds.), Failure of a Dream: Essays in the History of American Socialism. New York: Doubleday Anchor Books. (Originally published 1906.)

Sorokin, Pitirim, and Lundin, Walter. 1959. Power and Morality. Boston: Sargent

South, Scott, and Crowder, Kyle. 1997. "Escaping Distressed Neighborhoods: Individual, Community, and Metropolitan Influences." American Journal of Sociology 102:1040–1084.

Speare, Alden Jr., and Avery, Roger. 1993. "Who Helps Whom in Older Parent-Child Families." Journal of Gerontology 48:564–573.

Spilka, Bernard, Shaver, Phillip, and Kirkpatrick, Lee A. 1985. "A General Attribution Theory for the Psychology of Religion." The Journal for the Scientific Study of Religion 24 (1):1–20.

Spitz, René. 1945. "Hospitalism: An Inquiry into the Genesis of Psychiatric Conditions in Early Childhood." In Anna Freud, Heinz Hartman, and Ernst Kris (eds.), The Psychoanalytic Study of the Child. Vol 1. New York: International Universities Press.

Spitze, Glenna. 1986. "The Division of Task Responsibility in U.S. Households: Longitudinal Adjustments to Change." Social Forces 64 (March): 689–701.

Stanglin, Douglas. 1992. "Toxic Wasteland." U.S. News and World Report (April 13): 40–46.

Stark, Rodney, and Bainbridge, William Sims. 1979. "Of Churches, Sects, and Cults: Preliminary Concepts for a Theory of Religious Movements." Journal for the Scientific Study of Religion 18 (2): 117–133.

Stark, Rodney, and Finke, Roger. 1988. "American Religion in 1976: A Statistical Portrait." Sociological Analysis 49 (1):39–51.

Starr, Paul. 1982. The Social Transformation of American Medicine. New York: Basic Books.

State of World Population. 1992. New York. United Nations Population Fund.

Steffensmeier, Darrel J., Allan, Emilie, Harer, Miles, and Streifel, Cathy. 1989. "Age and the Distribution of Crime." American Journal of Sociology 94:803–831.

Stearn, Peter N. 1976. "The Evolution of Traditional Culture Toward Aging." In Jon Hendricks and C. Davis Hendricks (eds.), Dimensions of Aging: Readings. Cambridge, Mass.: Winthrop.

Stevenson, M. H. 1975. "Relative Wages and Sex Segregation by Occupation." Pp. 174–200 in C. B. Lloyd (ed.), Sex Discrimination and the Division of Labor. New York: Columbia University Press.

Stokes, Randall, and Anderson, Andy. 1990. "Disarticulation and Human Welfare in Less Developed Countries." American Sociological Review 55:63–74.

Stone, John. 1985. Racial Conflict in Contemporary Society. Cambridge, Mass.: Harvard University Press.

Straus, Murray, and Gelles, Richard. 1986. "Societal Change and Change in Family Violence from 1975 to 1985 as Revealed by Two National Surveys." Journal of Marriage and the Family 48:465–479.

Streib, Gordon. 1985. "Social Stratification and Aging." In Robert Binstock and Ethel Shanas (eds.), Handbook of Aging and the Social Sciences. (2nd ed.) New York: Van Nostrand Reinhold.

Stryker, Sheldon. 1981. "Symbolic Interactionism: Themes and Variations." In Morris Rosenberg and Ralph H. Turner (eds.), Social Psychology: Sociological Perspectives. New York: Basic Books.

Sudo, Phil. 1993. "The Faith and the Followers." Scholastic Update (October 22): 2–5.

Suomi, S. J., Harlow, H. H., and McKinney, W. T. 1972. "Monkey Psychiatrists." American Journal of Psychiatry 128 (February): 927–932.

Sutherland, Edwin H. 1961. White-Collar Crime. New York: Holt, Reinhart & Winston.

Suzuki, Bob. 1989. "Asian Americans as the Model Minority." Change (November–December):12–20.

Swidler, Ann. 1986. "Culture in Action: Symbols and Strategies." American Sociological Review 51 (April): 273–286.

Takagi, Dana Y. 1990. "From Discrimination to Affirmative Action: Facts in the Asian American Admissions Controversy." Social Problems 37:578–592.

Takaki, Ronald. 1993. A Different Mirror. Boston: Little, Brown.

Talero, Eduardo, and Gaudette, Philip. 1996. Harnessing Information for Development—A Proposal for a World Bank Group Strategy. New York: World Bank.

Tang, Joyce 1993. "The Career Attainment of Caucasian and Asian Engineers." The Sociological Quarterly 34:467–496.

Tannen, Deborah. 1990. You Just Don't Understand. New York: William Morrow and Co., Inc.

Tannen, Deborah. 1994. Talking from 9 to 5: How Women's and Men's Conversational Styles Affect Who Gets Heard, Who Gets Credit, and What Gets Done at Work. New York: William Morrow and Co., Inc.

Tannenbaum, Frank. 1979. "The Survival of the Fittest." In George Modelski (ed.), Transnational Corporations and the World Order. New York: W. H. Freeman. (Originally published 1968.)

Tarrow, Sidney. 1988. "National Politics and Collective Action: Recent Theory and Research in Western Europe and the United States." Annual Review of Sociology 14:421–440.

Taylor, Ronald E. 1994. Minority Families in the United States. Englewood Cliffs, N.J.: Prentice-Hall.

Taylor, Verta, and Raeburn, Nicole C. 1995. "Identity Politics as High-Risk Activism: Career Consequences for Lesbian, Gay, and Bisexual Sociologists." Social Problems 42: 252–273.

Teachman, Jay. 1987. "Family Background, Educational Resources, and Educational Attainment." American Sociological Review 52:548–557.

Tedeschi, James T., and Riess, Marc. 1981. "Identities, the Phenomenal Self, and Laboratory Research." In J. T. Tedeschi (ed.), Impression Management Theory and Social Psychological Research. Orlando, Fla.: Academic Press.

Teitelbaum, Michael S. 1987 "The Fear of Population Decline." Population Today 15 (March): 6–8.

Thomas, Darwin L., and Cornwall, Marie. 1990. "Religion and Family in the 1980's." Pp. 265–274 in Alan Booth (ed.), Contemporary Families. Minneapolis, Minn.: National Council on Family Relations.

Thomas, Melvin E. 1993. "Race, Class and Personal Income: An Empirical Test of the Declining Significance of Race Thesis, 1968–1988." Social Problems 40:328–342.

Thomas, W. I., and Thomas, Dorothy. 1928. The Child in America: Behavior Problems and Programs. New York: Knopf.

Thomis, Malcolm I. 1970. The Luddites: Machine-Breaking in Regency England. Hamden, Conn.: Archon Books.

Thomlinson, Ralph. 1976. Population Dynamics: Causes and Consequences of World Demographic Change. (2nd ed.) New York: Random House.

Thompson, Anthony P. 1983. "Extramarital Sex: A Review of the Research Literature." Journal of Sex Research (February): 1–21.

Thompson, Kevin, 1989. "Gender and Adolescent Drinking Problems: The Effects of Occupational Structure." Social Problems 36:30–47.

Thompson, Linda, and Walker, Alexis J. 1989. "Gender in Families: Women and Men in Marriage, Work and Parenthood." Journal of Marriage and the Family 51:845–872.

Thornberry, Terence P., and Farnworth, Margaret. 1982. "Social Correlates of Criminal Involvement: Further Evidence on the Relationship Between Social Status and Criminal Behavior." American Sociological Review 47 (August): 505–518.

Thorson, James A. 1995. Aging in a Changing Society. Belmont, Calif.: Wadsworth.

Thurow, Lester C. 1980. The Zero-Sum Society. New York: Basic Books.

Tillman, Robert, and Pontell, Henry N. 1992. "Is Justice Collar-Blind?: Punishing Medicaid Provider Fraud." Criminology 30:547–574.

Tilly, Charles. 1978. From Mobilization to Revolution. Reading, Mass.: Addison-Wesley.

Tittle, Charles R., and Meier, Robert F. 1990. "Specifying the SES/Delinquency Relationship." Criminology 28:271–299.

Tittle, Charles R., Villemez, Wayne, and Smith, Douglas. 1978. "The Myth of Social Class and Criminality: An Empirical Assessment of the Empirical Evidence." American Sociological Review 43 (October): 643–656.

Treas, Judith. 1995. "Older Americans in the 1990s and Beyond." Population Bulletin. Vol. 50. Washington, D.C.: Population Reference Bureau.

Troeltsch, Ernst. 1931. The Social Teaching of the Christian Churches. New York: Macmillan.

"Trying to Take Back the Planet." 1990. Newsweek (February 5): 24ff.

Turner, Jonathan H. 1972. Patterns of Social Organization. New York: McGraw-Hill.

Turner, Jonathan H. 1982. The Structure of Sociological Theory. (3rd ed.) Homewood, Ill.: Dorsey Press.

Turner, Jonathan, and Beegley, Leonard. 1981. The Emergence of Sociological Theory. Homewood, Ill.: Dorsey Press.

Turner, Jonathan, and Musick, David.1985. American Dilemmas. New York: Columbia University Press.

Turner, Ralph H. 1985. "Unanswered Questions in the Convergence Between Structuralist and Interactionist Role Theories." In S. N. Eisenstadt and H.J. Helle (eds.), Microsociological Theory: Perspectives on Sociological Theory. Vol. 2. Newbury Park, Calif.: Sage.

Udry, J. Richard. 1988. "Biological Predispositions and Social Control in Adolescent Sexual Behavior." American Sociological Review 53:709–722.

Ulbrich, Patricia, Warheit, George, and Zimmerman, Rick. 1989. "Race, Socioeconomic Status, and Psychological Distress: An Examination of Differential Vulnerability." Journal of Health and Social Behavior 30: 131–146.

United Nations 1988. 1986 Demographic Yearbook. New York: United Nations.

United Nations. 1994. "World Contraceptive Use 1994." New York: United Nations Department for Economic and Social Information and Policy Analysis, Population Division.

United Nations Development Programme. Human Development Report 1997. http://www.undp.org/undp/hdro/table2.htm

United Nations Population Fund. 1991. Population and the Environment: The Challenges Ahead. New York: United Nations Population Fund.

Unnever, James D., Frazier, Charles E., and Henretta, John C. 1980. "Race Differences in Criminal Sentencing." Sociological Quarterly 21 (Spring): 197–205.

U.S. Bureau of the Census. 1975a. Historical Statistics of the United States: Colonial Times to 1970 (Bicentennial ed., Part 1). Washington, D.C.: U.S. Government Printing Office.

U.S. Bureau of the Census. 1975b. Statistical Abstract of the United States, 1975. Washington, D.C.: U.S. Government Printing Office.

U.S. Bureau of the Census. 1982. "Money Income and Poverty Status of Persons in the United States: 1981." Current Population Reports, Series P-60, no. 132. Washington, D.C.: U.S. Government Printing Office.

U.S. Bureau of the Census. 1986. "Fertility of American Women: June 1985." Current Population Reports, Series P-20, No. 406. Washington, D.C.: U.S. Government Printing Office.

U.S. Bureau of the Census. 1989a. "The Black Population of the United States: March 1988." Current Population Reports, Series P-20, no. 442. Washington, D.C.: U.S. Government Printing Office.

U.S. Bureau of the Census. 1989b. "The Hispanic Population of the United States: March 1988." Current Population Reports, Series P-20, no. 438. Washington, D.C.: U.S. Government Printing Office.

U.S. Bureau of the Census. 1989c. "Household and Family Characteristics: March 1988." Current Population Reports, Series P-20, no. 437. Washington, D.C.: U.S. Government Printing Office.

U.S. Bureau of the Census. 1989d. "Marital Status and Living Arrangements: March 1988." Current Population Reports, Series P-20, no. 433. Washington, D.C.: U.S. Government Printing Office.

U.S. Bureau of the Census. 1989e. "Money Income and Poverty Status in the United States, 1988." Current Population Reports, Series P-60, no. 166. Washington, D.C.: U.S. Government Printing Office.

U.S. Bureau of the Census. 1989f. "Patterns of Metropolitan Area and County Population Growth: 1980 to 1987." Current Population Reports, Series P-25, no. 1039. Washington, D.C.: U.S. Government Printing Office.

U.S. Bureau of the Census. 1989g. "Fertility of American Women: June 1988." Current Population Reports, Series P-20, no. 436. Washington, D.C.: U.S. Government Printing Office.

U.S. Bureau of the Census. 1993a Statistical Abstract of the United States: 1993. Washington, D.C.: U.S. Government Printing Office.

U.S. Bureau of the Census. 1993b. "The Hispanic Population in the U.S.: March 1992." Current Population Reports, Series P-20, no. 465RV. Washington, D.C.: U.S. Government Printing Office.

U.S. Bureau of the Census. 1993c. "The Black Population in the U.S.: 1992." Current Population Reports, Series P-20, no. 471. Washington, D.C.: U.S. Government Printing Office.

U.S. Bureau of the Census. 1994a. "How We're Changing Demographic State of the Nation: 1994." Current Population Reports, Series P-23, no. 187. Washington, D.C.: U.S. Government Printing Office.

U.S. Bureau of the Census. 1994b. Statistical Abstract of the United States: 1994. Washington, D.C.: U.S. Government Printing Office.

U.S. Bureau of the Census. 1996. Statistical Abstract of the United States: 1996. Washington, D.C.: U.S. Government Printing Office.

U.S. Department of Justice. 1989. "Prisoners in 1988." Bureau of Justice Statistics Bulletin. Washington, D.C.: Bureau of Justice Statistics.

U.S. Department of Justice. 1993. Sourcebook of Criminal Justice Statistics: 1992. Washington, D.C.: U.S. Government Printing Office.

U.S. Department of Justice. 1994. "Violence Between Intimates." Bureau of Justice Statistics: Selected Findings. Publication # NCJ-149259 (November). Washington, D.C.: U.S. Department of Justice.

U.S. Department of Justice. 1996a. Uniform Crime Reports 1995. Washington, D.C.: U.S. Government Printing Office, Bureau of Justice Statistics.

U.S. Department of Justice. 1996b. Sourcebook of Criminal Justice Statistics, 1995. Washington, D.C.: U.S. Government Printing Office.

U.S. Department of Labor. 1985. The Impact of Technology on Labor in Four Industries. Bulletin 2263. Washington, D.C.: U.S. Government Printing Office.

U.S. Department of Labor. 1986a. Employment Projections for 1995: Data and Methods. Bureau of Labor Statistics Bulletin 2253. Washington, D.C.: U.S. Government Printing Office.

U.S. Department of Labor. 1986b. The Impact of Technology on Labor in Four Industries. Bulletin 2263. Washington, D.C.: U.S. Government Printing Office.

U.S. Department of Labor. 1992. Employment and Earnings: March 1992. Washington, D.C.: U.S. Government Printing Office.

U.S. Department of Labor. 1993. Employment and Earnings: October 1993. Washington, D.C.: U.S. Government Printing Office.

Useem, Bert. 1980. "Solidarity Model, Breakdown Model, and the Boston Anti-Busing Movement." American Sociological Review 45 (June): 357–369.

Vago, Steven. 1989. Law and Society. (2nd ed.) Englewood Cliffs, N.J.: Prentice-Hall.

Vallas, Steven P. 1987. "While Collar Proletarians? The Structure of Clerical Work and Levels of Class Consciousness." The Sociological Quarterly 28:523–540.

Vallas, Steven P., and Yarrow, Michael. 1987. "Advanced Technology and Worker Alienation." Working and Occupations 14 (February): 126–42.

van de Walle, Etienne, and Knodel, John. 1980. "Europe's Fertility Transition." Population Bulletin 34 (6): 1–43.

van den Berghe, Pierre L. 1978. Man in Society. New York: Elsevier North-Holland.

van den Hombergh, Heleen. 1993. Gender, Environment, and Development. Utrecht, Netherlands. International Books.

Veblen, Thorstein. 1919. The Vested Interests and the State of the Industrial Arts. New York: Huebsch.

Vega, W.A., and Amero, H. 1994. "Latino Outlook: Good Health, Uncertain Prognosis." American Review of Public Health 10: 333–361.

Vernon, JoEtta 1988. "The Grandparent/Grandchild Relationship: An Exploratory Study." Unpublished Manuscript, Department of Sociology, University of Nebraska–Lincoln.

Villemez, Wayne, and Bridges, William. 1988. "When Bigger Is Better: Differences in the Individual-Level Effect of Firm and Establishment Size." American Sociological Review 53:237–55.

Wagner, David G., Ford, Rebecca S, and Ford, Thomas W. 1986. "Can Gender Inequalities be Reduced?" American Sociological Review 51:47–61.

Wald, Kenneth D. 1987 Religion and Politics in the United States. New York: St. Martin's Press.

Waldholz, Michael. 1991. "Uninsured Infants Taken to Hospital Get Fewer Services." Wall Street Journal (December 18): B3.

Waldman, Steven. 1992. "Benefits 'R' Us." Newsweek (August 10): 56–58.

Waldron, Ingrid. 1983. "Sex Differences in Human Mortality: The Role of Genetic Factors." Social Science and Medicine 17 (6):321–333.

Wallace, Walter, 1969. Sociological Theory. Hawthorne, New York: Aldine.

Walshok, Mary Lindenstein. 1993. "Blue Collar Women." Pp. 256–264 in Albert H. Teich (ed.), Technology and the Future. (6th ed.) New York: St. Martin's Press.

Walster, Elaine, Arenson, V., Abrahams, D., and Rottman, L. 1966. "Importance of Physical Attractiveness in Dating Behavior." Journal of Personality and Social Psychology 4 (November): 508–516.

Watson, Russell. 1995. "When Words are the Best Weapon." Newsweek (February) 27: 35–40.

Weber, Max. 1954. Law in Economy and Society (Max Rheinstein, ed.: Edward Shils and Max Reinstein, trans.) Cambridge, Mass.: Harvard University Press. (Originally published 1914.)

Weber, Max. 1958. The Protestant Ethic and the Spirit of Capitalism. (Talcott Parsons, trans.) New York: Scribner's (Originally published 1904–5.)

Weber, Max. 1970a. "Bureaucracy." In H. H. Gerth and C. Wright Mills (trans.), From Max Weber Essays in Sociology. New York: Oxford University Press (Originally published 1910.)

Weber, Max. 1970b. "Class, Status, and Party." In H. H. Gerth and C. Wright Mills (trans.), From Max Weber Essays in Sociology. New York: Oxford University Press. (Originally published 1910.)

Weber, Max. 1970C. "Religion." In H. H. Gerth and C. Wright Mills (trans.), From Max Weber: Essays in Sociology. New York: Oxford University Press. (Originally published 1922).

Weber, Max. 1970e. "The Sociology of Charismatic Authority." In H. H. Gerth and C. Wright Mills (trans.), From Max Weber: Essays in Sociology. New York: Oxford University Press. (Originally published 1910.)

Wechsler, David. 1958. The Measurement and Appraisal of Adult Intelligence. (4th ed.) Baltimore, Md.: Williams & Wilkins.

Weibel-Orlando, Joan. 1986–1987. "Drinking Patterns of Urban and Rural American Indians." Alcohol Health and Research World II (2): 8–12, 54.

Weil, Frederick. 1985. "The Variable Effects of Education on Liberal Attitudes." American Sociological Review 50:458–74.

Weil, Frederick. 1989. "The Sources and Structure of Legitimation in Western Democracies." American Sociological Review 54:682–706.

Weitzman, Lenore. 1985. The Divorce Revolution: The Unexpected Social and Economic Consequences for Women and Children in America. New York. Free Press.

"Weitzman's Research Plays Key Role in New Legislation." 1985. Footnotes 13 (8): 1, 9.

Wellman, Barry. 1979. "The Community Question: The Intimate Networks of East Yorkers." American Journal of Sociology 84 (March): 1201–1231

Wellman, Barry, and Berkowitz, S. D. (eds.) 1988. Social Structures: A Network Approach. New York: Cambridge University Press.

Wellman, Barry, and Wortley, Scot. 1990. "Different Strokes from Different Folks: Community Ties and Social Support." American Journal of Sociology 96:558–588.

Werner, Emy E., and Smith, Ruth. 1982. Vulnerable but Invincible: A Longitudinal Study of Resilient Children and Youth. New York: McGraw-Hill.

Wessells, Michael G. 1990. Computer, Self, and Society. Englewood Cliffs, N.J.: Prentice Hall.

West, Candace. 1984. Routine Complications: Troubles with Talk Between Doctors and Patients. Bloomington, Ind.: Indiana University Press.

Western, Bruce. 1993. "Postwar Unionization in Eighteen Advanced Capitalist Countries." American Sociological Review 58:266–282.

Weston, Kath. 1991. Families We Choose: Lesbians, Gays, Kinship. New York: Columbia University Press.

White, Jr., Lynn. 1967. "The Historical Roots of Our Ecologic Crisis." Science 155:1203–1207.

White, Lynn. 1990. "Determinants of Divorce: A Review of Research in the Eighties." Journal of Marriage and the Family 52 (November).

White, Lynn, and Edwards, John. 1990. "Emptying the Nest and Parental Well-Being." American Sociological Review 55:235–242.

White, Lynn K. 1994. "Coresidence and Leaving Home: Young Adults and Their Parents." Annual Review of Sociology 20:81–102.

White, Lynn K., and Riedman, Agnes. 1992. "Ties Among Adult Siblings." Social Forces 71:85–102.

Whitt, J. Allen. 1979. "Toward a Class-Dialectical Model of Power." American Sociological Review 44 (February): 81–99.

Whorf, Benjamin L. 1956. Language, Thought, and Reality. Cambridge, Mass.: MIT Press.

Wilentz, Amy. 1993. "Love and Haiti." The New Republic 209:18–19.

Williams, Christine L. 1992. "The Glass Escalator: Hidden Advantages for Men in the 'Female' Professions." Social Problems 39:253–267.

Williams, J. Allen, Jr., Vernon, JoEtta, Williams, Martha C., and Malecha, Karen. 1987. "Sex Role Socialization in Picture Books: An Update." Social Science Quarterly 68:148–156.

Williams, Kirk, and Drake, Susan. 1980. "Social Structure, Crime, and Criminalization: An Empirical Examination of the Conflict Perspective." Sociological Quarterly 21 (Autumn): 563–575.

Williams, Rhys H. 1995. "Constructing the Public Good: Social Movements and Cultural Resources." Social Problems 42: 124–144.

Williams, Robin M., Jr. 1970. American Society: A Sociological Interpretation. (3rd ed.) New York: Knopf.

Wilson, Edward O. 1978. "Introduction: What Is Sociobiology?" In Michael S. Gregory, Anita Silvers, and Diane Sutch (eds.), Sociobiology and Human Nature. San Francisco: Jossey-Bass.

Wilson, George. 1997. "Pathways to Power: Racial Differences in the Determinants of Job Authority." Social Problems 44: 38–52.

Wilson, James Q. 1992. "Crime, Race, and Values." Society 30: 90–93.

Wilson, Thomas C. 1986. "Interregional Migration and Racial Attitudes." Social Forces 65 (September): 177–186.

Wilson, Thomas C. 1991. "Urbanism, Migration, and Tolerance: A Reassessment." American Sociological Review 56:117–123.

Wilson, William J. 1978. The Declining Significance of Race. Chicago: University of Chicago Press.

Wilson, William J. 1987. The Truly Disadvantaged. Chicago: University of Chicago Press.

Wilson, William J., and Aponte, Robert. 1985. "Urban Poverty." Annual Review of Sociology 11:231–258.

Wimberly, Dale. 1990. "Investment Dependence and Alternative Explanations of Third World Mortality: A Cross-National Study." American Sociological Review 55:75–91.

Wirth, Louis. 1938. "Urbanism as a Way of Life." American Journal of Sociology 44 (1): 1–24.

Wojtkiewicz, Roger, McLanahan, Sara, and Garfinkel, Irwin. 1990. "The Growth of Families Headed by Women: 1950–1980." Demography 27:19–30.

Wolfgang, Marvin E., and Reidel, Marc. 1973. "Race, Judicial Discretion, and the Death Penalty." Annals of the American Academy of Political and Social Science 407:119–133.

Woods, Kathryn, and Clouse, Meghan. 1994. Facts About Female Genital Mutilation. Santa Cruz, Calif.: Body Image Task Force.

Wright, Erik O. 1985. Classes. London: Verso.

Wright, Robert. 1995. "Hyperdemocracy." Newsweek 26 (January): 15–25.

Wrigley, E. A. 1969. Population in History. New York: McGraw-Hill.

Wrong, Dennis. 1961. "The Oversocialized Conception of Man in Modern Sociology." American Sociological Review 26 (April): 183–193.

Wrong, Dennis. 1979. Power. New York: Harper and Row.

Wuthnow, Robert, and Witten, Marsha. 1988. "New Directions in the Sociology of Culture." Annual Review of Sociology 8:49–67.

Yamane, David. 1994. "Professional Socialization for What?" Footnotes 22 (March): 7.

Yinger, J. Milton. 1957. Religion, Society, and the Individual. New York: Macmillan.

Yinger, J. Milton. 1985. "Ethnicity." Annual Review of Sociology 11:151–80.

Zavella, Patricia. 1987. Women's Work and Chicano Families. Ithaca, N.Y.: Cornell University Press.

Zhang, Lening, and Messner, Steven F. 1995. "Family Deviance and Delinquency in China." Criminology 33: 359–387.

Zuo, JiPing, and Benford, Robert. 1995. "Mobilization Processes and the 1989 Chinese Democracy Movement." The Sociological Quarterly 36:131–156.

Zygmunt, Joseph E. 1986. "Collective Behavior as a Phase of Societal Life: Blumer's Emergent Views and Their Implications." Research in Social Movements, Conflicts, and Change 9:25–46.

Author Index

Subject Index

Informal economy, 330
Informal social control, 113
Information technology
    definition of, 389
    and global culture, 176
    and social change, 389–392
    spread of, 391*m*
Innovation, as form of deviance, 115
Institution(s), 48–50
    basic, 49
    conflict view of, 50
    definition of, 48
    destruction of, effects of, 55–59
    economic, 323
    educational, 279
    interdependence of, 49
    medicine as, 242
    political, 314
    and power inequalities, 313
    and prejudice, 189–190
    structural-functional view of, 50
    total, 89
Institutionalized discrimination, against women, 211
Institutionalized racism, 200
Instrumental activities, 71
Instrumental rewards, 117
Intelligence, and social class, 149
Intelligence tests, 286, 287*f*
Intensive supervision probation (ISP), 132
Intergenerational bonds, 261–263
Intergenerational mobility, 137
Internalization, 113
International economic enterprises, 328–329
International Sociological Association, membership
    of, 11*m*
Internet, 391
    access to, 391*m*
Intragenerational mobility, 137
Intrinsic rewards, 333–334
IQ tests, 286, 287*f*
Iran, Islam in, 301
Islam, 301–302
Isolation
    and culture, 44
    social, effects of, 85–86
ISP. *See* Intensive supervision probation
Italy
    fertility in, 350, 350*t*
    nonmarital births in, 270*t*

Japan
    divorce rates in, 273*t*
    human development ranking of, 167*t*, 168
    nonmarital births in, 270*t*
    organizational model of, 80
Job(s)
    gendered, 213
    in sociology, 27–30
    U.S., protecting, 337
Job satisfaction
    versus alienation, 335
    factors determining, 333–334
Job search, social networks and, 74–75
Jurisdiction, state and, 314
Justice, 127–128
    sociological perspectives on, 392–393
    street-level, 128
Justifications, 105

Kenya
    fertility in, 348–349
    human development ranking of, 167*t*
Kinkeeping role, 265

Labeling theory of deviance, 118–119
Labor force
    displacement of, 335–336
    fertility rates and, 350
    in U.S., 328*f*, 331–332
Labor market
    projected changes in, 332–333, 333*t*
    segmented, 215, 218–220, 330–331
    and social class, 149–150
    women's participation in, 212–216, 213*f*
Labor unions
    Swedish, 326
    U.S., 322–323
Language, 37–39
    body, 102–103
    and culture, 36, 38
    as framework, 38–39
    as symbol, 39
Latent dysfunctions, 12
    of education, 280–281
Latent functions, 12
    of education, 280–281
Latinos. *See* Hispanic Americans
Latvia, fertility in, 350
Law(s), 41
    rational, 127
    sociology of, 127–133
    substantive, 127
    theories of, 127
Learning, socialization and, 90–93
Least-developed countries, 167
    and changes to status quo, 180–181
    demographic transition in, 345, 345*f*
    as market, 181–182
Leisure, of working class, 154
Lesbians. *See* Homosexuality
Less-developed countries, 166
    cities in, 363*f*, 363–364
    demographic transition in, 345, 345*f*
    population growth and poverty in, 353
Life chances, gender differences in, 211–218
Life expectancy, 230
    in Eastern Europe, 241–242
    income and, 239*t*
    race and, 239*t*, 240
    sex differential in, 217, 237, 239*t*, 239–240
    social class and, 240
Lifetime divorce probability, 259–260
Linguistic relativity hypothesis, 38
Lithuania, human development ranking of, 167*t*
Living standards
    and expectations, 375, 376*f*
    and life expectancy, 240–241
    and mortality rates, 344, 346
Looking-glass self, 94–95
Luddite uprising, 336

Macrosociology, 18
Mainline churches, 299
Majority group, 186–187
    interaction with minority groups, patterns of,
        187–188
Management, bottom-up, 80
Manifest dysfunctions, 12

Manifest functions, 12
Marijuana use, labeling as deviance, 118–119
Marriage
    choice of partners, 258–259
    versus cohabitation, 267
    definition of, 255
    economic determinism and, 6, 273
    gender roles in, 263
    homosexuality and, 268
    and parenthood, 260
    rates of, 259
    remarriage, 260
    sexual roles in, 263–264, 264*t*
    in U.S., 257–258
Marriage markets, local, 258–259
Martineau, Harriet, 5
Marx, Karl
    on class, 138
    on inequality, 145
    on religion, 297
    and sociology, 5–6
Mass media
    as agent of socialization, 92–93
    and social movements, 384–385
Material culture, 33
Mathematical skills, gender differences in, 86–87
Me, 84
Mead, George Herbert, and symbolic interaction
        theory, 95–96
Medicaid, 248
Medicare, 248
Melting pot, 194
Men. *See also* Gender differences; Sex differential
    disadvantages of, 217–218, 240
    and kinkeeping, 265
Mental ability, measuring, 286, 287*f*
Metropolitan area, 362. *See also* under Urban
Micromobilization, 379–380
Microsociology, 18
Middle age, 260–261
Middle class
    African Americans in, 196
    and democracy, 316
    U.S., 140, 140*f*, 151
Migration
    international, 358, 359*m*
    in U.S., 355–356, 357*m*
Mills, C. Wright, 9
Minority group(s), 187
    aging among, 237
    and gender inequality, 220–222
    intelligence tests and, 286
    interaction with majority group, patterns of,
        187–188
Mixed economies, 325
Mobility
    intergenerational, 137
    intragenerational, 137
    social, 137
    and urbanization, 362
Mobilization, 378–380
    bloc, 380
    definition of, 378
    micromobilization, 379–380
Modernization theory
    of aging, 235
    of development, 169
Modified extended family, 265
Moral entrepreneurs, 119

Photo Credits